Encyclopedia of Protein Engineering

.

Encyclopedia of Protein Engineering

Edited by **Anton Torres**

R Callisto
Reference

New York

Published by Callisto Reference,
106 Park Avenue, Suite 200,
New York, NY 10016, USA
www.callistoreference.com

Encyclopedia of Protein Engineering
Edited by Anton Torres

© 2015 Callisto Reference

International Standard Book Number: 978-1-63239-286-2 (Hardback)

Printed in the United States of America.

Contents

Preface

I am honored to present to you this unique book which encompasses the most up-to-date data in the field. I was extremely pleased to get this opportunity of editing the work of experts from across the globe. I have also written papers in this field and researched the various aspects revolving around the progress of the discipline. I have tried to unify my knowledge along with that of stalwarts from every corner of the world, to produce a text which not only benefits the readers but also facilitates the growth of the field.

In this book, a wide variety of data is enclosed by presenting a solid base in protein engineering. It provides readers with information crucial to the design and fabrication of proteins. This book provides debates on a range of techniques for protein engineering, featuring researches from experts practicing around the globe. A wide range of topics analyzing important features of techniques and applications in the composition of new proteins are presented. These comprise the use of unnatural amino acids, molecular progression and protein folding to construct helpful proteins with better properties.

Finally, I would like to thank all the contributing authors for their valuable time and contributions. This book would not have been possible without their efforts. I would also like to thank my friends and family for their constant support.

Editor

Structure-Functional Insight Into Transmembrane Helix Dimerization

Eduard V. Bocharov, Konstantin V. Pavlov, Pavel E. Volynsky,
Roman G. Efremov and Alexander S. Arseniev
Shemyakin-Ovchinnikov Institute of Bioorganic Chemistry RAS
Russia

1. Introduction

Membrane proteins, constituting ~30% of proteins encoded by whole genomes (Krogh et al., 2001), are heavily implicated in all fundamental cellular processes and, therefore, represent up to 60% of targets for all currently marketed drugs (Overington et al., 2006). Nevertheless, in spite of their significance, only few tens of spatial structures of membrane proteins have been obtained so far, while design of new types of drugs targeting membrane proteins requires precise structural information about this class of objects. Hydrophobic α-helices represent a dominant structural motif found in membrane-spanning domains of proteins, excluding membrane β-barrels. So, a membrane part a large variety of membrane proteins is formed by α-helical bundle (polytopic proteins) or just by single α-helix (bitopic proteins) (Fig. 1). Besides structural switching, oligomerization of helical membrane proteins forms the basis for various functions in the living cell including reception of extracellular signals, signal transduction, ion transfer, catalysis, energy conversion and so on (Ubarretxena-Belandia & Engelman, 2001). The mechanisms, by which helical membrane proteins fold into native structures and functionally oligomerize, are beginning to be understood from a confluence of structural and biochemical studies. Folding determinants of a membrane protein can be partially understood by dissecting its structure into pairs of interacting transmembrane (TM) helices, which, together with the connecting loops and extramembrane domains, comprise the overall structure. Obviously, the fold of helical membrane proteins along with their biological activity is largely determined by proper interactions of membrane-embedded helices. Either destroying or enhancing such helix-helix interactions can result in many diseases (developmental, oncogenic, neurodegenerative, immune, cardiovascular, and so on) related to dysfunction of different tissues in the human body.

Activity regulation of bitopic proteins that have only single-spanning TM domain is mostly associated with their lateral dimerization in cell membranes. Bitopic proteins are a broad class of biologically significant membrane proteins including the majority of receptor protein kinases, immune receptors and apoptotic proteins, which are involved in development regulation and homeostasis of multicellular organisms. Homo- and heterodimerization of bitopic proteins was earlier thought to involve mostly their extracellular and cytoplasmic domains, but recent studies have been making it increasingly clear that the single-spanning TM domains are also critical for their dimerization and modulation of biological function. Upon bitopic protein activation, ligand-dependent or not, significant intramolecular conformational transitions result in rearrangement of the receptor domains and following receptor dimerization or switching from one dimerization state to another, e.g. ligand-dependent transition from preformed inactive dimeric state into active dimer of ErbB receptor tyrosine kinase (Schlessinger, 2000; Moriki et al., 2001; Fleishman et al., 2002; Mendrola et al., 2002). The so-called "rotation-coupled" and "flexible rotation" activation mechanisms (Moriki et al., 2001; Fleishman et al., 2002; Mendrola et al., 2002), which were initially proposed for receptor tyrosine kinases and imply active involvement of TM domains in dimerization and activation of the receptors via proper TM helix-helix packing and rearranging, are possibly widespread among bitopic proteins. However, if biological functions are carried out using only one homo- or heterodimeric state of bitopic protein TM domains, the TM helix-helix interaction can be strong, as in the case of permeabilization of the outer mitochondrial membrane by proapoptotic protein BNip3 in the course of hypoxia-acidosis induced cell death. Furthermore, amino acid polymorphisms and mutations in the TM domain of bitopic proteins have been implicated in numerous human pathological states, including many types of cancers, Alzheimer's disease, tissue dysplasias and abnormalities (Li & Hristova, 2006; Selkoe, 2001). It was shown that the mutations affect both the behavior of the isolated TM domains in model lipid bilayers, and the behavior of the full length receptors in the plasma membrane. Most probably, the effects are exerted via yet unknown mutation-induced changes in dimeric structure of the TM domains. Importantly, it was found that isolated TM domains revealed ability not only to homo- and heterodimerize in membrane-like environment, but also to specifically inhibit biological activity of bitopic proteins in cell membrane (Li & Hristova, 2006; Bennasroune et al., 2004; Rath et al., 2007). So, membrane-spanning segments of bitopic proteins represent a novel class of pharmacologically important targets, whose activity can be modulated by natural or specially designed molecules. Among the most perspective candidates for these purposes are artificial hydrophobic helical peptides, the so-called peptide "interceptors" (Bennasroune et al., 2004) or "computer helical antimembrane proteins" (CHAMPs) (Caputo et al., 2008), which are capable of specifically recognizing the target wild-type TM segments of bitopic proteins and interfering with their lateral association in cell membrane. Therefore, understanding the factors that drive packing of α-helices in membranes has attracted considerable interest of researchers from both scientific and medical communities. Nevertheless, in spite of their significance, only few spatial structures of the homo- and heterodimeric single-span TM domains have been obtained so far, notwithstanding that design of new types of drugs targeting bitopic proteins requires precise structural information about this class of objects.

At the present stage of development of the structural biology methods, obtaining high-resolution structure of a full-length bitopic protein is a scientific challenge. Issues with crystallization of membrane proteins are inherent to X-ray techniques, whereas NMR cannot effectively handle large protein-lipid complexes. The crystallographic methods, which recently allowed obtaining high-resolution structure of such multi-span TM receptors as G-protein coupled receptors (Cherezov et al., 2007), cannot be directly translated to multiple-domains flexible receptors like receptor kinases and immune system receptors. Therefore, the structural-dynamic properties of the extracellular, cytoplasmic and intramembrane parts of such bitopic proteins are still studied separately. Extensive structural studies of extracellular and cytoplasmic domains in different functional states of the bitopic proteins are closely followed by detailed analysis of their TM domain dimerization in membrane environment. Apparently, thorough understanding of all the aspects of TM helix-helix interactions in bitopic proteins can only be achieved with multi-disciplinary approach based on a comprehensive set of modeling, biochemical and biophysical tools. The already available information about structural-dynamic properties of the dimeric TM domains of bitopic proteins along with the biophysical and biochemical data provides useful insights into the protein functioning in the human organism on the atomistic scale. This review will discuss the applicable methods, from purely theoretical approaches to direct experimental techniques, which recently allowed describing high-resolution dimeric TM domain structure for several bitopic proteins and understanding some aspects of structure-function relations and their biological activity.

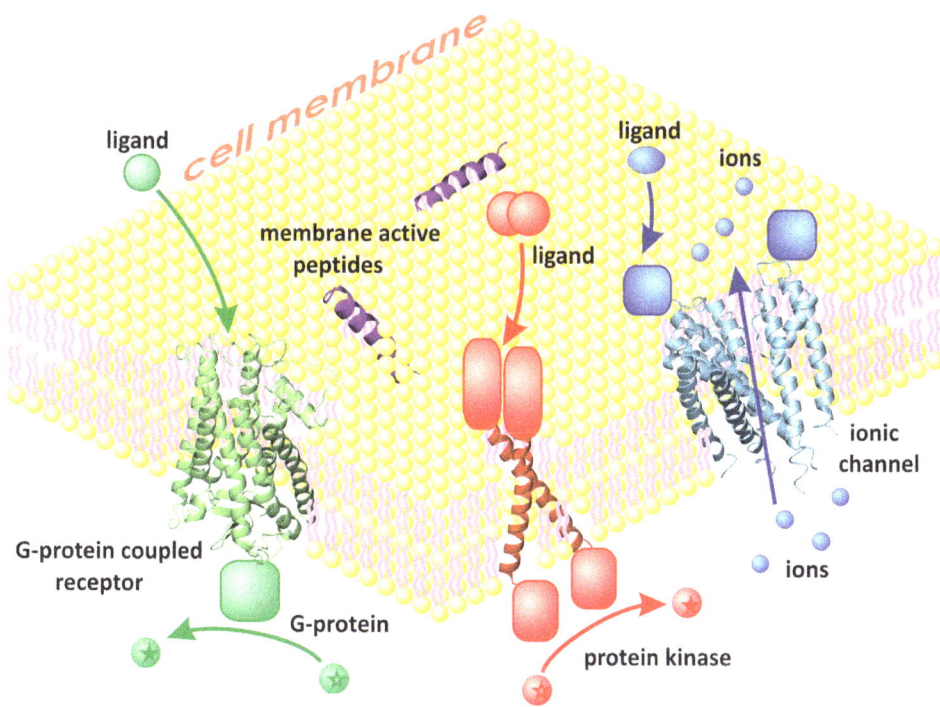

Fig. 1. Representatives of bitopic and polytopic helical TM proteins.

2. Thermodynamical aspects of helix-helix interaction in membrane

The balance of forces driving association of proteins in lipid membranes, in particular the helix-helix interaction of transmembrane domains, differs fundamentally from the case of protein interaction in aqueous solutions to the such degree that in some cases might seem counterintuitive. For the sake of clarity, folding of an α-helical membrane protein can be conceptualized as a process that occurs in two thermodynamically distinct steps, involving the formation of independently stable TM helices and the subsequent specific TM helix-helix interactions giving rise to higher-order structures (Popot & Engelman, 1990), in which TM helices are usually more or less tilted with respect to the membrane plane. The former step, similarly to the case of water soluble proteins, is controlled by hydrophobic interactions, profile of which changes essentially once the helices are inserted into the lipid bilayer and the hydrophobic side chains can be exposed to hydrophobic environment without energy penalties. Lipid-protein interactions are also most likely involved, though indirectly, in driving the association of TM helices in the form of the entropy term (Helms, 2002; Schneider, 2004). Although the formation of higher ordered helix oligomers decreases the entropy of the proteins, the entropy of the lipids is greatly increased. Every TM helix is surrounded by a "coat" of lipids more or less tightly associated with it. After interaction of individual TM helices, a part of these "frozen" or anyhow correlated lipids (Morrow et al., 1985) is released into the membrane lipid pool. Therefore, TM helix oligomerization would decrease the area of protein-lipid interface and thereby increase the overall entropy of the system, thus contributing to stabilization of the protein-protein complex. In addition, adjustments of local lipid composition of the membrane and matching the hydrophobic thickness of lipid bilayer with the hydrophobic length of TM proteins can regulate lipid-protein and protein-protein interactions, e.g. resulting in cooperative lipid-mediated protein-protein lateral association into signaling platforms in biomembranes (Lee, 2004; Nyholm et al., 2007; Sparr et al., 2005; Marsh, 2008; Vidal & McIntosh, 2005; de Meyer et al., 2008).

Specific helix-helix interactions require precise mutual orientation of TM helices, imposing certain restrictions on their tilt angle and tilt direction between dimer axis and normal to the membrane, therefore proper hydrophobic matching may influence the specific TM domain association. Depending on the specifics of the protein, this would result in sorting different biologically relevant states of dimeric bitopic proteins between lipid phases and microdomains of cell membrane or in shifting the occupancies of the resultant conformation of the pair of TM helices and proteins depending on the surrounding lipid phase or microdomain (Nyholm et al., 2007; Sparr et al., 2005). Even when helices do not exhibit any tendency for specific association (Lee, 2004; Nyholm et al., 2007), helix-helix association could still occur as a result of poor packing between the lipids and helices, or of a favorable change in entropy due to the release of helix-bound lipids upon helix association. In these cases, helix association is primarily driven by lipid-protein interactions rather than strongly favorable protein-protein interactions. However, while entropy considerations and hydrophobic matching or mismatching could partly explain the formation of higher ordered TM structures in the membrane, it cannot serve the sole explanation of the specificity of TM helix interactions.

Protein-lipid interaction is not the only noncovalent force involved in the formation of TM helix oligomers, van-der-Waals forces and polar interactions also play important roles (Senes et al., 2004; Curran & Engelman, 2003). Association of TM helices often proceeds through a "ridge-into-groove" or a "knob-into-hole" packing (Langosch & Heringa, 1998; Walther et al., 1996). The ridges or knobs on the surface of one TM helix fit well into grooves or holes on the complementary helical surface. Such a complementarity of contacting adjacent TM ensures most favorable polar and van-der-Waals interactions. Electrostatic interactions also cannot be excluded from consideration despite relative rarity of occurrence of charged residues in the TM segments and play a specific role in membrane protein folding (Zhou et al., 2000, Zhou et al., 2001; Choma et al., 2000; Adamian et al., 2003; Gratkowski et al., 2001), since the strength of such interactions increases with a decreasing dielectric constant of the environment. Electrostatic interactions stabilize folded membrane structures via polar backbone-backbone, backbone-side chain, or side chain-side chain interactions resulting in hydrogen bond formation between adjacent TM helices. Contribution of amino acid residues into interaction energy in the hydrophobic environment is a function of their polarity. Weakly polar amino acids, like glycine, alanine, serine, and threonine are characterized by a relatively small electrostatic component of the interaction energy and a complex nature of interaction. In addition to forming electrostatic interactions, these polar residues with small side chains also allow two TM helices to come into close contact and to tightly pack without significant entropy loss of side chain rotamers upon dimer formation (MacKenzie et al., 1997). This does not only facilitate the interhelical hydrogen bonding with participation of polar side chains of serine or threonine, but also enables van-der-Waals interactions between surrounding residues. In addition to polar side chains, the CαH groups of such tightly packed residues are capable of participating in non-canonical hydrogen bonding, e.g. with the opposite carbonyl groups across the helix-helix interface (Senes et al., 2001). In other words, the marginal polarity of the Cα proton might be sufficient to serve as a hydrogen bond donor in a highly hydrophobic environment. However, although the slightly polar residues could form hydrogen bonds with an adjacent TM helix, they are able to contribute significantly to the specific helix-helix interactions only consisting in an amino acid context, which promotes association of TM helices, e.g. by proper packing (Gratkowski et al., 2001; Dawson et al., 2002; Schneider & Engelman, 2004; Arbely & Arkin, 2004; Mottamal & Lazaridis, 2005).

Presence of highly polar residues, like histidine, asparagine, aspartic acid, glutamine, glutamic acid, arginine or lysine in the membrane environment can apparently drive noncovalent association of TM helices through more specific strong hydrogen bonding and salt bridge formation, resulting in very stable helix oligomers. These residues are rarely found in membrane proteins (Arkin & Brunger, 1998), but it has been shown that the presence of a single asparagine, aspartic acid, glutamine, or glutamic acid in a TM helix is sufficient to drive stable oligomerization (Gratkowski et al., 2001; Zhou et al., 2000; Zhou et al., 2001). While highly polar residues can contribute significantly to the stability of the helix-helix interaction, several problems arise when these residues are present in a membrane. Transfer of highly polar residues into a membrane is thermodynamically unfavorable, and only very few of these residues can be tolerated in a single TM helix. Furthermore, in membrane environment, the ionizable side chains of these residues prefer uncharged state and their pKa values can vary substantially depending on numerous parameters, such as local hydrogen bond network, membrane composition, transmembrane potential, and juxtamembrane environment (Smith et al., 1996; Bocharov et al., 2008a). Since highly polar residues could interact with any potential binding partner for hydrogen bonding or salt-bridge formation, which create the danger of non-specific helix-helix association and misfolding (Schneider, 2004), the polar substitutions are apparently the most common pathogenic mutations in membrane proteins that cause different human diseases (Li & Hristova, 2006; Moore et al., 2008). On the other hand, for the polar residues located at the level of the lipid headgroups where solubility of charged groups is higher than in the hydrophobic core but the electrostatic shielding is accordingly more effective, the individual interactions are not so formidable and can be modulated by external ligands (Lau et al.,

2009). In addition, arginine and lysine residues are frequently found at the ends of TM helices, where they have a tendency to participate in direct or water-mediated polar–polar interactions with phospholipid headgroups (Arkin & Brunger, 1998; Wallin et al., 1997; Adamian et al., 2005) and can modulate the helix-helix dimerization strength (Peng et al., 2009).

A separate important class of participants of specific TM helix association processes are π-π and cation-π aromatic interactions arising either between two aromatic residues or between a basic and an aromatic residue, respectively (Johnson et al., 2007; Unterreitmeier et al., 2007; Sal-Man et al., 2007). Interactions of aromatic rings of tryptophan, phenylalanine, tyrosine, and histidine residues and their self-association or interaction with protonated cation side chains of arginine, lysine, and histidine residues have been proposed to consist of van-der-Waals and electrostatic forces complemented by correct packing geometry and interactions with the aromatic ring quadrupole moment. Besides, the indole, phenol, and imidazole group of the aromatic residues can participate in hydrogen bonding across TM helix packing interface. Even though weak, CαH-π interactions enhanced in the low dielectric membrane environment can be considered as additional interactions supporting specific TM helix association (Unterreitmeier et al., 2007). In addition, aromatic residues have a strong propensity to face phospholipids in the headgroup region and are thought to act as anchors for a membrane protein, influencing on helix tilting and hydrophobic matching in the membrane (Adamian et al., 2005). Cation-π interactions occurring at the headgroup levels are often contributed or mediated by additional interaction with water molecules.

3. Common motifs employed for helix-helix interaction in membrane

The helical configuration of TM segments imposes certain limitations and regularities on the amino acid sequences that are suitable for forming intermolecular contacts. The TM helix-helix association modes can be roughly grouped on the basis of sequence patterning and interhelical geometry. Since N- and C-termini of α-helical TM domains of bitope proteins are usually exposed to extracellular and cytoplasmic sides of membrane respectively, such proteins specifically associate into homo- and heterodimers in a parallel manner, in the so-called "head-to-head" orientation. Both right- and left-handed variants of parallel helix-helix dimers with most frequently occurring helix-helix crossing angles near -40° and 20°, respectively, and the distance of 7-9 Å between helix axes appear to be quite common for TM helix packing in membrane (Walters & DeGrado, 2006). The interfaces of TM helices crossing at negative angles are often formed by [abcd]n tetrad repeats, in which a and b correspond to interfacial residues (Langosch et al., 2002). Right-handed packing of helix pairs is most often characterized by an i, i+4 separation of "small" residues, such as glycine, alanine, serine and threonine, along the TM sequence, which is alternately termed "small-xxx-small" or GG4-like motif first exemplified by self-assembling TM domain of glycophorin A (MacKenzie et al., 1997). Small residues in this motif create a shallow weakly polar groove that complements the surface of an adjacent helix and allows the helices to approach closely. The association is stabilized by van-der-Waals contacts resulting from the excellent geometric fit and weak polar interactions, which can contribute to non-canonical hydrogen bonding between CαH and carbonyl groups across helix-helix interface (Senes et al., 2001). Two GG4-like motifs in tandem form the so-called "glycine zipper" motif, which is statistically overrepresented in membrane proteins (Kim et al., 2005). The geometry of left-handed pairs of TM helices characterized by positive crossing angles requires longer [abcdefg]n heptad minimal repeat motifs, where e and g positions are located at the periphery of these helix–helix interfaces and side-chains at a and d positions interdigitate repeatedly (Langosch & Heringa, 1998). Such a heptad pattern was originally identified in water soluble "leucine zipper" interaction domains and gives rise to "knobs-into-holes" packing of side-chains (Lupas, 1996). The left-handed TM helix pairings are mostly stabilized along heptad repeats by van-der-Waals contacts of large side chains of valine, leucine and isoleucine residues, while slightly polar interactions of interfacial residues having small side chains, like glycine, alanine, and serine, are also important for left-handed oligomerization (Lear et al., 2004; Ruan et al., 2004; North et al., 2006). In addition, the TM helix-helix dimerization via both tetrad and heptad repeat motifs can be enhanced by π-π, cation-π and CαH-π interactions across helix packing interface with participation of aromatic side chains (Johnson et al., 2007; Unterreitmeier et al., 2007). Furthermore, interhelical hydrogen bonding with participation of polar residues can work in concert with other helix packing interactions to strongly stabilize both right- and left-handed motifs, which appear to be essential for proper alignment of the polar side chains required for formation of hydrogen bonds (Moore et al., 2008).

TM helix interactions are mostly driven and stabilized by a broad spectrum of forces caused by protein-protein interactions via such motifs as well as interactions of the helices with the membrane environment. The precise interplay of all these forces is unique for each system and warrants individual detailed analysis since it often defines the functionality of interacting membrane proteins. Currently, many unique sequence motifs that are responsible for specific helix-helix association have been identified on the basis of tetrad and heptad repeats, which play primarily a permissive role for close helix-helix interactions (for a review see refs. Moore et al., 2008; Walters & DeGrado, 2006; Langosch & Arkin, 2009; Mackenzie, 2006). The relative importance of the sequence motifs in stabilizing helix-helix interactions depends on the specific combination of residues and location of the interacting surfaces relative to the N- and C-termini of α-helical TM segments (Johnson, 2006). Besides, the affinity of TM helix association can be modulated by flanking and non-interfacial residues (Zhang & Lazaridis, 2009).

One or a few potential dimerization motifs can be usually identified in each TM region of bitopic proteins that participate in two broad categories of helix-helix interactions (Moore et al., 2008). In the first of them, the TM domains form relatively static contacts that might be necessary e.g. for the assembly of a functional protein complex or for proper folding and export from endoplasmic reticulum. In other cases, the TM domains can undergo dynamic conformational changes between alternative dimerization modes important e.g. for signaling process that can involve a change in association state and/or lateral, vertical, and rotational motions in the membrane. Such triggering interactions cannot play a thermodynamically dominant role in overall protein conformational transitions, but are quite capable of fine-tuning the system energetics, leveraging TM coupling and restricting the pool of the allowable conformations of the full length bitope proteins in the course of their biological activity.

4. Predicting spatial structure of dimeric transmembrane helices by molecular modeling

Molecular modeling is a reasonably quick and efficient tool for quantitative assessment of the possible modes of helix association in membranes, especially when direct structural methods fail to provide the necessary insights or are prohibitively resource-consuming. Moreover, relative simplicity and stability of homo- and heterodimers of TM domains of bitopic proteins facilitates development and application of computational techniques for assessing the helix-helix interactions in membranes. Though only a few experimental spatial structures of TM helical dimers are available so far, molecular modeling offers quite reasonable atomic-scale models of dimeric structures.

Adequate molecular modeling of TM protein-protein interactions is impossible without a proper representation of the membrane. Three generic techniques have been developed for representing the membrane environment for the purpose of membrane protein simulations. The simplest option is to model the effect of heterogeneous membrane environment implicitly by means of some potential of mean force. This is commonly achieved by adding special terms to the potential energy function of a protein in the framework of so-called implicit or "hydrophobic slab" membrane models (Efremov et al., 2004; Feig & Brooks, 2004). Though this kind of representation can not provide atomistic details of protein-membrane interactions, it adequately mimics the basic membrane properties, such as membrane transversal hydrophobicity, thickness, curvature, and transmembrane voltage. These approaches are quite computationally effective and allow fast sampling of the protein configurational space and reasonably guessing the key trends of protein behavior in membrane (spatial structure in the membrane-bound state, geometry of binding, etc.). The second group of modeling techniques employs explicit membrane representation. The simulations are carried out for full-atom hydrated lipid bilayers or detergent micelles with imposed periodic boundary conditions (Forrest & Sansom, 2000). This class of models is capable of providing the most reliable dimeric structures of TM peptides. Unfortunately, due to large size of the systems (up to 106 particles), such calculations are very time- and resource-consuming. Finally, the third class of membrane models, so-called "coarse-grain" (CG) models, is a reasonable trade-off between the simplicity of the former and accuracy of the latter approach (Sansom et al., 2008). In CG-models, standard groups of atoms are replaced with "grains", thus reducing considerably the number of degrees of freedom in the protein-membrane systems.

The approaches commonly employed for such studies can be subdivided into three major categories: molecular docking, Monte Carlo and molecular dynamics simulations. A group of docking techniques is intended for fast identification of homo- and heterodimeric states of bitopic protein TM domains based on their amino acid sequence (Casciari et al., 2006). Usually, one of the TM monomers is considered as a target, and the other as a ligand, the conformational lability being limited for one or both of the monomers defined with the parameters of the backbone and side chains typical for α-helical TM segments. The membrane is either ignored or modeled implicitly. This method allows quick scanning for spatially complementary surfaces with optimally matched geometrical, hydrophobic/hydrophilic, and electrostatic properties of the interacting TM helices, and thus predicts potential dimerization interfaces and intermonomer hydrogen bonds. However, due to restrictions imposed on the TM helix mobilities and due to many physical factors of protein-protein and protein-lipid interactions being ignored, docking methods are typically used only for initial characterization of the specific helix-helix packing, to be subsequently supplemented by other methods.

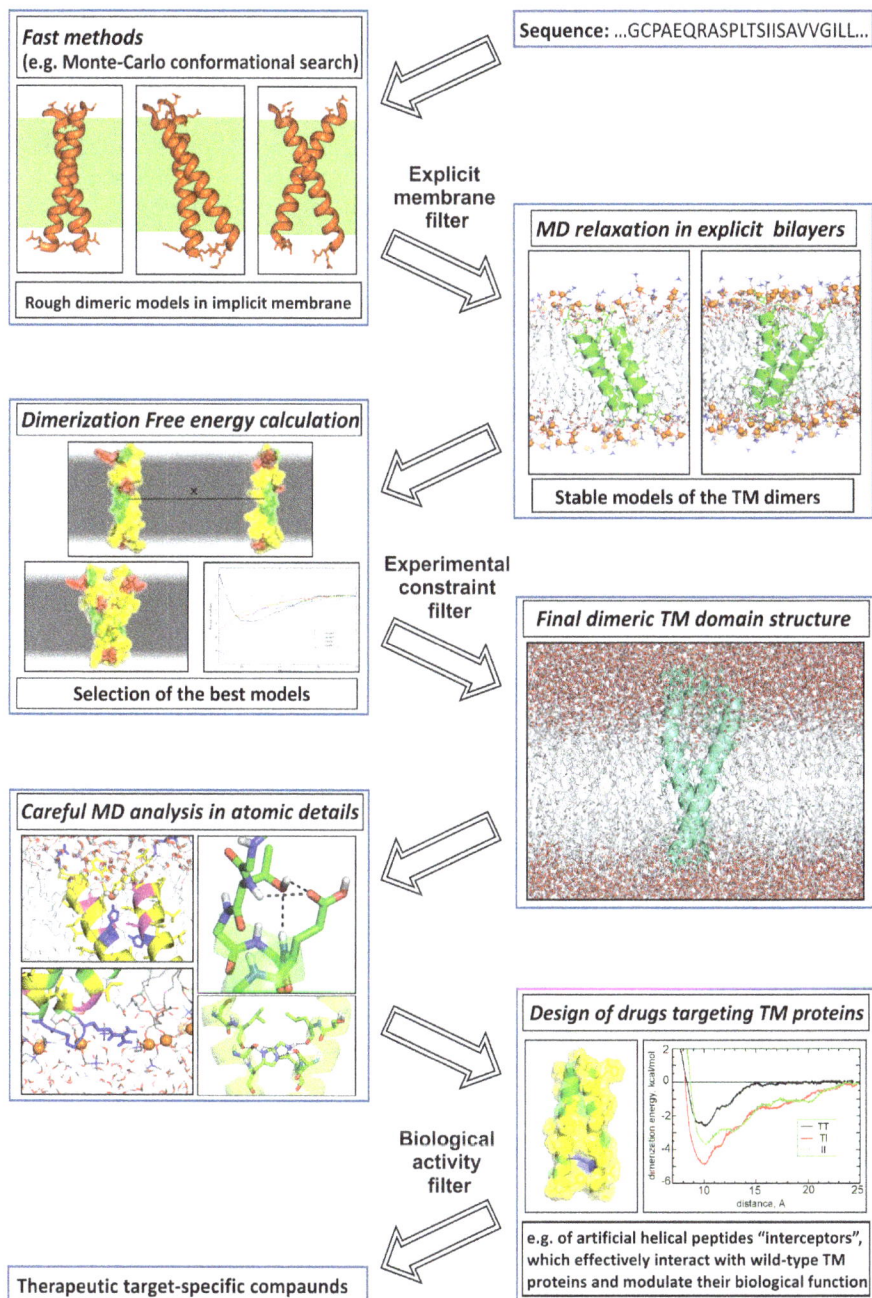

Fig. 2. Scheme of the spatial structure elucidation of homo- and heterodimeric TM domains of bitopic proteins and the subsequent molecular design of drugs targeting TM proteins with the aid of computer simulations techniques.

In Monte Carlo conformational search, both monomers are flexible permitting more careful scanning of the conformational space and thus potentially yielding more credible calculated structures. Clearly, these approaches allow the membrane to be more accurately taken into account, using either implicit or explicit representation. With implicit membrane representation, more extensive scanning of conformational space becomes feasible due to its lower computational cost, and therefore the chance of missing a realistic helix-helix configuration decreases greatly. On the other hand, the predicted dimeric structures can be graded more accurately with explicit membrane models. For acceleration of the Monte Carlo conformational search it was often assumed a priori that the TM helices adopt a proper TM orientation and their backbones were considered "rigid", and hence, common occurrence of local distortions in TM helices, like kinks and bends, was not taken into account. Under such assumptions, the effects of membrane environment on the secondary structure formation and/or stabilization, along with the events accompanying insertion of the peptides, also can not be assessed. However, Monte Carlo protocols without imposing any restraints on the secondary structure and a priori knowledge of the mode of membrane binding for the peptides were recently developed (Efremov et al., 2006; Vereshaga et al, 2005). Often, Monte Carlo algorithms operate in dihedral angles space, thus reducing dimensionality of the computational task. Usually, Monte Carlo simulations help in delineation of a limited number of low-energy conformational states of TM helical dimers (Vereshaga et al, 2005). Subsequent analysis of these families of conformers results in very few "native-like" structures, thus facilitating selection of the final models.

Molecular dynamics (MD) is one of the most informative methods, since besides providing the spatial structure it allows estimation of dynamic parameters of interaction, identification of the most important residues, etc. Membrane models of any degree of complexity can be used in MD calculations. It comes at a price of great computational intensity, therefore selection of the starting state becomes a real issue due to limited capabilities for scanning conformational space, making it virtually impossible to obtain correct structure starting from an essentially wrong one. This problem is especially significant in case of calculations in the explicit bilayer. One of the ways to resolve it is based on generating a set of initial states with different geometries of the dimer packing. Though providing most detailed scanning of the conformational space, this method is often impractical due to unacceptable computational resource requirements, and is essentially limited to implicit membrane calculations. For explicit membranes, the starting structures can be obtained as a result of Monte Carlo search in an implicit membrane (or docking) with subsequent relaxation in the explicit bilayers. An alternative approach consists in preliminary investigation of the dimerization by the CG representation. In this case, the molecules are represented by "grains" (e.g. each of which roughly corresponds to 4 heavy atoms) that substantially improves the calculation time, so the intervals of up to ~1 microsecond can be investigated. As was shown Psachoulia et al., 2009, this time scale is sufficient for obtaining a realistic model of the TM dimer, which after MD relaxation in the full atomic representation correlates well with the NMR structure.

There is a number of examples of extensive application of computer modeling methods for investigation of specific TM dimerization of several bitopic proteins, including the wild type and mutated TM domains of glycophorin A (Lemmon et al., 1992), bacteriophage M13 major coat protein (Melnyk et al., 2002), proapoptotic protein BNip3 (Sulistijo et al., 2003), erythropoietin receptor (Constantinescu et al., 2001), amyloid precursor protein APP (Scheuermann et al., 2001), and ErbB receptor tyrosine kinases (Schlessinger, 2000). Most of the methods of molecular modeling of the TM helix specific dimerization have been developed and successfully tested on the TM domain of glycophorin A protein, homodimeric conformation of which was first obtained with high resolution (MacKenzie et al., 1997). Although a number of successful in silico predictions of TM helix-helix complexes have been reported, the uncertainty of the energy estimate of the final state, which is a measure of certainty of selection of the correct conformation, is still relatively high. Therefore, without employment of additional data it is usually very difficult to choose between several alternative models with close energies, having substantially different geometries. Moreover, if several dimerization modes are actually realized for a protein, computational methods provide little or no information about population and relative stability of the possible modes of helix-helix associates, which can be affected by modeling assumptions in silico as well as by variations of membrane environment and ligand binding in vivo. Partially, such a hypothesis is corroborated by somewhat vague results of mutagenesis studies (Lemmon et al., 1992), as well as by NMR (Gratkowski et al., 2002) and MD (Im et al., 2003; Petrache et al., 2000) data that demonstrate the importance of media effects for stability of helical oligomers and provide examples of their multi-state equilibrium in lipid bilayers and membrane mimics. In real biological membranes, the situation may be more complex due to inhomogeneous composition of lipid bilayers, their domain structure, variations of physico-chemical characteristics, presence of small molecules (e.g., cholesterol), etc.

Some conformations obtained by modeling are artificially introduced by computational assumptions, and they cannot be readily discriminated from those really occurring in cellular membrane without additional experimental information, in particular about the TM dimerization interface, see Fig. 2. Such information can be obtained by solid state NMR, site-specific infrared dichroism, mutagenesis in combination with the techniques permitting assessment of dimerization degree (SDS electrophoresis, bioassays in ToxR systems, FRET), Cys scanning (insertion of cysteine residues and analysis of the extent of disulphide bridges formations), and so on (for a review see refs. Rath et al., 2007; MacKenzie, 2006; Li et al., 2009; Schneider et al., 2007). Experimental limitation can be either imposed at the stage of calculations, e.g. in the form of limitations on the distances between atoms in different monomers, or used for assessing appropriateness of the predicted structures after completion of calculations. Such a combination of experimental and modeling techniques provides important advantages, substantially narrowing the search of dimeric TM structures and simplifying membrane representation and hence significantly accelerating the analysis. Compared to direct structural methods that usually identify only one conformation, this approach gives better credit for a conformational diversity of homo- and heterodimeric TM domain structures, which can occur in vivo during biological activity of a bitopic proteins.

Effectiveness of such a combination of computational methods with various biophysical and biochemical techniques was proved by its successful applications in a number of studies several of which are presented below. Selection of the proper dimeric structure of glycophorin A TM helix in Adams et al., 1996, was done based on mutagenesis data superposed on the set of structures obtained by global conformational search in vacuum.

The proposed model of the dimer was in good agreement with the spatial structure obtained by means of NMR spectroscopy in detergent micelles (MacKenzie et al., 1997). This method was later used for analysis of the glycophorin A TM domain dimerization in lipid bilayers, where the conformational search was done with the distance restraints from solid-state NMR spectroscopy (Smith et al., 2001; Smith et al., 2002a). Beevers et al., 2006, obtained spatial structure of the TM domain of the oncogenic mutant of rat receptor tyrosine kinase Neu by MD calculations in the explicit bilayer with different possible orientations of the monomers. Correctness of the resulting 'consensus' structures was assessed based on the information about orientation of the CO groups determined by site-specific infrared dichroism. Vereshaga et al., 2007, calculated the spatial structure of TM segment dimer of human proapoptotic protein Bnip3. In this case, Monte Carlo conformational search in an implicit membrane with subsequent MD relaxation of the best models in the full-atom DMPC bilayer was used for identification of the potential structures. Dynamically unstable models were screened out at the stage of MD relaxation. Correctness of the remaining models was assessed via comparison with the mutagenesis data. As a result, one of the final models consistent with the mutagenesis data was also in good agreement with the NMR-derived structure of dimeric Bnip3 TM domain in lipid bicelles (Bocharov et al., 2007). Volynsky et al., 2010, used modeling methods in combination with ToxR assays to study dimerization of TM segments of ephrin receptor EphA1. A set of spatial structures of the dimer proposed based on Monte Carlo simulations in implicit membrane followed by MD relaxation in explicit lipid bilayer were employed for rational design of wild-type and mutant genetic constructions for ToxR assays. Such a combined, self-consistent, application of modeling and experimental techniques allowed defining the factors important for dimerization of the TM segment of the EphA1 receptor, providing unambiguous spatial model consistent with the NMR-derived structure (Bocharov et al., 2008a) of the EphA1 TM dimer in lipid bicelles. Moreover, alternative conformations of the dimer were proposed. Metcalf et al., 2009, reported the models of integrin $\alpha IIb\beta 3$ TM heterodimers obtained using a Monte Carlo algorithm that selects conformations by a geometrical filter based on mutagenesis data. The Monte Carlo search for integrin $\alpha IIb\beta 3$ TM heterodimers was also carried out with an additional energy term using distance restraints obtained from cysteine-scanning mutagenesis bioassay data (Zhu et al., 2009). In both cases the proposed heterodimeric structures were in good agreement with recently obtained NMR structure of heterodimeric integrin $\alpha_{IIb}\beta 3$ TM complex embedded in lipid bicelles (Lau et al., 2009).

5. Determination of high-resolution structure of dimeric transmembrane helices by NMR spectroscopy

Over a number of recent years, structural biology has witnessed a race of rapidly developing experimental methods matched closely by increase of complexity of the experimental objects. Nowadays, obtaining high resolution structure of entire membrane proteins or functionally essential fragments thereof has become a reality. Isolation, purification, and handling of membrane proteins in their "native-like" conformations are still associated with enormous difficulties and often require expanding the limits of the modern experimental techniques. Besides, tertiary and quaternary structures of membrane proteins are only moderately stabilized and transitions are often observed between conformational substates. Multiple conformations and dynamics considerably complicate characterizing the structure of membrane proteins and their oligomers. For this reasons, despite recent increases in the number of high resolution structures of membrane proteins solved annually, the gap between soluble and membrane protein structures continues to increase. Even among the membrane proteins of known structure, specific oligomeric complexes of small membrane-spanning proteins such as TM domains of bitopic proteins are underrepresented.

Heteronuclear NMR spectroscopy proved an effective tool for investigating the systems of oligomeric α-helical TM domains of membrane proteins. Solid-state NMR has been successfully employed to obtain highly resolved spectra of membrane-bound peptides and proteins in lipid bilayer model systems, such as liposomes, which can have composition, thickness, surface tension and curvature similar to those of native lipid bilayers and thus adequately mimic cell membranes. Solid-state NMR techniques for membrane protein samples are rapidly evolving, and the structures of several small proteins in lipid bilayers have been already obtained with the aid of these methods (Opella & Marassi, 2004; Andronesi et al., 2005). There are two ways of obtaining high-resolution solid-state NMR spectra, either by performing magic angle spinning (MAS) in order to mimic the rapid tumbling that would naturally occur for a small molecule in solution for averaging the anisotropic interactions in solid-state, or by observing uniformly aligned molecules. Smith and co-workers have used ^{13}C–^{13}C rotational resonance and ^{13}C–^{15}N rotational echo double resonance MAS experiments to measure interhelical distances in the α-helical TM domain dimers of human glycophorin A (Smith et al., 2001; Smith et al., 2002a), human amyloid precursor protein (Sato et al., 2008) and rat receptor tyrosine kinase Neu (homologue of human ErbB2 receptor) with its constitutively activate Val664Glu mutant (Smith et al., 2002b). That allowed developing the structural models for the helix–helix packing interactions in lipid bilayer for these bitopic proteins. The tilt angle and rotational angle of TM helices can be estimated by analysis of the position, shape, and size of the so-called PISA wheels obtained from polarization inversion with spin exchange at the magic angle (PISEMA) experiment acquiring for oriented ^{15}N-labeled membrane proteins (Opella & Marassi, 2004).

Solution NMR became a major method to determine structures of water-soluble proteins and their complexes (Wüthrich, 1986). In addition to elucidation of their structures, NMR also offers unique opportunities to probe dynamical processes in them. However, membrane proteins embedded into lipid bilayers cannot be studied by means of solution NMR techniques because their rotations in these environments are slow and highly anisotropic, which leads to unfavorable relaxation and very wide or undetectable resonance lines. An alternative approach to solving high-resolution spatial structures and obtaining dynamical information on membrane proteins is to extract the proteins from their host membranes and disperse them in non-denaturing membrane-mimicking detergent/lipid systems such as micelles, bicelles, and nanodiscs, which tumble fast enough to give well-resolved resonance lines when using solution NMR methods. Since resulting supramolecular membrane protein–detergent/lipid complexes are usually still large on the scale of protein structures that are routinely solved by NMR, the most advanced solution NMR techniques and spectrometers operating at high magnetic fields and equipped with highly sensitive cryoprobes are typically employed to solve high-resolution structure of the membrane proteins. These include labeling the proteins with two or three low-abundant isotopes ^{2}H, ^{13}C and ^{15}N, deuterating of detergents and lipids at least on hydrophobic tails, using transverse relaxation-optimized spectroscopy (TROSY) (Pervushin et al., 1997), and obtaining structural restraints in addition to those typically obtained from nuclear Overhauser effects (NOE) and chemical shifts, such as restraints obtained from residual dipolar couplings (RDC) and paramagnetic relaxation enhancements (PRE), which can drastically improve both quality and throughput of membrane protein structure determination (for comprehensive review see ref. Kim et al., 2009). The accuracy of determining the protein structure is controlled by many factors, including the dynamical properties of the protein itself, as well as the nature and quantity of the experimentally obtained restraints. In case of dimeric TM α-helical proteins, if a well defined structure of monomers is known (particularly the side chain conformations and helix bending), just a few restrains can fully determine the structure provided that they are, in a broad sense, independent enough. However, since every restraint has an experimental error associated with the precision of measurements and with the accuracy of assignment in case of NOE contacts, having larger number of independently derived consistent restraints greatly increases confidence in the structure of individual TM helices and of the dimer as a whole. In case of underdetermined structures where there are substantial ambiguities in the NMR-derived structural information with only few reliable restraints defining global dimer structure, molecular modeling can allow making a choice in favor of the most physically justifiable model of the dimer. Obviously, this process directly depends on the accuracy of the underlying physical assumptions, i.e. the force fields used in the modeling of the membrane proteins. Given the limited amount of structures obtained in the membrane-mimicking environments, each new experimental structure is of utmost practical and methodological importance.

The smallest among membrane mimicking particles – micelles, which are formed of soft detergents, short-chain lipids or lysolipids, are optimal from the standpoint of NMR relaxation, allowing recording spectra with narrow lines and rather good chemical shift dispersion (Gautier et al., 2008; Krueger-Koplin et al., 2004) A lot of membrane-penetrating peptides, membrane associated peptides and fragments of membrane proteins were studied in micellar solutions by NMR spectroscopy (Kim et al., 2009; Krueger-Koplin et al., 2004; Sanders & Sönnichsen, 2006). Most of the structures of helical membrane proteins resolved with NMR spectroscopy were determined in micelles of different types, indicating that there is no universal detergent, applicable for every membrane protein. Therefore, extensive detergent screening is usually made to find a proper environment (Krueger-Koplin et al., 2004; Page et al., 2006; Maslennikov et al., 2007). Although majority of the membrane proteins maintain native-like structures in micelles and some retain activity, sometimes the detergent providing the best appearance of NMR spectra does not provide proper folding, and the protein dissolved in it remains inactive. Micelles have some disadvantages associated with high curvature of their spherical surfaces. Curvature effects are occasionally observed with small peptides, and the absence of specific phospholipids or mixtures of phospholipids may cause amphiphillic peptides interacting with the membrane surface to have distorted structures in micelles environment (Lindberg, 2003, Chou, 2002). Integral membrane proteins, especially those having structural element in the lipid headgroup region, can also have distorted structure and poor spectrum appearance in micellar solutions. Both the headgroup region and the hydrocarbon core in a highly curved micelle are packed less orderly and exhibit greater dynamics than in a planar or near-planar lipid bilayer (Lindberg, 2003, Chou, 2002). The shielding effect of the interfacial headgroup region is less pronounced, and water molecules can penetrate more easily into the micellar core, resulting in distortions of TM helix structure (Bordag & Keller, 2010). Importantly, addition of a very modest amounts of phospholipids to micelles can result in dramatic enhancements of NMR spectral quality for some integral membrane proteins (Sulistijo & Mackenzie, 2010). This lipid dependence appears to reflect the requirement of some membrane proteins for specific or semi-specific lipid-protein interactions, which cannot be satisfied by detergents only. So, detergent micelles with some amounts of phospholipids offer a viable compromise for investigating TM peptides in membrane-mimetic systems, combining ease of use and good dissolving properties with anisotropic environment. Nevertheless, many detergents exert a denaturing effect on membrane proteins and peptides by abrogating helix-helix interactions (Melnyk et al., 2001; Therien & Deber, 2002). These problems could be overcome by using membrane mimicking particles with elements of flat surface, such as bicelles and nanodiscs.

Nanodiscs are similar to high-density lipoprotein particles and consist of fairly large patches of planar lipid bilayers (~160 lipid molecules) surrounded by the rim formed by apolipoprotein A-I (Borch & Hamann, 2009; Nath et al., 2007; Ritchie et al., 2007). The particles have the diameter of about 12 nm and thickness of 4 nm with the overall rotational correlation time of about 80 ns (Lyukmanova et a., 2008), which is rather high for structural NMR studies, but with TROSY (Pervushin et al., 1997) and CRINEPT (Riek et al., 1999) techniques one can record a readable heteronuclear spectrum and compare it to the one recorded in micelles or bicelles. Nanodiscs have only been applied in NMR spectroscopy for a couple of years and but few membrane protein were studied in this environment so far. However, they proved useful for verifying that other membrane mimicking media provide proper tertiary structure of membrane proteins (Shenkarev et al., 2010); they also have high potential for various bioassay applications (Borch & Hamann, 2009).

A reasonable compromise between micelles and nanodiscs – small isotropic bicelles are binary mixed micelles, consisting of two types of molecules: long-chain lipids (with long hydrophobic tails) and short-chain lipids or detergents, e.g. dimyristoylphosphatidylcholine (DMPC) mixed with dihexanoylphosphatidylcholine (DHPC) or zwitterionic bile sole derivative CHAPSO (Kim et al., 2009). As such, they represent the most convenient environment with excellent bilayer-mimicking properties for NMR structural studies of small membrane protein and their complexes (Kim et al., 2009; Poget & Girvin, 2007). A number of bicelle systems have been developed and characterized for their unique liquid-crystal phase behavior. It was shown that bicelles at some conditions have discoidal shape with a bilayer formed by long-chain lipids and a rim of short-chain lipids (Vold et al., 1997; Lee et al., 2008; van Dam et al., 2004; Glover et al., 2001; Luchette et al., 2001). The shape of the particles is controlled by three parameters: the molar ratio q of long- and short-chain lipid (or detergent) concentrations (adjusted for concentrations of free lipids in the bicellar suspension), total lipid concentration c_L, and temperature T; and it can be either disc or perforated bilayer the dependence being rather complex (Vold et al., 1997). At q between 0.25 and 0.5 bicelles are tumbling fast, are almost isotropic and can be used for high-resolution structure determination (Kim et al., 2009; Prosser et al., 2006). The hydrophobic thickness of the aggregates can be controlled by the choice of long-chained lipids, and it was also shown that charged lipids, e.g. with either negative serine or glycerol headgroups, can be incorporated into such particles without loss of stability (Lind et al., 2008; Struppe et al.,

2000). A number of publications report smaller distorting effect of bicellar media on the structure of membrane proteins (Kim et al., 2009). Recent determination of the structure of the heterodimeric TM domain of the platelet integrin $a_{IIb}\beta_3$ in bicelles provides an elegant example of using this medium to solve an important structural biology problem that proved elusive when conventional micelles were used (Lau et al., 2008; Lau et al., 2009). Detergent micelles destabilize the heterodimer to the point where interaction cannot be detected, while the environment provided by bicelles allows at least partial retention of native-like heterodimer avidity. Typical size of the particles consisting of fast-tumbling bicelles (e.g. DMPC/DHPC bicelle of ~80 lipid molecules, q of 0.25, c_L of 3%, at 40 °C) with two embedded bitope protein TM fragments (~40 residues including hydrophobic TM segment flanked by polar N- and C-terminal regions) is ~5 nm corresponding to overall rotational correlation time of ~18 ns and the effective molecular weight of ~50 kDa. Therefore, extensive capabilities of solution heteronuclear NMR technique can be readily employed for investigating structural-dynamic properties of membrane proteins (Bocharov et al., 2008a).

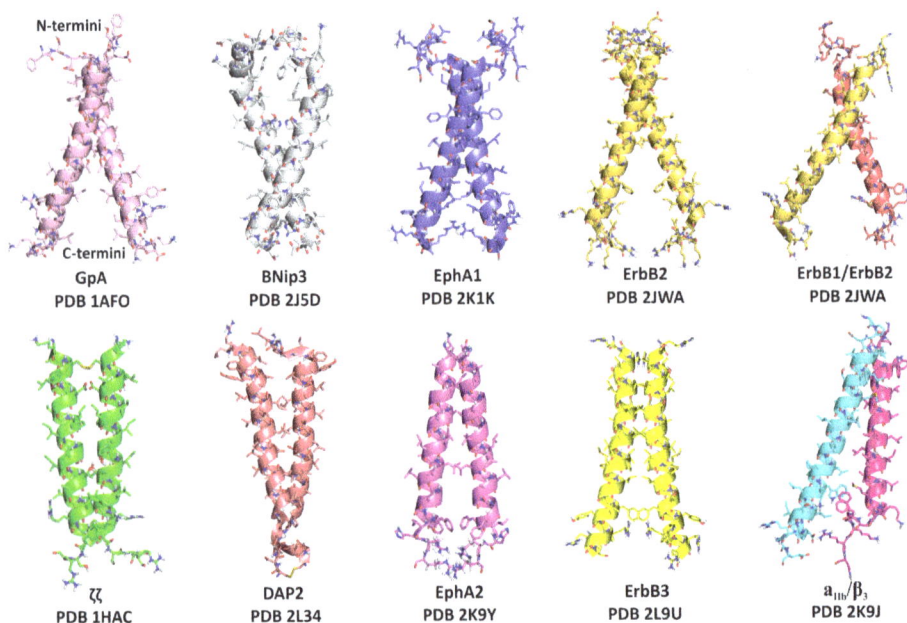

Fig. 3. High-resolution spatial structures of homo- and heterodimeric TM domains of bitopic protein obtained at present time.

A useful property of such systems is low effective ratio of detergent/lipid to protein and restricted protein mobility that can make homo- or heterodimerization effective enough even if specific interaction of TM helices are weak (e.g. in the case of receptor tyrosine kinase TM domains). Moreover, typical size of micelles and bicelles allows detecting intermolecular NOE contact network (up to ~6 Å) along TM helix-helix interface that is crucial for obtaining high-resolution structures of homo- and heterodimeric TM domains of bitopic protein. Nevertheless, one of the main problems encountered in structure determination of molecular complexes by NMR spectroscopy is to distinguish between intra- and intermolecular NOE contacts. In case of self-association of bitopic protein TM domains, if the dimers of α-helical TM segments are symmetrical on the NMR time scale, their two monomer chains display similar chemical shifts so that inter- and intramonomeric NOE contacts are indistinguishable in the NMR spectra. Furthermore, small chemical shift dispersion inherent to α-helical structure as well as line broadening owing to large size of the supramolecular system and slow conformational exchange widespread in oligomeric complexes are additional unfavorable factors complicating unambiguous identification of intermonomeric NOE contacts also in the cases of TM heterodimers or asymmetric homodimers.

This symmetry degeneracy problem can be tackled analytically, with the aid of the so-called "ambiguous distance restraints" method (Nilges & O'Donoghue, 1998), according of which spatial structure of a symmetrical dimer is calculated in two stages, involving an initial stage the structure refinement of the monomer subunit before proceeding to the dimer. Experimentally identified NOE contacts are interpreted in a conservative manner and only those that are clearly inconsistent with the global fold of the monomer could be assigned as unambiguous intermonomeric NOE contacts. All other NOE contacts are treated as having arisen from either intra- or intermonomer cross-relaxation. Back in 1997 MacKenzie et al., successfully used this strategy in the pioneering work of determining high-resolution structure of homodimeric TM domain of glycophorin A (GpA), which was solubilized in DPC micellar media (PDB 1AFO) (Fig. 3). Glycophorin A, a surface protein marker of human erythrocytes, is widely used as a model protein in developing the experimental and theoretical methods to study the specific dimerization of TM domains of bitopic proteins. In detergent micelles, the membrane-spanning α-helices of glycophorin A self-associate in a parallel right-handed manner with crossing angle of -40° *via* tetrad repeat dimerization pattern L[75]IxxG[79]VxxG[83]VxxT[87] including the so-called tandem GG4-like motif (also known as 'glycine zipper' (Kim et al., 2005)) composed of residues with small side chains allowing close approach of the helices. Along with numerous van-der-Waals interactions, four close polar CαH⋯O helix-helix contacts, which can be described as non-canonical hydrogen bonds across the dimer interface afforded by GG4-like motif, occur between CαH$_1$ of Gly79 and Gly83 and opposite backbone carbonyls of Ile76 and Val80. The dimer structure also revealed the intramolecular hydrogen bonding of hydroxyl group of Thr87 with backbone carbonyl group of Gly79. As shown recently, the structure of the TM domain dimer of glycophorin A embedded into DMPC/DHPC lipid bicelles is similar (PDB 2KPF, Mineev et al., 2011a). Nevertheless, the formation of an intermonomeric hydrogen bond between side chain hydroxyl group of Thr87 and backbone carboxyl group of Val84 was proposed based on several dipolar interaction observed with solid state NMR using dry DMPC and POPC lipid bilayers (Smith et al., 2001; Smith et al., 2002a). The work of MacKenzie et al., 1997, was an important early accomplishment both for technical reasons and because of the insight that the glycophorin A TM domain structure provides into membrane protein folding and stability.

There is a more straightforward, experimental approach to circumventing the symmetry degeneracy problem through a direct search of intermolecular NOE contacts in dimer interface. For this purpose, an isotopic "heterodimer, consisting of ^{2}H, ^{13}C, ^{15}N isotope labeled and natural abundance monomers, is to be prepared for the NMR experiments to select NOE contacts between isotopically bound and nonisotopically bound protons. Besides the case of symmetrical homodimerization, such experiments are useful for directly obtaining interhelical spatial restraints for asymmetric TM dimers (or oligomers) as well for identifying close intermolecular protein-lipid contacts. A simple method to distinguish intermonomer NOE contacts is to produce a ^{2}H/^{15}N-isotopic "heterodimer", in which one subunit is ^{15}N-labeled and fully deuterated (except NH groups) whereas the other subunit is unlabelled (^{1}H/^{12}C/^{14}N). This method allows directly obtaining interhelical proton-proton restrains from side chain and backbone groups of one subunit to backbone amide groups of the other. Such strategy was successfully used for determination of high-resolution NMR structure of a constitutively disulfide-linked TM domains of the T cell receptor $\zeta\zeta$-chain homodimer embedded into mixed 5:1 DPC/SDS micelles (PDB 2HAC, Call et al., 2006) (Fig. 3). In detergent micelles the TM $\zeta\zeta$-chain helices form a left-handed dimer with a crossing angle +23° *via* extended heptad repeat dimerization pattern $C^2xxL^5D^6xxL^9xxY^{12}xxxL^{16}T^{17}xxF^{20}xxV^{23}$ encompassing almost entire TM segment and making numerous interhelical side chain contacts, several of which are polar. It was shown that the side-chain hydroxyls of Tyr12 and Thr17 form a pair of interhelical hydrogen bonds that create "brackets" defining the lateral edges of the dimer interface. Structural and mutagenesis analysis revealed that two aspartic acid Asp6 situated near intersubunit Cys2-Cys2 bridge, which are required for receptor assembly, can form extensive hydrogen-bonding network with several hydrogen-bond donors and acceptors including at least one water molecule, the cysteine carbonyls, the carboxyl side chain and amide groups of aspartic acids themselves. So, the structure of the TM $\zeta\zeta$-chain dimer nicely demonstrated how multiple hydrogen bonding can establish a left-handed TM homodimer. A more recent study provided the structure of another, functionally homologous TM-signaling dimer, DAP12, both alone (PDB 2L34) (Fig. 3) and with a receptor TM domain, NKG2C (PDB 2L35), in an assembled trimeric complex using mixed-label (^{15}N^{2}H+^{13}C^{1}H)-isotopic "heterodimer" samples (Call et al., 2010). In detergent TDPC/SDS micelles the TM DAP12 helices, linked covalently through a native disulfide bond in extracellular stalk region, form a left-handed dimer with a crossing angle +18° *via* extended heptad repeat dimerization pattern $L^9xxI^{12}V^{13}xxD^{16}xxL^{19}T^{20}xxI^{23}xxxV^{27}$ making numerous interhelical side chain contacts, several of which are polar but without inter-helical hydrogen bonding. Assembled in immunoreceptor complex with the TM domain of type II, C-type lectin-like receptor NKG2C the DAP12 TM dimer formation of an extensive membrane-embedded electrostatic

The strategy of ILV-methyl-selective protonation (Tugarinov & Kay, 2005) was employed for high-resolution structure determination of the heterodimeric TM domain of intact $a_{IIb}\beta_3$ integrin in POPS/POPC/DHPC (q = 0.32) and deuterated DMPC/DHPC (q = 0.30) lipid bicelles (PDB 2K9J, Lau et al., 2009) (Fig. 3). The ^{1}H^{13}C$_3$-Ile,Leu,Val;^{2}H/^{13}C/^{15}N-labeled and unlabelled 1:1 mixtures of the a_{IIb} and β_3 integrin TM subunits were used for partial side-chain assignments and for identification of intermonomeric proton-proton NOE contacts between methyl groups of one subunits and any groups of the second subunit. Guided by packing interaction with three distinct glycine residues, the integrin TM helices cross at an angle of -25° and connect through tetrad repeat patterns $G^{972}xxxG^{976}xxL^{979}L^{980}xxxL^{984}$ and $V^{700}M^{701}xxI^{704}L^{705}xxG^{708}xxxL^{712}$ of a_{IIb} and β_3, respectively, forming a TM heterodimer of unique structural complexity. The assembly enables strong electrostatic interactions (as detected by mutagenesis) between side chains of Arg995 and Asp723 of a_{IIb} and β_3, respectively, within the relatively low dielectric environment of lipid headgroups. The reported heterodimeric TM structure along with structure-based side-directed mutagenesis of $a_{IIb}\beta_3$ integrin provides important insights into the structural basis for integrin signaling in cell membrane, revealing the structural events that underlie the transition from associated to dissociated states upon receptor activation (Lau et al., 2009).

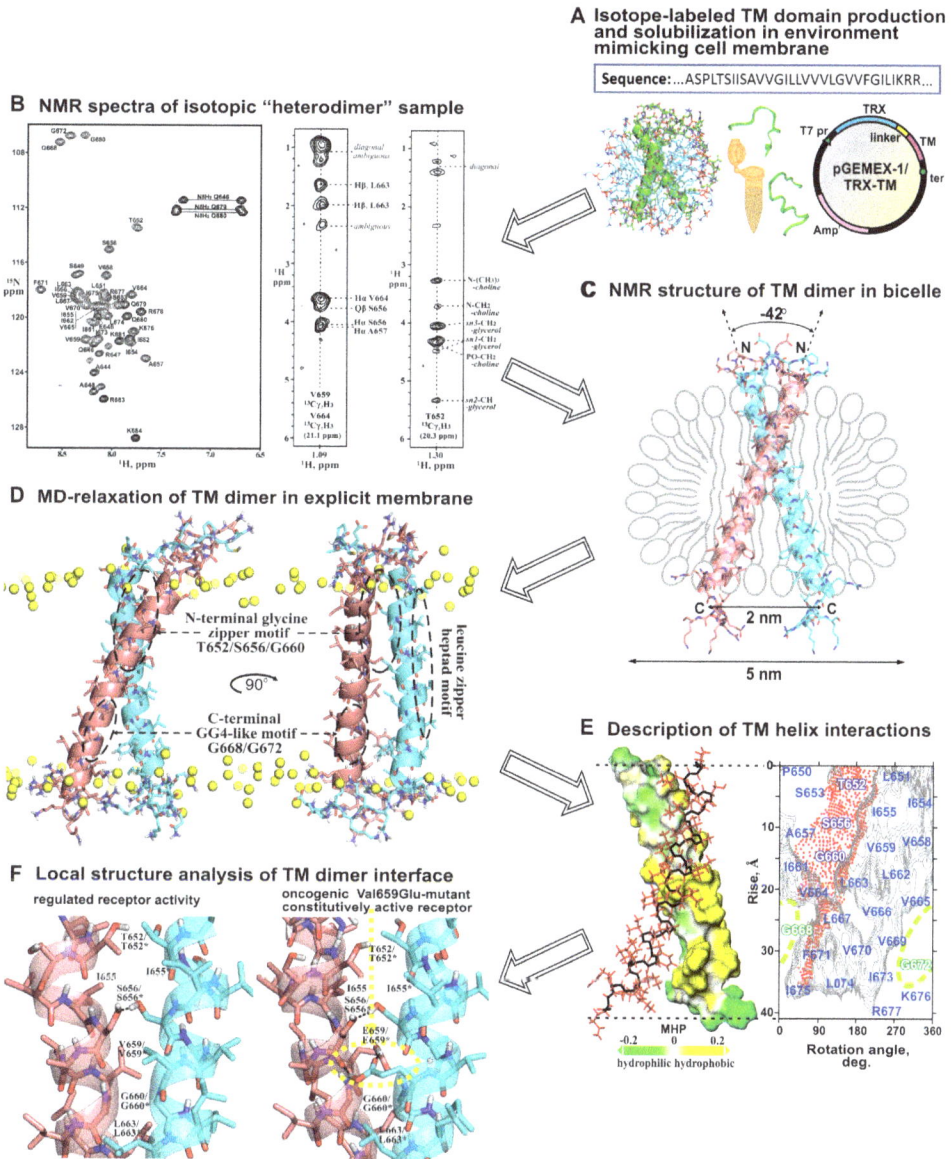

A Isotope-labeled TM domain production and solubilization in environment mimicking cell membrane

Sequence: ...ASPLTSIISAVVGILLVVVLGVVFGILIKRR...

B NMR spectra of isotopic "heterodimer" sample

C NMR structure of TM dimer in bicelle

D MD-relaxation of TM dimer in explicit membrane

N-terminal glycine zipper motif — T652/S656/G660

leucine zipper heptad motif

C-terminal GG4-like motif — G668/G672

E Description of TM helix interactions

F Local structure analysis of TM dimer interface

regulated receptor activity

oncogenic Val659Glu-mutant constitutively active receptor

Fig. 4. Spatial structure elucidation of dimeric TM domains of bitopic proteins with the aid of heteronuclear NMR spectroscopy combined with MD-relaxation (exemplified by receptor tyrosine kinase ErbB2; Bocharov et al., 2008a).

Figure was adapted from Bocharov et al., 2010. (A) Production of the isotope-labeled TM fragments of the protein (e.g. chemical synthesis, bacterial or cell-free expression) and their subsequent solubilization in membrane mimicking environment (e.g. in detergent micelles or lipid bicelles). (B) Acquisition of NMR spectra of isotopic "heterodimer", consisting of

^{13}C/^{15}N-isotope labeled and natural abundance ErbB2 TM fragments (residues 641-685) embedded into DMPC/DHPC lipid bicelles. From *left* to *right*, ^{1}H-^{15}N HSQC spectrum with amide backbone resonance assignments, two representative 2D strips from the 3D ^{13}C F1-filtered/F3-edited-NOESY spectrum with intermolecular protein-protein and protein-lipid NOE contacts are presented. (C) Determination of high-resolution spatial structure of the right-handed ErbB2 TM homodimer in lipid bicelle using NMR-derived restraints. The obtained N-terminal association mode of the ErbB2 TM dimer via N-terminal dimerization motif corresponds to the receptor active state. (D) MD-relaxation of the ErbB2 TM homodimer in hydrated explicit DMPC lipid bilayer with imposed NMR-derived constraints. *Yellow* balls show phosphorus atoms of lipid heads. The spatial locations of the three characteristic dimerization motifs of ErbB2tm are marked by dashed ovals. (E) Analysis of interacting surfaces of the ErbB2 TM helices. In *left*, hydrophobic and hydrophilic (polar) surfaces of one TM helix in the homodimer colored in *yellow* and *green* according to the molecular hydrophobicity potential (MHP) (Efremov & Vergoten, 1995). The second monomer of the dimer is shown with *red* side chains. In *right*, hydrophobicity map for ErbB2 TM helix surface with contour isolines encircling hydrophobic regions with high values of MHP is presented with red-point area indicating the helix packing interface via N-terminal glycine zipper motif T^{652}xxxS^{656}xxxG^{660}. The residues composing C-terminal unemployed dimerization GG4-like motif G^{668}xxxG^{672} are highlighted in green. (F) Local structure analysis of intra- and intermolecular interactions in the ErbB2 TM dimer. Comparison of intermonomeric hydrogen bonding (*black* dotted lines) in the TM helix-helix interface of ErbB2 and its constitutively active Val659Glu-mutant is presented.

A robust strategy to distinguish intermonomeric NOE contacts in protein dimers was based on producing a ^{13}C/^{15}N-isotopic "heterodimer", in which one subunit is ^{13}C/^{15}N-labeled and the other subunit is unlabelled. For the direct detection of the intermolecular NOE contacts in such isotopic "heterodimer", NMR pulse sequences were developed (Zwahlen et al., 1997; Stuart et al., 1999), employing so-called X-filtering elements to select NOE contacts arising between nonisotopically and isotopically bound protons. Due to fast transverse magnetic relaxation as consequence of relatively big overall correlation time of the studied supramolecular systems, the intermonomeric proton-proton contacts in the ^{13}C/^{15}N-isotopic "heterodimer" are mainly detected from methyl groups (having smallest relaxation rates) to other groups. This approach was successfully applied in our lab for elucidation of structural-dynamic properties of homo- and heterodimeric α-helical TM domains of several biologically different human proteins, including proapoptotic protein BNip3 and representatives of receptor tyrosine kinase ErbB and Eph subfamilies. The high-resolution NMR structures of dimeric TM domains of these bitopic proteins were obtained using DMPC/DHPC (q = 0.25) bicelles consisting of lipids with deuterated hydrophobic tails and lipid/protein molar ratios of ~35. The resulting NMR structures of the TM domain dimers were subjected to energy relaxation using MD during several ns of MD trajectory in hydrated explicit lipid bilayers with the imposed NMR-derived constraints and then without constraints to study the conformational stability of the dimer in the membrane. The MD relaxation procedure provided a detailed atomistic picture of the intra- and intermolecular (protein-protein, protein-membrane and protein-water) interactions and allowed estimating the influence of amino acid substitution, including pathogenic TM mutations, on the structural-dynamic properties of bitopic proteins, see Fig. 4.

BNip3 is a prominent representative of apoptotic Bcl-2 proteins with unique properties initiating an atypical programmed cell death pathway (Chen et al., 1999). Investigation of spatial structure and internal dynamics of the homodimeric TM domain of human protein BNip3 (PDB 2J5D, Bocharov et al., 2007) revealed that in the lipid bicelles the central membrane-spanning α-helices of BNip3 cross at the angle of -45° and form a right-handed parallel symmetric dimer via tetrad repeat pattern $S^{172}H^{173}xxA^{176}xxxG^{180}xxxG^{184}$ (Fig. 3). In addition, labile *Phe*-ring hydrophobic cluster with numerous intermonomeric stacking interactions between six phenylalanine residues $(Phe^{157}/Phe^{161}/Phe^{165})_2$ was identified in the interface between short mobile N-terminal helices, flanking the central helices. According to the obtained NMR data supported by MD relaxation, a hydrophilic motif $(Ser^{172}/His^{173})_2$ in the centre of dimerization interface of BNip3 TM domain forms a water-accessible *His-Ser* node of inter- and intramonomeric hydrogen bonds decreasing apparent pKa of the imidazole group below 4. The C-terminal TM part of the BNip3tm dimer is stabilized by van-der-Waals side chain contacts and by weakly hydrophilic backbone contacts of the helices tightly self-associated through a glycine zipper motif, which appears to be essential for proper alignment of the side chains in the *His-Ser* node required for hydrogen bonding. In the DMPC/DHPC bicelles the *His-Ser* node undergoes slow conformational exchange with ~10% occupancy of the minor state probably associated with alternative hydrogen bonding and water permeability. Nevertheless, it was shown that an addition of long chain DPPC lipid to DPC micelles (lipid/detergent ratio of 1:50) with embedded dimeric BNiP3 TM domain allows to eliminate the conformational inhomogenity in the dimer interface (Sulistijo & Mackenzie, 2009). The revealed structural-dynamic properties of the BNip3 TM domain with a potentially switchable network of hydrogen bonds and water accessibility up to the middle of the membrane appear to enable the protein to form ion-conducting pathway across the membranes. Indeed, the TM domain was shown to induce conductivity of artificial bilayer lipid membrane in a pH-dependent manner (Bocharov et al., 2007). These findings and currently available information about phenomenology of programmed cell death allowed us to propose a mechanism of triggering necrosis-like cell death by BNip3 in case of hypoxia-acidosis of human tissues.

Receptor tyrosine kinases conducting biochemical signals across plasma membrane *via* lateral dimerization play an important role in normal and in pathological conditions of human organism by providing cell signaling, maintaining cellular homeostasis and controlling cell fate (Schlessinger, 2000). Eph receptors are found in a wide variety of cells in developing and mature tissues and represent the largest family of receptor tyrosine kinases regulating cell shape, movement, and attachment (Pasquale, 2005). Because all Eph receptors and their ligand ephrins are cell surface-associated proteins, a direct cell-cell contact is required for receptor activation resulting in cytoskeletal remodeling that underlies cell adhesion, repulsion and motility in both communicating cells. Although the Eph TM segments reveal relatively low amino acid sequence homology, several dimerization motifs, including at least one explicit GG4-like motif, can be identified in each Eph TM region. Structural-dynamic properties of the homodimeric TM domains of the EphA1 and EphA2 receptors were investigated with the aid of solution NMR in lipid bicelles and MD relaxation in explicit lipid bilayers of different composition. High-resolution spatial structures of homodimeric TM domains of EphA1 (PDB 2K1K and 2K1L, Bocharov et al., 2008a) and EphA2 (PDB 2K9Y, Bocharov et al., 2010a) embedded into DMPC/DHPC bicelles (q = 0.25) revealed a right- and left-handed parallel packing of the α-helical TM domains with crossing angle of -45° and +15°, respectively (Fig. 3). The EphA1 TM segment self-associates through the N-terminal glycine zipper motif $A^{550}xxxG^{554}xxxG^{558}$ whereas the C-terminal GG4-like dimerization motif $A^{560}X_3G^{564}$ is not employed. And vice versa, the EphA2 TM helices interact through the extended heptad repeat motif $L^{535}xxxG^{539}xxA^{542}xxxV^{546}xxxL^{549}$ assisted by intermolecular stacking interactions of aromatic rings of $(FF^{557})_2$, whereas the N-terminal glycine zipper motif $A^{536}X_3G^{540}X_3G^{544}$ remains vacant. Thus, our studies of the Eph1 and EphA2 receptors demonstrated that the TM domains of different representatives of the same receptor tyrosine kinase family can use alternative dimerization motifs in the same bicellar system, the different motifs possibly being corresponding to active and inactive dimeric state of the receptor. Moreover, in the case of EphA1 TM domain, variations of external pH and lipid composition of the bicelles initiated triggering between the alternative motifs, which can be viewed as an argument in favor to the so-called "rotation-coupled" mechanism of the receptor tyrosine kinase activation (Moriki et al., 2001; Fleishman et al., 2002; Mendrola et al., 2002). The obtained results indicated also that alternative dimeric conformations of the TM domains can influence the receptor localization in plasma membrane microdomains and signaling platform, such as rafts and caveolae (Bocharov et al., 2010a).

Four human ErbB members of epidermal growth factor receptor family form numerous homo- and heterodimer combinations, recognizing different EGF-related ligands and performing diverse functions in a complex signaling network (Warren & Landgraf, 2006). All the species of the ErbB family are activated by proper ligand-induced dimerization or by reorientation of monomers in preformed receptor dimers after ligand binding that can be widespread among receptor tyrosine kinase family (Schlessinger, 2000; Tao & Maruyama, 2008). The TM segments of all four human ErbB receptors have at least one such motif, and all except ErbB3 have two of them, located in the N- and C-terminal parts of the TM helices. So, two possible dimeric conformations of the α-helical ErbB TM segments with interfaces located either at N- or C-terminus were proposed to associate different receptor active states (Moriki et al., 2001; Fleishman et al., 2002; Mendrola et al., 2002). According to high-resolution spatial structure of homodimeric ErbB2 TM domain embedded into DMPC/DHPC lipid bicelles (PDB 2JWA, Bocharov et al., 2008b), the α-helical TM segments of ErbB2 interact with right-handed crossing angle of -42° through the N-terminal glycine zipper motif $T^{652}xxxS^{656}xxxG^{660}$ (Fig. 4). Polar contact area of this motif is shielded from lipid tails by the side chains of leucine, isoleucine, and valine residues, while slightly polar concave surface of the C-terminal GG4-like motif $G^{668}xxxG^{672}$ is exposed to hydrophobic lipid environment. In the C-terminal part of the dimeric interface, aromatic rings of the opposite Phe671 residues participate in intermolecular edge-face stacking interaction. Constrained MD relaxation of the ErbB2tm dimer structure revealed that the $(Thr652/Ser656)_2$ hydrophilic motif in the N-terminal part of the dimerization interface forms a node of switching inter- and intramonomeric hydrogen bonds mediating the ErbB2 TM helix packing. Based on the NMR-derived structure it was also shown by molecular modeling that pro-oncogenic Val659Glu mutation leads to overstabilization of the described ErbB2 TM domain conformation which was ascribed to the active state of the tyrosine kinase. Spatial structure of the heterodimeric complex formed by TM domains of ErbB1 and ErbB2 receptors was also obtained using the bicellar environment, in which the domains associate in a right-handed α-helical bundle with crossing angle of -46° through their N-terminal double GG4-like motif $T^{648}G^{649}X_2G^{652}A^{653}$ and glycine zipper motif $T^{652}X_3S^{656}X_3G^{660}$, respectively (PDB 2KS1, Mineev et al., 2010) (Fig. 3). The described heterodimer conformation is believed to support the juxtamembrane and kinase domain configuration corresponding to the receptor active state. The capability for multiple polar interactions along with hydrogen bonding between TM segments correlates with the observed highest affinity of the ErbB1/ErbB2 heterodimer, implying an important contribution of the TM helix-helix interaction to signal transduction. Recently, an alternative left-handed homodimeric conformation was described for the ErbB3 TM domain embedded in DPC micelles (PDB 2L9U, Mineev et al., 2011b) (Fig. 3). The tight association of ErbB3tm α-helices with crossing angle of +24° is accomplished via the extended heptad-like motif $I^{649}xxL^{652}VxI^{655}FxxL^{659}xxxF^{663}LxxR^{667}$, which is similar to the motif, implemented in the dimerization of TM segments of EphA2 tyrosine kinase receptor (Bocharov et al., 2008a).

The assumption that the N-terminal association mode of the ErbB TM dimer corresponds to the receptor active state has been supported by recent structural studies of the juxtamembrane segment and kinase domain dimerization upon kinase activation of the ErbB1 receptor (Jura et al., 2009). It was shown that folding of the juxtamembrane regions of both monomers in the receptor dimer into an antiparallel helical structure, requiring the spacing between the C-termini of the TM helices to be about 20 Å, is essential for the kinase domain activation (Jura et al., 2009). The homodimeric ErbB2 TM structure we obtained has exactly the required distance between the C-termini of the TM helices (Fig. 4C), and is thus allowing proper kinase domain activation. Overall these findings enhance understanding of the functional conformational changes of receptor tyrosine kinases during activation of the signaling ligand-receptor complex in cell membranes in normal and pathologic states of human organism.

6. Conclusion

Information about structure and dynamic of non-covalently bonded protein oligomers in the membrane is very challenging to obtain. To date, there are only a few experimentally solved dimeric structures of the TM domains of bitopic proteins. Several strategies based on various theoretical and physicochemical methods and their combination are currently available, providing structural-dynamic information about atomic-scale details of TM helix-helix and helix-membrane interactions. Experimental high-resolution structure obtained in a particular membrane mimicking environment usually corresponds to only one of homo- or heterodimeric states of TM domains, which are apparently realized in vivo in the course of bitopic protein activity. Even if special selection of environment allows obtaining an alternative conformation for some proteins, it is impossible to stabilize every conformation of interest to live long enough for comprehensive experimental investigation by mere choice of the external conditions. Molecular modeling, in its turn, predicts all possible alternative dimerization interfaces of the bitopic protein TM domains, existence of which in vivo should be verified in experiment wherever possible. Many aspects of the specific helix-helix interactions in membranes are yet far from being completely understood and are awaiting detailed investigation, which is only possible through concerted use of various physical-chemical and biological methods supported by molecular modeling. Theoretical and experimental methods to study protein-protein and protein-lipid interactions in membrane are rapidly evolving in a correlated manner. Molecular modeling is used to support interpretation of data about specific TM helix association, whereas the theoretical modeling parameters are refined based on the experimentally obtained information. This will likely result, within a few years to come, in detailed description of a large variety of intra- and intermolecular interactions in membranes and elucidation of the roles of the TM domains in normal and abnormal functioning of the proteins and in their proper localization in cell membranes. The most important practical implications of these studies are primarily related to molecular design of pharmaceutical compositions that can affect specific helix-helix association in cell membrane, providing a novel form of therapy of many human diseases related with abnormal activity of the bitopic proteins. Naturally, that does not diminish the current importance and topicality of obtaining the structure of full-length bitope proteins both separately and in complexes. However, at the present state of development of structural biology this remains quite an ambitious undertaking.

7. Acknowledgments

Financial support was received from the Russian Foundation for Basic Research, the Program of the Russian Academy of Science "Molecular and Cellular Biology", the Russian Funds Investment Group, the Federal Target Program "Research and development in priority fields of Russian scientific and technological complex in 2007-2013". E.V. Bocharov thanks personally K.A. Beirit for financial support.

8. References

Adamian, L., Jackups, R., Binkowski, T.A. & Liang, J. (2003). Higher-order interhelical spatial interactions in membrane proteins. *J. Mol. Biol.*, Vol. 327, pp. 251-272

Adamian, L., Nanda, V., DeGrado, W.F. & Liang, J. (2005). Empirical lipid propensities of amino acid residues in multispan alpha helical membrane proteins. *Proteins*, Vol. 59, pp. 496-509

Adams, P.D., Engelman, D.M. & Brünger, A.T. (1996). Improved prediction for the structure of the dimeric transmembrane domain of glycophorin A obtained through global searching. *Proteins*, Vol. 26, pp. 257-261

Andronesi, O.C., Becker, S., Seidel, K., Heise, H., Young, H.S. & Baldus, M. (2005). Determination of membrane protein structure and dynamics by magic-angle-spinning solid-state NMR spectroscopy. *J. Am. Chem. Soc.*, Vol. 127, pp. 12965-12974

Arbely, E. & Arkin, I.T. (2004). Experimental measurement of the strength of a C alpha-H...O bond in a lipid bilayer. *J. Am. Chem. Soc.*, Vol. 126, pp. 5362-5363

Arkin, I.T. & Brunger, A.T. (1998). Statistical analysis of predicted transmembrane alpha-helices. *Biochim. Biophys. Acta*, Vol. 1429, pp. 113-128

Beevers, A.J. & Kukol, A. (2006). The transmembrane domain of the oncogenic mutant ErbB-2 receptor: a structure obtained from site-specific infrared dichroism and molecular dynamics. *J. Mol. Biol.*, Vol. 361, pp. 945-953

Bennasroune, A., Fickova, M., Gardin, A., Dirrig-Grosch, S., Aunis, D., Cremel, G. & Hubert, P. (2004). Transmembrane peptides as inhibitors of ErbB receptor signaling. *Mol. Biol. Cell.*, Vol. 15, pp.. 3464-3474

Bocharov, E.V., Pustovalova, Y.E., Pavlov, K.V., Volynsky, P.E., Goncharuk, M.V., Ermolyuk, Y.S, Karpunin, D.V., Schulga, A.A., Kirpichnikov, M.P., Efremov, R.G., Maslennikov, I.V. & Arseniev, A.S. (2007). Unique dimeric structure of BNip3 transmembrane domain suggests membrane permeabilization as a cell death trigger. *J. Biol. Chem.*, Vol. 282, pp. 16256-16266

Bocharov, E.V., Mayzel, M.L., Volynsky, P.E., Goncharuk, M.V., Ermolyuk, Y.S., Schulga, A.A., Artemenko, E.O., Efremov, R.G. & Arseniev, A.S. (2008a). Spatial structure and pH-dependent conformational diversity of dimeric transmembrane domain of the receptor tyrosine kinase EphA1. *J. Biol. Chem.*, Vol. 283, pp. 29385-29395

Bocharov, E.V., Mineev, K.S., Volynsky, P.E., Ermolyuk, Y.S., Tkach, E.N., Sobol, A.G., Chupin, V.V., Kirpichnikov, M.P., Efremov, R.G. & Arseniev, A.S. (2008b). Spatial structure of the dimeric transmembrane domain of the growth factor receptor ErbB2 presumably corresponding to the receptor active state. *J. Biol. Chem.*, Vol. 283, pp. 6950-6956

Bocharov, E.V., Mayzel, M.L., Mineev, K.S., Tkach, E.N., Ermolyuk, Y.S., Schulga, A.A. & Arseniev, A.S. (2010a). Left-handed dimer of EphA2 transmembrane domain: helix packing diversity among receptor tyrosine kinases. *Biophys. J.*, Vol. 98, pp. 881-889

Bocharov, E.V., Volynsky, P.E., Pavlov, K.V., Efremov, R.G. & Arseniev, A.S. (2010b). Structure elucidation of dimeric transmembrane domains of bitopic proteins. *Cell Adh. Migr.*, Vol. 4, pp. 284-298

Borch, J. & Hamann, T. (2009). The nanodisc: a novel tool for membrane protein studies. *Biol. Chem.*, Vol. 390, pp. 805-814

Bordag, N. & Keller, S. (2010). Alpha-helical transmembrane peptides: a "divide and conquer" approach to membrane proteins. *Chem. Phys. Lipids*, Vol. 163, pp. 1-26

Call, M.E., Schnell, J.R., Xu, C., Lutz, R.A., Chou, J.J. & Wucherpfennig, K.W. (2006). The structure of the zetazeta transmembrane dimer reveals features essential for its assembly with the T cell receptor. *Cell*, Vol. 127, pp. 355-368

Call, M.E., Wucherpfennig, K.W. & Chou, J.J. (2010). The structural basis for intramembrane assembly of an activating immunoreceptor complex. *Nat. Immunol.*, Vol. 11, pp. 1023-1029

Caputo, G.A., Litvinov, R.I., Li, W., Bennett, J.S., Degrado, W.F. & Yin, H. (2008). Computationally designed peptide inhibitors of protein-protein interactions in membranes. *Biochemistry*, Vol. 47, pp. 8600-8606

Casciari, D., Seeber, M. & Fanelli, F. (2006). Quaternary structure predictions of transmembrane proteins starting from the monomer: a docking-based approach. *BMC Bioinformatics*, Vol. 7, pp. 340

Chen, G., Cizeau, J., Vande Velde, C., Park, J.H., Bozek, G., Bolton, J., Shi, L., Dubik, D. & Greenberg, A. (1999). Nix and Nip3 form a subfamily of pro-apoptotic mitochondrial proteins. *J. Biol. Chem.*, Vol. 274, pp. 7-10.

Cherezov, V., Rosenbaum, D.M., Hanson, M.A., Rasmussen, S.G., Thian, F.S., Kobilka, T.S., Choi, H.J., Kuhn, P., Weis, W.I., Kobilka, B.K. & Stevens, R.C. (2007). High-resolution crystal structure of an engineered human beta2-adrenergic G protein-coupled receptor. *Science*, Vol. 318, pp. 1258-1265

Choma, C., Gratkowski, H., Lear, J.D. & DeGrado, W.F. (2000). Asparagine-mediated self-association of a model transmembrane helix. *Nat. Struct. Biol.*, Vol. 7, pp. 161-166

Chou, J.J., Kaufman, J.D., Stahl, S.J., Wingfield, P.T. & Bax, A. (2002). Micelle-induced curvature in a water-insoluble HIV-1 Env peptide revealed by NMR dipolar coupling measurement in stretched polyacrylamide gel. *J. Am. Chem. Soc.*, Vol. 124, pp. 2450-2451

Constantinescu, S.N., Keren, T., Socolovsky, M., Nam, H., Henis, Y.I. & Lodish, H.F. (2001). Ligand-independent oligomerization of cell-surface erythropoietin receptor is mediated by the transmembrane domain. *Proc. Natl. Acad. Sci. USA*, Vol. 98, pp. 4379-4384

Curran, A.R. & Engelman, D.M. (2003). Sequence motifs, polar interactions and conformational changes in helical membrane proteins. *Curr. Opin. Struct. Biol.*, Vol. 13, pp. 412-417

Dawson, J.P., Weinger, J.S. & Engelman, D.M. (2002). Motifs of serine and threonine can drive association of transmembrane helices. *J. Mol. Biol.*, Vol. 316, pp. 799-805

de Meyer, F.J., Venturoli, M. & Smit, B. (2008). Molecular simulations of lipid-mediated protein-protein interactions. *Biophys. J.*, Vol. 95, pp. 1851-1865

Efremov, R.G. & Vergoten, G. (1995). Hydrophobic nature of membrane-spanning α-helical peptides as revealed by Monte Carlo simulations and molecular hydrophobicity potential analysis. *J. Phys. Chem.*, Vol. 99, pp. 10658-1066

Efremov, R.G., Nolde, D.E., Konshina, A.G., Syrtcev, N.P. & Arseniev, A.S. (2004). Peptides and proteins in membranes: what can we learn via computer simulations? *Curr. Med. Chem.*, Vol. 11, pp. 2421-2442

Efremov, R.G., Vereshaga, Y.A, Volynsky, P.E., Nolde, D.E. & Arseniev, A.S. (2006). Association of transmembrane helices: what determines assembling of a dimer? *J. Comput. Aided. Mol. Des.*, Vol. 20, pp. 27-45

Feig, M. & Brooks, C.L. 3rd. (2004). Recent advances in the development and application of implicit solvent models in biomolecule simulations. *Curr. Opin. Struct. Biol.*, Vol. 14, pp. 217-224

Fleishman, S.J., Schlessinger, J. & Ben-Tal, N. (2002). A putative molecular-activation switch in the transmembrane domain of ErbB2. *Proc. Natl. Acad. Sci. USA*, Vol. 99, pp. 15937-15940

Forrest, L.R. & Sansom, M.S. (2000). Membrane simulations: bigger and better? *Curr. Opin. Struct. Biol.*, Vol. 10, pp. 174-181

Gautier, A., Kirkpatrick, J.P. & Nietlispach, D. (2008). Solution-state NMR spectroscopy of a seven-helix transmembrane protein receptor: backbone assignment, secondary structure, and dynamics. *Angew. Chem. Int. Ed. Engl.*, Vol. 47, pp. 7297-7300

Glover, K.J., Whiles, J.A., Wu, G., Yu, N., Deems, R., Struppe, J.O., Stark, R.E., Komives, E.A. & Vold, R.R. (2001). Structural evaluation of phospholipid bicelles for solution-state studies of membrane-associated biomolecules. *Biophys. J.*, Vol. 81, pp. 2163-2171

Gratkowski, H., Lear, J.D. & DeGrado, W.F. (2001). Polar side chains drive the association of model transmembrane peptides. *Proc. Natl. Acad. Sci. USA*, Vol. 98, pp. 880-885

Gratkowski, H., Dai, Q., Wand, A.J., DeGrado, W.F. & Lear, J.D. (2002). Cooperativity and specificity of association of a designed transmembrane peptide. *Biophys. J.*, Vol. 83, pp. 1613-1619

Helms, V. (2002). Attraction within the membrane. Forces behind transmembrane protein folding and supramolecular complex assembly. *EMBO Rep.*, Vol. 3, pp. 1133-1138

Im, W., Feig, M. & Brooks, C.L. 3rd. (2003). An implicit membrane generalized born theory for the study of structure, stability, and interactions of membrane proteins. *Biophys. J.*, Vol. 85, pp. 2900-2918

Johnson, R.M., Rath, A., & Deber, C.M. (2006). The position of the Gly-xxx-Gly motif in transmembrane segments modulates dimer affinity. *Biochem. Cell Biol.*, Vol. 84, pp. 1006–1012

Johnson, R.M., Hecht, K. & Deber, C.M. (2007). Aromatic and cation-pi interactions enhance helix-helix association in a membrane environment. *Biochemistry*, Vol. 46, pp. 9208-9214

Jura, N., Endres, N.F., Engel, K., Deindl, S., Das, R., Lamers, M.H., Wemmer, D.E., Zhang, X. & Kuriyan, J. (2009). Mechanism for activation of the EGF receptor catalytic domain by the juxtamembrane segment. *Cell*, Vol. 137, pp. 1293-1307

Kim, H.J., Howell, S.C., Van Horn, W.D., Jeon, Y.H. & Sanders, C.R. (2009). Recent advances in the application of solution NMR spectroscopy to multi-span integral membrane proteins. *Prog. Nucl. Magn. Reson. Spectrosc.*, Vol. 55, pp. 335-360

Kim, S., Jeon, T.-J., Oberai, A., Yang, D., Schmidt, J.J. & Bowie. J.U. (2005). Transmembrane glycine zippers: physiological and pathological roles in membrane proteins. *Proc. Natl. Acad. Sci. USA*, Vol. 102, pp. 14279-14283

Krogh, A.; Larsson, B.; von Heijne, G. & Sonnhammer, E.L. (2001). Predicting transmembrane protein topology with a hidden Markov model: application to complete genomes. *J. Mol. Biol.*, Vol. 305, pp. 567-580

Krueger-Koplin, R.D., Sorgen, P.L., Krueger-Koplin, S.T., Rivera-Torres, I.O., Cahill, S.M., Hicks, D.B., Grinius, L., Krulwich, T.A. & Girvin, M.E. (2004). An evaluation of detergents for NMR structural studies of membrane proteins. *J. Biomol. NMR*, Vol. 28, pp. 43-57

Langosch, D. & Heringa, J. (1998). Interaction of transmembrane helices by a knobs-into-holes packing characteristic of soluble coiled coils. *Proteins*, Vol. 31, pp. 150-159

Langosch, D., Lindner, E. & Gurezka, R. (2002). In vitro selection of self-interacting transmembrane segments--membrane proteins approached from a different perspective. *IUBMB Life*, Vol. 54, pp. 109-113

Langosch, D. & Arkin, I.T. (2009). Interaction and conformational dynamics of membrane-spanning protein helices. *Protein Sci.*, Vol. 18, pp. 1343-1358

Lau, T.L., Partridge, A.W., Ginsberg M.H. & Ulmer, T.S. (2008). Structure of the integrin beta3 transmembrane segment in phospholipid bicelles and detergent micelles. *Biochemistry*, Vol. 47, pp. 4008-4016

Lau, T.L., Kim, C., Ginsberg, M.H. & Ulmer, T.S. (2009). The structure of the integrin aIIbβ3 transmembrane complex explains integrin transmembrane signalling. *EMBO J.*, Vol. 28, pp. 1351-1361

Lear, J.D., Stouffer, A.L., Gratkowski, H., Nanda, V. & Degrado, W.F. (2004). Association of a model transmembrane peptide containing gly in a heptad sequence motif. *Biophys. J.*, Vol. 87, pp. 3421–3429

Lee, A.G. (2004). How lipids affect the activities of integral membrane proteins. *Biochim. Biophys. Acta*, Vol. 1666, pp. 62–87

Lee, D., Walter, K.F., Brückner, A.K., Hilty, C., Becker, S. & Griesinger, C. (2008). Bilayer in small bicelles revealed by lipid-protein interactions using NMR spectroscopy. *J. Am. Chem. Soc.*, Vol. 130, pp. 13822-13823

Lemmon, M.A., Flangan, J.M., Treutlein, H.R., Zhang, J. & Engelman, D.M. (1992). Sequence specificity in the dimerization of transmembrane alpha-helices. *Biochemistry*, Vol. 31, pp. 12719-12725

Li, E. & Hristova, K. (2006). Role of receptor tyrosine kinase transmembrane domains in cell signaling and human pathologies. *Biochemistry*, Vol. 45, pp. 6241-6251

Li, E., Merzlyakov, M., Lin, J., Searson, P. & Hristova, K. (2009). Utility of surface-supported bilayers in studies of transmembrane helix dimerization. *J. Struct. Biol.*, Vol. 168, pp. 53-60

Lind, J., Nordin, J. & Mäler, L. (2008). Lipid dynamics in fast-tumbling bicelles with varying bilayer thickness: effect of model transmembrane peptides. *Biochim. Biophys. Acta*, Vol. 1778, pp. 2526-2534

Lindberg, M., Biverståhl, H., Gräslund, A. & Mäler, L. (2003). Structure and positioning comparison of two variants of penetratin in two different membrane mimicking systems by NMR. *Eur. J. Biochem.*, Vol. 270, pp. 3055-3063

Luchette, P.A., Vetman, T.N., Prosser, R.S., Hancock, R.E., Nieh, M.P., Glinka, C.J., Krueger, S. & Katsaras, J. (2001). Morphology of fast-tumbling bicelles: a small angle neutron scattering and NMR study. *Biochim. Biophys. Acta*, Vol. 1513, pp. 83–94

Lupas, A. (1996). Coiled coils: New structures and new functions. *Trends Biochem. Sci.*, Vol. 21, pp. 375–382

Lyukmanova, E.N., Shenkarev, Z.O., Paramonov, A.S., Sobol, A.G., Ovchinnikova, T.V., Chupin, V.V., Kirpichnikov, M.P., Blommers, M.J. & Arseniev, A.S. (2008). Lipid-protein nanoscale bilayers: a versatile medium for NMR investigations of membrane proteins and membrane-active peptides. *J. Am. Chem. Soc.*, Vol. 130, pp. 2140-2141

MacKenzie, K.R., Prestegard, J.H. & Engelman, D.M. (1997). A transmembrane helix dimer: structure and implications. *Science*, Vol. 276, pp. 131-133

Mackenzie, K.R. (2006). Folding and stability of alpha-helical integral membrane proteins. *Chem. Rev.*, Vol. 106, pp. 1931-1977

Marsh, D. (2008). Protein modulation of lipids, and vice-versa, in membranes. *Biochim. Biophys. Acta*, Vol. 1778, pp. 1545-1575

Maslennikov, I., Kefala, G., Johnson, C., Riek, R., Choe, S. & Kwiatkowski, W. (2007). NMR spectroscopic and analytical ultracentrifuge analysis of membrane protein detergent complexes. *BMC Struct. Biol.*, Vol. 7, pp. 74

Melnyk, R.A., Partridge, A.W. & Deber, C.M. (2001). Retention of native-like oligomerization states in transmembrane segment peptides: application to the Escherichia coli aspartate receptor. *Biochemistry*, Vol. 40, pp. 11106-11113

Melnyk, R.A., Partridge, A.W. & Deber, C.M. (2002). Transmembrane domain mediated self-assembly of major coat protein subunits from Ff bacteriophage. *J. Mol. Biol.*, Vol. 315, pp. 63-72

Mendrola, J.M., Berger, M.B., King, M.C. & Lemmon, M.A. (2002). The single transmembrane domains of ErbB receptors self-associate in cell membranes. *J. Biol. Chem.*, Vol. 277, pp. 4704–4712

Metcalf, D.G, Kulp, D.W., Bennett, J.S., DeGrado, W.F. (2009). Multiple approaches converge on the structure of the integrin alphaIIb/beta3 transmembrane heterodimer. *J. Mol. Biol.*, Vol. 392, pp. 1087-1101

Mineev, K.S., Bocharov, E.V., Pustovalova, Y.E., Bocharova, O.V., Chupin, V.V. & Arseniev, A.S. (2010). Spatial structure of the transmembrane domain heterodimer of ErbB1 and ErbB2 receptor tyrosine kinases. *J. Mol. Biol.*, Vol. 400, pp. 231-243

Mineev, K.S., Bocharov, E.V., Volynsky, P.E., Goncharuk, M.V., Tkach, E.N., Ermolyuk, Ya.S., Schulga, A.A., Chupin, V.V., Maslennikov, I.V., Efremov, R.G. & Arseniev, A.S. (2011a). Dimeric structure of transmembrane domain of glycophorin A in lipidic and in detergent environments. *Acta Naturae*, Vol. 3, pp. 85-93

Mineev, K.S., Khabibullina, N.F., Lyukmanova, E.N., Dolgikh, D.A., Kirpichnikov, M.P. & Arseniev, A.S. (2011b). Spatial structure and dimer-monomer equilibrium of the ErbB3 transmembrane domain in DPC micelles. *Biochim. Biophys. Acta*, Vol. 1808, pp. 2081-2088

Moore, D.T., Berger, B.W. & DeGrado, W.F. (2008). Protein-protein interactions in the membrane: sequence, structural, and biological motifs. *Structure*, Vol. 16, pp. 991-1001

Moriki, T., Maruyama, H. & Maruyama, I. N. (2001). Activation of preformed EGF receptor dimers by ligand-induced rotation of the transmembrane domain. *J. Mol. Biol.*, Vol. 311, pp. 1011-1026

Morrow, M.R., Huschilt, J.C. & Davis, J.H. (1985). Simultaneous modeling of phase and calorimetric behavior in an amphiphilic peptide/phospholipid model membrane. *Biochemistry*, Vol. 24, pp. 5396-5406

Mottamal, M. & Lazaridis, T. (2005). The contribution of C alpha-H...O hydrogen bonds to membrane protein stability depends on the position of the amide. *Biochemistry*, Vol. 44, pp. 1607-1613

Nath, A., Atkins, W.M. & Sligar, S.G. (2007). Applications of phospholipid bilayer nanodiscs in the study of membranes and membrane proteins. *Biochemistry*, Vol. 46, pp. 2059-2069

Nilges, M., & O'Donoghue, S.I. (1998). Ambiguous NOEs and automated NOE assignment. *Prog. Nucl. Magn. Reson. Spectrosc.*, Vol. 32, pp. 107-139

North, B., Cristian, L., Fu Stowell, X., Lear, J.D., Saven, J.G., & Degrado, W.F. (2006). Characterization of a membrane protein folding motif, the Ser zipper, using designed peptides. *J. Mol. Biol.*, Vol. 359, pp. 930–939

Nyholm, T.K., Ozdirekcan, S. & Killian, J.A. (2007). How protein transmembrane segments sense the lipid environment. *Biochemistry*, Vol. 46, pp. 1457-1465

Opella, S.J. & Marassi, F.M. (2004). Structure determination of membrane proteins by NMR spectroscopy. *Chem. Rev.*, Vol. 104, pp. 3587-3606

Overington, J.P.; Al-Lazikani, B. & Hopkins, A.L. (2006). How many drug targets are there? *Nat. Rev. Drug Discov.*, Vol. 5, pp. 993-996

Page, R.C., Moore, J.D., Nguyen, H.B, Sharma, M., Chase, R., Gao, F.P., Mobley, C.K, Sanders, C.R., Ma, L., Sönnichsen, F.D., Lee, S., Howell, S.C., Opella, S.J. & Cross, T.A. (2006). Comprehensive evaluation of solution nuclear magnetic resonance spectroscopy sample preparation for helical integral membrane proteins. *J. Struct. Funct. Genomics.*, Vol. 7, pp. 51-64

Pasquale, E.B. (2005). Eph receptor signalling casts a wide net on cell behaviour. *Nat. Rev. Mol. Cell Biol.*, Vol. 6, pp. 462-475.

Peng, W.C., Lin, X. & Torres, J. (2009). The strong dimerization of the transmembrane domain of the fibroblast growth factor receptor (FGFR) is modulated by C-terminal juxtamembrane residues. *Protein Sci.*, Vol. 18, pp. 450-459

Pervushin, K., Riek, R., Wider, G. & Wüthrich, K. (1997). Attenuated T2 relaxation by mutual cancellation of dipole-dipole coupling and chemical shift anisotropy indicates an avenue to NMR structures of very large biological macromolecules in solution. *Proc. Natl. Acad. Sci. USA*, Vol. 94, pp. 12366-12371

Petrache, H.I., Grossfield, A., MacKenzie, K.R., Engelman, D.M. & Woolf, T.B. (2000). Modulation of glycophorin A transmembrane helix interactions by lipid bilayers: molecular dynamics calculations. *J. Mol. Biol.*, Vol. 302, pp. 727-746

Poget, S.F. & Girvin, M.E. (2007). Solution NMR of membrane proteins in bilayer mimics: small is beautiful, but sometimes bigger is better. *Biochim. Biophys. Acta.*, Vol. 1768, pp. 3098-3106

Popot, J.L. & Engelman, D.M. (1990). Membrane protein folding and oligomerization: the two-stage model. *Biochemistry*, Vol. 29, pp. 4031-4037

Prosser, R.S., Evanics, F., Kitevski, J.L. & Al-Abdul-Wahid, M.S. (2006). Current applications of bicelles in NMR studies of membrane-associated amphiphiles and proteins. *Biochemistry*, Vol. 45, pp. 8453-8465

Psachoulia, E., Marshall, D.P. & Sansom, M.S. (2010). Molecular dynamics simulations of the dimerization of transmembrane alpha-helices. *Acc. Chem. Res.*, Vol. 43, pp. 388-396

Rath, A., Johnson, R.M. & Deber, C.M. (2007). Peptides as transmembrane segments: decrypting the determinants for helix-helix interactions in membrane proteins. *Biopolymers*, Vol. 88, pp. 217-232

Riek, R., Wider, G., Pervushin, K. & Wüthrich, K. (1999). Polarization transfer by cross-correlated relaxation in solution NMR with very large molecules. *Proc. Natl. Acad. Sci. USA*, Vol. 96, pp. 4918-4923

Ritchie, T.K., Grinkova, Y.V., Bayburt, T.H., Denisov, I.G., Zolnerciks, J.K., Atkins, W.M. & Sligar, S.G. (2009). Reconstitution of membrane proteins in phospholipid bilayer nanodiscs. *Methods Enzymol.*, Vol. 464, pp. 211-231

Ruan, W., Becker, V., Klingmüller, U. & Langosch, D. (2004). The interface between self-assembling erythropoietin receptor transmembrane segments corresponds to a membrane-spanning leucine zipper. *J. Biol. Chem.*, Vol. 279, pp. 3273-3279

Sal-Man, N., Gerber, D., Bloch, I. & Shai, Y. (2007). Specificity in transmembrane helix-helix interactions mediated by aromatic residues. *J. Biol. Chem.*, Vol. 282, pp. 19753-19761

Sanders, C.R. & Sönnichsen, F. (2006). Solution NMR of membrane proteins: practice and challenges. *Magn. Reson. Chem.*, Vol. 44, pp. S24-40

Sansom, M.S., Scott, K.A. & Bond, P.J. (2008). Coarse-grained simulation: a high-throughput computational approach to membrane proteins. *Biochem. Soc. Trans.*, Vol. 36, pp. 27-32

Sato, T., Tang, T., Reubins, G., Fei, J.Z., Taiki F., Kienlen-Campar, P., Constantinescu, S.N., Octave, J.-N., Aimoto, S. & Smith S.O. (2008). A helix-to-coil transition at the ε-cut site in the transmembrane dimer of the amyloid precursor protein is required for proteolysis. *Proc. Natl. Acad. Sci. USA*, Vol. 106, pp. 1421–1426

Scheuermann, S., Hambsch, B., Hesse, L., Stumm, J., Schmidt, C., Beher, D., Bayer, T.A., Beyreuther, K. & Multhaup, G. (2001). Homodimerization of amyloid precursor protein and its implication in the amyloidogenic pathway of Alzheimer's disease. *J. Biol. Chem.*, Vol. 276, pp. 33923-33929

Schlessinger, J. (2000). Cell signaling by receptor tyrosine kinases. Cell, Vol. 281, pp. 211-225

Schneider, D. (2004). Rendezvous in a membrane: close packing, hydrogen bonding, and the formation of transmembrane helix oligomers. *FEBS Lett.*, Vol. 577, pp. 5-8

Schneider, D. & Engelman, D.M. (2004). Motifs of two small residues can assist but are not sufficient to mediate transmembrane helix interactions. *J. Mol. Biol.*, Vol. 343, pp. 799-804

Schneider, D., Finger, C., Prodöhl, A. & Volkmer, T. (2007). From interactions of single transmembrane helices to folding of alpha-helical membrane proteins: analyzing transmembrane helix-helix interactions in bacteria. *Curr. Protein Pept. Sci.*, Vol. 8, pp. 45-61

Selkoe, D.J. (2001). Alzheimer's disease: genes, proteins, and therapy. *Physiol. Rev.*, Vol. 81. pp. 741-766

Senes, A., Engel, D.E. & DeGrado, W.F. (2004). Folding of helical membrane proteins: the role of polar, GxxxG-like and proline motifs. *Curr. Opin. Struct. Biol.*, Vol. 14, pp. 465-479

Senes, A., Ubarretxena-Belandia, I. & Engelman, D.M. (2001). The Cα-H...O hydrogen bond: a determinant of stability and specificity in transmembrane helix interactions. *Proc. Natl. Acad. Sci. USA*, Vol. 98, pp. 9056-9061

Shenkarev, Z.O., Lyukmanova, E.N., Paramonov, A.S., Shingarova, L.N., Chupin, V.V., Kirpichnikov, M.P., Blommers, M.J. & Arseniev, A.S. (2010). Lipid-Protein Nanodiscs as Reference Medium in Detergent Screening for High-Resolution NMR Studies of Integral Membrane Proteins. *J. Am. Chem. Soc.*, Vol. 132, pp. 5628-5629

Smith, S.O., Smith, C.S., & Bormann, B.J. (1996). Strong hydrogen bonding interactions involving a buried glutamic acid in the transmembrane sequence of the neu/erbB-2 receptor. *Nat. Struct. Biol.*, Vol. 3, pp.252-258

Smith, S.O., Song, D., Shekar, S., Groesbeek, M., Ziliox, M. & Aimoto, S. (2001). Structure of the transmembrane dimer interface of glycophorin A in membrane bilayers. *Biochemistry*, Vol. 40, pp. 6553-6558

Smith, S.O, Eilers, M., Song, D., Crocker, E., Ying, W., Groesbeek, M., Metz, G., Ziliox, M. & Aimoto, S. (2002a). Implications of threonine hydrogen bonding in the glycophorin A transmembrane helix dimer. *Biophys J.*, Vol. 82, pp. 2476-2486

Smith, S.O., Smith, C., Shekar, S., Peersen, O., Ziliox, M. & Aimoto, S. (2002b). Transmembrane interactions in the activation of the Neu receptor tyrosine kinase. *Biochemistry*, Vol. 41, pp. 9321-9332

Sparr, E., Ash, W.L., Nazarov, P.V., Rijkers, D.T., Hemminga, M.A., Tieleman, D.P. & Killian, J.A. (2005). Self-association of transmembrane alpha-helices in model membranes: importance of helix orientation and role of hydrophobic mismatch. *J. Biol. Chem.*, Vol. 280, pp. 39324-39331

Struppe, J., Whiles, J.A. & Vold, R.R. (2000). Acidic phospholipid bicelles: a versatile model membrane system. *Biophys. J.*, Vol. 78, pp. 281-289

Stuart, A.C., Borzilleri, K.A., Withka, J.M., Palmer III, A.G. (1999). Compensating for variations in 1H−13C scalar coupling constants in isotope-filtered NMR experiments. *J. Am. Chem. Soc.*, Vol. 121, pp. 5346–5347

Sulistijo, E.S., Jaszewski, T.M. & MacKenzie, K.R. (2003). Sequence-specific dimerization of the transmembrane domain of the "BH3-only" protein BNIP3 in membranes and detergent. *J. Biol. Chem.*, Vol. 278, pp. 51950-51956

Sulistijo, E.S. & Mackenzie, K.R. (2009). Structural basis for dimerization of the BNIP3 transmembrane domain. *Biochemistry*, Vol. 48, pp. 5106-5120

Tao, R.H. & Maruyama, I.N. (2008). All EGF (ErbB) receptors have preformed homo- and heterodimeric structures in living cells. *J. Cell. Sci.*, Vol. 121, pp. 3207-3217

Therien, A.G. & Deber, C.M. (2002). Oligomerization of a peptide derived from the transmembrane region of the sodium pump gamma subunit: effect of the pathological mutation G41R. *J. Mol. Biol.*, Vol. 322, pp. 583-550

Tugarinov, V. & Kay, L.E. (2005). Methyl groups as probes of structure and dynamics in NMR studies of high-molecular-weight proteins. *Chembiochem*, Vol. 6, pp. 1567-1577

Ubarretxena-Belandia, I. & Engelman, D.M. (2001). Helical membrane proteins: diversity of functions in the context of simple architecture. *Curr. Opin. Struct. Biol.*, Vol. 11, pp. 370-376

Unterreitmeier, S., Fuchs, A., Schäffler, T., Heym, R.G., Frishman, D. & Langosch, D. (2007). Phenylalanine promotes interaction of transmembrane domains via GxxxG motifs. *J. Mol. Biol.*, Vol. 374, pp. 705-718

Van Dam, L., Karlsson, G. & Edwards, K. (2004). Direct observation and characterization of DMPC/DHPC aggregates under conditions relevant for biological solution NMR. *Biochim. Biophys. Acta*, Vol. 1664, pp. 241-256

Vereshaga, Y.A., Volynsky, P.E, Nolde, D.E., Arseniev, A.S. & Efremov, R.G. (2005). Helix interactions in membranes: lessons from unrestrained Monte Carlo simulations. *J. Chem. Theory Comput.*, Vol. 1, pp. 1252-1264

Vereshaga, Y.A., Volynsky, P.E., Pustovalova, J.E., Nolde, D.E., Arseniev, A.S. & Efremov, R.G. (2007). Specificity of helix packing in transmembrane dimer of the cell death factor BNIP3: a molecular modeling study. *Proteins*, Vol. 69, pp. 309-325

Vidal, A. & McIntosh, T.J. (2005). Transbilayer peptide sorting between raft and nonraft bilayers: comparisons of detergent extraction and confocal microscopy. *Biophys. J.*, Vol. 89, pp. 1102-1108

Vold, R.R., Prosser, R.S. & Deese, A.J. (1997). Isotropic Solutions of phospholipid micelles: A new membrane mimetic for high-resolution NMR studies of polypeptides. *J. Biomol. NMR*, Vol. 9, pp. 329-335

Volynsky, P.E., Mineeva, E.A., Goncharuk, M.V., Ermolyuk, Y.S, Arseniev, A.S. & Efremov, R.G. (2010). Computer simulations and modeling-assisted ToxR screening in deciphering 3D structures of transmembrane α-helical dimers: ephrin receptor A1. *Phys. Biol.*, Vol. 7, pp. 16014

Wallin, E., Tsukihara, T., Yoshikawa, S., von Heijne, G. & Elofsson, A. (1997). Architecture of helix bundle membrane proteins: an analysis of cytochrome c oxidase from bovine mitochondria. *Protein Sci.*, Vol. 6, pp. 808-815

Walters, R.F.S., & DeGrado, W.F. (2006). Helix-packing motifs in membrane proteins. *Proc. Natl. Acad. Sci. USA*, Vol. 103, pp. 13658-13663

Walther, D., Eisenhaber, F. & Argos, P. (1996). Principles of helix-helix packing in proteins: the helical lattice superposition model. *J. Mol. Biol.*, Vol. 255, pp. 536-553

Warren, C.M. & Landgraf, R. (2006). Signaling through ERBB receptors: multiple layers of diversity and control. *Cell Signal.*, Vol. 18, pp. 923-933

Wüthrich, K. (1986). *NMR of Proteins and Nucleic Acids*, ISBN 0-471-82893-9, Wiley, NewYork, USA.

Zhang, J., & Lazaridis, T. (2009). Transmembrane helix helix association affinity can be modulated by flanking and noninterfacial residues. *Biophys. J.*, Vol. 96, pp. 4418-4427

Zhou, F.X., Cocco, M.J., Russ, W.P., Brunger, A.T. & Engelman, D.M. (2000). Interhelical hydrogen bonding drives strong interactions in membrane proteins. *Nat. Struct. Biol.*, Vol. 7, pp. 154-160

Zhou, F.X., Merianos, H.J., Brunger, A.T. & Engelman, D.M. (2001). Polar residues drive association of polyleucine transmembrane helices. *Proc. Natl. Acad. Sci. USA*, Vol. 98, pp. 2250-2255

Zhu, J., Luo, B.H., Barth, P., Schonbrun, J., Baker, D. & Springer, T.A. (2009). The structure of a receptor with two associating transmembrane domains on the cell surface: integrin alphaIIbbeta3. *Mol. Cell*, Vol. 34, pp. 234-249

Zwahlen, C., Legault, P., Vincent, S.J.F., Greenblatt, J., Konrat, R. & Kay, L.E. (1997). Methods for Measurement of Intermolecular NOEs by Multinuclear NMR Spectroscopy: Application to a Bacteriophage λ N-Peptide/boxB RNA Complex. *J. Am. Chem. Soc.*, Vol. 119, pp. 6711-6721

Protein Engineering Methods and Applications

Burcu Turanli-Yildiz[1,2], Ceren Alkim[1,2] and Z. Petek Cakar[1,2,]
[1]Istanbul Technical University (ITU), Dept. of Molecular Biology and Genetics,
[2]ITU Dr. Orhan Ocalgiray Molecular Biology, Biotechnology and Genetics
Research Center (ITU-MOBGAM), Istanbul,
Turkey

1. Introduction

Protein engineering is the design of new enzymes or proteins with new or desirable functions. It is based on the use of recombinant DNA technology to change amino acid sequences. The first papers on protein engineering date back to early 1980ies: in a review by Ulmer (1983), the prospects for protein engineering, such as X-ray crystallography, chemical DNA synthesis, computer modelling of protein structure and folding were discussed and the combination of crystal structure and protein chemistry information with artificial gene synthesis was emphasized as a powerful approach to obtain proteins with desirable properties (Ulmer, 1983). In a later review in 1992, protein engineering was mentioned as a highly promising technique within the frame of biocatalyst engineering to improve enzyme stability and efficiency in low water systems (Gupta, 1992). Today, owing to the development in recombinant DNA technology and high-throughput screening techniques, protein engineering methods and applications are becoming increasingly important and widespread. In this Chapter, a chronological review of protein engineering methods and applications is provided.

2. Protein engineering methods

Many different protein engineering methods are available today, owing to the rapid development in biological sciences, more specifically, recombinant DNA technology. These methods are chronologically reviewed in this section, and summarized in Table 1.

The most classical method in protein engineering is the so-called "rational design" approach which involves "site-directed mutagenesis" of proteins (Arnold, 1993). Site-directed mutagenesis allows introduction of specific amino acids into a target gene. There are two common methods for site-directed mutagenesis. One is called the "overlap extension" method. This method involves two primer pairs, where one primer of each primer pair contains the mutant codon with a mismatched sequence. These four primers are used in the first polymerase chain reaction (PCR), where two PCRs take place, and two double-stranded DNA products are obtained. Upon denaturation and annealing of them, two heteroduplexes are formed, and each strand of the heteroduplex involves the desired mutagenic codon. DNA polymerase is then used to fill in the overlapping 3' and 5' ends of each heteroduplex and the second PCR takes place using the nonmutated primer set to amplify the mutagenic

DNA. The other site-directed mutagenesis method is called "whole plasmid single round PCR". This method forms the basis of the commercial "QuikChange Site-Directed Mutagenesis Kit" from Stratagene. It requires two oligonucleotide primers with the desired mutation(s) which are complementary to the opposite strands of a double-stranded DNA plasmid template. Using DNA polymerase PCR takes place, and both strands of the template are replicated without displacing the primers and a mutated plasmid is obtained with breaks that do not overlap. DpnI methylase is then used for selective digestion to obtain a circular, nicked vector with the mutant gene. Upon transformation of the nicked vector into competent cells, the nick in the DNA is repaired, and a circular, mutated plasmid is obtained (Antikainen & Martin, 2005).

Rational design is an effective approach when the structure and mechanism of the protein of interest are well-known. In many cases of protein engineering, however, there is limited amount of information on the structure and mechanisms of the protein of interest. Thus, the use of "evolutionary methods" that involve "random mutagenesis and selection" for the desired protein properties was introduced as an alternative approach. Application of random mutagenesis could be an effective method, particularly when there is limited information on protein structure and mechanism. The only requirement here is the availability of a suitable selection scheme that favours the desired protein properties (Arnold, 1993). A simple and common technique for random mutagenesis is "saturation mutagenesis". It involves the replacement of a single amino acid within a protein with each of the natural amino acids, and provides all possible variations at that site. "Localized or region-specific random mutagenesis" is another technique which is a combination of rational and random approaches of protein engineering. It includes the simultaneous replacement of a few amino acid residues in a specific region, to obtain proteins with new specificities. This technique also makes use of overlap extension, and the whole-plasmid, single round PCR mutagenesis, as in the case of site-directed mutagenesis. However, the major difference here is that the codons for the selected amino acids are randomized, such that a mixture of 64 different forward and 64 different reverse primers are used, based on a statistical mixture of four bases and three nucleotides in a randomized codon (Antikainen & Martin, 2005).

In 1994, important fields for protein engineering were also discussed in a review article by Anthonsen and co-workers (Anthonsen et al., 1994). The challenge in protein sequence deduction from DNA sequence, resulting from post-transcriptional and post-translational modifications and splicing, was emphasized. Homology modelling of protein structures, NMR of large proteins, molecular dynamics simulations of protein structures, and simulation of electrostatic effects (such as pH-dependent effects) were mentioned as important scientific areas to provide additional key information to protein engineering studies.

Another important method that finds applications in protein engineering is "peptidomimetics". It involves mimicking or blocking the activity of enzymes or natural peptides upon design and synthesis of peptide analogs that are metabolically stable. Peptidomimetics is an important approach for bioorganic and medical chemistry. It includes a variety of synthesis methods such as the use of a common intermediate, solid phase synthesis and combinatorial approaches (Venkatesan & Kim, 2002).

"*In vitro* protein evolution systems" are also important methods in protein engineering. They are based on the hierarchical evolution principle of genes. It was suggested that modern genes developed from small genetic units upon hierarchical and combinatorial processes. An example is MolCraft, an *in silico* evolved microgene which was then tandemly polymerized, including insertion or deletion mutations at the junctions between microgene units. The junctional perturbations allowed molecular diversity and the formation of combinatorial peptide polymers, whereas the repetitiousness allowed the formation of ordered structures (Shiba, 2004).

In a review article by Antikainen and Martin, (2005), the major protein engineering methods were described in detail. These methods were classified as rational methods that involve site-directed mutagenesis, random methods including random mutagenesis and evolutionary methods which involve "DNA shuffling". In DNA shuffling method, a group of genes with a double-stranded DNA and similar sequences is obtained from various organisms or produced by error-prone PCR. Digestion of these genes with *DNase*I yields randomly cleaved small fragments, which are purified and reassembled by PCR, using an error-prone and thermostable DNA polymerase. The fragments themselves are used as PCR primers, which align and cross-prime each other. Thus, a hybrid DNA with parts from different parent genes is obtained. Variations of DNA shuffling method such as the use of a mixture of restriction endonucleases instead of *DNase*I, or the "staggered extension process" that does not require parental gene fragmentation were also discussed (Antikainen & Martin, 2005). Additionally, the development of efficient screening methods to screen large libraries of proteins/enzymes such as "cell surface libraries coupled with fluorescence activated cell sorting (FACS)", or "phage display technology" were discussed (Antikainen & Martin, 2005). The combination of cell surface libraries with FACS can be used to screen very large libraries. The system is based on a scissile bond of the substrate, such as an Arg-Val linkage, which can be cleaved by a surface-displayed enzyme or not. The scissile bond on the designed substrate links a fluorophore and a quencher. If the scissile bond of the substrate is not cleaved by the surface-displayed enzyme, the fluorophore emission is then quenched by the quenching fluorophore. Thus, no fluorescence emission occurs. However, if the enzyme cleaves the scissile bond of the substrate, the fluorophore and the quenching fluorophore are then separated, and fluorescence occurs. Fluorescence of the clones with cleavage of the scissile bond is then detected by FACS (Antikainen & Martin, 2005). Phage display technology is another powerful technique for screening large libraries of proteins. The method requires degenerate reverse primers to be used in a PCR for random mutagenesis of the starting cDNA throughout a target region. The PCR products are then subcloned into a bacteriophage vector coding for a phage coat protein. Each phage of the mutant pool expresses a different protein displayed on the coat protein of the phage surface. Elution experiments help screen and identify the variants that bind tightly to a substrate of interest. Thus, the identified mutants are purified and sequenced (Antikainen & Martin, 2005).

"Flow cytometry", a powerful method for single cell analysis, is also used in protein engineering studies. A variety of examples are available where the sorting was done according to ligand binding in antibody and peptide surface display studies, or enzyme engineering of intra- and extracellular enzymes (Mattanovich & Borth, 2006).

The advantages and disadvantages of random mutagenesis methods used in protein engineering were also determined and compared to each other in detail. Based on the nucleotide substitution method used, these random mutagenesis methods were divided into four major groups: enzyme-based methods, synthetic chemistry-based methods, whole cell methods and combined methods. Their comparison was made according to a variety of parameters such as controllable mutation frequency, technical robustness, cost-effectiveness, etc. (Wong et al., 2006).

"Cell-free translation systems" were also described as important tools for protein engineering and production. They are an alternative to in vivo protein expression. When template DNA or mRNA is added to a reaction mixture, proteins are produced upon incubation in the absence of cells. PCR products can be used, and proteins are synthesized from cDNA rapidly. Cell-free translation systems are based on the ribosomal protein system of cells, which is provided as a cell extract from Escherichia coli etc. obtained as a supernatant upon centrifugation at 30'000 g. This supernatant contains necessary compounds for protein synthesis, such as ribosomes, t-RNAs, translation factors and aminoacyl-tRNA synthetases. Potential applications involve production of biologically active proteins, synthesis of membrane proteins for minimal cells, and artificial proteins. With further development, cell-free translation systems could be a strong alternative to in vivo protein expression, due to their high level of controllability and simplicity. The limitations of recombinant protein expression in living cells, such as protein degradation and aggregation will also be avoided (Shimizu et al., 2006).

Green fluorescent protein (GFP) is a very important protein that is widely used for biological and medical research purposes. It is a 238-residue protein from the jellyfish Aequorea victoria. GFP has unique spectroscopic characteristics, undergoes an autocatalytic post-translational cyclization and oxidation of the polypeptide chain around Ser65, Tyr66, and Gly67 residues, to form an extended and rigidly encapsulated conjugated Π system, the chromophore, that emits green fluorescence. Additionally, no cofactors are required for the formation or the function of the chromophore. GFP has high structural stability and high fluorescence quantum yield, which are other important properties for its widespread use. GFP has been modified extensively to be used as a marker for gene expression, protein localization and protein-protein interactions, as well as a biosensor. The proper folding of GFP is critical for its functional efficiency. Thus, protein engineering methods such as random mutagenesis and screening, DNA shuffling, as well as computational methods and X-ray crystallography improved the folding of GFP and emphasized the importance of the use of different methods such as biophysical techniques in improving protein properties (Jackson et al., 2006).

"Designed divergent evolution" is also an important protein engineering method that is used in redesigning enzyme function. The method is based on the theories of divergent molecular evolution. According to these theories, firstly, enzymes with more specialized and active functions have evolved from those enzymes with promiscuous functions. Secondly, this process is driven by a few amino acid substitutions; and finally, the effects of double/multiple mutations are usually additive. Thus, the method allows the selection of combinations of mutations that would confer the desired functions and their introduction into the enzymes (Yoshikuni & Keasling, 2007).

"Stimulus-responsive peptide systems" are based on both naturally existing peptides and rationally engineered systems. These systems exploit the fact that the peptides and proteins are able to change their conformations as a response to external stimulants such as pH, temperature or some specific molecules. There is a broad range of applications of these systems in research fields such as biosensors, bioseparations, drug delivery, nanodevices and tissue engineering. Directed evolution of stimulus-responsive peptides, however, requires an appropriate selection or screening scheme. Thus, protein-based conformational change sensors (CCSs) were developed using immunofluorescence and recombinant DNA technology (Chockalingam *et al.*, 2007).

Avidin and streptavidin are proteins that are structurally and functionally analogous. Because of their ability to bind biotin very tightly, they are widely used in (strept)avidin-biotin binding technology that is a common tool in life sciences and nanotechnology. To further improve these protein tools and obtain genetically engineered (strept)avidins, protein engineering methods were applied including simple amino acid substitutions to change physico-chemical properties, or more complex changes, such as chimeric (strept)avidins, topology rearrangements and non-natural amino acid stitching into the active sites (Laitinen *et al.*, 2007).

"Receptor-based QSAR methods" are also valuable for protein engineering studies. These methods are based on a computational combination of structure-activity relationship analysis and receptor structure-based design. They provide valuable pharmacological information on therapeutic targets. The Comparative Binding Energy (COMBINE) analysis, for example, probes bioactivity changes with respect to amino acid variations in a series of homologous protein receptors and with respect to conformational changes within a protein of interest (Lushington *et al.*, 2007).

As mentioned previously, phage display technology is one of the most commonly known molecular display technologies which relates phenotypes with their corresponding genotypes. Phage display technique is used particularly in "synthetic binding protein engineering", where libraries of 'synthetic' binding proteins were developed with antigen-binding sites constructed from man-made diversity. It was suggested that the combination of phage display and synthetic combinatorial libraries will be preferred for synthetic binding protein engineering (Sidhu & Koide, 2007). Similar to phage display technology, "yeast surface display" is also a useful method for protein engineering and characterization. Using this method, many different proteins can be displayed on yeast surface, and the yeast secretory biosynthetic system promotes efficient N-linked glycosylation and oxidative protein folding. Rapid and quantitative library screening by FACS analysis and easy characterization of mutants without requiring their soluble expression and purification are among the major advantages of this method. Yeast surface display has recently been suggested as an important methodology for protein characterization, and identifying protein-protein interactions (Gai & Wittrup, 2007). In a later review article, library creation methods and display technologies related with enzyme evolution and protein engineering were also discussed in detail (Chaput *et al.*, 2008).

An interesting protein tool that was obtained by protein engineering methods is "anticalin". It offers a variety of applications in biochemical research as well as in medical therapy as potential drugs. Anticalins are a combination of antibodies and lipocalins. Lipocalins are a protein family with a binding site that has high structural plasticity. By applying protein engineering methods such as site-directed random mutagenesis and selection by phage display technology, artificial lipocalins with novel ligand specificities, i.e. "anticalins" were obtained. Anticalins have many advantages such as being significantly smaller than antibodies, not requiring post-translational modifications, having robust biophysical properties and the ability to be produced in microbial expression systems (Skerra, 2008).

Method name	Reference(s)
Rational design	(Arnold, 1993)
Site-directed mutagenesis	(Arnold, 1993), (Antikainen & Martin, 2005)
Evolutionary methods/directed evolution	(Arnold, 1993)
Random mutagenesis	(Antikainen & Martin, 2005), (Wong *et al.*, 2006), (Jackson *et al.*, 2006), (Labrou, 2010)
DNA shuffling	(Antikainen & Martin, 2005), (Jackson *et al.*, 2006)
Molecular dynamics	(Anthonsen *et al.*, 1994)
Homology modeling	(Anthonsen *et al.*, 1994)
'MolCraft' *in vitro* protein evolution systems	(Shiba, 2004)
Computational methods (computational protein design)	(Jackson *et al.*, 2006), (Van der Sloot *et al.*, 2009), (Golynskiy & Seelig, 2010)
Receptor-based QSAR methods	(Lushington *et al.*, 2007)
NMR	(Anthonsen *et al.*, 1994)
X-ray crystallography	(Jackson *et al.*, 2006)
Peptidomimetics	(Venkatesan & Kim, 2002)
Phage display technology	(Antikainen & Martin, 2005), (Sidhu & Koide, 2007), (Chaput *et al.*, 2008)
Cell surface display technology	(Antikainen & Martin, 2005), (Gai & Wittrup, 2007), (Chaput *et al.*, 2008)
Flow cytometry / Cell sorting	(Mattanovich & Borth, 2006)
Cell-free translation systems	(Shimizu *et al.*, 2006)
Designed divergent evolution	(Yoshikuni & Keasling, 2007)
Stimulus-responsive peptide systems	(Chockalingam *et al.*, 2007)
Mechanical engineering of elastomeric proteins	(Li, 2008)
Engineering extracellular matrix variants	(Carson & Barker, 2009)
Traceless Staudinger ligation	(Tam & Raines, 2009)
De novo enzyme engineering	(Golynskiy & Seelig, 2010)
mRNA display	(Golynskiy & Seelig, 2010)

Table 1. A summary of different methods used in protein engineering

In a recent review by Goodey and Benkovic (2008), the allosteric regulation of proteins was discussed in detail. Ligand binding or an amino acid mutation at an allosteric site can significantly change enzymatic activity or binding affinity at another site such as the active site. Thus, this site-to-site communication of allosteric regulation is an important concept to be considered for protein engineering studies. Particularly, if the allosteric mechanisms are well understood, new proteins with switch-like properties could be designed for drug delivery, etc. (Goodey & Benkovic, 2008).

Recently, engineering of elastomeric proteins has been discussed as a new approach to improve the mechanical properties for the construction of biomaterials. Elastomeric proteins are important in regulating the mechanical properties in biological machineries. Using a combination of protein engineering methods and single molecule atomic force microscopy, the molecular basis of the mechanical stability of elastomeric proteins could be understood, and the mechanical properties of elastomeric proteins could be further improved by their 'mechanical engineering' (Li, 2008).

Protein and catalytic promiscuity are also important concepts for protein engineering. Catalytic promiscuity is defined as the ability of a single active site to catalyse more than one chemical reaction (Kazlauskas, 2005). Understanding protein and catalytic promiscuity is important for optimizing protein engineering applications (Nobeli *et al.*, 2009).

In a recent review, the advances in mammalian cell and protein evolution were discussed, which would have important applications in commercial mammalian cell biotechnology. As mutagenesis and selection of mammalian cells is quite elaborate, the improvement of mammalian protein evolution systems would be crucial for obtaining new diagnostic tools and designer polypeptides (Majors *et al.*, 2009).

Another recent concept in protein engineering research is the "engineering of extracellular matrix variants" to direct cell behaviour, particularly differentiation, as a response to biomaterials, in regenerative medicine applications. Extracellular matrix-derived peptides, such as Arg-Gly-Asp, are useful in supporting cell adhesion and specific integrin-signalling scaffolds and growth factor-receptor signalling are required for directing cell phenotype. Thus, by making use of this information, engineering of extracellular matrix variants could be a promising protein engineering approach (Carson & Barker, 2009).

Manipulation of proteins in a controlled way is a key requirement for many protein engineering studies. To facilitate that, "the traceless Staudinger ligation" method was recently introduced. It is based on the Staudinger reaction, where a phosphine is used to reduce an azide to an amide. The reaction occurs by means of a stable intermediate, an iminophosphorane, that has a nucleophilic nitrogen which can be acylated in inter- and intramolecular ligations. In peptide synthesis, the Staudinger reaction is applied by using a phosphinothiol for uniting an azide and a thioester. This method allows convergent chemical synthesis of proteins, and can ligate peptides at noncysteine residues. Thus, it overcomes a limitation of other strategies, and can be used as a powerful method for protein engineering (Tam & Raines, 2009).

In addition to the traditional methods of protein engineering, such as 'classical' rational design and directed-evolution methods, computational protein design tools are becoming increasingly important. In a recent review by Van der Sloot *et al.* (2009), "computational protein design principles and applications" were discussed. Computational protein design principles are based on the combination of a force field and a search algorithm to identify the amino acid sequence that is most compatible with a given protein three-dimensional backbone structure. At selected positions, the computational protein design algorithm 'mutates' or changes the original amino acid to all other natural amino acids and results in new conformations. The energy of the structure is determined after simultaneous optimization of the side-chain and/or backbone conformations of the substituted amino acid and the interacting amino acids. Thus, low energy substitutions which are favorable are retained (Van der Sloot *et al.*, 2009). Another recent review article on random mutagenesis methods used in protein engineering/enzyme evolution also discussed different methods such as "error-prone" PCR mutagenesis, chemical mutagenesis, rolling circle error-prone PCR, saturation mutagenesis and novel methodologies (Labrou, 2010). The potential of *"de novo* enzyme engineering" method was also emphasized recently (Golynskiy & Seelig, 2010). *De novo* means that the enzymes are not based on a related parent protein regarding substrate or reaction mechanism. Obtaining *de novo* enzymes from scratch has been possible by i) *in silico* rational design; ii) utilizing the understanding of a reaction mechanism and the diversity of the immune system by means of catalytic antibodies; and iii) empirical search of large protein libraries by using mRNA display. mRNA display is a powerful new technique that can select *de novo* proteins from libraries that are several orders of magnitude larger than most other selection methods such as phage display and cell surface display. The proteins obtained by mRNA display method are covalently attached to the mRNA encoding them. Thus, each protein becomes directly amplifiable. The key feature of this method is the presence of the antibiotic puromycin that mimics a charged tRNA. Thus, puromycin is added into the growing polypeptide chain by the ribosome. The transcription of a synthetic DNA library into mRNA and its modification with puromycin is usually followed by *in vitro* translation, where a covalent link is made between each protein and the mRNA encoding that protein. This step is followed by reverse transcription of the library of mRNA-displayed proteins with a substrate-modified primer, and attachment of the substrate to the cDNA/RNA/protein complex. Proteins catalyzing the substrate reaction change their encoding cDNA with the product, and selected cDNA sequences are amplified by PCR and used for the next selection step (Golynskiy & Seelig, 2010). The future of protein engineering will definitely involve many new technologies and combinational use of existing methods.

3. Protein engineering applications

A variety of protein engineering applications have been reported in the literature. These applications range from biocatalysis for food and industry to environmental, medical and nanobiotechnology applications (as summarized in Table 2), and will be discussed in this section.

3.1 Food and detergent industry applications

Early reports on the importance of protein engineering methods to design new enzymes for enzyme biotechnological industries date back to 1993 (Wiseman, 1993). Particularly, the enzymes used in food industry were emphasized as an important group of enzymes, the industrially important properties of which could be further improved by protein engineering. Those properties include thermostability, specificity and catalytic efficiency. Additionally, the design and production of new enzymes for food industry by using protein engineering was discussed to produce new food ingredients (James & Simpson, 1996). In a later review, new application areas of enzymes were discussed, resulting from significant developments in biotechnology, such as protein engineering and directed evolution. Successful combinations of rational protein engineering with directed evolution (Voigt *et al.*, 2000; Altamirano *et al.*, 2000) have also been mentioned and it was emphasized that the combined use of rational design, directed evolution and the diversity of the nature would be much more powerful than the use of a single technique (Kirk *et al.*, 2002).

Application name	Example reference(s)
Food industry applications	(James & Simpson, 1996), (Kirk *et al.*, 2002), (Akoh *et al.*, 2008)
Detergent industry applications (proteases)	(Gupta *et al.*, 2002)
Environmental applications	(Wiseman, 1993), (Cirino & Arnold, 2002), (Le Borgne & Quintero, 2003), (Ayala *et al.*, 2008), (Cao *et al.*, 2009)
Medical applications	(Buckel, 1996), (Filpula & McGuire, 1999), (Paques & Duchateau, 2007), (Nuttall & Walsh, 2008), (Liu *et al.*, 2009), (Lam *et al.*, 2003), (Zafir-Lavie *et al.*, 2007), (Vazquez *et al.*, 2009), (Olafsen & Wu, 2010)
Biopolymer production applications	(Chow *et al.*, 2008), (Rehm, 2010), (Banta *et al.*, 2010)
Nanobiotechnology applications	(Hamada *et al.*, 2004), (Banta *et al.*, 2007) (Sarikaya *et al.*, 2003) (Tamerler *et al.*, 2010)
Applications with redox proteins and enzymes	(Saab-Rincon & Valderrama, 2009), (Kumar, 2010)
Applications with various industrially important enzymes	(Martinkova & Kren, 2010), (Clapes *et al.*, 2010), (Jordan & Wagschal, 2010), (Rao *et al.*, 2009), (Marcaida *et al.*, 2010).
Other new applications	(Lofblom *et al.*, 2010), (Elleuche & Poggeler, 2010), (Klug, 2010), (Guven *et al.*, 2010),(Nagahara *et al.*, 2009), (Henriques & Craik, 2010)

Table 2. A general summary of selected protein engineering applications

An important application area of protein engineering regarding food industry is the wheat gluten proteins. Their heterologous expression and protein engineering has been studied using a variety of expression systems, such as *E.coli*, yeasts or cultured insect cells. Wild-type and mutant wheat gluten proteins were produced to compare them to each other for protein structure-function studies. Generally, *E.coli* expression systems were suggested as suitable systems for many applications, because of their availability, rapid and easy use, as well as high expression levels (Tamas & Shewry, 2006). Food industry makes use of a variety of food-processing enzymes, such as amylases and lipases, the properties of which are improved using recombinant DNA technology and protein engineering. The deletion of native genes encoding extracellular proteases, for example, increased enzyme production yields of microbial hosts. In fungi, for example, the production of toxic secondary metabolites has been reduced to improve their productivity as enzyme-producing hosts (Olempska-Beer et al., 2006).

Some large groups of enzymes like proteases, amylases and lipases are important for both food and detergent industries, as they have a broad range of industrial applications. Proteases, for example, are used in several applications of food industry regarding low allergenic infant formulas, milk clotting and flavors. They are also important for detergent industry for removing protein stains (Kirk et al., 2002). The improvement of proteases for industry to have, for example, high activity at alkaline pH and low temperatures, or improved stability at high temperatures is a challenge for protein engineering. Microbial protease production is industrially suitable because of low costs, high production yields, and easy genetic manipulation. Microbial protease genes have also been investigated for protein engineering of the enzymatic properties, clarifying the role of proteases in pathogenicity, as well as for overproduction purposes (Rao et al., 1998). There are some protein engineering applications to improve proteases: cold adaptation of a mesophilic subtilisin-like protease was performed using laboratory evolution (Wintrode et al., 2000); and DNA shuffling was applied to isolate new proteases with improved properties from an initial material of 26 subtilisin proteases (Ness et al., 1999).

Among different proteases, bacterial alkaline proteases are a commercially important group. They are particularly important for detergent industry and commercial products include subtilisin Carlsberg, subtilisin BPN and Savinase. The use of protein engineering techniques resulted in improvement of their catalytic efficiency, stability against high temperatures, oxidation and changes in washing conditions. Site directed mutagenesis and/or random mutagenesis resulted in new alkaline proteases, such as Durazym, Maxapem and Purafect, whereas new subtilisin products with improved stability and specificity were also obtained by directed evolution. The recent "metagenomic" approaches to discover natural and molecular diversities were also suggested as new technologies to isolate new microbial sources with better alkaline protease activities (Gupta et al., 2002). Among many bacteria, Bacillus species play an important role in microbial commercial enzyme production. The fact that some Bacillus species are classified as GRAS (generally regarded as safe) organisms, and have the ability to produce and secrete high amounts of extracellular enzymes, makes them valuable hosts for industrial enzyme production. Classical mutation and selection techniques, as well as protein engineering methods resulted in high-efficiency production of new enzymes with improved properties (Schallmey et al., 2004).

Amylases are also important for both food and detergent industries. In food industry, they are used for liquefaction and saccharification of starch, as well as in adjustment of flour and bread softness and volume in baking. The detergent industry makes use of amylases in removal of starch stains (Kirk et al., 2002). Recently, the production of "functional foods" is becoming increasingly important for food industry. Particularly, the production of industrial products and functional foods from cheap and renewable raw agricultural materials is desirable. Conversion of starch to bioethanol or to functional ingredients like fructose, wine, glucose and trehalose, for example, has been studied. Such a conversion requires microbial fermentation in the presence of biocatalysts such as amylases to liquefy and saccharify starch. To improve the industrially important properties of amylases, such as high activity, high thermo- and pH-stability, high productivity, etc.; recombinant enzyme technology, protein engineering and enzyme immobilization have been used. In a recent review article, rice was given as a typical example for biocatalytical production of useful industrial products and functional foods from cheap agricultural raw materials and transgenic plants (Akoh et al., 2008).

Another major group of enzymes utilized by food and detergent industries is constituted by lipases. They are used in many applications of food industry such as for the stability and conditioning of dough (as an in situ emulsifier), and in cheese flavor applications. Lipases are also crucial for the detergent industry, as they are used in removal of lipid stains (Kirk et al., 2002). As lipases are commonly used in food industrial applications, having toxicologically safe lipases is an important requirement of food industry. The commercial lipase isoform mixtures prepared from Candida rugosa meet this requirement. Obtaining pure and different C. rugosa lipase isoforms is possible by means of computer modelling of lipase isoforms, and protein engineering methods such as lid swapping and DNA shuffling (Akoh et al., 2004). A recent review on microbial lipases focused on non-aqueous microbial lipase catalysis and major factors affecting esterification/transesterification processes in organic media. Additionally, protein engineering, directed evolution, metagenomics and application of these strategies on lipase catalysis were discussed (Verma et al., 2008). Similarly, lipases from other organisms such as mammals and fishes were also reviewed (Kurtovic et al., 2009).

3.2 Environmental applications

Environmental applications of enzyme and protein engineering are also another important field. Early reports on enzyme and cell applications in industry and in environmental monitoring, such as environmental biosensors, date back to 1993 (Wiseman, 1993). One year later, recent genetic methods and strategies for designing microorganisms to eliminate environmental pollutants were discussed in detail. Those methods and strategies included gene expression regulation to provide high catalytic activity under environmental stress conditions, such as the presence of a toxic compound, rational changes introduced in regulatory proteins that control catabolic activities, creation of new metabolic routes and combinations thereof etc. (Timmis et al., 1994).

In a later review in 2000, the importance of microbial strains and their enzymes in bioremediation and biotransformation applications was discussed, pointing out the utilization of modern strategies such as protein engineering or pathway engineering to improve microbial processes. Molybdenum hydroxylases, enzymes that catalyze the initial bacterial hydroxylation of a N-heteroaromatic compound, and ring-opening 2,4-dioxygenases that play a role in the bacterial quinaldine degradation, were investigated in detail, to study and improve the enzymes involved in aerobic bacterial degradation of N-heteroaromatic compounds (Fetzner, 2000). Protein engineering of oxygenases, an important group of enzymes with high selectivity and specificity, which enable the microbial utilization and biodegradation of organic, toxic compounds, was also discussed. The potential application of oxygenases in chemical synthesis and bioremediation was also emphasized (Cirino & Arnold, 2002). Apart from oxygenases, other oxidative enzymes such as peroxidases and laccases are also important for the treatment of organic pollutants. These enzymes have broad substrate specificities and can catalyze the oxidation of a wide range of toxic organic compounds. Many organic pollutants such as phenols, azo dyes, organophosphorus pesticides and polycyclic aromatic hydrocarbons can be detoxified using enzymatic oxidation. However, there are some limitations of enzymatic treatment which should be overcome. These include enzyme denaturation by the use of organic solvents used in enzymatic reactions, inhibition/stabilization of enzyme-substrate complexes, low reaction rates of laccases, toxicity of mediators, high costs and limited availability of the enzymes, etc. Chemical modification or protein engineering of oxidative enzymes to have robust enzymes with high activity was suggested (Torres *et al.*, 2003). Another review article published in 2004 focused on the environmental applications with enzymes, such as the use of enzymes in waste management and pollution control. Protein engineering, rational enzyme design and recombinant DNA technology were mentioned as important research areas that would influence environmental enzyme applications. Utilization of new technologies such as gene shuffling, high throughput screening, and nanotechnology was suggested as future prospects of environmental enzyme applications (Ahuja *et al.*, 2004).

Petroleum biorefining is also an important environmental application area, where new biocatalysts are required. Protein engineering, isolation and study of new extremophilic microorganisms, genetic engineering developments are all promising advances to develop new biocatalysts for petroleum refining. Petroleum biorefining applications such as fuel biodesulfurization, denitrogenation of fuels, heavy metal removal, depolymerisation of asphaltenes, etc. were discussed (Le Borgne & Quintero, 2003).

Microbial bioplastics, or polyhydroxyalkanoates (PHAs), are also an important research area in environmental biotechnology. They are storage polymers produced by many bacteria and archea, and their properties are similar to those of petroleum-derived plastics. PHAs are, however, biodegradable and thus, environment-friendly. Thus, microbial large-scale and low-cost production of PHAs is a challenge for biotechnologists. PHAs are deposited in cells as water-soluble, cytoplasmic granules of nano-size. Protein engineering of polyester synthases and phasins, the two proteins involved in PHA polyester formation, and structural issues, respectively, was used to understand the genetics and biochemistry of PHA granule self-assembly. This information would also be used for medical applications involving biocompatible and biodegradable biomaterials (Rehm, 2006). The biogenesis of microbial polyhydroxyalkanoate granules, and protein engineering of polyester synthases and phasins to functionalize the polyester particle surface allowed microbial and biocatalytic production of particles with controlled size, polyester care composition and surface functionality. This would allow a platform technology for the production of tailor-made bioparticles, particularly for medical applications (Rehm, 2007).

In a recent review, microbial surface display applications for environmental bioremediation and biofuels production were discussed. Yeast and bacterial cell systems where proteins or peptides are expected on the cell exterior were reported to be used as biocatalysts, biosorbents and biostimulants (Wu *et al.*, 2008).

Another important environmental application of protein engineering involves fungal enzymes. Particularly peroxidases isolated from fungi can transform xenobiotics and many pollutants. For the development of applications, the enzyme stability and availability need to be improved. Thus, many protein engineering strategies were identified such as improvement of hydrogen peroxide stability, increasing the redox potential to broaden the substrate range, heterologous expression and industrial production development (Ayala *et al.*, 2008).

In recent reviews on environmental applications of protein engineering, recent 'omics' technologies have also been discussed. Metagenomic libraries, which identify and analyze genetic resources of complex microbial communities were suggested to help identify microbial enzymatic diversity, with implications in medicine, environmental issues, agriculture etc. Thus, contributions in renewable energy sources, decrease in pollutant burdens and process energies were expected with metagenomics applications in the future (Ferrer *et al.*, 2009). Similarly, in a review on biodegradation of aromatic compounds such as benzene, toluene, ethylbenzene and xylene, the importance of metabolic engineering, protein engineering, and "omics" technologies were emphasized (Cao *et al.*, 2009).

3.3 Medical applications

Medical applications of protein engineering are also diverse. The use of protein engineering for cancer treatment studies is a major area of interest. Pretargeted radioimmunotherapy has been discussed as a potential cancer treatment. By pretargeting, radiation toxicity is minimized by separating the rapidly cleared radionuclide and the long-circulating antibody. Advances in protein engineering and recombinant DNA technology were expected to increase the use of pretargeted radioimmunotherapy (Lam *et al.*, 2003). The use of novel antibodies as anticancer agents is also an important field of application, where the ability of antibodies to select antigens specifically and with high affinity is exploited, and protein engineering methods are used to modify antibodies to target cancer cells for clinical applications (Zafir-Lavie *et al.*, 2007). Recently, the term "modular protein engineering" has been introduced for emerging cancer therapies. Treatment strategies based on targeted nanoconjugates to be specifically directed against target cells are becoming increasingly important. Additionally, multifunctional and smart drug vehicles can be produced at the nanoscale, by protein engineering. These strategies could be combined to identify and select targets for protein-based drug delivery (Vazquez *et al.*, 2009).

Protein engineering applications for therapeutic protein production is an important area, particularly for medicine. In 1996, recombinant protein production for therapeutic purposes was reviewed. It was stated that protein engineering resulted in a second generation of therapeutic protein products with application-specific properties obtained by mutation, deletion of fusion. The third generation of such products were mentioned as "gene therapy" protein products to be produced by the patients, upon gene transfer (Buckel, 1996). Other studies on therapeutic protein production include single-chain Fv designs for protein, cell and gene therapy (Filpula & McGuire, 1999). DNA shuffling and recursive genetic recombination studies to improve therapeutic proteins (Kurtzman *et al.*, 2001); development of secreted proteins such as insulin, interferon, erythropoietin as biotherapeutics agents (Bonin-Debs *et al.*, 2004), combinatorial protein biochemistry for therapeutics and proteomics (Lowe & Jermutus, 2004), meganucleases and DNA double-strand break-induced recombination for gene therapy (Paques & Duchateau, 2007), the use of protein cationization techniques for future drug discovery and development (Futami *et al.*, 2007), protein display scaffolds for protein engineering of new therapeutics (Nuttall & Walsh, 2008), and polymer-based therapeutics for drug delivery and tissue regeneration (Liu *et al.*, 2009).

Protein engineering applications with antibodies are also diverse. Owing to advances in recombinant DNA technology, "antibody engineering" is possible. Improvements such as minimal recognition units and antigenized antibodies were described. Combinational approaches such as bacteriophage display libraries have been introduced as a strong alternative to hybridoma technology for antibody production with desired antigen binding characteristics (Sandhu, 1992). Studies on genetic manipulation of mouse monoclonals for producing humanized antibodies and bacteriophage display libraries for Ig repertoires have been reported (Zaccolo & Malavasi, 1993). Phage display has become a powerful technique in protein engineering, immunology, oncology, etc. Phage display of antibody fragments, particularly the production of artificial epitopes by phage antibodies is an important application (Pini & Bracci, 2000). "Antibody modeling" studies to engineer antibody-like molecules and increase their stability and specificity are also common, particularly for humanization of antibodies of animal origin (Morea *et al.*, 2000). Recently, the use of antibodies as vectors for molecular imaging has become popular. Pharmacokinetic properties of antibodies have been improved by protein engineering and antibody variants of different size and antigen binding sites have been produced for the ultimate use as imaging probes specific to target tissues. A variety of examples include antibody fragments which have been conjugated to bioluminescence, fluorescence, quantum dots for optical imaging, as well as iron oxide nanoparticles for magnetic resonance imaging. It is obvious that molecular imaging tools based on antibodies will find more applications in the future regarding diagnosis and treatment of cancer and other complex diseases (Olafsen & Wu, 2010).

3.4 Applications for biopolymer production

Protein engineering applications for biopolymer production are also promising. Particularly, peptides are becoming increasingly important as biomaterials because of their specific physical, chemical and biological properties. Protein engineering and macromolecular self-assembly are utilized to produce peptide-based biomaterials, such as elastin-like polypeptides, silk-like polymers, etc. (Chow *et al.*, 2008). Similarly, biosynthesis, modification and applications of bacterial polymers have also been discussed recently (Rehm, 2010).

The ability of protein engineering to create and improve protein domains can be utilized for producing new biomaterials for medical and engineering applications. One such example is the use of protein engineering to make new protein and peptide domains which enable advanced functional hydrogel formation. These domains include leucine zipper coiled-coil domains, the EF-band domains and elastin-like polypeptides (Banta *et al.*, 2010).

3.5 Nanobiotechnology applications

Nanobiotechnology applications of protein engineering are becoming increasingly important. The synthesis and assembly of nanotechnological systems into functional structures and devices has been difficult and limiting their potential applications for a long time. However, when biomaterials are investigated, it can be realized that they are highly organized from molecular to the nano- and macroscales, hierarchically. Biological macromolecules, such as proteins, carbohydrates and lipids are used in the synthesis of biological tissues in aqueous environments and mild physiological conditions, where this biosynthetic process is under genetic regulation. Particularly proteins are crucial elements of biological systems, based on their roles in transport, regulation of tissue formation, physical performance and biological functions. Thus, they are suitable components for controlled synthesis and assembly of nanotechnological systems. Combinatorial biology methods commonly applied in protein engineering studies, such as phage display and bacterial cell surface display technologies, are also used to select polypeptide sequences which selectively bind to inorganic compound surfaces, for ultimate applications of nanobiotechnology. Biopanning procedures that involve washing cycles of the phages or the cells to remove nonbinders from the surface reveal individual clones that strongly bind to a given inorganic surface. Those clones are then sequenced to identify the amino acid sequences of the polypeptides which bind strongly to the inorganic target compound surface, such as (noble) metals, semiconducting oxides and other important compounds for nanotechnology. The so-called "genetically engineered proteins for inorganics" (GEPIs) were suggested as important tools for the self-assembly of molecular systems in nanobiotechnology (Sarikaya et al., 2003). Since then, many genetically engineered peptides have been selected that specifically bind a variety of inorganic materials such as platinum, gold, and quartz; and their binding characteristics were investigated (Seker et al., 2009; Oren et al., 2010). Combining experimental approaches with computational tools allows engineering of the peptide binding and assembly characteristics. Thus, higher generation function-specific peptides can be obtained for applications in tissue engineering, therapeutics, and nanotechnology where inorganic, organic and biological materials are used (Tamerler et al., 2010). Engineering protein and peptide building blocks to be used as molecular motors, transducers, biosensors, and structural elements of nanodevices, and the importance of proteins and peptides for the development of biocompatible nanomaterials, as well as the impact of computational techniques in this field have been well recognized (Banta et al., 2007).

Another interesting nanotechnology application is the use of amyloid fibrils as structural templates for nanowire construction. This application is based on the fact that some proteins form well-ordered fibrillar aggregates that are called amyloid fibrils. As the self-organization and assembly of small molecules are crucial for nanotechnology, the self-association of well-ordered growth fibrils through noncovalent bonds under controlled conditions was suggested to have a high potential to be used for nanobiotechnology. The use of amyloid fibrils as structural templates for nanowire construction was explained as a typical example of potential applications (Hamada et al., 2004).

3.6 Applications with redox proteins and enzymes

Improvement of redox proteins and enzymes by protein engineering is also an important application field. Such proteins and enzymes can be modified to be used in nanodevices for biosensing, as well as for nanobiotechnology applications (Gilardi & Fantuzzi, 2001). The electrochemistry of redox proteins particularly draws attention for applications in biofuel cells, chemical synthesis and biosensors. Thus, protein engineering applications using rational design, directed evolution and combination thereof are found for bioelectrocatalysis

(Wong & Schwaneberg, 2003). A recent review on protein engineering of redox-active enzymes pointed out two emerging areas of protein engineering of redox-active enzymes: novel nucleic acid-based catalyst construction, and intra-molecular electron transfer network remodelling (Saab-Rincon & Valderrama, 2009). A variety of studies focused on cytochrome P450 superfamily of enzymes, such as heme monooxygenases which are involved in biosynthesis and biodegradation of metabolic compounds and in the oxidation of xenobiotics. Thus, protein engineering of P450 enzymes for degradation of xenobiotics is a biotechnological challenge (Wong et al., 1997). Additionally, the fact that heterologous expression of P450s in bacteria resulted in blue pigment formation required detailed studies of intermediary metabolism, toxicology, further protein engineering studies and suggested potential applications in dye industry (Gillam & Guengerich, 2001). More recently, review articles were published on cytochrome P450 monooxygenases (Urlacher & Eiben, 2006) and protein engineering of cytochrome P450 biocatalysts for medical, biotechnological and bioremediation applications (Kumar, 2010).

3.7 Applications with various industrially important enzymes

Protein engineering applications with a variety of industrially important enzymes can be found in the literature. These include nitrilases (Martinkova & Kren, 2010), aldolases (Clapes et al., 2010), microbial beta-D-xylosidases (Jordan & Wagschal, 2010) etc. Nitrilases are important enzymes for biotransformation, but the enzymatic reactions require improvement for higher industrial process efficiencies. For this purpose, new enzymes were screened from new isolates, medium and protein engineering methods were applied (Martinkova & Kren, 2010). Aldolases are also important enzymes for stereoselective synthesis reactions regarding carbon-carbon bond formation in synthetic organic chemistry. Protein engineering or screening methods improved aldolases for such synthesis reactions. De novo computational design of aldolases, aldolase ribozymes etc. are promising applications (Clapes et al., 2010). Microbial beta-D-xylosidases are also an industrially important group of enzymes, particularly for baking industry, animal feeding, D-xylose production for xylitol manufacturing and deinking of recycled paper. As they catalyse hydrolysis of non-reducing end xylose residues from xylooligosaccharides, they could be used for the hydrolysis of lignocellulosic biomass in biofuel fermentations to produce ethanol and butanol. Thus, improving the catalytic efficiency of beta-D-xylosidases is crucial for many industrial applications (Jordan & Wagschal, 2010). As the use of organic solvents is industrially suitable for enzymatic reactions, but has adverse effects on enzyme activity and/or stability, protein engineering of organic solvent tolerant enzymes (Gupta, 1992; Doukyu & Ogino, 2010) has become an important research area. Screening organic solvent-tolerant bacteria or extremophiles has been preferred to isolate and improve naturally solvent-stable enzymes (Gupta & Khare, 2009; Doukyu & Ogino, 2010). Other protein engineering examples with industrially and/or pharmacologically important enzymes include studies on cholesterol oxidase (Pollegioni et al., 2009), cyclodextrin glucanotransferases (Leemhuis et al., 2010), human butyrylcholinesterase (Masson et al., 2009), microbial glucoamylases (Kumar & Satyanarayana, 2009), lipases of different origins (Akoh et al., 2004; Verma et al., 2008; Kurtovic et al., 2009), phospholipases (Song et al., 2005; De Maria et al., 2007; Simockova & Griac, 2009) and phytases (Rao et al., 2009). Studies on extremozymes, enzymes isolated from extremophilic species, revealed their different structural and functional characteristics which could be exploited for biotechnological applications and improved further by protein engineering (Bjarnason et al., 1993; Hough & Danson, 1999; Georlette et al., 2004). Homing endonucleases are another important group of enzymes with application potential in gene therapy of monogenic diseases. They are double-stranded DNases with extremely rare recognition sites, and are used as templates for engineering genetic tools to cleave DNA sequences different from the wild-type targets (Marcaida et al., 2010).

3.8 Other new applications

Recently, novel types of proteins have been developed, using combinatorial protein engineering techniques. These binding proteins of non-Ig origin are called "affibody binding proteins". With their high affinity, these proteins have been used in many different applications such as diagnostics, bioseparation, functional inhibition, viral targeting, and in vivo tumor imaging or therapy (Nygren, 2008). More recently, comprehensive reviews on engineered affinity proteins (Gronwall & Stahl, 2009), and affibody molecules (Lofblom et al., 2010) were published, where their therapeutic, diagnostic and biotechnological applications were discussed in detail.

Inteins are protein splicing elements that are involved in a variety of applications such as protein purification, protein semisynthesis, in vivo and in vitro protein modifications. The use of intein tags for protein purification in plants with high protein production could potentially enable industrial production of pharmaceutically important proteins (Evans et al., 2005). The proteolytic cleavage and ligation activities of inteins have been understood, which resulted in novel intein applications in protein engineering, enzymology, microarray production, target detection and transgene activation in plants. The conversion of inteins into molecular switches was introduced by intein-mediated protein attachment to solid supports for microarray and western blot studies and by linking nucleic acids to proteins and controlled splicing (Perler, 2005). Recent intein-mediated protein engineering applications like protein purification, ligation, cyclization and selenoprotein production have been discussed in detail lately (Elleuche & Poggeler, 2010).

"Insertional protein engineering" applications are also becoming important, particularly for biosensor studies. The applications of insertional protein engineering for analytical molecular sensing have been reviewed by Ferraz and coworkers (Ferraz et al., 2006).

"Zinc finger protein engineering" is another approach that has been used in gene regulation applications. The zinc finger design and principle is used to design DNA binding proteins to control gene expression. Examples include a three-finger protein to block the expression of an oncogene that was transformed into a mouse cell line. Fusion of zinc finger peptides to repression or activation domains allows selective gene switching off and on (Klug, 2010).

Applications of protein engineering in enzymatic biofuel cell design is also becoming increasingly important. Particularly, obtaining biofuels from lignocellulosic resources is a challenge, as the enzyme hydrolysis efficiency of lignocellulose is low which increases the costs of biofuels. Thus, protein engineering methods have been used to improve the performance of lignocellulose-degrading enzymes, and biofuels-synthesizing enzymes (Wen et al., 2009). Protein engineering is also applied to obtain an efficient electrical communication between biocatalyst(s) and the electrode by rational design and directed evolution, within the frame of biocatalyst engineering (Guven *et al.*, 2010).

"Virus engineering" is another emerging field, where the virus particles are modified by protein engineering. Viruses have many promising applications in medicine, biotechnology and nanotechnology. They could be used as new vaccines, gene therapy and targeted drug delivery vectors, molecular imaging agents and as building blocks for electronic nanodevices or nanomaterials construction. Thus, the improvement of the physical stability of viral particles is crucial for efficient applications with them. Protein engineering methods are employed to improve physical stability of viral particles (Mateu, 2011).

"Protein cysteine modifications" are also important protein engineering applications. As cysteine modifications in proteins cause diversities in protein functions, cysteine thiol chemistry has been applied for *in vitro* glycoprotein synthesis. This method could be potentially used for development of new protein-based drugs, improving their half-life, reducing their toxicity and preventing multidrug resistance development (Nagahara *et al.*, 2009).

Cyclotides are important proteins that have recently been popular for protein engineering applications. They are plant proteins made up from small disulfide-rich peptides and are exceptionally stable to thermal, chemical or enzymatic degradation. This property of cyclotides makes them valuable molecular templates for many protein engineering and drug design applications (Craik *et al.*, 2007; Daly *et al.*, 2009; Henriques & Craik, 2010).

4. Conclusion

The modification of natural enzymes and proteins by protein engineering is an increasingly important scientific field. The well-known methods of rational design and directed evolution, as well as new techniques will enable efficient and easy modification of proteins. New technologies such as computational design, catalytic antibodies and mRNA display would be crucial for *de novo* engineering of enzymes and also for new areas of protein engineering.

Protein engineering applications cover a broad range, including biocatalysis for food and industry, as well as medical, environmental and nanobiotechnological applications. With advances in recombinant DNA technology tools, "omics" technologies and high-throughput screening facilities, improved methods for protein engineering will be available, which would enable easy modification or improvement of more proteins/enzymes for further specific applications.

5. References

Ahuja, SK., Ferreira, GM. & Moreira, AR. (2004). Utilization of enzymes for environmental applications. *Critical Reviews in Biotechnology*, Vol.24, No.2-3, pp.125-154, ISSN: 0738-8551

Akoh, CC., Chang, SW., Lee, GC. & Shaw, JF. (2008). Biocatalysis for the production of industrial products and functional foods from rice and other agricultural produce. *Journal of Agricultural and Food Chemistry*, Vol.56, No.22, (November 2008), pp.10445-10451, ISSN: 0021-8561

Akoh, CC., Lee, GC. & Shaw, JF. (2004). Protein engineering and applications of *Candida rugosa* lipase isoforms. *Lipids*, Vol.39, No.6, (June 2004), pp.513-526, ISSN: 0024-4201

Altamirano, MM., Blackburn, JM., Aguayo, C. & Fersht, AR. (2000). Directed evolution of a new catalytic activity using the α/β-barrel scaffold. *Nature*, Vol.403, (February 2000), pp.617-622, ISSN: 0028-0836

Anthonsen, HW., Baptista A., Drablos, F., Martel, P. & Petersen SB. (1994). The blind watchmaker and rational protein engineering. *Journal of Biotechnology*, Vol. 36, No. 3, (August 1994), pp. 185-220, ISSN: 0168-1656

Antikainen, NM. & Martin, SF. (2005). Altering protein specificity: techniques and applications. *Bioorganic & Medicinal Chemistry*, Vol. 13, No. 8, (April 2005), pp.2701-2716, ISSN: 0968-0896

Arnold, FH. (1993), Engineering proteins for non-natural environments. *The FASEB Journal*, Vol. 7, No. 9, (June 1993), pp.744-749, ISSN: 0892-6638

Ayala, M., Pickard, MA. & Vazquez-Duhalt, R. (2008). Fungal enzymes for environmental purposes, a molecular biology challenge. *Journal of Molecular Microbiology and Biotechnology*, Vol.15, No.2-3, pp.172-180, ISSN: 1464-1801

Banta, S., Megeed, Z., Casali, M., Rege, K. & Yarmush, ML. (2007). Engineering protein and peptide building blocks for nanotechnology. *Journal of Nanoscience and Nanotechnology*, Vol.7, No.2, (February 2007), pp.387-401, ISSN: 1533-4880

Banta, S., Wheeldon, IR. & Blenner, M. (2010). Protein engineering in the development of functional hydrogels. *Annual Review of Biomedical Engineering*, Vol.12, No.12, pp.167-186, ISSN: 1523-9829

Bjarnason, JB., Asgeirsson, B. & Fox, JW. (1993). Psychrophilic proteinases from Atlantic cod. *ACS Symposium Series*, Vol.516, No.5, (December 1993), pp.69-82, ISSN: 0097-6156

Bonin-Debs, AL., Boche, I., Gille, H. & Brinkmann, U. (2004). Development of secreted proteins as biotherapeutic agents. *Expert Opinion on Biological Therapy*, Vol.4, No.4, (April 2004, pp.551-558, ISSN: 1471-2598

Buckel, P. (1996). Recombinant proteins for therapy. *Trends in Pharmacological Sciences*, Vol.17, No.12, (December 1996), pp.450-456, ISSN: 0165-6147

Cao, B., Nagarajan, K. & Loh, KC. (2009). Biodegradation of aromatic compounds: current status and opportunities for biomolecular approaches. *Applied Microbiology & Biotechnology*, Vol.85, No.2, (November 2009), pp.207-228, ISSN: 0175-7598

Carson, AE. & Barker, TH. (2009). Emerging concepts in engineering extracellular matrix variants for directing cell phenotype. *Regenerative Medicine*, Vol. 4, No. 4, (July 2009), pp.593-600, ISSN: 1746-0751

Chaput, JC., Woodbury, NW., Stearns, LA. & Williams, BAR. (2008). Creating protein biocatalysts as tools for future industrial applications. *Expert Opinion on Biological Therapy*, Vol. 8, No. 8, (August 2008), pp.1087-1098, ISSN: 1471-2598

Chockalingam, K., Blenner, M. & Banta, S. (2007). Design and application of stimulus-responsive peptide systems. *Protein Engineering Design & Selection*, Vol. 20, No. 4, (April 2007), pp.155-161, ISSN: 1741-0126

Chow, D., Nunalee, ML., Lim, DW., Simnick, AJ. & Chilkoti, A. (2008). Peptide-based biopolymers in biomedicine and biotechnology. *Materials Science & Engineering Reports*, Vol.62, No.4, (September 2008), pp.125-155, ISSN: 0927-796X

Cirino, PC. & Arnold, FH. (2002). Protein engineering of oxygenases for biocatalysis. *Current Opinion in Chemical Biology*, Vol.6, No.2, (April 2002), pp.130-135, ISSN: 1367-5931

Clapes, P.,Fessner, WD., Sprenger, GA. & Samland, AK. (2010). Recent progress in stereoselective synthesis with aldolases. *Current Opinion in Chemical Biology*, Vol.14, No.2, pp.154-167, ISSN: 1367-5931

Craik, DJ., Cemazar, M. & Daly, NL. (2007). The chemistry and biology of cyclotides. *Current Opinion in Drug Discovery & Development*, Vol.10, No.2, (March 2007), pp.176-184, ISSN: 1367-6733

Daly, NL., Rosengren, KJ. & Craik, DJ. (2009). Discovery, structure and biological activities of cyclotides. *Advanced Drug Delivery Reviews*, Vol.61, No.11, (September 2009), pp.918-930, ISSN: 0169-409X

De Maria, L., Vind, J., Oxenboll, KM., Svendsen, A. & Patkar, S. (2007). Phospholipases and their industrial applications. *Applied Microbiology and Biotechnology*, Vol.74, No.2, (February 2007), pp.290-300, ISSN: 0175-7598

Doukyu, N. & Ogino, H. (2010). Organic solvent-tolerant enzymes. *Biochemical Engineering Journal*, Vol.48, No.3, (February 2010), pp.270-282, ISSN: 1369-703X

Elleuche, S. & Poggeler, S. (2010). Inteins, valuable genetic elements in molecular biology and biotechnology. *Applied Microbiology and Biotechnology*, Vol.87, No.2, (June 2010), pp.479-489, ISSN: 0175-7598

Evans, TC., Xu, MQ. & Pradhan, S. (2005). Protein splicing elements and plants: From transgene containment to protein purification. *Annual Review of Plant Biology*, Vol.56, pp. 375-392, ISSN: 1040-2519

Ferraz, RM., Vera, A., Aris, A. & Villaverde, A. (2006). Insertional protein engineering for analytical molecular sensing. *Microbial Cell Factories*, Vol.5, No.15, (April 2006), ISSN: 1475-2859

Ferrer, M., Beloqui, A., Timmis, KN. & Golyshin, PN. (2009). Metagenomics for mining new genetic resources of microbial communities. *Journal of Molecular Microbiology & Biotechnology*, Vol.16, No.1-2, pp.109-123, ISSN: 1464-1801

Fetzner, S. (2000). Enzymes involved in the aerobic bacterial degradation of N-heteroaromatic compounds: Molybdenum hydroxylases and ring-opening 2,4-dioxygenases. *Naturwissenschaften*, Vol.87, No.2, (February 2000), pp.59-69, ISSN: 0028-1042

Filpula, D. & McGuire, J. (1999). Single-chain Fv designs for protein, cell and gene therapeutics. *Expert Opinion on Therapeutic Patents*, Vol.9, No.3, (March 1999), pp.231-245, ISSN: 1354-3776

Futami, J., Kitazoe, M., Murata, H. & Yamada, H. (2007). Exploiting protein cationization techniques in future drug development. *Expert Opinion on Drug Discovery*, Vol.2, No.2, (February 2007), pp.261-269, ISSN: 1746-0441

Gai, SA. & Wittrup, KD. (2007). Yeast surface display for protein engineering and characterization. *Current Opinion in Structural Biology*, Vol. 17, No. 4, (August 2007), pp.467-473, ISSN: 0959-440X

Georlette, D., Blaise, V., Collins, T., D'Amico, S., Gratia, E., Hoyoux, A., Marx, JC, Sonan, G., Feller, G. & Gerday, C. (2004). Some like it cold: biocatalysis at low temperatures. *FEMS Microbiology Reviews*, Vol.28, No.1, (February 2004), pp.25-42, ISSN: 0168-6445

Gilardi, G. & Fantuzzi, A. (2001). Manipulating redox systems: application to nanotechnology. *Trends in Biotechnology*, Vol.19, No.11, (November 2011), pp.468-476, ISSN: 0167-7799

Gillam, EMJ. & Guengerich, EP. (2001). Exploiting the versatility of human cytochrome P450 enzymes: the promise of blue roses from biotechnology. *IUBMB Life*, Vol.52, No.6, (December 2001), pp.271-277, ISSN: 1521-6543

Golynskiy, MV. & Seelig, B. (2010). De novo enzymes: from computational design to mRNA display. *Trends in Biotechnology*, Vol. 28, No. 7, (July 2010), pp.340-345, ISSN: 0167-7799

Goodey, NM. & Benkovic, SJ. (2008), Allosteric regulation and catalysis emerge via a common route. *Nature Chemical Biology*, Vol. 4, No. 8, (August 2008), pp.474-482, ISSN: 1552-4450

Gronwall, C. & Stahl, S. (2009). Engineered affinity proteins – Generation and applications. *Journal of Biotechnology*, Vol.140, No.3-4, (March 2009), pp. 254-269, ISSN: 0168-1656

Gupta, A. & Khare, SK. (2009). Enzymes from solvent-tolerant microbes: Useful biocatalysts for non-aqueous enzymology. *Critical Reviews in Biotechnology*, Vol.29, No.1, (March 2009), pp.44-54, ISSN: 0738-8551

Gupta, MN. (1992). Enzyme function in organic-solvents. *European Journal of Biochemistry*, Vol. 203, No. 1-2, (January 1992), pp.25-32, ISSN: 0014-2956

Gupta, R., Beg, QK. & Lorenz P. (2002). Bacterial alkaline proteases: molecular approaches and industrial applications. *Applied Microbioology and Biotechnology*, Vol.59, No.1, (June 2002), pp.15-32, ISSN: 0175-7598

Guven, G., Prodanovic, R. & Schwaneberg, U. (2010). Protein engineering – an option for enzymatic biofuel cell design. *Electroanalysis*, Vol.22, No.7-8, (April 2010), pp.765-775, ISSN: 1040-0397

Hamada, D., Yanagihana, I. & Tsumoto, K. (2004). Engineering amyloidogenicity towards the development of nanofibrillar materials. *Trends in Biotechnology*, Vol.22, No.2, (February 2004), pp.93-97, ISSN: 0167-7799

Henriques, ST. & Craik, DJ. (2010). Cyclotides as templates in drug design. *Drug Discovery Today*, Vol.15, No.1-2, (January 2010), pp.57-64, ISSN: 1359-6446

Hough, DW. & Danson, MJ. (1999). Extremozymes. *Current Opinion in Chemical Biology*, Vol.3, No.1, (February 1999), pp.39-46, ISSN: 1367-5931

Jackson, SE., D Craggs, T. & Huang, JR. (2006). Understanding the folding of GFP using biophysical techniques. *Expert Review of Proteomics*, Vol. 3, No. 5, (October 2006), pp.545-559, ISSN: 1478-9450

James, J. & Simpson, BK. (1996). Application of enzymes in food processing. *Critical Reviews in Food Science and Nutrition*, Vol.36, No.5, pp.437-463, ISSN: 1040-8398.

Jordan, DB. & Wagschal, K. (2010). Properties and applications of microbial beta-D-xylosidases featuring the catalytically efficient enzyme from *Selenomonas ruminantium*. *Applied Microbiology & Biotechnology*, Vol.86, No.6, (May 2010), pp.1647-1658, ISSN: 0175-7598

Kazlauskas, RJ. (2005). Enhancing catalytic promiscuity for biocatalysis. *Current Opinion in Chemical Biology*, Vol. 9, No. 2, (April 2005), pp.195-201, ISSN: 1616-301X

Kirk, O., Borchert, TV. & Fuglsang, CC. (2002). Industrial enzyme applications. *Current Opinion in Biotechnology*, Vol.13, No.4, (August 2002), pp.345-351, ISSN: 0958-1669

Klug, A. (2010). The discovery of zinc fingers and their applications in gene regulation and genome manipulation. *Annual Review of Biochemistry*, Vol.79, No.79, (July 2010), pp.213-231, ISSN: 0066-4154

Kumar, P. & Satyanarayana, T. (2009). Microbial glucoamylases: characteristics and applications. *Critical Reviews in Biotechnology*, Vol.29, No.3, (September 2009), pp.225-255, ISSN: 0738-8551

Kumar, S. (2010). Engineering cytochrome P450 biocatalysts for biotechnology, medicine and bioremediation. *Expert Opinion on Drug Metabolism & Toxicity*, Vol.6, No.2, (February 2010), pp.115-131, ISSN: 1742-5255

Kurtovic, I., Marshall, SN., Zhao, X. & Simpson, BK. (2009). Lipases from mammals and fishes. *Reviews in Fisheries Science*, Vol.17, No.1, pp.18-40, ISSN: 1064-1262

Kurtzman, AL., Govindarajan, S., Vahle, K., Jones, JT., Heinrichs, V. & Patten, PA. (2001). Advances in directed protein evolution by recursive genetic recombination: applications to therapeutic proteins. *Current Opinion in Biotechnology*, Vol.12, No.4, (August 2001), pp.361-370, ISSN: 0958-1669

Labrou, NE. (2010). Random mutagenesis methods for *in vitro* directed enzyme evolution. *Current Protein & Peptide Science*, Vol. 11, No. 1, (February 2010), pp.91-100, ISSN 1389-2037

Laitinen, OH., Nordlund, HR., Hytonen, VP. & Kulomaa, MS. (2007). Brave new (strept)avidins in biotechnology. *Trends in Biotechnology*, Vol. 25, No. 6, (June 2007), pp.269-277, ISSN: 0167-7799

Lam, L., Liu, XY. & Cao, Y. (2003). Pretargeted radioimmunotherapy, a potential cancer treatment. *Drugs of the Future*, Vol.28, No.2, (February 2003), pp.167-173, ISSN: 0377-8282

Le Borgne, S. & Quintero, R. (2003). Biotechnological processes for the refining of petroleum. *Fuel Processing Technology*, Vol.81, No.2, (May 2003), pp.155-169, ISSN: 0378-3820

Leemhuis, H., Kelly, RM. & Dijkhuizen, L. (2010). Engineering of cyclodextrin glucanotransferases and the impact for biotechnological applications. *Applied Microbiology and Biotechnology*, Vol.85, No.4, (January 2010), pp.823-835, ISSN: 0175-7598

Li, HB. (2008). 'Mechanical engineering' of elastomeric proteins: Toward designing new protein building blocks for biomaterials. *Advanced Functional Materials*, Vol. 18, No. 18, (September 2008), pp.2643-2657, ISSN: 1616-301X

Liu, S., Maheshwari, R. & Kiick, KL. (2009). Polymer-based therapeutics. *Macromolecules*, Vol.42, No.1, (January 2009), pp.3-13, ISSN: 0024-9297

Lofblom, J., Feldwisch, J., Tolmachev, V., Carlsson, J., Stahl, S. & Frejd, FY. (2010). Affinity molecules. Engineered proteins for therapeutic, diagnostic and biotechnological applications. *FEBS Letters*, Vol.584, No.12, (June 2010), pp. 2670-2680, ISSN: 0014-5793

Lowe, D. & Jermutus, L. (2004). Combinatorial protein biochemistry for therapeutics and proteomics. *Current Pharmaceutical Biotechnology*, Vol.5, No.1, (February 2004), pp.17-27, ISSN: 1389-2010

Lushington, GH., Gu, JX. & Wang, JL. (2007). Whither combine? New opportunities for receptor-based QSAR. *Current Medicinal Chemistry*, Vol. 14, No. 17, pp.1863-1877, ISSN: 0929-8673

Majors, BS., Chiang, GG. & Betenbaugh, MJ. (2009). Protein and genome evolution in mammalian cells for biotechnology applications. *Molecular Biotechnology*, Vol. 42, No. 2, (June 2009), pp.216-223, ISSN: 1073-6085

Marcaida, MJ., Munoz, IG., Blanco, FJ., Prieto, J. & Montoya, G. (2010). Homing endonucleases: from basics to therapeutic applications. *Cellular and Molecular Life Sciences*, Vol.67, No.5, (March 2010), pp.727-748, ISSN: 1420-682X

Martinkova, L. & Kren, V. (2010). Biotransformations with nitrilases. *Current Opinion in Chemical Biology*, Vol.14, No.2, (April 2010), pp.130-137, ISSN: 1367-5931

Masson, P., Carletti, E. & Nachon, F. (2009). Structure, activities and biomedical applications of human butyrylcholinesterase. *Protein and Peptide Letters*, Vol.16, No.10, pp. 1215-1224, ISSN: 0929-8665

Mateu, MG. (2011). Virus engineering: functionalization and stabilization. *Protein Engineering Design & Selection*, Vol.24, No.1-2, (January-February 2011), pp.53-63, ISSN: 1741-0126

Mattanovich, D. & Borth, N. (2006). Applications of cell sorting in biotechnology. *Microbial Cell Factories*, Vol. 5, No.12, (March 2006), ISSN: 1475-2859

Morea, V., Lesk, AM. & Tromantono, A. (2000). Antibody modelling: Implications for engineering and design. *Methods*, Vol.20, No.3, (March 2000), pp. 267-279, ISSN: 1046-2023

Nagahara, N., Matsumura, T., Okamoto, R. & Kajihara, Y. (2009). Protein cysteine modifications: (2) Reactivity specificity and topics of medicinal chemistry and protein engineering. *Current Medicinal Chemistry*, Vol.16, No.34, (December 2009), pp.4490-4501, ISSN: 0929-8673

Ness, JE., Welch, M., Giver, L., Bueno, M., Cherry, JR., Borchert, TV., Stemmer, WPC & Minshull, J. (1999). DNA shuffling of subgenomic sequences of subtilisin. *Nature Biotechnology*, Vol.17, pp.893-896, ISSN: 1087-0156

Nobeli, I., Favia, A.D. & Thornton, JM. (2009). Protein promiscuity and its implications for biotechnology. *Nature Biotechnology*, 27, 2, (February 2009), pp.157-167, ISSN: 1087-0156

Nuttall, SD. & Walsh, RB. (2008). Display scaffolds: protein engineering for novel therapeutics. *Current Opinion in Pharmacology*, Vol.8, No.5, (October 2008), pp.609-615, ISSN: 1471-4892

Nygren PA. (2008). Alternative binding proteins: Affibody binding proteins developed from a small three-helix bundle scaffold. *FEBS Journal*, Vol.275, No.11, (June 2008), pp. 2668-2676, ISSN: 1742-464X

Olafsen, T. & Wu, AM. (2010). Antibody vectors for imaging. *Seminars in Nuclear Medicine*, Vol.40, No.3, (May 2010), pp.167-181, ISSN. 0001-2998

Olempska-Beer, ZS., Merker, RI., Ditto, MD. & DiNovi, MJ. (2006). Food-processing enzymes from recombinant microorganisms – a review. *Regulatory Toxicology and Pharmacology*, Vol.45, No.2, (July 2006), pp.144-158, ISSN: 0273-2300.

Oren, EE., Notman, R., Kim, IW., Evans, JS., Walsh, TR., Samudrala, R., Tamerler, C. & Sarikaya, M. (2010). Probing the molecular mechanisms of quartz-binding peptides. *Langmuir*, Vol.26, No.13, (July 2010), pp. 11003-11009, ISSN: 0743-7463.

Paques, F. & Duchateau, P. (2007). Meganucleases and DNA double-strand break-induced recombination: Perspectives for gene therapy. *Current Gene Therapy*, Vol.7, No.1, (February 2007), pp.49-66, ISSN: 1566-5232

Perler FB. (2005). Protein splicing mechanisms and applications. *IUBMB Life*, Vol.57, No.7, (July 2005), pp.469-476, ISSN: 1521-6543

Pini, A. & Bracci, L. (2000). Phage display of antibody fragments. *Current Protein & Peptide Science*, Vol.1, No.2, (September 2000), pp.155-169, ISSN: 1389-2037

Pollegioni, L., Piubelli, L. & Molla, G. (2009). Cholesterol oxidase: biotechnological applications. *FEBS Journal*, Vol.276, No.23, (December 2009), pp.6857-6870, ISSN: 1742-464X

Rao, DECS., Rao, KV., Reddy, TP. & Reddy, VD. (2009). Molecular characterization, physicochemical properties, known and potential applications of phytases: an overview. *Critical Reviews in Biotechnology*, Vol.29, No.2, (June 2009), pp.182-198, ISSN: 0738-8551

Rao, MB., Tanksale, AM., Ghatge, MS. & Deshpande, W. (1998). Molecular and biotechnological aspects of microbial proteases. *Microbiology and Molecular Biology Reviews*, Vol.62, No.3, (September 1998), pp.597-635, ISSN: 1092-2172

Rehm, BHA. (2006). Genetics and biochemistry of polyhydroxyalkanoate granule self-assembly: the key role of polyester synthases. *Biotechnology Letters*, Vol.28, No.4, (February 2006), pp.207-213, ISSN: 0141-5492

Rehm, BHA. (2007). Biogenesis of microbial polyhydroxyalkanoate granules: a platform technology for the production of tailor-made bioparticles. *Current Issues in Molecular Biology*, Vol.9, pp.41-62, ISSN: 1467-3037.

Rehm, BHA. (2010). Bacterial polymers: biosynthesis, modifications and applications. *Nature Reviews Microbiology*, Vol.8, No.8, (August 2010), pp.578-592, ISSN: 1740-1526

Saab-Rincon, G. & Valderrama, B. (2009). Protein engineering of redox-active enzymes. *Antioxidants & Redox Signalling*, Vol.11, No.2, (February 2009), pp.167-192, ISSN: 1523-0864

Sandhu, JS. (1992). Protein engineering of antibodies. *Critical Reviews in Biotechnology*, Vol.12, No.5-6, pp.437-462, ISSN: 0738-8551

Sarikaya, M., Tamerler, C., Jen, AKY., Schulten, K. & Baneyx, F. (2003). Molecular biomimetics: nanotechnology through biology. *Nature Materials*, Vol. 2, No.9, (September 2003), pp. 577-585, ISSN: 1476-1122

Schallmey, M., Singh, A. & Ward, OP, (2004). Developments in the use of *Bacillus* species for industrial production. *Canadian Journal of Microbiology*, Vol.50, No.1, (January 2004), pp.1-17, ISSN: 0008-4166

Seker, UOS., Wilson, B., Sahin, D., Tamerler, C. & Sarikaya, M. (2009). Quantitative affinity of genetically engineered repeating polypeptides to inorganic surfaces. *Biomacromolecules*, Vol.10, No.2, (February 2009), pp. 250-257, ISSN: 1525-7797

Shiba, K. (2004). MolCraft: a hierarchical approach to the synthesis of artificial proteins. *Journal of Molecular Catalysis B: Enzymatic*, Vol. 28, No. 4-6, (June 2004), pp.145-153, ISSN: 1381-1177

Shimizu, Y., Kuruma, Y., Ying, BW., Umekage, S. & Ueda, T. (2006). Cell-free translation systems for protein engineering. *FEBS Journal*, Vol. 273, No. 18, (September 2006), pp.4133-4140, ISSN: 1742-464X

Sidhu, SS. & Koide, S. (2007). Phage display for engineering and analysing protein interaction interfaces. *Current Opinion in Structural Biology*, Vol. 17, No. 4, (August 2007), pp.481-487, ISSN: 0959-440X

Simockova, M. & Griac, P. (2009). Phospholipid degradation: making new from the old. *Chemicke Listy*, Vol.103, No.9, (April 2009), pp.704-711, ISSN: 0009-2770

Skerra, A. (2008). Alternative binding proteins: Anticalins-harnessing the structural plasticity of the lipocalin ligand pocket to engineer novel binding activities. *FEBS Journal*, Vol. 275, No. 11, (June 2008). pp.2677-2683, ISSN: 1742-464X

Song, JK., Han, JJ. & Rhee, JS. (2005). Phospholipases: Occurrence and production in microorganisms, assay for high-throughput screening, and gene discovery from natural and man-made diversity. *Journal of the American Oil Chemists Society*, Vol.82, No.10, (October 2005), pp.691-705, ISSN: 0003-021X

Tam, A. & Raines, RT. (2009). Protein engineering with the traceless Staudinger ligation. *Methods in Enzymology: Non-natural amino acids*, Vol. 462, pp.25-44, ISSN: 0076-6879

Tamas, L. & Shewry, PR. (2006). Heterologous expression and protein engineering of wheat gluten proteins. *Journal of Cereal Science*, Vol.43, No.3, (May 2006), pp.259-274, ISSN: 0733-5210

Tamerler, C., Khatayevich, D., Gungormus, M., Kacar, T., Oren, EE., Hnilova, M. & Sarikaya, M. (2010). Molecular biomimetics: GEPI-based biological routes to technology. *Biopolymers*, Vol.94, No.1, (January 2010), pp. 78-94, ISSN: 0006-3525

Timmis, KN., Steffan, RJ. & Unterman, R. (1994). Designing microorganisms for the treatment of toxic wastes. *Annual Review of Microbiology*, Vol.48, pp.525-557, ISSN: 0066-4227

Torres, E., Bustos-Jaimes, I. & Le Borgne, S. (2003). Potential use of oxidative enzymes for the detoxification of organic pollutants. *Applied Catalysis B-Environmental*, Vol.46, No.1, (October 2003), pp.1-15, ISSN: 0926-3373

Ulmer, KM. (1983). Protein engineering. *Science*, Vol. 219, No. 4585, (February 1983), pp.666-671, ISSN: 0036-8075

Urlacher, VB. & Eiben, S. (2006). Cytochrome P450 monooxygenases: perspectives for synthetic application. *Trends in Biotechnology*, Vol.24, No.7, (July 2006), pp.324-330, ISSN: 0167-7799

Van der Sloot, AM., Kiel, C., Serrano, L. & Stricher, F. (2009). Protein design in biological networks: from manipulating the input to modifying the output. *Protein Engineering Design & Selection*, Vol. 22, No. 9, (September 2009), pp.537-542, ISSN: 1741-0126

Vazquez, E., Ferrer-Miralles, N., Mangues, R., Corchero, JL., Schwartz, S. & Villaverde, A. (2009). Modular protein engineering in emerging cancer therapies. *Current Pharmaceutical Design*, Vol.15, No.8, (March 2009), pp. 893-916, ISSN: 1381-6128

Venkatesan, N. & Kim, BH. (2002). Synthesis and enzyme inhibitory activities of novel peptide isosteres. *Current Medicinal Chemistry*, Vol. 9, No. 24, (December 2002), pp.2243-2270, ISSN: 0929-8673

Verma, ML., Azmi, W. & Kanwar, SS. (2008). Microbial lipases: At the interface of aqueous and non-aqueous media – A review. *Acta Microbiologica et Immunologica Hungarica*, Vol.55, No.3, (September 2008), pp.265-294, ISSN: 1217-8950

Voigt, CA., Kauffman, S. & Wang, ZG. (2000). Rational evolutionary design: the theory of *in vitro* protein engineering evolution. *Advanced Protein Chemistry*, Vol.55, pp.79-160, ISSN: 0065-3233

Wen, F., Nair, NU. & Zhao, HM. (2009). Protein engineering in designing tailored enzymes and microorganisms for biofuels production. *Current Opinion in Biotechnology*. Vol.20, No.4, (August 2009), pp.412-419, ISSN: 0958-1669

Wintrode, PL., Miyazaki, K. & Arnold, FH. (2000). Cold adaptation of a mesophilic subtilisin-like protease by laboratory evolution. *Journal of Biological Chemistry*, Vol.275, No.41, (October 2000), pp.31635-31640, ISSN: 0021-9258

Wiseman, A. (1993). Designer enzyme and cell applications in industry and in environmental monitoring. *Journal of Chemical Technology and Biotechnology*, Vol.56, No.1, pp.3-13, ISSN: 0268-2575

Wong, LL., Westlake, ACG. & Nickerson, DP. (1997). Protein engineering of cytochrome P450 (cam). *Metal sites in proteins and models*, Vol.88, pp.175-207, ISSN: 0081-5993

Wong, TS. & Schwaneberg, U. (2003). Protein engineering in bioelectrocatalysis. *Current Opinion in Biotechnology*, Vol.14, No.6, (December 2003), pp.590-596, ISSN: 0958-1669

Wong, TS., Zhurina, D. & Schwaneberg, U. (2006). The diversity challenge in directed protein evolution. *Combinatorial Chemistry & High Throughput Screening*, Vol. 9, No. 4, (May 2006), pp.271-288, ISSN: 1386-2073

Wu CH., Mulchandani A. & Chen W. (2008). Versatile microbial surface-display for environmental remediation and biofuels production. *Trends in Microbiology*, Vol. 16, No. 4, (April 2008), pp. 181 188, ISSN: 0966-842X

Yoshikuni, Y. & Keasling, JD. (2007). Pathway engineering by designed divergent evolution. *Current Opinion in Chemical Biology*, Vol. 11, No. 2, (April 2007), pp.233-239, ISSN: 1367-5931

Zaccolo, M. & Malavasi, F. (1993). From cells to genes – how to make antibodies useful in human diagnosis and therapy. *International Journal of Clinical & Laboratory Research*, Vol.23, No.4, (November 1993), pp.192-198, ISSN: 0940-5437

Zafir-Lavie, I., Michaeli, Y. & Reiter, Y. (2007). Novel antibodies as anticancer agents. *Oncogene*, Vol.26, No.25, (May 2007), pp.3714-3733, ISSN: 0950-9232

3

Protein Engineering Applications on Industrially Important Enzymes: *Candida methylica* FDH as a Case Study

Emel Ordu[1] and Nevin Gül Karagüler[2]
[1]Yıldız Technical University
[2]Istanbul Technical University
Turkey

1. Introduction

Global warming and the resultant problems it could cause has made the scientific community and general public realize that there is a need to decrease dependency on oil for energy. It has also shown the need to increase the efficient use of energy in manufacturing and industrial processes. Green processes would be an alternative way of decreasing oil dependency and biotechnology is promoted as a way both to decrease oil dependency and as a source for renewable bio-based products. Biocatalysis (using enzymes or whole systems) can provide a valuable alternative to traditional chemical processes because it has many advantages such as; efficiency in enhancing the rate of chemical reactions and for their ability to discriminate between potential substrates. Therefore there is the possibility of using biological molecules to catalyse any reaction or modify any product of interest to industry. Biological catalysts from animal, plant and bacterial sources have evolved to perform most types of organic reactions, producing chirally pure and complex molecules with interesting biological properties. Biocatalysts have thus become important tools in medicine, the chemical industry, food processing and in agriculture. Industrial processes often require extreme conditions such as high pressure, temperature and extreme pH which require a large amount of energy to achieve and may produce unwanted toxic waste. Biological enzymes do not require such conditions and produce chirally pure products often without the disadvantages of unwanted toxic by-products. They offer a number of advantages over conventional chemical catalysts (Davies, 2003; Kirk et al., 2002; Perez, 2010; Tao, 2009; Wojtasiak, 2006): i) Most enzymes catalyze their reactions under mild conditions such as physiological temperature and pH (6-8) and so are therefore often compatible with one another. Compatible enzymes can be used together either in sequence or cooperatively to catalyze multistep reactions. ii) Enzymes are regioselective, and also stereospecific and this allows the production of exact chiral products from racemic mixtures. Enantiomerically pure compounds are specially demanded by the pharmaceutical, food and cosmetics industries. iii) They may be cheap and easy to use because many enzymes are commercially available. iv) They are regarded as environmentally friendly because catalysis is achieved without organic solvents or the heavy metal toxic waste.

Biotransformations can be carried out using pure enzymes in solution or inside intact organisms. It remains an unanswered question as to which method is better as each has advantages and disadvantages. Enzymatic transformation usually gives a single product so they can be specific for chosen reactions. However, they are expensive and many potentially useful enzymes need cofactors. These co-factors may also be expensive, therefore it would be useful to develop a co-factor recycling system in these cases. This has been achieved for oxidoreductases, but it is not always possible and with high development costs it may not be cost effective. Whole organisms do not have this disadvantage because enzyme co-factors and their recycling systems are already present in cells. However, growing and harvesting whole cells is very labor intensive and the end product is not always pure.

2. Protein engineering methods

Protein engineering, which is the design and construction of novel proteins, usually by manipulation of their genes, is a promising approach which can be used to create enzymes with the desired properties. Proteins are engineered with the goal of better understanding the molecular basis for their functions and also so they will be able to synthesize novel products in non-native environments. Success would greatly expand the possible applications of enzymes in industrial processes.

Protein engineering methods comprise three main strategies; rational design, directed evolution and a combination of both methods, semi rational design (site saturation mutagenesis) (Figure 1).

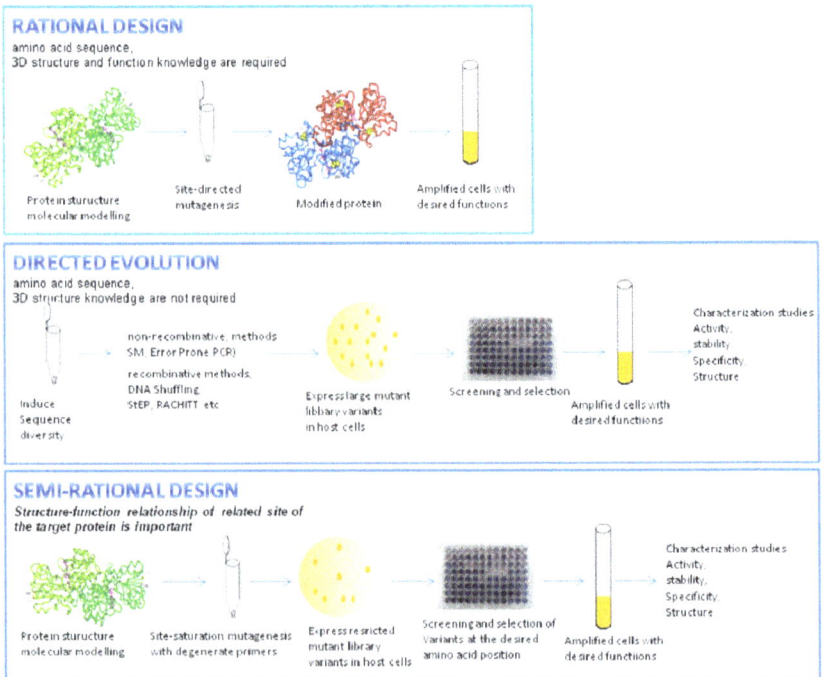

Fig. 1. Protein engineering methods

2.1 Rational design

Rational design in other words computational design of proteins requires the amino acid sequence, 3D structure and function knowledge of the protein of interest. This method provides controllable amino acid sequence changes (insertion, deletion or substitution). Controlled changes are important to determine the effect of individual residue changes on the protein structure, folding, stability or function.

Knowledge of three-dimensional structure is a key for understanding the biological function. Although understanding of 3D structure of proteins is crucial in terms of their function, only about 1 % of proteins (68.812 proteins with known structures have accumulated in the PDB database in the date of July 2011) for which the amino acid sequence is known, had their 3D structure determined because of the time consuming nature and difficulty of crystallographic experimental methods (Sanchez & Sail, 1997). As a result, the gap between the numbers of known sequences and structures continuously grows. In addition to enlarging databases, improvements in sequence comparison, fold recognition and protein modelling algorithms have supported the enhancement of protein structure prediction studies based on computer modelling methods to bridge this gap (Hillisch, 2004).

In the absence of 3D structure of the interested protein, homology modelling, which is used to predict the 3D structure of a target protein by using an experimentally (x-ray crystallography or NMR) determined protein structures as a template, is already the most promising and easiest technique among the computer based structure prediction methods (Sali & Blundel, 1993). Therefore the importance of homology or comparative modelling which can provide a useful 3D models for many proteins is steadily increasing (Suarez, 2009).

When the mutants obtained and characterized from both computational and random mutagenesis methods have been compared, it is often found that the best mutants obtained from both methods have the same residue changes (Binay et al., 2009). On the other hand strongly destabilized mutants obtained from the computational method can not be found by random mutagenesis. This explains the advantages of rational design in terms of either increasing stability or determination of individual residue effect on the protein stability, folding or function (Wunderlich et al., 2002).

The first step in rational design is the development of a molecular model by using an appropriate algorithm. This is followed by experimental construction and analysis of the properties of the designed protein. Besides the improvement of several enzyme properties like coenzyme and substrate specifity (Chul Lee et al., 2009), stability towards to oxidative stress (Slusarczyk et al., 2000), rational protein design has also been applied to improving the thermostability of several cases (Annaluru et al., 2006; Spadiut et al, 2009; Voutilainen et al., 2009; Wei et al., 2009).

Mechanisms for altering these properties include manipulation of the primary structure. Just a single point mutation may cause significant structural or functional changes in the protein. There are many rational strategies to change protein characteristics such as introducing disulfide bridges, optimization of electrostatic interactions, improved core packing, shorter and/or tighter surface loops etc. These changes are put in practice by site-directed mutagenesis. In this technique, mutations are created at computationally defined sites in the gene sequence via PCR using primers containing nucleic acids changes which correspond to the desired amino acid changes (Walker & Rapley, 2008).

2.2 Directed evolution

Unlike rational design, directed evolution (in vitro evolution or random mutagenesis) does not require any knowledge about sequence, structure or function of proteins. Directed evolution mimicks natural evolution *in vitro* by reducing the time scale of evolution from millions of years to months or weeks. This method has been used since 1980s to enhance or alter various enzyme functions. It has become a powerful technology through the work of Arnold and Stemmer in the 1990's which enhance the existing methods (Arnold, 2001; Stemmer, 1994a, 1994b). Today, directed evolution methods can be divided into two classes; (i) non-recombinative, random mutagenesis of genes (e.g. Sequence saturation mutagenesis (SeSaM), Error Prone PCR (epPCR)) and (ii) recombinative methods, recombination of gene fragments of homologous enzymes from different sources (e.g. DNA Shuffling, Family DNA Shuffling, Random Chimeragenesis on Transient Templates (RACHITT)) (Bornscheuer & Pohl, 2001; Williams et al., 2004).

Directed evolution requires two essential steps; one is the generation of random genetic libraries and the other one is screening and selection of variant enzymes that possess the desired characteristics, for example increased catalytic activity, enhanced selectivity or improved stability. Choice of the right strategy for both steps is very important to achieve the desired goal. In order to select a target protein from a large pool of mutant proteins, an efficient screening strategy, such as high-throughput solid phase digital imaging, phage display and other different screening techniques, is the most important requirement for the success of this method. The disadvantage of this method is the time-consuming process of screening and the selection of desired mutants and generally it requires robotic equipment to screen large libraries of enzyme variants (Turner, 2003). Screening of libraries on the order of 10^{3}-10^{4} variants seems sufficient for reliable selection (Tao & Cornish, 2002).

2.3 Semirational design-site saturation mutagenesis

Although either rational design or directed evolution gives effective results, to overcome the time consuming screening and selection process of directed evolution and the necessity of amino acid sequence and 3D information for rational design, a new approach would be useful. A combination of both strategies represented the new route to improve the properties and function of an enzyme (Bommarius, 2006).

Site saturation mutagenesis (SSM) method has some advantages when it is compared to other directed evolution methods. In directed evolution methods such as DNA shuffling (Stemmer, 1994) or error-prone PCR (Wong et al., 2004), random or targeted mutations in the whole sequence coding for the protein generates a large mutant library which is very time consuming to screen. With saturation mutagenesis, it is possible to create a library of mutants containing all possible combination of 22 different amino acids at one or more predetermined target positions in a gene. Choice of the correct mutagenesis, positions that can be responsible for desired changes is determined by homology modelling which requires 3D information (Lehmann & Wyss, 2001).

Because of the rational approach of this method, it is possible to obtain more effective results by combining it with high thoughput screening methods. Site saturation mutagenesis has been successfully used to improve several enzymatic properties as well as thermostability (Andreadeli et al., 2008; Reetz et al. 2006; Wu et al., 2009; Zheng, 2004).

3. *Candida methylica* NAD⁺-dependent formate dehydrogenase

A wide range of organisms use formate in a variety of metabolic pathways. From aerobic to anaerobic organisms formate dehydrogenase is the last enzyme in the metabolic pathway which catalyses the oxidation of formate to CO_2 and water. The use of formate and formate dehydrogenase (FDH) have been extensively studied and reviewed (Thiskov & Popov, 2004, 2006).

There are three families of FDH namely; complex non- Nicotinamide Adenine Dinucleotide (NAD⁺) dependent FDH, complex, soluble NAD⁺-dependent type and simple, soluble NAD⁺-dependent FDH types. Two of them are complex and use heavy metals such as molybdenum, selenium, iron, etc. The third one which is the simplest and is called NAD⁺-dependent FDH because it only requires NAD⁺ as a coenzyme. It is also the slowest class of FDH with regard to catalytic rate. The third class is represented by proteins devoid of any prosthetic groups. The molecular properties of these FDHs from prokaryotes reveal that their FDHs belong to the same family as the NAD⁺-dependent FDHs from yeasts and higher plants. In other words, the molecular masses, affinities for formate and the substrate specificities of the enzymes from bacteria, yeast and plants all resemble one another. This has been confirmed in many cases, by comparison of gene-derived amino acid sequences (Popov & Lamzin, 1994)

In the last decades active sequencing of genomes resulted in the discovery of FDH genes in various organisms such as *Staphylococcus aureus*, *Mycobacterium avium subsp. paratuberculosis* str.k10, different strains of *Bordetella*, and *Legionella*, *Francisella tularensis subsp. tularensis* SCHU S4, *Histoplasma capsulatum*, *Cryptococcus neoformans var. neoformans* JEC21, *A. thaliana*, potato, rice, barley, cotton plant, English oak, *Mesembryanthemum crystallinum*, (*S. cerevisiae*, *C. boidinii*, *C. methylica*, *Hansenula polymorpha*, and *Pichia pastoris*, *A. nidulans*, *N. crassa*, *G. zeae* PH-1, *M. grisea*, *M. graminicola*, *U. Maydis* (Figure 2) (The detailed information about organisms can be found in the web page of The National Center for Biotechnology Information). Among the large number of microorganisms that have formate dehydrogenase, attention has been mainly focused on the yeasts. In yeast, the ability to utilize methanol as the sole carbon source is limited to members of 4 genera, namely *Candida*, *Hansenula*, *Pichia* and *Torulopsis*. An investigation of FDHs led to the selection of *Candida* species NAD⁺-dependent FDH as the best candidate for the NADH regeneration system because it is stable and it has relatively good activity. However, while several FDH enzymes have been isolated, crystal structures of FDHs from the bacterium *Pseudomonas sp.*101 and yeast *Candida boidini* have been solved (Lamzin et al., 1992; Schirwitz et al., 2007).

NAD+-dependent FDH is a dimeric enzyme with two identical subunits each has an independent active site, containing no metal ions or prosthetic groups. They are unable to use one-electron carriers as oxidizers and are highly specific to both formate and NAD⁺. FDH catalyzes the oxidation of formate to carbon dioxide coupled with reduction of NAD⁺ to NADH (Thiskov & Popov, 2004, 2006):

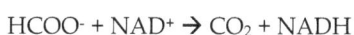

$$HCOO^- + NAD^+ \rightarrow CO_2 + NADH$$

```
                                               1                              50
c-methyl1  ..................  MKIVLVLYDA  GKHAADE...  .......EKLY  GCTENKLGIA
h-polymo   ..................  .KVVLVLYDA  GKHAADE...  .......ERLY  CCTENALGIR
a-nidula   .............MVLLDG  GSHAKDQ...                 ...PGLL  GTTENELGTR
n-crassa   ...........M  VKVLAVLYDG  GKHGEEV...          ...PELL  GTIQNELGLR
s-tubero   MSKVASTAAR AITSPSSLVT TRELQASFGP KKIVGVFKA NEYAEMN...  PNFL  GCAENALGIR
p-sp.101   AKVLCVLYDD PVDGYPKTYA RDDLPKIDHY PGGQTLPTFK AIDFTPGQLL GSVSGELGLR

                    100                          120
c-methyl1  NWLKDQGHEL  IITSDKEET  SELDKHIPDA
h-polymo   DWLEKQGHDV  VVTSDKEQN  SVLEKNISDA
a-nidula   KWILEQGHTL  VTTSDKDEN  STFDKELVDA
n-crassa   KWLEQGHTL   VTTCDKDEN  STFDKELEDA
s-tubero   EWLESKGHQY  IVTPSKGPD  CELEKHIPDL
p-sp.101   KYLESNGHTL  VVTSDKSGPD SVTERELVDA

            121                              150
c-methyl1  DIIITTPFHP  AYITKERLDK  AENIKSVVVA  GVGSDHIDLD  YINQTGKKIS
h-polymo   DVIISTPFHP  AYIIKERIDK  AEKIKILVVA  GVGSDHIDLD  YINQSGREIS
a-nidula   EVIITTPFHP  GVLEAERLAK  ANNIKLAVTA  GIGSDHVDLD  AANKTNGGIT
n-crassa   EIIITTPFHP  GVLEARLAR   AEKILAVTA   GIGSDHVDLN  AANKTNGGIT
s-tubero   EVLISTPFHP  AYVAERIKK   ANNIQLLLTA  GIGSDHVDLK  AA..AAAGLT
p-sp.101   DVVISQPFWP  AYLPERIAK   AENIKLALTA  GIGSDHVDLQ  SA..IDRNVV

                    200
c-methyl1  VLEVTGSNVV  SVAEHVVMTM  HEQIINHDWE  VAAIAKDAYD  IIGKTIATIG  AGRIGYRVIE
h-polymo   VLEVTGSNVV  SVAEHVVMTM  HEQIISGG.N  VAEIAKDSFD  IIGKVIATIG  AGRIGYRVIE
a-nidula   VAEVTGSNVV  SVAEHVVMTI  HDQIRNGD.N  VVAVAKNEFD  LNKVVGVVG   VGRIGERVIR
n-crassa   VAEVTGSNVV  SVAEHVLMTI  HEQIQEGRD   VEAAKNEFD   LGKVVGVVG   VGRIGERVIR
s-tubero   VAEVTGSMTV  SVAEDELMRI  HHQVINGE.N  VIAIBRAYD   LGKTVGTVG   AGRIGRLLLQ
p-sp.101   TAFVVTYNSI  SLVRNYLFS   HEWARKGG.N  IADCVSHAYD  LAMEVGTVA   AGRIGLAVIR

            241                              250        300
c-methyl1  RLLPFNFKEL  LYDYQALPK   EAEKVGARR   VENIEELVAQ  ADIVTVVAPL  HAGTKGLINK  ELLSKFK....
h-polymo   RVAANFKEL   LYDYQOSLSK  EAEKVGARR   VHDIKELVAQ  ADIVTINCPL  HAGSKGLVNA  LLLKHFK....
a-nidula   RLKPFDCKEL  EVKEIGARR   VDSLEEMVSQ  CDVTINCPL   HEKTRGLFNK  ELISKMKPGK  SALLYLIIPM  LMYHKGSWLV
n-crassa   RLKPFDCKEL  LYDYQPLSA   EKCAEIGCRR  VADLEEMIAQ  CDVTINCPL   HEKTQGLFNK  ELISRMK....
s-tubero   RLKPFNC.NL  LYHRLKMDS   ELENQIGAKF  EEDLDKMLSK  CDIVVINTPL  TEKTKGMFDK  ERIAKLK....
p-sp.101   RLAPFD.HVL  HYTRHRLPE   SVEKELNLTW  HATREDMYPV  CDVVLNCPL   HPETEHMIND  ETLKLFK....

                    350                          360
c-methyl1  ..KGAWLV    NTARGAICVA  EDVAAALESG  QLRGYGGDVW
h-polymo   ..KGAWLV    NTARGAICVA  EDVAAAVKSG  QLRGYGGDVW
a-nidula   SALLYLIIPM  NTARGAIVVK  EDVAEALKSG  HLRGYGGDVW
n-crassa   ..KGSWLV    NTARGAIVVK  EDVAEALKSG  HLRGYGGDVW
s-tubero   ..KGVLIV    NNARGAIMDT  QAVVDJCNSG  HIAGYSGDVW
p-sp.101   ..RGAYIV    NTARGELCDR  DAVARALESG  RLAGYAGDVW

            361                              400        451
c-methyl1  FFQPAEKDHP  WRDMRNKYGA  GNAMPEYSG   TTILDAQTRYA  EQTKNILESF  FTGKPDYRPQ  DIILLNGEYV  TKAYGKHDKK
h-polymo   FFQPAEKDHP  WRSMANKYGA  GRAMPEYSG   SVIDAQVRYA   QGTKNILESF  FTQKPDYRPQ  DIILLNGKYK  TKSYGADK..
a-nidula   FFQPAEKEHP  LRYAEHPWGG  GNATVPEMSG  TSLAAQIRYA   NGTKAILDSY  FSGRPDYQPQ  DLIVHGDYA   TKAYGQREKK
n-crassa   FFQPAEKEHP  GNAMVPEMSG  TSLDACKRYA  AGTKAIIHSY   LSGKHDYRPE  DLIVYG-DYA  TKSYGERERA  KAAAAAAKS.
s-tubero   YFQPAEKDHP  WRIMPN....  .QAMPHISG   TTIDALQRYA   AGTKDMLDRY  FKGE.DFPAE  NYIVKD-ELA  PQYR......
p-sp.101   FFQPAEKDHP  WRIMPY....  .DGMTPHISG  TTILTAQAYA   AGTREILECF  FEGR.PIRDE  YLIVQG-ALA  GTGAHSYSKG NATGGSEEAA KVFKKAV.....

G     xGxxG  ------

-(17/18x)-  ---D
NAD-motif
```

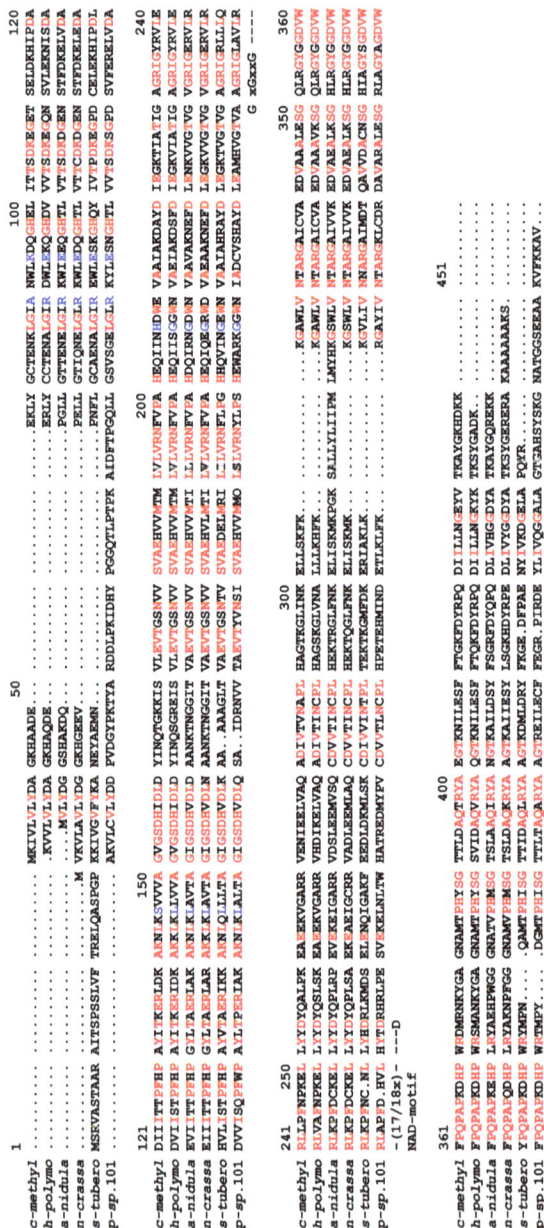

Fig. 2. Sequence alignment of *C. methylica* FDH with other FDHs from *hp*fdh (*Hansenula Polymorpha*), *nc*fdh (*Neurospora Crassa*), *st*fdh (*Solanum Tuberosum*), *an*fdh (*Aspergillus Nidulans*), *ps*fdh (*Pseudomonas sp.* 101). Residues conserved in all FDHs, including *C. methylica* FDH are in red and residues conserved in all FDHs but not in *C. methylica* FDH are in blue.

The FDH enzyme was first discovered in 1950 (Uversky, 2003) but it has attracted attention in recent years due to its practical application in the regeneration of NAD(P)H in the enzymatic processes of chiral synthesis. FDH is widely used for coenzyme regeneration with enzymes used for optically pure product synthesis in the pharmaceutical, food, cosmetic and agriculture industries (Patel, 2004; Jormakka et al., 2003). The FDH catalysed reaction is also a suitable model for investigating the general mechanism of hydride ion transfer because of direct transfer of hydride ion from the substrate onto the C4-atom of the nicotinamide moiety of NAD+ without stages of acid-base catalysis (Serov et al., 2002a).

NAD+-dependent FDH from *Candida methylica* (*cm*FDH) was previously isolated by Allen & Hollbrook, 1995). Its N-terminal amino acid sequence was determined and it was cloned into pKK223-3 and overproduced in *Escherichia coli*. *cm*FDH in pKK223-3 vector has been used in several studies but purification of enzyme was a time consuming and costly process. Therefore, in order to eliminate difficulties in the purification of FDH and to produce quick and highly purified-homogeneous-recombinant protein, *cm*FDH was subcloned into pQE-2 expression vector and the amount of purified protein improved about 3 times. It was observed that the N-terminally His tagged FDH has similar activity to the FDH enzyme without the His-tag after digestion with exopeptidases (Ordu and Karagüler, 2007). Since then, the recombinant FDH from *Candida methylica* has been intensively studied to improve the properties for the NAD(P)H regeneration by using protein engineering techniques.

3.1 Kinetic and thermodynamic properties of the folding and assembly of *cm*FDH

Although there has been much empirical work on stabilizing FDHs versus increasing temperatures and other environmental factors such as oxidation (Thiskov & Popov, 2006), the thermodynamic and kinetic properties of its folding and unfolding pathway have not been dissected in detail. Whereas, in order to control the stability of proteins by genetic modification of sequence we need to understand the full mechanism of folding and unfolding of the active system. This, ideally, requires elucidation of the kinetic and equilibrium properties of each step and to target certain critical steps in the process, so that the sequence engineering would be more rationally directed.

While many of the proteins of interest in biotechnology are oligomeric, as are many of the structures in biological systems where we want to understand the dynamics of assembly and disassembly, our understanding of folding and assembly processes in multi-chain proteins is less comprehensive than kinetics and thermodynamics of folding in single-chain proteins. For these reasons it is useful to examine mechanisms such as formate dehydrogenase enzyme containing two identical subunit with a view to provide a framework for their analysis. The native form of *cm*FDH is a dimer and each subunit has 364 residues folded into two distinct domains, each comprising a parallel β-sheet core surrounded by α-helices arranged in a Rossmann-type fold. While one domain is functionally defined by co-enzyme binding, the other domain is defined for catalysis; coenzyme and the substrate are encapsulated in the inter-domain cleft during the catalytic reaction. The molecule dimerizes by 2-fold symmetrical interactions between the co-enzyme-binding domains while the catalytic domains are distal to the dimer interface (Schirwitz et al., 2007).

In this section, we define the rates of steps in the minimal model of folding and assembly reaction and deduce the equilibrium properties of the system with respect to its thermal and denaturant sensitivities. These results act as a basis for understanding the effects of site-specific engineering or forced evolution on the stability of this molecule.

Equilibrium denaturation data of *cm*FDH yielded a dissociation constant of about 10^{-13} M. Findings showed that homodimeric *cm*FDH unfolds by two state single transition model without intermediates in equilibrium and in the equilibrium one dimer is equal to two unfolded monomers including both folding and dissociation processes.

Thermodynamics of the folding-unfolding transition showed that, at the reference temperature of 25 °C the enthalpy change on folding (ΔH) is unfavourable (approximately +27 ± 18 kcal mol^{-1}), while the entropy change is favourable (-TΔS = -46 ± 18 kcal mol^{-1}). The heat capacity change (ΔC_p) is -10.5 ± 1.8 kcal mol^{-1} K^{-1}; a value that is dominated by the degree of desolvation of non-polar surface during the folding process (Ordu et al, 2009).

The refolding process of *cm*FDH enzyme is rate-limited by at least two steps occurring on the same time scale. Further, one of these steps must be multi-molecular since the half-time of refolding is clearly sensitive to the protein concentration. The intermediate state (or states) lie between the two rate-limiting steps is devoid of activity. The simplest model that accounts for the data is an essentially irreversible uni-bi reaction: 2U $\rightarrow$ 2M $\rightarrow$ D, where U is the unfolded chain, M the folded monomer and D the active dimer.

According to this kinetic model, rate constant values yielded as yielded values of 1.9 ± 0.4 x10^{-3} s^{-1} for the unimolecular folding step and 1.6 ± 0.5 x 10^4 M^{-1}s^{-1} for the bimolecular association of subunits (Ordu et al, 2009). The monomeric intermediate of *cm*FDH forms active dimers at a rate which is slower than expected for a process limited by subunit diffusion in solution. Considerations based on orientational constraints and Brownian diffusion for protein associations have suggested that the basal rate should lie in the range of 10^5–10^6 M^{-1}s^{-1} (Northrup & Erickson, 1992; Schlosshauer & Baker, 2004; Zhou et al., 1997).

The relatively slow bimolecular rate constant is combined with a slow rate of dissociation to yield a high stability for the native dimer (ΔG° = -14.6 kcal mol^{-1}), in keeping with other dimeric proteins of high molecular weight. This level of stability (K_d ~ 10^{-13} M) means that at a micromolar concentration, less than one millionth of the enzyme is in the inactive and unstable monomeric state. It is interesting to note that the rate constant for FDH dissociation is so slow that, once formed, the dimer has a dissociative half-life of one and a half years. The heat inactivation of *cm*FDH, as for most proteins, is not reversible and follows a first-order decay. As a result of this, denaturation cannot be formally treated as an equilibrium system to which the orthodox analysis can be applied, rather it should be thought of as being defined by a temperature-sensitive rate constant for the irreversible step (Ordu et al, 2009).

3.2 Coenzyme regeneration

Production of optically pure compounds is important for product quality and customer safety in industry. In traditional industrial chemical synthesis of chiral compounds, the products are usually racemic mixtures which contain both forms of optically active compounds. In the pharmaceutical industry, when only one enantiomer has the appropriate physiological activity, problems from side effects of another enantiomer can arise like the case of thalidomide. While the R-enantiomer of thalidomide has an analgesic activity, the S-enantiomer causes defects in the fetus (Muller, 1997). In food industry, chiral molecules are indicator of quality and purity of products. Natural food components are optically pure but extreme industrial process like high pH, temperature and irridiation cause racemic mixture in the food. Chirality is also important in taste and odour applications in food industry. For example, while the L-form of asparagine, tryptophan, tyrosine and isoleucine are characterized by bitter taste, they are characterized as sweet taste in the D-form (Wojtasiak , 2006).

Enzymatic reactions catalyzed by oxidoreductases (e.g. lactate dehydrogenase, hydroxyisocaproate dehydrogenase, xylitol dehydrogenase, mannitol dehydrogenase and limone monooxygenase) are highly sterospecific and very important for the production of chirally pure products. Lactate dehydrogenase and hydroxyisocaproate dehydrogenase can be used in the production of optically pure hydroxyacids which are used for the production of semi synthetic antibiotic (S - α-hydroxyisocaproic acid) and medical diagnosis (S – phenyl pyruvate in diagnosis of phenylketonuria and S – ketoisocaproic acid for some urine disease) (Nakamura, 1988; Van der Donk & Zhao, 2003;). Xylitol dehydrogenase and mannitol dehydrogenase are used in D-xylitol and D-mannitol synthesis, respectively and limone monooxygenase used in aroma synthesis also provides sterospecific compounds in the food industry (Kaup et al, 2005; Mayer et al, 2002). The reduction of a carbonyl group particularly to generate a new chiral center, is one of the most widely used biotransformation in industry. Enzymatic synthesis of the final product may reach 100 %, while the processes based on chemical synthesis of racemic mixtures can only provide the theoretical yield of 50 %. Enzymatic reactions take place under mild conditions, this minimizes problems from side reactions and they have a regio and stereochemical specificity that can be difficult to achieve chemically. However use of these enzymes and similars is still limited because of the requirement for stoichiometric amounts of the very expensive NAD(P)H coenzyme. Enzymes remain unchanged after the reaction is completed, but coenzymes react with the substrate chemically and they are usually more expensive than desired product. It is possible to recover used coenzymes via recycling reactions, but existing methods for regenerating NAD(P)H are still a significant expense and are not cost-effective in the manufacturing process. Therefore, there is a need for a low-priced method of coenzyme regeneration (Liu and Wang, 2007; Patel, 2004; Vrtis, 2002).

Several approaches including chemical, electrochemical, photochemical, microbial and enzymatic synthesis have been investigated for coenzyme regeneration. Enzymatic regeneration is the most promising one in the industrial process due to its high selectivity, efficiency, aqueous solvent as operational medium and environmentally safe waste (Van der Donk and Zhao, 2003).

There are two different approaches for enzymatic regeneration. One of them is substrate coupled reaction systems in which one enzyme that reacts with both the reduced and oxidized forms of coenzyme to catalyze both the desired synthesis of the product from one substrate and the coenzyme regeneration with a second substrate. But in this system it is difficult to find thermodynamically favourite conditions for both reactions in the same medium. The other approach, enzyme coupled, is the usage of second enzyme to catalyze the coenzyme regeneration (Eckstain et al., 2004; Popov & Thiskov, 2003). In this way second substrate must be very cheap or can be regenerated easily.

Several enzymes have been studied for NAD(P)H regeneration based on enzyme coupled approach. Formate dehydrogenase (FDH) (Bolivar et al., 2007; Gül-Karagüler et al., 2001;

Seelbach et al., 1996;), phosphite dehydrogenase (PTDH) (Johannes et al., 2007; Relyea et al., 2005; Woodyer et al., 2006), alcohol dehydrogenase (ADH), glucose-6 phosphate dehydrogenase, glucose dehydrogenase(GDH) (Endo &Koizumi, 2001; Xu et al., 2007) are currently available systems.

PTDH, catalyzes the oxidation of phosphite to phosphate by reducing the NAD^+ to NADH, works in very narrow pH range (pH 7.0-7.6) and destabilize above 35 °C. Altough the cost of phosphite is cheaper than formate, which is the substrate of FDH, 90 g of phosphite is necessary to regenerate 1 mol of NADH by PTDH while formate dehydrogenase needs only 45 g of formate to produce 1 mol of NADH. GDH which catalyzes the hydrolization of gluconolactone to gluconic acid has both coenzyme specifity, NAD^+ and $NADP^+$ (Nicotinamide Adenine Dinucleotide Phosphate). Although this enzyme has higher specific activity than FDH (FDH, 2,5-10U/mg; GDH, 20-100U/mg), 172 g of glucose is needed for reduction of 1mol of NAD(P) to 1 mol of NAD(P)H. Moreover, gluconic acid should be removed from the reaction medium to produce pure chiral product.

When FDH is compared to other studied dehydrogenases it offers several advantages. The enzyme is commercially available at low cost and has a favourable thermodynamic equilibrium. It is used for the industrial scale regeneration of NADH in bioreactors with recycle numbers for the cofactor in the order of 130.000 during the production of 640 $g^{-1}l^{-1}d^{-1}$ of L-leucine (Wandrey, 1986) at present. Its reaction results in a 99–100 % yield of the final product. Since the reaction of FDH is essentially irreversible and the product (CO_2) can be easily removed from the reaction system and formate or CO_2 does not inhibit the other reaction and does not interfere with the purification of the final product. FDH also has a wide range pH optima so that it can work with lots of different enzymes (Popov & Thiskov, 2003). Because of its advantages FDH is very suitable for NAD(P)H regeneration. Therefore use of a cheap reductant is possible. Wilks et. al. (1986) adapted the general scheme for cofactor recycling using formate dehydrogenase (Figure 3).

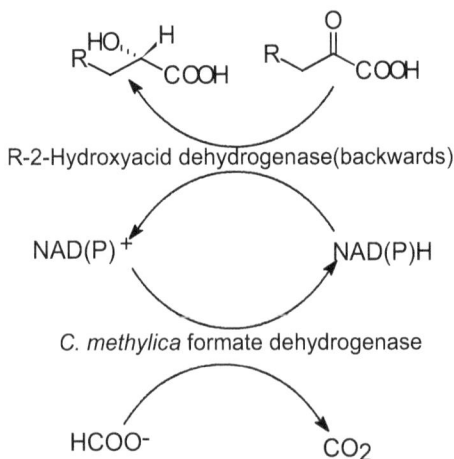

Fig. 3. General scheme for cofactor recycling using formate dehydrogenase. This diagram is the conventional Wandrey-Kuhla route for the bulk synthesis with reducing equivalents from formic acid (Wilks et. al., 1986).

In laboratory scale, FDH was used for NADH regeneration with several enzymes like lactate dehydrogenase, xylitol dehydrogenase and mannitol dehydrogenase in the production of hydroxyacids, xylitol and mannitol, respectively, (Kaup, 2005; Mayer et al., 2002; Van der Donk & Zhao, 2003). FDH from the yeast Candida boidinii was used in the first commercial scale process of chiral synthesis of tert-L-leucine with leucine dehydrogenase by German Degussa company (Popov & Thiskov, 2003). NADP+-dependent enzymes are less common than their NAD+-dependent counterparts. Engineered FDH is also a favourite enzyme to regenerate NADPH among other investigated enzymes such as ADH and glucose-6 phosphate dehydrogenase because of its cheap substrate. Engineered FDH that can accept NADP+ and it is preferred over glucose-6 phosphate dehydrogenase given the expense of its substrate (Andreadeli et al., 2008; Wu et al., 2009). The number of studies on FDH and its application for coenzyme regeneration in the processes of chiral synthesis with NAD(P)H-dependent enzymes is getting larger year by year (Holbrook et al., 2000; Tishkov & Zaitseva, 2008).

Unfortunately, native FDHs have some disadvantages. These are low k_{cat}, high K_M, its limited coenzyme specificity solvent tolerance and lack of thermostability. Hence it is important to improve the stability of FDH to cope with the harsh conditions like high temperature, pressure or pH required for the most of manufacturing processes in food or pharmaceutical industries. To make FDH a more suitable enzyme for cofactor regeneration these disadvantages should be removed. Protein engineering is a promising approach to improve the industrial parameters of FDH.

4. Protein engineering studies on *cm*FDH

During the evolution, organisms adapted to extremes of pH, salinity, pressure and temperature by means of mutations that affect stability, catalytic activity or substrate specificity of enzymes (Bloom et al., 2006). Proteins are the best candidates to evolve because they may change their biochemical function with just a few mutations (Razvi & Scholtz, 2006). These molecules have evolved by different strategies to protect their stability and function under unusual conditions.

4.1 Increasing the thermostability

Studies in the literature show that thermophilic proteins typically have increased number of van der Waals interactions, hydrogen bonds, salt bridges, dipole-dipole interactions, disulphide bridges, hydrophobic interactions, aromatic stacking interactions, improved core packing, shorter and/or tighter surface loops, enhanced secondary structure propensities, decreased conformational entropy of the unfolded state and oligomerization at the molecular level (Karshikoff & Ladenstain, 1998; Kumar et al., 2000; Robinson-Rechavi et al., 2006). Thermophilic enzymes are already used in industrial applications, e.g. the α-amylase from *Bacillus licheniformis* is among the most thermostable natural enzymes used in biotechnological processes. However, the use of thermophilic enzymes depends on their availability. Identification and purification of new industrial enzymes suitable to harsh conditions is not always straightforward. Therefore, improving of wild type enzymes for a desired function or increased stability by using protein engineering is an important approach to overcome this restriction. Several efforts have been attempted to identify the most efficient method to enhance thermostability of a mesophilic protein by using protein engineering methods (Annaluru et al., 2006; Rodriguez et al., 2000; Yokoigawa et al., 2003). Unfortunately, there is no general rule by which a mesophilic protein can be converted into a thermophilic one.

In the laboratory, protein stability studies have been performed against to several denaturating agents like extremes of pH, salinity, pressure and temperature. Among these extreme conditions, high temperature stability (thermostability) is particularly important in the industrial and biotechnological enzymatic production process where the enzymes are often inactived due to high temperature. For example in the starch sector, which is one of the largest users of enzymes, conversion of starch include liquefaction and saccharification steps. The temperature has to be 105-110 °C during these processes. Otherwise, below 105 °C, gelatinisation of starch granules is not achieved successfully and causes filtration problems in the other steps (Synowiecki, 2006; Haki, 2003). This causes an economical challenge for the manufacturer. There are many advantages for industry by using thermostable enzymes in commercial applications . Higher temperatures can lead to i) Increased activity, reaction rates typically increase two to three fold for each 10 °C increase of temperature, ii) increased yield of product because many substrates are more soluble at higher temperatures, and by a shift of the thermodynamic equilibrium for endothermic reactions, iii) increased storage and operational stability, iv) increased stability against other denaturing conditions, for example the possibility of sterilizing products in the course of the production. If such problems can be solved by high temperature, any FDH enzyme that would be used for NAD(P)H regeneration in this kind of high temperature process would need to be thermostable.

The lack of thermostability is the most important disadvantage of FDH. Studies in the literature with the aim of improving the thermostability of FDH have been focused on *Pseudomonas* sp 101 which has the best thermostability among the known FDHs and the FDH from *Candida bodiini* by using rational design or directed evolution. The FDH enzymes from these organisms have the advantage of solved crystal structures.

In the case of FDH from *Candida methylica*, the first attempment was made to produce a more thermostable enzyme by the DNA shuffling method because the 3D structure of FDH from *C. methylica* is not yet solved. In DNA shuffling, the ability to select an improved protein from a large pool of protein variants depend on sensitive screening or selection methods. In this study host proteins could not be inactivated through a heat step screening method. In the light of this problem a homology model of *cm*FDH was generated to design rational mutations for optimizing the surface electrostatic interaction on the protein surface and introduction of disulphide bridges into protein structure.

After the determination of importance of hydrophobic interaction by Langmuir in 1938, hydrophobic interactions are one of the widely studied and observed driving forces behind the folding and stability of globular proteins (Dill, 1990; Folch, et al., 2008; Reiersen & Rees, 2000). Recent results show that two other strategies have attracted attention to increase protein thermostability. One of them is improving the electrostatic interactions on the protein surface to optimize the surface electrostatic interactions (Eijsink, 2004; Kumar & Nussinov, 2001; Roca et al., 2007; Takita et al., 2008;). The second one is the introduction of disulphide bonds between Cys residues (Chu et al., 2007; Hamza & Engel, 2007; Yang et al., 2007). Although the contribution of surface electrostatics or disulphide bridges to protein stability is still not fully understood, it is clear that these interactions are important in protein thermostability.

4.1.1 Homology modelling of *cm*FDH

In order to generate homology model of *Candida methylica,* crystal structures of FDH from *Pseudomonas sp.*101 (Lamzin et al., 1992) which has 49 % amino acid sequence identity with *cm*FDH and *Candida boidinii* which has about 90% identity with *cm*FDH (Schirwitz et al., 2007) were used as the basis of model. Richard Sessions within the University of Bristol, helped with the computer simulations presented in this section. Amino acid changes were performed using InsightII 2005 (Accelrys). All the differences (insertions and/or deletions) between template (*Pseudomonas sp.*101 and *Candida boidini*) and target (*Candida methylica*) sequences in the alignment appeared at the surface of the structure, the loop building facility in InsightII was used to model these differences in the protein backbone. Energy minimization of modelled native *cm*FDH and various mutant FDHs were carried out using the steepest descents and conjugate gradients methods to minimize the energy because of the large number of atoms in FDH protein from *Candida methylica*. All energy minimizations were calculated using Discover, Version 2.95 (Accelrys) on a Linux workstation, which can be used for minimization and molecular dynamic simulations.

Gratifyingly, we find that the two models are closely similar with the following Root Mean Squared Deviations of the Cα postions: complete dimer, 2.7 Å; NAD⁺-binding domain, 1.7 Å; catalytic domain, 1.8 Å. Likewise, this result validates the designs and interpretations presented in our previous mutagenesis (Karaguler et al., 2007).

The all mutants designed either to optimize the surface electrostatic interactions or to introduce a disulphide bridge were applied to *cm*FDH by site directed mutagenesis to generate a thermostable *cm*FDH.

4.1.2 Optimization of electrostatic interactions on the surface of *cm*FDH

Compared with positions buried in the core, stabilizing surface mutations are less likely to disrupt the tertiary structure, which may be considered as the evidence of evolutionary selection (Alsop et al., 2003; Robinson-Rechavi, 2006). Electrostatic interactions also affect the protein flexibility that is significant for movement of residues with respect to each other and their environment. The changes in electrostatic interaction of charged side chains of residues are critical for protein folding and stability thus biological activity of the protein.

In order to increase the thermostability of formate dehydrogenase enzyme, Fedorchuk et al. (2002) tried to optimize electrostatic interactions by engineering residues in positions 43 and 61 of *ps*FDH by comparison to the *Mycobacterium vaccae* N10 FDH that only differs in two amino acid residues but has a lower stability than *ps*FDH. They showed the thermostability effect of loop regions of bacterial formate dehydrogenases which are absent in analogous eukaryotic enzyme and the replacement of Asp43 and in the *ps*FDH molecule does not result in an increase in stability. Rojkova et al. (1999) applied the hydrophobization of alpha helix strategy. They selected the 5 serine residues occupying positions 131, 160, 163, 184 and 222 of *ps*FDH. Their results showed that a combination of mutations had an additive effect to FDH stabilization and obtained a four-point mutant FDH which has a thermal stability 1.5 times higher, compared to the wild type enzyme. Serov et al. (2005) tried to optimize the polypeptide chain conformation to increase thermostability of *ps*FDH.

In our study, according to model mentioned in the section 4.1.1, a set of surface-charge mutations, simply based on inspection of the *cm*FDH homology model and chemical intuition, was designed. Comparison of the thermodynamic properties of mutants and native *cm*FDH shows that additional electrostatic interactions have different effects on the enthalpy and entropy changes at room temperature. Based on the thermodynamic results, it is suggested that three of nine catalytically active mutants resulted in increased protein stability by the measures of free energy of folding, Tm values and persistence of activity. Except relatively improved mutants, melting temperature increased between 2 and 6 °C while the folding and unfolding patterns of native *cm*FDH was not altered.

This result corresponds to a 33 % accuracy of designing mutant to increase thermostability. On the other hand, all the other mutation positions were significant such that they all had a measurable effect on the *cm*FDH folding and activity, and hence contributed to our understanding of the interactions. Another point of interest concerns the effect of double mutants on stability. It is known that, stabilization effect is generally additive allowing significant stabilization to be achieved with a number of single-point mutations (Serov et al., (2005). In contrast, a stabilizing effect of double mutants was not observed even though the corresponding single mutants had stabilizing effects.

4.1.3 Introduction of disulphide bridges into the structure of *cm*FDH

Disulphide bridges are also thought to stabilize the three-dimensional structure of proteins (Thornton, 1981). This hypothesis is based on observations that the disruption of disulphide bridges leads to a decrease in the thermodynamic stability of proteins (Anfinsen & Scheraga, 1975; Creighton , 1978). Although the mechanism by which disulphide bonds confer stability is not known yet in detail, one major aspect is that introducing disulphide bridges into proteins is the most obvious way to decrease the number of conformations available to the unfolded protein. Since part of the driving force to denature a protein is conformational entropy (the unfolded state has many more configurations than the folded state) reducing the number of conformations available to the unfolded state should stabilise the folded state.

FDH enzyme is one of several enzymes which have been subjected to disulphide bridge or Cys residue engineering to stabilize it against several factors or to overcome atmospheric oxidation of Cys residues. Odintseva et al. (2002) have engineered Cys residues in an attempt to increase the stability of *ps*FDH, obtaining mutants that showed the same kinetic parameters as the wild type enzyme, but its thermal stability dropped four-fold. Slusarczyk et al. (2000) have engineered the Cys residues of *cb*FDH and they showed that mutations affecting Cys23 are more effective than Cys262 on the stability. The first attempt to introduce a disulphide bridge into the FDH enzyme from *Candida methylica* to improve the thermostability was applied by Karagüler et al, in 2007b. Three pairs of cysteine residues were introduced into the *cm*FDH gene to construct three different disulphide bridged (T169/T226, V88/V112, M156/L159) mutants by site directed mutagenesis. The wild type *cm*FDH contain two cysteine residues buried in separate hydrophobic pockets but does not contain disulphide bridges, hence it is a good candidate for attempted stabilization by disulphide bridge approach.

No formate-dehydrogenase activity could be measured for the mutants V88C/ V112C (N-domain bridge) and M156C/ L159C (inter-subunit bridges) in either reducing or oxidising conditions. Presumably these mutations have distorted the structure of the FDH dimer to such an extent that the catalytic machinery is rendered ineffective. In contrast, the mutations T169C and T226C in the C-domain of *cm*FDH are tolerated. The catalytic efficiency (k_{cat}/K_M)

of the mutant protein in both its oxidised and reduced form is about four fold lower than the wild type *cm*FDH. Interestingly, the individual k_{cat} and K_M data show that the introduction of the cysteine residues in the reduced mutant raises K_M by about two fold and reducing k_{cat} by about two fold, with respect to the wild type. Oxidising the mutant and forming a cystine bridge restores K_M close to that of the wild type, consistent with forming a more native-like substrate binding site but reduces k_{cat} a further two fold (Karagüler et al, 2007b).

Another attempt to introduce a disulphide bridge into *cm*FDH on the positions of A153 and I239 by mutating these residues to cysteine have been performed by the same group (unpublished data). The disulphide bridge mutants investigated in this work and these mutants did not increase the thermostability of *cm*FDH. Morever, only one of them could be tolerated to give a functional enzyme. Hence it is clear that choosing sterically ideal sites for the insertion of disulphide bridges is crucial.

In all previous studies, to generate homology model of *Candida methylica*, crystal structures of FDH from *Pseudomonas sp.*101 (Lamzin et al., 1992) which has 49 % amino acid sequence identity with *cm*FDH was used as the basis of model. In our latest attempt, according to homology modelling of *cm*FDH based on *Candida boidinii* FDH crystal structure (Schirwitz et al., 2007) which has amino acid sequence similarity 97 % to *cm*FDH, two residues selected to change by Cys residue lie on the start point of two different β-strands. This is a good position to introduce a disulphide bridge in the N-terminal of catalytic domain of *cm*FDH and investigate the effect of Met to Cys replacement for stabilizing the structure.

Altough the mutant *cm*FDH which has disulphide bridge did not show the expected improvement in the stability, characterization studies of individual mutants showed that Met to Cys change is related to temperature stability. Catalytic efficiency (k_{cat}/K_M) of this single mutant was 63 % better than that of the native *cm*FDH and T_m value was measured as 2 °C higher than that of the native *cm*FDH. Substitution of methionine to cysteine, which converts a hydrophobic residue into a more hydrophilic one, can markedly alter the properties of a protein (Daia et al., 2007). Methionine and cysteine are two sulfur-containing amino acids but side chain of Cys is shorter than Met. Therefore, probably, change of Met to Cys affects the catalytic efficiency of protein and increase the thermostability by providing a more sterically compact structure.

The Far –UV CD spectra of native *cm*FDH and Met to Cys mutant exhibits similar shape at 25 °C. This result shows that Met to Cys mutant and native *cm*FDH have a nearly identical secondary structure content and indicates that the presence of the disulphide bridge does not affect significantly the enzyme secondary structure. During the heating to different temperatures, the CD spectra of both enzymes suggest that the replacement of methionine to cysteine leads to an increase of the thermostability of the secondary structure of *cm*FDH. On the other hand, it is known that introduction of disulphide bonds is one of the strategies to increase stability of a protein, which arises from a decrease of the conformational entropy (Mårtensson et al., 2002; Ordu, 2011, in preperation).

4.2 Alteration of coenzyme specificity

In biological systems, the majority of redox enzymes involved in anabolic processes use the coenzyme NADPH whereas those of catabolic processes generally use the coenzyme NADH (Stryer, 1988). NADPH is distinguished from NAD^+ by the presence of phosphate group esterified to the 2′ hydroxyl group of its adenosine moiety.

The majority of NAD+-dependent FDHs are highly specific towards NAD+ and do not utilise NADP+ as a coenzyme (Popov & Lamzin, 1994). Since many enzymatic synthesis of chemical compounds in industry require cofactors like NAD(H) or NADP(H) as discussed in section 3.2., it has been problem that the known FDHs only work with NAD+ as a cofactor but not NADP+. However, *Pseudomonas sp.* 101 FDH is exceptional in that it turns over NADP+ at 25 % of the rate of NAD+ under optimal reaction conditions (Popov & Lamzin, 1994). A number of positively charged amino acid side chains (Arg222, His223 and His379) are located close to the region which binds the 2'-O-PO$_4$$^{-2}$ of NADP+ and may provide a suitable electrostatic stabilisation for the bulky negatively charged phosphate group of NADP+.

One challenge is to convert NAD+-specific formate dehydrogenase to an NADPH-specific enzyme. It would be an advantage of NAD+-dependent FDHs if they could be modified to work with NADP+ as well because many of enzymatic synthesis of chemical compounds in industry require NADPH. Therefore, the dual activity enzyme could be used to input cheap reducing equivalent in a commercial process, normally using an NADP+ enzyme, as seen in figure 3.

Many attempts using both rational design and site saturation mutagenesis approaches have been made to change the coenzyme specificity of FDH from different sources (Andreadeli et al., 2008; Karaguler et al., 2001; Serov et al., 2002b; Wu et al., 2009) In our previous experiments we achieved *cm*FDH to use NADP+ by the single mutation Asp195Ser (Karaguler et al., 2001). However, this mutant binds NADP+ weakly and we suggested that *cm*FDH possesses an aspartic acid residue (195) which binds the hydroxyl groups of the adenine ribose moiety of NAD+, in common with many NAD+-dependent dehydrogenases containing the Rossmann fold. D195S shows similar catalytic constants to wild type in the reaction with NAD+. In contrast with wild type, the reaction of NADP+ catalysed by the mutant is clearly discernible. However, the accessible concentrations of NADP+ are well below K_M, hence only $(k_{cat}/K_M)_{NADP}$ can be determined. The ratio of the catalytic efficiencies for NAD+ versus NADP+ for the mutant protein is 40:1 in favour of NAD+. Likewise, the D195S enzyme is at least 8300-fold more efficient at turning over NADP+ than the wild type. These results demonstrate that D195S is a major determinant of cofactor specificity in *cm*FDH. Wu et al., (2009) revealed that site saturation mutagenesis application on residues Asp195, Tyr196 and Gln197 of *Candida boidinii* FDH produce more mutants with significant NADP+ specificity, which indicate the critical roles of these residues in determining the enzyme`s cofactor specificity. In the light of these foundations, we have also explored further mutations (Ozgun et al., submitted in 2011) in the coenzyme binding domain to improve the K_M of *cm*FDH for NADP+. The single mutations at D195, Y196 and Q197 in the coenzyme binding domain are introduced by using site saturation mutagenesis. In the library, two single mutants exhibit the highest catalytic efficiency (k_{cat} / K_M) in the presence of NADP+. Mutation on the residue 195, which has a proven role for the substrate specifity of FDH, is selected from the first generation mutant library as a template to construct the secondary generation mutant library. Second generation mutation is introduced into Y196 and Q197 residues. 2 double mutants are selected as promising NADP+ specific FDH mutants from the library. These mutants increased the overall catalytic efficiency of NADP+ to 56000 and 50000- fold, respectively. Results emphasize that; SSM is an efficient method for creating 'smarter libraries' for improving the properties of *cm*FDH.

5. Conclusion

In order to make FDH more suitable enzyme for its industrial applications, protein engineering is a promising approach. The work presented in this chapter demonstrate the difficulty of engineering hyper-stability into highly evolved enzyme structures using these methods. Each method has its own advantages and disadvantages. Although a combination of site-directed mutagenesis and model building is now being used to engineer novel proteins the rational design still can be inefficient because it is so time-consuming and only a limited number of amino acid variants can ever be evaluated for function. The results of DNA shuffling experiments showed that a very efficient screening technique is necessary to make this approach possible. Screening limits the search for beneficial mutations, because the basic rule of DNA shuffling is you get what you screen for, therefore it is vital to know exactly what the screen is actually selecting. The application of site saturation mutagenesis which is a combination of both strategies represent the new route to obtain the biocatalysts with the desired properties

6. Acknowledgment

Parts of this work are supported by Turkish State Planning Organisation in Advanced Technologies Program, Turkish State Planning Organisation (Project No: 90188), TUBITAK (Project No:107T684) and ITU Research Funds (Project no: 33309).

7. References

Allen, S. & Holbrook, J. (1995). Isolation, sequence and overexpression of the gene encoding NAD+-dependent formate dehydrogenase from the methylotrophic yeast *Candida methylica. Gene,* 162, 99-104.

Alsop, E.; Silver, M. & Livesay, D. R. (2003). Optimized electrostatic surfaces paralel increased thermostability: a structural bioinformatic analysis. *Protein Eng.,* 16,pp. 871-874.

Andreadeli, A.; Platis, D.; Tishkov, V.; Popov, V. & Labrou, N. E. (2008). Structure-guided alteration of coenzyme specificity of formate dehydrogenase by saturation mutagenesis to enable efficient utilization of NADP+. *FEBS Journal,* 275, pp.3859-3869.

Anfinsen, C. & Scheraga, H. (1975). Experimental and theoretical aspects of protein folding. *Adv Protein Chem,* 29, pp. 205-300.

Annaluru, N.; Watanabe, S., Saleh, A. A.; Kodak, T. & Makino, K. (2006). Site-directed mutagenesis of a yeast gene for improvement of enzyme thermostability. *Nucleic Acids Symposium Series,* 50, 1, pp. 281-282.

Annaluru, N.; Watanabe, S.; Saleh, A. A.; Kodak, T. & Makino, K. (2006). Site-directed mutagenesis of a yeast gene for improvement of enzyme thermostability. *Nucleic Acids Symposium Series,* 50, 1, pp. 281-282.

Arnold, F. H. (1998). Design by Directed Evolution. *Accounts of Chemical Research.,* 31, pp.125-131.

Arnold, F. H. (2001). Combinatorial and computational challenges for biocatalyst design *Nature* , 409, pp. 253-257.

Binay, B.; Shoemark, D.K.; Sessions, R.B.; Clarke, A.R. & Karaguler, N. G. (2009). Increasing the substrate specificity of *Bacillus stearothermophillus* lactate dehydrogenase by DNA shuffling. *Biochemical Engineering Journal*, 48, pp.118–123.

Bloom, J. D.; Labthavikul, S. T., Otey, C. R. & Arnold, F. H. (2006). Protein stability promotes evolvability *Proceedings of the National Academy of Science of the USA*, 103, 15, pp.5869–5874.

Bolivar, J. M.; Wilson, L.; Ferrarotti, S. A.; Fernandez-Lafuente, R.; Guisan, J. M. & Mateo, C. (2007). Evaluation of different immobilization strategies to prepare an industrial biocatalyst of formate dehydrogenase from *Candida boidinii*. *Enzyme and Microbial Technology*, 40, 4, pp. 540-546.

Bommarius, A. S.; Broering, J. M.; Chaparro-Riggers, J. F. & Polizzi, K. M. (2006). High-throughput screening for enhanced protein stability. *Current Opinion in Biotechnology*, 17, pp.606–610.

Bornscheuer, U. T. & Pohl, M. (2001). Improved biocatalysts by directed evolution and rational protein design. *Current Opinion in Chemical Biology*, 5, pp.137–143.

Chu, X.; Yu, W.; Wu, L.; Liu, X.; Li, N. & Li, D. (2007). Effect of a disulfide bond on mevalonate kinase. *Biochimica et Biophysica Acta - Proteins & Proteomics*, 1774, 12, pp. 1571-1581.

Chul Lee, S.; Chang, Y. J.; Min Shina, D.; Hana, J.; Seoa, M. H., Fazelini, H.; Maranasb, C. D. & Kima, H. S. (2009). Designing the substrate speci.city of d-hydantoinase using a rational approach. *Enzyme and Microbial Technology*, 44, pp.170–175.

Creighton, T. (1978). Experimental studies of protein folding and unfolding. *Progress in Drug Research*, 33, pp. 231-297.

Daia, X. L.; Sunb, Y. X. & Jiang, Z. F. (2007). Attenuated cytotoxicity but enhanced b.bril of a mutant amyloid b-peptide with a methionine to cysteine substitution. *FEBS Lett.* 581, pp. 1269–1274.

Davies, N. M. & Teng, X. V. (1997). Importance of Chirality in Drug Therapy and Pharmacy Practice: Implications for Psychiatry. *Advances in Pharmacy*, I, 3, pp. 242–252.

Davies, N. M. & Teng, X. W. (2003). Importance of Chirality in Drug , Therapy and Pharmacy Practice: Implications for Psychiatry. *Advances in Pharmacy*, 1, 3, pp. 242–252.

Dill, K. A. (1990). The meaning of hydrophobicity. *Science*, 250, pp.297–298.

Eckstein, M.; Daußmann, T. & Kragl, U. (2004). Recent Developments in NAD(P)H Regeneration for Enzymatic Reductions in One- and Two-Phase Systems. *Biocatalysis and Biotransformation*, 22, 2, pp.89-96.

Eijsink, V.G.H.; Bjørkb, A.; Gaseidnes, S.; Sirevag, R.; Synstad, B.; Van den Burgc, B. & Vriend, G. (2004). Rational engineering of enzyme stability. *Journal of Biotechnology*, 113, pp.105–120.

Endo, T. & Koizumi, S. (2001). Microbial conversion with cofactor regeneration using genetically engineered bacteria. *Advanced Synthesis and Catalysis*, 343, pp.521-526.

Fedorchuk, V.V.; Galkin, A.G.; Yasny, I.E.; Kulakova, L.B.; Rojkova, A.M.; Filippova, A.A. & Tishkov, V.I. (2002). Influence of interactions between amino acid residues 43 and 61 on thermal stability of bacterial formate dehydrogenases. *Biochemistry (Mosc.)*, 67, pp. 1145–1151.

Folch, B.; Rooman, M. & Dehouck, Y. (2008). Thermostability of Salt Bridges versus Hydrophobic Interactions in Proteins Probed by Statistical Potentials. *Journal of Chemical Information and Modeling*, 48, 1, pp. 119-127.

Gül-Karagüler, N.; Sessions, R. B.; Clarke, A. R. & Holbrook, J. J. (2001). A single mutation in the NAD-specific formate dehydrogenase from *Candida methylica* allows enzyme to use NADP. *Biotechnology Letters*, 23, pp. 283-287.

Haki, G.D. & Rakshit, S.K. (2003). Developments in industrially important thermostable enzymes. *Biosource Technology*, 89, pp.17-34.

Hamza, M. A. & Engel, P. C. (2007). Enhancing long-term thermal stability in mesophilic glutamate dehydrogenase from *Clostridium symbiosum* by eliminating cysteine residues. *Enzyme and Microbial Technology*, 41, 6-7, pp.706-710.

Hillisch, A.; Pineda, L. F. & Hilgenfeld, R. (2004). Utility of homology models in the drug discovery process. *Drug Discovery Today*, 9, 15, pp. 659-69.

Holbrook, J. J.; Louise, C.; Keyji J. & Hateley, John M. (2000). Chiral synthesis of 2-hydroxy carboxylic acids with a dehydrogenase United States Patent US6033882.

Johannes, T. W. & Woodyer, R., Zhao, H. (2007). Efficient Regeneration of NADPH using an Engineered Phosphite Dehydrogenase. *Biotechnology and Bioengineering*, 96, pp.18-26.

Jormakka, M., Byrney, B. & Iwata, S. (2003). Formate dehydrogenase – a versatile enzyme in changing environments. *Current Opinion in Structural Biology*, 13, pp.418-423.

Karagüler, N. G.; Sessions R. B. & Clarke, A. R. (2007b). Effects of disulphide bridges on the activity and stability of the formate dehydrogenase from *Candida methylica*. *Biotechnology Lett.* 29, pp.1375-1380.

Karaguler, N.G., Sessions, R. B.; Clarke A.R. &. Holbrook, J.J. (2001).A single mutation in the NAD-specific formate dehydrogenase from *Candida methylica* allows the enzyme to use NADP. *Biotechnology Letters*, 23, pp. 283-287.

Karaguler, N.G.; Sessions, R.B.; Binay, B.; Ordu, E.B. & Clarke, A.R. (2007). Protein engineering applications of industrially exploitable enzymes: *Bacillus stearothermophilus* LDH and *Candida methylica* FDH. *Biochemical Society Transactions*, 35, pp.1610-1615

Karshikoff, A. & Ladenstein, R. (2000). Protein from thermophilic and mesophilic organisms essentally do not differ in packing. *Protein Engineering*, 11, 10, pp. 867-872.

Kaup, B.; Meyer, S. B. & Sahm, H. (2005). D-mannitol formation from D-glucose in a whole cell biotransformation with recombinant E. coli. *Applied Microbiology and Biotechnology*, 69, pp. 397-403.

Kirk, O.; Borchert, T. W. & Fuglsang, C. C. (2002). Industrial enzyme applications. *Current Opinion in Biotechnology*, 13, pp. 345–351.

Kumar, S.; Ma, B.; Tsai, C. J. & Nussinov, R. (2000). Electrostatic Strengths of Salt Bridges in Thermophilic and Mesophilic Glutamate Dehydrogenase Monomers, *PROTEINS: Structure, Function, and Genetics*, 38, pp. 368–383.

Kumar, S.; Wolfson, H.J. & Nussinov, R. (2001). Protein flexibility and elecrostatic interactions. *IBM Journal of Research and Development*, 45, 3-4, pp. 499-512.

Lamzin, V. S.; Aleshin, A. L.; Stroskopytov, B. V.; Yukhnevich, M. G.; Popov, V. O., Harutyunyan, H. R. & Wilson, K. S. (1992). Crystal structure of NAD-dependent formate dehydrogenase. *European Journal of Biochemistry*, 206, pp. 441 -452.

Lehmann, M. & Wyss, M. (2001). Engineering proteins for thermostability: the use of sequence alignments versus rational design and directed evolution. *Current Opinion in Biotechnology*, 12, pp. 371–375.

Liu, W. & Wang, P. (2007). Cofactor regeneration for suitable enzymatic biosynthesis. *Biotechnology Advances*, 25, pp.369-384.

Mårtensson, L. G.; Karlsson, M. M. & Carlsson, U. (2002). Dramatic stabilization of the native state of human carbonic anhydrase II by an engineered disulphide bond. *Biochemistry*. 41, pp. 15867–15875.

Mayer, G.; Kulbe, K.D. & Nidetzky, B. (2002). Utilization of xylitol dehydrogenase in a combined microbial/enzymatic process for production of xylitol from D-glucose. *Applied Biochemistry*, 99, 3, pp.577-590.

Muller, G. W. (1997). Thalidomide: From tragedy to new drug discovery. *ChemTech*, 27, pp. 21-25.

Nakamura, K.; Aizawa, M. & Miyawaki, O. (1988). In electro-enzymology coenzyme regeneration., pp. 87-161, Springer-Verlag, Berlin and Heidelberg.

Northrup, S.H. & Erickson, H.P. (1992). Kinetics of protein–protein association explained by Brownian dynamics computer simulation. *Proc. Nalt. Acad. Sci. USA, Biochem.* 89, pp. 3338–3342.

Odintseva, E.R.; Popova, A.S.; Rojkova, A.M. & Tishkov, V.I. (2002). Role of cysteine residues in stability of bacterial formate dehydrogenase. *Bull. Moscow Univ., Ser. 2 Chem.* 43, pp. 356–359.

Ordu, E. B.; Cameron, G.; Clarke, A. R.; Karagüler, N. G. (2009). Kinetic and thermodynamic properties of the folding and assembly of formate dehydrogenase. *FEBS Letters*, 583, pp. 2887-2892.

Ozgun, G. P.; Ordu, E. B.; Bıyık, E. H.; Yelboga, E.; Sessions, R. B. & Karagüler, n. G. (submitted in 2011). Site Saturation Mutagenesis Applications on *Candida methylica* Formate Dehydrogenase.

Patel, R. N. (2004). Biocatalytic Synthesis of Chiral Intermediates. *Food Technology and Biotechnology*, 42, 4, pp. 305–325.

Péreza, M., Sinisterraa, J. V. & Hernáiz, M. J. (2010). Hydrolases in Green Solvents. *Current Organic Chemistry*, 14, pp. 2366-2383.

Popov, V., & Lamzin, V. (1994). NAD+-dependent formate dehydrogenase. *Biochemic J.*, 301, pp. 625-643.

Popov, V.O. & Thiskov, V.I. (2003). NAD+-dependent formate dehydrogenase. From a model enzyme to a versatile biocatalyst. *Protein Structures: Kaleidoscope of Structural Properties and Functions*, pp. 441-443.

Razvi, A &Scholtz, J. M. (2006). Lessons in stability from thermophilic proteins. *Protein Science*, 15, pp.1569-1578.

Reetz, M. F.; Carballeira, j. D. & Vogel, A. (2006). Iterative Saturation Mutagenesis on the Basis of B Factors as a Strategy for Increasing Protein Thermostability. *Angewandte Chemie International Edition*, 45, pp.7745 –7751.

Reiersen, H. & Rees, A. R. (2000). Trifluoroethanol may form a solvent matrix for assisted hydrophobic interactions between peptide side chains. *Protein Engineering*, 13, 11, pp.739-743.

Relyea, H.A., Vrtis, J. M., Woodyer, R.; Rimkus S. A. & Van der Donk, W. A. (2005). Inhibition and pH dependence of phosphite dehydrogenase. *Biochemistry*, 44, pp. 6640-6649.

Robinson-Rechavi, M. (2006). Contribution of electrostatic interactions, compactness and quaternary structure to protein thermostability: Lessons from Structural Genomics of *Thermotoga maritim.*, *J. Mol. Biol.*, 356, pp. 547-557.

Robinson-Rechavi, M.; Alibés, A. & Godzik, A. (2006). Contribution of Electrostatic Interactions, Compactness and Quaternary Structure to Protein Thermostability: Lessons from Structural Genomics of *Thermotoga maritime. Journal of Molecular Biology*, 356, 2, pp. 547-557.

Roca, M.; Messer, B. & Warshel A. (2007). Electrostatic contributions to protein stability and folding energy. *FEBS Letters*, 581, 10, pp.2065-2071.

Rodriguez, E.; Wood, Z. A., Karplus, P. A. & Lei, X. G. (2000). Site-Directed Mutagenesis Improves Catalytic Efficiency and Thermostability of *Escherichia coli* pH 2.5 Acid Phosphatase/Phytase Expressed in *Pichia pastoris. Archives of Biochemistry and Biophysics*, 382, 1, pp. 105-112.

Rojkova, A.M.; Galkin, A.G.; Kulakova, L.B.; Serov, A.E.; Savitsky, P.A.; Fedorchuk, V.V. & Tishkov, V.I. (1999). Bacterial formate dehydrogenase. Increasing the enzyme thermal stability by hydrophobization of alpha Helices. *FEBS Letters*, 445, pp. 183-188.

Rudra, S.G.; Shivhare, U.S.; Basu, S. & Sarkar, B.C. (2008). Thermal inactivation kinetics of peroxidase in coriander leaves. *Food Bioprocess. Technol.* , 1, pp. 187- 195.

Sali, A. & Blundell, T. L. (1993). Comparative protein modelling by satisfaction of spatial restraints. *Journal of Molecular Biology*, 234, pp.779-815.

Sanchez, R. & Sail, A. (1997). Advances in comparative protein-structure modelling. *Current Opinion in Structural Biology*, 7, pp.206-214.

Sarmentoa, A.C.; Oliveiraa, C.S.; Pereiraa, A.; Esteves, V.I.; Moirc, A.J.G.; Saraivad, J.; Pirese, E. & Barros, M. (2009). Unfolding of cardosin A in organic solvents and detection of intermediaries. *J. Mol. Catal. B: Enzym.*, 57, pp. 115–122.

Schirwitz, K.; Schmidt, A. & Lamzin, V. S. (2007). High-resolution structures of formate dehydrogenase from *Candida boidinii. Protein Science*, 16, pp. 1146-1156.

Schlosshauer, M. & Baker, D. (2004). Realistic protein–protein association rates from a simple diffusional model neglecting long-range interactions, free energy barriers, and landscape ruggedness. *Protein Sci.* 13, pp. 1660–1669.

Seelbach, K.; Riebel, B.; Hummel W., Kula, M. R., Tishkov, V. I., Egorov, A. M.; Wandrey, C. & Kragl, U. (1996). A novel, efficient regenerating method of NADPH using a new formate dehydrogenase. *Tetrahedron Letters*, 37, 9, pp. 1377-1380.

Serov, A. E.; Popova, A. S & Tishkov, V. I. (2002a). The Kinetic Mechanism of Formate Dehydrogenase from Bakery Yeast. *Doklady Biochemistry and Biophysics*, 382, 26–30.

Serov, A. E.; Popova, A. S.; Fedorchuk, V. V. & Tishkov, V. I. (2002b). Engineering of coenzyme specificity of formate dehydrogenase from Saccharomyces cerevisiae. *Biochem J.*, 1, 367, pp. 841–847.

Serov, A.E.; Odintzeva, E.R.; Uporov, I.V. & Tishkov, V.I. (2005). Use of Ramachandran plot for increasing thermal stability of bacterial formate dehydrogenase. *Biochemistry (Mosc.)*, 70, pp. 804–808.

Slusarczyk, H.; Felber, S., Kula, M.R. & Pohl, M. (2000). Stabilization of NAD⁺-dependent formate dehydrogenase from *Candida boidinii* by site-directed mutagenesis of cysteine residues. *European Journal of Biochemistry*, 267, pp.1280–1289.

Slusarczyk, H.; Felber, S.; Kula, M.R. & Pohl, M. (2000). Stabilization of NAD⁺-dependent formate dehydrogenase from Candida boidinii by site-directed mutagenesis of cysteine residues. *European Journal of Biochemistry*, 267, pp.1280–1289.

Spadiut, O.; Leitner, C.; Salaheddin, C.; Varga, B.; Vertessy, B. G.; Tan, T. C.; Divne, C. & Haltrich, D. (2009). Improving thermostability and catalytic activity of pyranose 2-oxidase from Trametes multicolor by rational and semi-rational design. *FEBS Journal*, 276, 3, pp. 776-792.

Stemmer, W. P. (1994). DNA shuffling by random fragmentation and reassembly: in vitro recombination for molecular evolution. *Proc Natl Acad Sci U S A*, 25, 91, 22, pp. 10747-51.

Stemmer, W. P. (1994). Rapid evolution of a protein in vitro by DNA shuffling. *Nature*, 370, pp.389 – 391.

Stryer, L. (1988). Chapter 13: Design of metabolis,. In: Biochemistry, 3rd edition, pp. 321, W.H. Freeman and Company, New York:

Suárez, M. & Jaramillo, A. (2009). Challenges in the computational design of proteins. Journal of the Royal *Society Interface*, pp.1-15.

Synowiecki, J.; Grzybowska, B. & Zdzieblo, A. (2006). Sources, properties and suitability of new thermosable enzymes in food processing. *Critical Reviews in Food Science and Nutrition*, 46, pp. 197-205.

Ordu, E. B. and Karagüler N. G. (2007). Improving the Purification of NAD1-Dependent Formate Dehydrogenase from Candida methylica, Prep. Biochem. Biotechnol., 37, 333-334.

Takita, T.; Aono, T.; Sakurama, H.; Itoh, T.; Wada, T.; Minoda, M.; Yasukawa, K. & Inouye, K. (2008). Effects of introducing negative charges into the molecular surface of thermolysin by site-directed mutagenesis on its activity and stability. *Biochimica et Biophysica Acta - Proteins & Proteomics*, 1784, 3, pp.481-488.

Tao, H. & Cornish, V. W. (2002). Milestones in directed enzyme evolution. *Current Opinion in Chemical Biology*, 6, pp.858–864.

Tao, H. & Cornish, V. W. (2002). Milestones in directed enzyme evolution. *Current Opinion in Chemical Biology*, 6, pp.858–864.

Tao, J. & Xu, J. H. (2009). Biocatalysis in development of green pharmaceutical processes. *Current Opinion in Chemical Biology*, 13, pp. 43–50.

Thornton, J. (1981). Disulphide bridges in globular proteins. *J Mol Biology*, 151, pp. 261-287.

Tishkov, V. I & Zaitseva, E. A. (2008). Modern Trends in Biocatalytic Synthesis of Chiral Compounds. *Moscow University Chemistry Bulletin.*, 63, 2, pp. 111–113.

Tishkov, V. I. & Popov, V.O. (2006). Protein engineering of formate dehydrogenase. *Biomolecular Engineering*, 23, pp.89–110.

Turner, N. J. (2003). Directed evolution of enzymes for applied biocatalysis. *Trends in Biotechnology*, 21, 11, pp.474-8.

Uversky, V. N. (2003). Protein folding revisited. A polypeptide chain at the folding – misfolding – nonfolding cross-roads: which way to go?, *Cellular and Molecular Life Sciences*, 60, pp.1852–1871.

Van der Donk, W. A. & Zhao, H. (2003). Recent developments in pyridine nucleotide regeneration. *Current Opinion in Biotechnology,* 14, pp. 421–426.

Voutilainen, S. P.; Boer, H.; Alapuranen, M.; Jänis, J.; Vehmaanperä, J. & Koivula, A. (2009). Improving the thermostability and activity of Melanocarpus albomyces cellobiohydrolase Cel7B. *Applied Microbiology and Biotechnology,* 83, pp.261-272.

Vrtis, J. M.; White, A. K.; Metcalf, W. W. & Van der Donk, W.A. (2002). Phosphite dehydrogenase: versatile cofactor-regeneration enzyme. *Angewandte Chemie International Edition,* 41, pp. 3257–3259.

Walker, J. M. &_Rapley, R. (2008). Chapter 35, In: *Molecular Biomethods,* pp. 587, Humana Press, Totowa-USA.

Wandrey, C. (1986). Synthesis of L-amino acids by isolated enzymes and microorganisms. In: Enzymes as catalysts in organic synthesis. Schneider M, ed. Holland: D. Reidel Publishing Company, pp. 263-284.

Wei, J.; Zhou, J.; Xu T. & Lu, B. (2009). Rational Design of Catechol-2, 3-dioxygenase for Improving the Enzyme Characteristics. *Applied Biochemistry and Biotechnology,*162, pp. 116-126.

Wilks, H.;Cortes, A.; Emery, D.; Halsall, D. Clarke, A. & Holbrook, J. (1992). Opportunities and limits in creating new enzymes: experiences with the NAD-dependent lactate dehydrogenase frameworks of humans and bacteria. *Annals of the New York Academy of Sciences,* 672, pp. 80-93.

Williams, G.J.; Nelson, A.S. & Berry, A. (2004). Directed evolution of enzymes for biocatalysis and the life sciences. *Cellular and Molecular Life Science,* 61, pp. 3034-3046.

Wojtasiak, R. W. (2006). Chirality and the nature of food authensiticy of aroma. *Acta Sci Po, Technol. Aliment.,* 5, 1, pp. 21-36.

Wong, T. S.; Tee, K. L.; Hauer, B. & Schwaneberg, U. (2004). Sequence saturation mutagenesis (SeSaM): a novel method for directed evolution. *Nucleic Acids Research,* 32, 3, pp.26.

Woodyer, R.; Van der Donk, W. A. & Zhao, H. (2006). Optimizing a biocatalyst for improved NAD(P)H regeneration: directed evolution of phosphite dehydrogenase. *Combinatorial Chemistry & High Throughput Screening,* 9, pp. 237-245.

Wu, W.; Zhu, D. & Hua, L. (2009). Site-saturation mutagenesis of formate dehydrogenase from Candida bodinii creating effective NADP⁺-dependent FDH enzymes. *Journal of Molecular Catalysis B: Enzymatic,* 61, 3-4, pp.157-161.

Wunderlich, M.; Martin, A.; Staab, C. A. & Schmid, F. X. (2005). Evolutionary Protein Stabilization in Comparison with Computational Design. *Journal of Molecular Biology,* 351, pp. 1160–1168.

Xu, Z., Jing, K.; Liu, Y. & Cen, P. (2007). High-level expression of recombinant glucose dehydrogenase and its application in NADPH regeneration, J. Ind. Microbiology and Biotechnology, 34, 1, pp.83-90.

Yan, J.; Yan, Y; Liu, S.; Hu, J. & Wang, G. (2011). Preparation of cross-linked lipase-coated micro-crystals for biodiesel production from waste cooking oil. *Bioresource Technology,* 102, pp. 4755–4758.

Yang, H. M.; Yao, B.; Meng, K.; Wang, Y. R.; Bai, Y. G. & Wu, N. F. (2007). Introduction of a disulfide bridge enhances the thermostability of a *Streptomyces olivaceoviridis* xylanase mutant. *Journal of Industrial Microbiology & Biotechnology,* 34, 3, pp. 213-218.

Yokoigawa, K.; Okubo, Y.; Soda, K. & Misono, H. (2003). Improvement in thermostability and psychrophilicity of psychrophilic alanine racemase by site-directed mutagenesis. *Journal of Molecular Catalysis. B, Enzymatic*, 23, 2-6, pp. 389-395.

Zheng, l.; Baumann, U. & Reymond, J. L. (2004). An efficient one-step site-directed and site-saturation mutagenesis protocol. *Nucleic Acids Research*, 32, pp. 14, 115.

Zhou, H.X.; Wong, K.Y. & Vijayakumar, M. (1997). Design of fast enzymes by optimizing interaction potential in active site. *Proc. Natl. Acad. Sci. USA, Biophys.* 94, pp. 12372–12377.

4

Evolutionary Engineering of Artificial Proteins with Limited Sets of Primitive Amino Acids

Junko Tanaka, Hiroshi Yanagawa and Nobuhide Doi
Department of Biosciences and Informatics, Keio University
Japan

1. Introduction

Because present-day proteins are composed of 20 kinds of amino acids, the number of possible amino-acid sequences in a 100-residue protein is 20^{100} (approximately 10^{130}), which is larger than the total number of atoms in the universe ($\sim 10^{80}$). The number of proteins that may have existed in nature throughout the history of life on the Earth has been estimated to be less than 10^{50} molecules (Mandecki, 1998) or 10^{43} molecules (Dryden et al., 2008). Thus, the vast sequence space available remains to be explored further, and the sequence space that remains unexplored provides an opportunity to create valuable proteins with novel structures and functions for biomedical and environmental applications. Evolutionary protein engineering or directed protein evolution has been used to create artificial proteins with novel functions (Bloom et al., 2005; Hoogenboom, 2005; Leemhuis et al., 2005; Romero & Arnold, 2009) by repeated mutation, selection and amplification, mimicking Darwinian evolution in the laboratory (Figure 1).

Using evolutionary engineering, several researchers have recently demonstrated in the laboratory how the steps of protein evolution might occur in nature. For example, Peisajovich et al. (2006) explored the plausibility of the permutation-by-duplication model of the evolution of the DNA-methyltransferase superfamily and indicated that new protein topologies can evolve gradually through multistep gene rearrangements while maintaining the function of the parent domain. Tokuriki and Tawfik (2009a) investigated the random mutational drift of several enzymes in the presence of overexpressed chaperonin and revealed that protein stability is a major constraint in protein evolution and is a buffering mechanism by which chaperonin can alleviate this constraint. Huang et al. (2008) indicated that new protein functions can be generated by combining unrelated domains and subsequently optimizing the domain interface. However, these studies have mainly focused on the relatively recent evolutionary pathways of modern proteins; none of the hypotheses regarding the early evolution of primitive proteins has yet been tested.

In this chapter, we focused on the hypothesis that proteins consisted of fewer amino acid types during the early stage of protein evolution. Although modern proteins consist of 20 amino acid types, it has been proposed that primordial proteins consisted of a smaller set of "primitive" amino acids that could have been abundantly formed on the prebiotic Earth. Additional, "new" amino acids were then gradually recruited into the genetic code (Section 2). To test this hypothesis, we used the powerful tool "mRNA display" (Section 3) and examined the rate at which folding ability (Section 4) and function (Section 5) occurred in artificial proteins consisting of limited sets of amino acids. An improved understanding of protein evolutionary pathways can provide more efficient tools for the creation of artificial proteins with novel functions and structures.

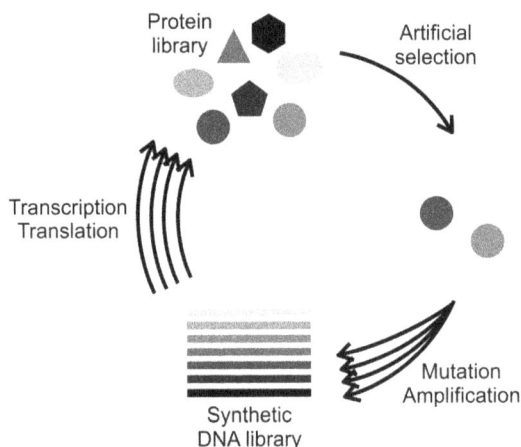

Fig. 1. The cycle used in the Darwinian evolution of proteins. A pool of genotype molecules (a synthetic DNA library) is converted to a pool of phenotype molecules (a protein library), from which proteins with a desired property are selected (artificial selection). The genotype molecules corresponding to the selected proteins are then amplified and mutated for further selection cycles.

2. A hypothesis regarding the origin and early evolution of proteins and the genetic code

According to Oparin's chemical evolutionary hypothesis regarding the Origin of Life (Oparin, 1961), a soup of nutrient organic compounds was available to the first organism on the primitive Earth. Once a self-replicating molecule formed from the primordial soup, this early replicator could have evolved to a primordial cell. Since the discovery of RNA enzymes (ribozymes) (Altman, 1981; Cech et al., 1981), RNA molecules have been postulated to be the first self-replicating molecule to play the following two roles: storing genetic information and catalyzing a chemical reaction. Later, RNA acquired various catalytic activities (in the RNA world), and protein synthesis could have been established. Finally, the location of genetic information moved from RNA to DNA because DNA is more stable than RNA. At present, proteins are synthesized based on the information contained in DNA and have particular structures and functions that provide various kinds of biological activities.

How were present-day proteins and the genetic code generated in the RNA world? It has been proposed that primordial proteins consisted of a small set of amino acids such as Ala, Gly, Asp and Val, which could have been abundantly formed early during chemical evolution (Miller, 1987). Interestingly, the codons for these amino acids all have guanosine (G) as the first nucleotide; for this reason, the codons GNC and GNN, where N denotes U, C, A or G, were proposed to have formed the early genetic code (Eigen & Schuster, 1978).

Thereafter, according to the coevolution theory (Wong, 1975, 1988, 2005), the genetic code coevolved with amino acid biosynthetic pathways, and additional amino acids were introduced after production through their synthetic pathways. Recently, Trifonov (2004) deduced a list of the consensus order in which amino acids were incorporated into the genetic code on the basis of 60 criteria (Figure 2A). This list revealed that the amino acids synthesized in Miller's spark discharge experiments (Gly, Ala, Asp, Val, Pro, Ser, Glu, Thr, Leu and Ile) appeared first, and that the amino acids associated with codon capture events (His, Cys, Phe, Tyr, Met and Trp) came last (Trifonov, 2004). Jordan et al. (2005) verified this list using comparative genome sequence analysis of orthologous proteins in the genomes of bacteria, archaea and eukaryotes. These authors clarified that the frequencies of Gly, Ala, Glu and Pro consistently decrease in proteins while the frequencies of Ser, His, Cys, Met and Phe increase during protein evolution (Figure 2B). They took into consideration the concept that the amino acids with decreasing frequencies are thought to have been the first amino acids incorporated into the genetic code; conversely, all amino acids with increasing frequencies, except Ser, are probably late recruitments (Jordan et al., 2005). The trend of transitioning amino acid composition (Figure 2B) corresponds well to Trifonov's list of the order of incorporation of amino acids into the genetic code (Figure 2A).

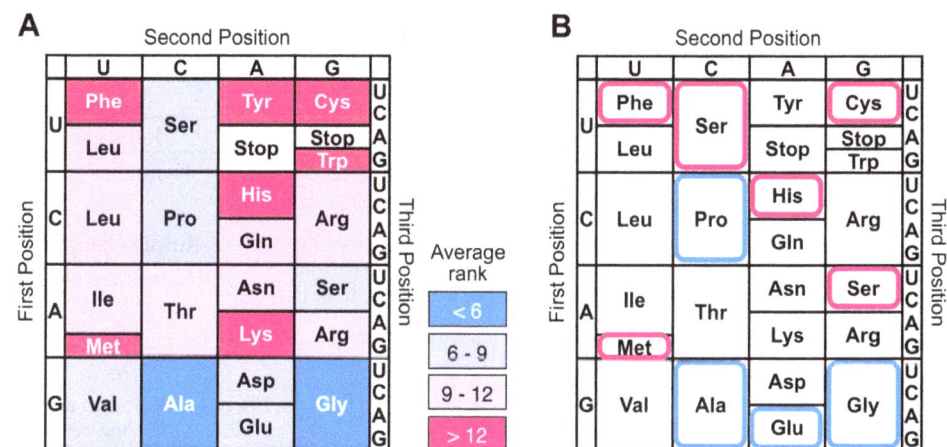

Fig. 2. The universal genetic code. (A) The average rank represents the chronological order of amino acid addition to the genetic code (Trifonov, 2004). The ranking values calculated based on 60 criteria were Gly, 3.5; Ala, 4.0; Asp, 6.0; Val 6.3; Pro, 7.3; Ser, 7.6; Glu, 8.1; Thr, 9.4; Leu, 9.9; Arg, 11.0; Asn, 11.3; Ile, 11.4; Gln, 11.4; His, 13.0; Lys, 13.3; Cys, 13.8; Phe, 14.2; Tyr, 15.2; Met, 15.4 and Trp, 16.5. (B) The trend of amino acid gain and loss during protein evolution (Jordan et al., 2005). The amino acids for which the frequencies consistently decrease (i.e., primitive amino acids) are highlighted in blue, and the amino acids for which the frequencies consistently increase (i.e., amino acids that were probably recruited late into the genetic code) are highlighted in red.

Can native protein structure and function be achieved with such reduced alphabets? Several researchers have demonstrated that the amino acid usage of various natural globular proteins and enzymes can be restricted to 5–9 members while retaining their structures and functions (Table 1). For example, Riddle et al. (1997) simplified SH3 domains in which 90%

of the sequence employed just five types of amino acids. Silverman et al. (2001) restricted 78% of the sequence of the prototypical $(\beta/\alpha)_8$ barrel enzyme, triosephosphate isomerase, to seven types of amino acids. Akanuma et al. (2002) generated variants of orotate phosphoribosyl transferase in which 88% of the sequence used just nine types of amino acids. Finally, Walter et al. (2005) created an active enzyme, chorismate mutase, which was constructed entirely from nine types of amino acids. Other researchers have attempted to produce *de novo* proteins from designed combinatorial libraries. Hecht's group created four helix bundle proteins based on binary patterning using five types of nonpolar amino acids (Val, Met, Ile, Phe and Leu) and six kinds of polar amino acids (Asp, Glu, Asn, Lys, Gln and His) (Go et al., 2008; Kamtekar et al., 1993; Patel et al., 2009). Jumawid et al. (2009) produced *de novo* proteins with an α3β3 structure using a simplified binary combination of hydrophobic amino acids (Val, Ile and Leu) and hydrophilic amino acids (Ala, Glu, Lys and Thr). These experiments support the hypothesis that the full amino-acid alphabet set is not essential for the structure and biological function of proteins. However, these experiments were not focused on whether the limited sets of amino acids used are primitive or not. Thus, the hypothesis that primordial proteins originally consisted of a small repertoire of primitive amino acids that gradually increased by coevolution with amino acid biosynthetic pathways has been insufficiently supported by experimental data thus far. In the following section, we summarize molecular display technologies that can be used to experimentally demonstrate the hypothesis regarding the existence of primordial proteins.

Protein	Amino Acids (variety of amino acids)	Reference
Four α-helix bundle *de novo* protein	Asp, Val, Glu, Leu, Asn, Ile, Gln, His, Lys, Phe, Met (11)	Kamtekar et al., 1993
SH3 domain	Gly, Ala, Glu, Ile, Lys (5)	Riddle et al., 1997
Triosephosphate isomerase	Ala, Val, Glu, Leu, Gln, Lys, Phe (7)	Silverman et al., 2001
Orotate phosphoribosyl transferase	Gly, Ala, Asp, Val, Pro, Thr, Leu, Arg, Tyr (9)	Akanuma et al., 2002
Chorismate mutase	Asp, Glu, Leu, Arg, Asn, Ile, Lys, Phe, Met (9)	Walter et al., 2005
α3β3 *de novo* protein	Ala, Val, Glu, Thr, Leu, Ile, Lys (7)	Jumawid et al., 2009

Table 1. Proteins constructed using reduced sets of amino acids with retention of their biological functions or structures.

3. mRNA display for *in vitro* selection of proteins

In the selection of targeted functional biomolecules by directed evolution, the most important consideration is the ability to link genotype and phenotype. "Phenotype" refers to biological functions, whereas "genotype" refers to the nucleic acids coding for replication. The nucleic acid portions of RNA aptamers and ribozymes have roles in both function and replication. Proteins, however, have only functional roles and cannot be replicated. Therefore, the development of a molecular display technique that physically links genotype with phenotype is essential for directed protein evolution. As shown in Figure 3, various molecular display techniques have been developed (Doi & Yanagawa, 2001; Matsumura et al., 2006). In 1985, Smith discovered that exogenous peptides could be displayed on a filamentous phage by fusing peptides of interest to the coat protein of a filamentous phage (Smith, 1985). This technology has been developed into the best-known display technique, phage display (Figure 3A). Phage display is a cell-based method in which proteins are expressed in *Escherichia coli*. Another display technique using living cells is cell-surface display (Figure 3B) in which proteins are displayed on the surface of living cells, such as yeast (Georgiou et al., 1997; Murai et al., 1997) or mammalian cells (Wolkowicz et al., 2005). These cell-based display techniques have some weaknesses; the library size is limited by the number of cells and transformation efficiency (typically below 10^9), and some proteins that are toxic to the cell are excluded from the library. To overcome such weaknesses, completely *in vitro* techniques have been developed, such as ribosome display (Hanes & Plückthun, 1997; Figure 3C), mRNA display (Nemoto et al., 1997; Roberts & Szostak, 1997; Figure 3D) and DNA display (Doi & Yanagawa, 1999; Figure 3E). Each display technique has been improved and applied to functional selection for peptides and proteins (Matsumura et al., 2006).

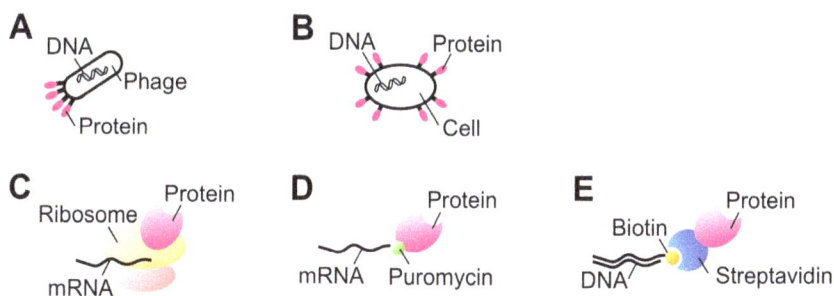

Fig. 3. Molecular display techniques. (A) Phage display (Smith, 1985). Proteins are displayed on a filamentous phage by fusing them to coat proteins of the phage. (B) Cell-surface display (Georgiou et al., 1997; Murai et al., 1997; Wolkowicz et al., 2005). Proteins are displayed on the surface of living cells. (C) Ribosome display (Hanes & Plückthun, 1997). Individual nascent proteins are coupled to their corresponding mRNA through ribosomes. (D) mRNA display (Nemoto et al., 1997; Roberts & Szostak, 1997). The protein is covalently linked with its corresponding mRNA *via* puromycin. (E) DNA display (Doi & Yanagawa, 1999). The protein is linked with its corresponding DNA by streptavidin-biotin interaction in water-in-oil emulsion.

Techniques for mRNA display have been developed in our laboratory and independently in that of Szostak (Nemoto et al., 1997; Roberts & Szostak, 1997). In this technique, each cell-free translated polypeptide (phenotype) in a library is covalently linked with its corresponding mRNA (genotype) *via* puromycin. This antibiotic is an analogue of the 3′ end of aminoacyl-tRNA (Figure 4A) and causes premature termination of translation by binding to the C-terminus of the nascent polypeptide chain. When its concentration is very low, puromycin is transferred to the C-terminus of the full-length protein (Miyamoto-Sato et al., 2000). Based on this property of puromycin, when mRNA lacking a stop codon is ligated with puromycin at the 3′ end and translated using a cell-free translation system, an mRNA (genotype) and full-length protein (phenotype) conjugate is produced (Figure 4B).

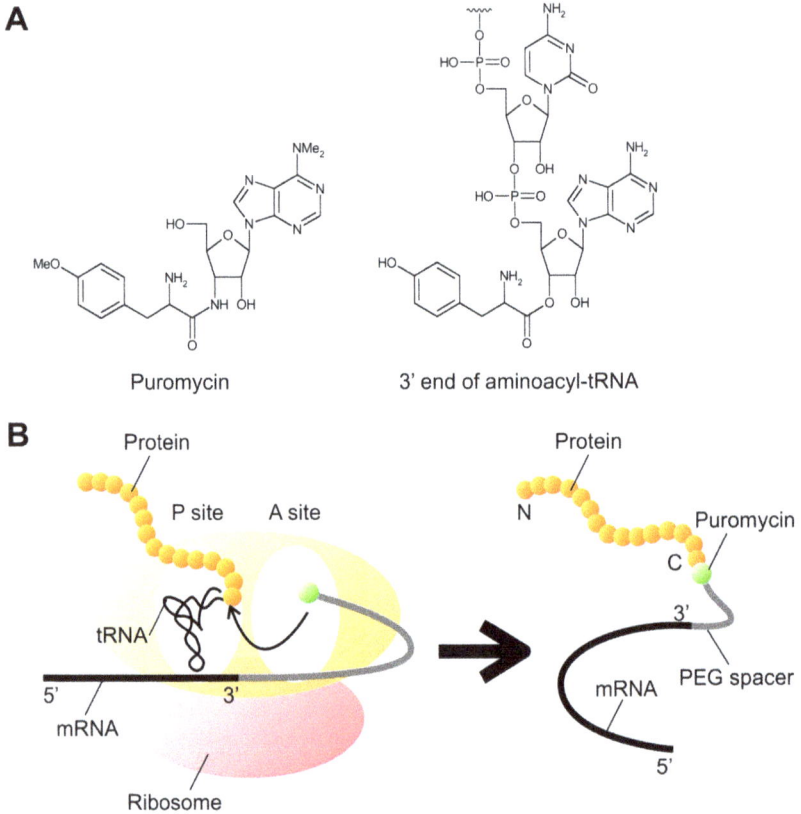

Fig. 4. The principle behind the formation of an mRNA-displayed protein. (A) The structure of puromycin and the 3' end of aminoacyl-tRNA. (B) Puromycin ligated to the 3'-terminal end of an mRNA *via* a polyethyleneglycol (PEG) spacer can enter the ribosomal A site to bind covalently to the C-terminal end of the protein that it encodes (Nemoto et al., 1997).

In mRNA display, a larger number of molecules (approximately 10^{12-13}) can be handled than is possible using other cell-based display techniques such as phage display. This enables the enrichment of active sequences with low abundance from libraries with high diversity and complexity.

The typical scheme of *in vitro* selection using mRNA display is shown in Figure 5. Proteins are displayed on mRNA by cell-free translation of modified mRNA as described above. After affinity selection *via* the protein portion of an mRNA-displayed protein from the library, selected proteins can be easily identified by amplification and sequencing of the mRNA portion. Moreover, targeted proteins with low-copy numbers can be also detected by iterative selection. In the following sections, we describe the application of mRNA display to the construction of random-sequence protein libraries with a limited set of amino acids (Section 4) (Tanaka et al., 2010) and to the selection of functional proteins from partially randomized libraries with a limited set of amino acids (Section 5) (Tanaka et al., 2011).

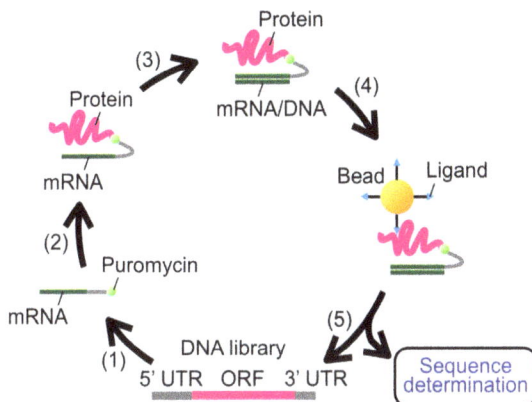

Fig. 5. Schematic representation of the mRNA-display selection of ligand-binding proteins. (1) A DNA library is transcribed and ligated with a polyethyleneglycol-puromycin spacer. (2) The modified mRNA library is translated using a cell-free translation system. (3) The resulting mRNA-protein conjugates are purified and reverse-transcribed. (4) The mRNA/DNA-protein conjugates are incubated with the ligand-immobilized beads, washed and competitively eluted with the free ligand. (5) The DNA portion of the eluted molecules is amplified using PCR to form a DNA library for the next round.

4. Random-sequence proteins with primitive amino acids

How frequently did functional or folded proteins occur in the RNA world? To answer this question, Keefe & Szostak (2001) selected novel ATP binding proteins from a random-sequence protein library based on the 20-amino acid alphabet using mRNA display. The authors roughly estimated that the frequency of occurrence of functional proteins is 1 in 10^{11}. Over the last decade, no functional protein has been obtained from random-sequence libraries. One of the difficulties in functional selection is that random-sequence proteins that use 20 types of amino acids tend to aggregate (Mandecki, 1990; Prijambada et al., 1996; Watters & Baker, 2004). Because primordial proteins presumably consisted of a smaller set of amino acids that could have been abundantly formed during early chemical evolution as mentioned above, random-sequence proteins that use 20 types of amino acids may have different physical properties from primordial proteins.

As shown in Table 2, random-sequence proteins that use a limited set of amino acids reportedly have different properties from random-sequence proteins that use 20 kinds of amino acids. Although random-sequence proteins based on three kinds of amino acids (QLR proteins, which consist of Gln, Leu and Arg) tend to strongly aggregate (Davidson & Sauer, 1994; Davidson et al., 1995), random-sequence proteins with the primitive amino acids Ala, Gly, Val, Asp and Glu, which are encoded by codons of the form GNN (N = T, C, A or G), demonstrated extremely high solubility (Doi et al., 2005). Using mRNA display, we constructed three classes of random-sequence libraries consisting of limited sets of amino acids (Tanaka et al., 2010); these libraries were encoded using the codons GNN, RNN (R = A or G, encoding a 12-amino acid alphabet) and NNN (encoding the full set of amino acids). When proteins that were arbitrarily chosen from these libraries were expressed in *Escherichia coli*, all proteins from the GNN library were present in the soluble fraction, all of the proteins from the NNN library were present in the insoluble fraction, and the proteins from the RNN library were intermediate in character, *i.e.*, one out of 14 RNN proteins was expressed only in the soluble fraction, 11 RNN proteins were expressed only in the insoluble fraction, and two were expressed in both fractions (Tanaka et al., 2010).

Protein	Variety of Amino Acids	Solubility	Secondary Structure Content	Reference
QLR protein	3	Quite low	Strong	Davidson & Sauer, 1994; Davidson et al., 1995
GNN protein	5	High	Low	Doi et al., 2005; Tanaka et al., 2010
RNN protein	12	Medium	Low	Tanaka et al., 2010
NNN protein	20	Low	Low	Yamauchi et al., 1998; Tanaka et al., 2010

Table 2. Biophysical properties of random-sequence proteins constructed using reduced alphabets.

What causes such difference in solubilities? To investigate this question, we examined the relationship between the solubility of random-sequence proteins and several properties of the amino acid sequences (Tanaka et al., 2010). It has been suggested that protein solubility is strongly affected by net charge and the fraction of turn-forming residues (Gly, Asp, Pro, Ser and Asn) and is weakly affected by hydrophobicity and protein size (Wilkinson & Harrison, 1991). We found no relation between solubility and the fraction of turn-forming residues, hydrophobicity [calculated based on the index of Kyte and Doolittle (1982)], or protein size for GNN, RNN and NNN proteins (Tanaka et al., 2010). The high solubility of GNN proteins could be attributed to net charge because all GNN proteins lack positively charged amino acids. Soluble RNN proteins have higher net charge and lower hydrophobicity than insoluble RNN proteins. However, the low solubility of NNN proteins with high net charge and low hydrophobicity cannot be easily explained.

Random-sequence proteins with limited sets of amino acids have been structurally characterized. QLR proteins, which have three kinds of amino acids, exhibited strong α-helical content in aqueous solution but tended to aggregate, and the addition of a denaturing agent is necessary for solubilization (Davidson & Sauer, 1994; Davidson et al., 1995). Soluble RNN proteins largely adopted random coil conformations but formed α-helical structures in a hydrophobic environment (Tanaka et al., 2010). Hence, these results indicate that RNN proteins have the potential to form at least partial secondary structures, similar to random-sequence proteins based on 20 amino acid types (Yamauchi et al., 1998). Thus, there may be a trade-off between secondary structure formation and high solubility among random-sequence proteins (Table 2). Other experiments showed the presence of hydrophobic clusters in GNN and RNN proteins (Doi et al., 2005; Tanaka et al., 2010). Furthermore, GNN and RNN proteins formed monomeric structures with more compact shapes than the random-coil structures adopted by denatured proteins of similar molecular weight but had more extended shapes than the globular structures of natural proteins. That is, they probably form molten globule-like structures (Figure 6). However, random-sequence proteins based on 3- and 20-amino acid alphabets have been reported to form oligomeric structures due to their tendency toward aggregation (Davidson & Sauer, 1994; Davidson et al., 1995; Yamauchi et al., 1998).

Recently, a large number of intrinsically unstructured domains that become structured only during binding to the target (*i.e.*, induced fit) have been identified in nature (Wright & Dyson, 1999). Moreover, artificial proteins that form well-folded structures after interaction with their target were produced (Walter et al., 2005; Vamvaca et al., 2004; Chaput & Szostak, 2004). Such partially structured polypeptides might have been the first evolutionary intermediates, and their functions and structures would have coevolved (Tokuriki & Tawfik, 2009b). Thus, random-sequence proteins based on the set of amino acids encoded by the codon RNN may include such evolutionary intermediates because these proteins contain partial secondary structures and hydrophobic clusters.

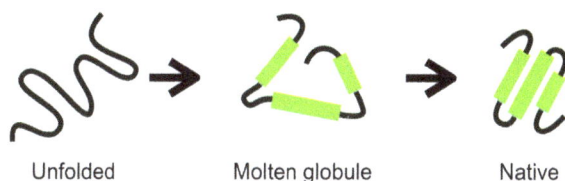

Unfolded	Molten globule	Native

Fig. 6. Protein folding. In the unfolded state, the polypeptide chain adopts an entirely random conformation. In the folded state, the protein takes on a unique conformation. The protein folds into the compact native structure through an intermediate state, *i.e.*, a molten globule state (Ohgushi & Wada, 1983), in which much of the secondary structure (light green) is present.

5. Functional proteins consisting of primitive amino acids

As described in the previous section, random-sequence proteins constructed with subsets of the putative primitive amino acids (5 and 12-amino acid alphabets) have higher solubility than those constructed using the natural 20-member alphabet, although other biophysical properties remain very similar. Because the solubility of globular proteins is an important factor in the exertion of their function, it is of interest to test whether functional proteins occur more frequently in a library based on a limited set of primitive amino acids than in a library based on the 20-amino acid alphabet or other non-primitive alphabets.

To address this question, we attempted to compare the frequencies with which functional proteins occur in libraries based on various sets of amino acids (Tanaka et al., 2011). First, we designed randomized *src* SH3 gene libraries in which approximately half the residues of the SH3 gene were replaced by various kinds of randomized codons (Figure 7A). We utilized three limited sets of amino acids: (1) the set coded by the lower half of the genetic code (RNN) contains mainly putative primitive amino acids (*e.g.*, Gly and Ala); (2) the set coded by the upper half of the genetic code (YNN, where Y = T or C) contains many putative new amino acids (*e.g.*, Cys, Phe, Tyr and Trp); and (3) the set coded using all bases (NNN) contains all 20 kinds of amino acids, used as a control. Subsequently, functional SH3 sequences that can bind to the SH3 ligand peptide were selected from each library using mRNA display as described in Section 3. After three rounds of *in vitro* selection, the contents of active SH3 domains in each round were analyzed using an enzyme-linked immunosorbent assay (ELISA) (Figure 7B). Functional SH3 sequences were enriched from the natural NNN library and the RNN library rich in "primitive" amino acids but not from the YNN library rich in "new" amino acids (Figure 7B).

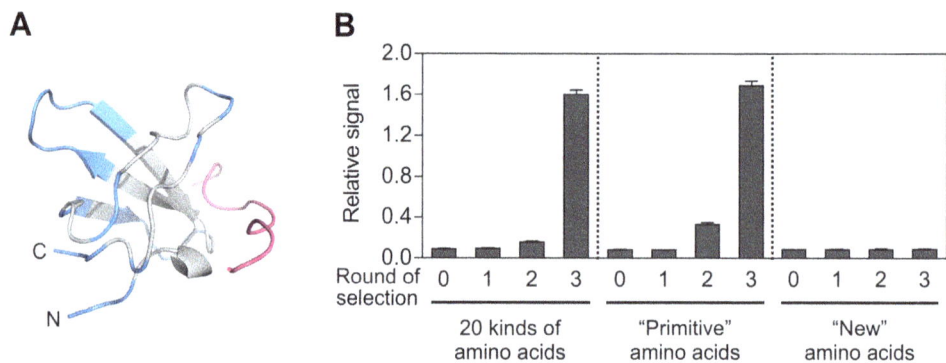

Fig. 7. *In vitro* selection of functional SH3 proteins using mRNA display. (A) The three-dimensional structure of the *src* SH3 domain (blue and gray) complexed with its peptide ligand VSL12 (red). The SH3 domain was partially randomized (blue). The structure was visualized using PyMol (PDBid, 1QWF; Feng et al., 1995). (B) The fraction of functional SH3 sequences at each round of mRNA-display selection (see Figure 5). The total amount of the three libraries [*i.e.*, those based on 20 kinds of amino acids (NNN), putative "primitive" amino acids (RNN) and putative "new" amino acids (YNN)] that bound to the peptide ligand before (0) and after 1–3 rounds of selection were quantified using ELISA. Error bars indicate the s.d. of four samples (Tanaka et al., 2011).

This result experimentally supports, for the first time, *in silico* simulations showing that modern proteins might be simplified more easily using a set of putative primitive amino acids than by a set of putative new amino acids (Babajide et al., 1997). We propose that this result cannot be explained based on differences in the typical biophysical properties (*e.g.*, charge and hydrophobicity) of individual amino acids coded by RNN versus those coded by YNN for the following reasons. First, we reconstructed a randomized SH3 domain in which highly conserved positions [*e.g.*, the ligand-binding region, the hydrophobic core and the polar surface region (Larson & Davidson, 2000)] were fixed. Second, the amino acid compositions of the randomized regions were designed to roughly equalize the biophysical properties (*i.e.*, the proportion of hydrophobic residues present and β-sheet propensity) among three kinds of random codons and resemble those of modern proteins. Thus, the reason behind the utility of primitive amino acids remains unknown but may reflect the evolutionary constraint that primordial proteins consisted of a small set of primitive amino acids and gradually acquired new amino acids during the course of neutral evolution.

Functional SH3 sequences were enriched during the second round in the library containing primitive amino acids but not during the second round in the library based on the 20-amino acid alphabet (Figure 7B). Because the biophysical properties (in particular, β-sheet propensity) were considered to be almost equal, it is not reasonable to suggest that particular amino acids that were not included in the library containing primitive amino acids, such as Pro, prevented the formation of secondary structure and peptide binding. Further study showed that the proteins selected from both libraries have similar biophysical properties, including ligand specificity, ligand affinity and thermostability (Tanaka et al., 2011). Therefore, the library rich in putative primitive amino acids included a slightly larger number of functional SH3 sequences than the randomized library based on the full set of amino acids.

Interestingly, proteins selected from the library based on the primitive amino acids were more likely to be expressed in the soluble fraction in E. coli than those selected from the library based on the 20-amino acid alphabet, in agreement with the results obtained using random-sequence proteins mentioned in Section 4. Thus, increasing the content of primitive amino acids in proteins may improve not only the frequency at which folded and functional proteins occur but also their solubility.

Recently, it has been reported that such limited sets of amino acids are effective for functional selection from randomized libraries (Reetz et al., 2008; Wu et al., 2010; Zheng & Reetz, 2010), although only a few amino acids were randomized in the active sites in these studies. Reetz et al. compared the quality of randomized libraries in which five amino acid residues around the active site of epoxide hydrolase were replaced by 20 kinds of amino acids encoded by NNK (K = T or G) or 12 kinds of amino acids (Gly, Asp, Val, Ser, Leu, Arg, Asn, Ile, His, Cys, Phe and Tyr) encoded by NDT (D = T, A or G). The NDT library produced many more variants with high activity than the NNK library (Reetz et al., 2008). Moreover, the authors succeeded in modifying other enzymes using a library based on the randomized codon NDT, for example, by inducing allosteric effects into Baeyer–Villiger monooxygenase (Wu et al., 2010) and manipulating the stereoselectivity of limonene epoxide hydrolase (Zheng & Reetz, 2010). Fellouse et al. (2004, 2005) demonstrated that the performance of a randomized antibody library was maintained when the number of amino acid types constituting part of a randomized complementarity-determining region (CDR) in the library was reduced to just four (Ala, Ser, Asn and Tyr) or even two (Ser and Tyr).

Although protein engineering using a limited set of primitive amino acids might improve protein folding ability and the frequency of occurrence of functional proteins, the need remains to determine the most appropriate subset of amino acids for functional selection because our study (Tanaka et al., 2011) and those of Reetz's group (Reetz et al., 2008; Wu et al., 2010; Zheng & Reetz, 2010) and Fellouse et al. (2004, 2005) simultaneously compared only a few subsets of amino acids. Some putative new amino acids may be essential for some structures and functions. For example, Cys, which might be a late recruit into the amino acid repertoire, improves structural stability by forming intra- and intermolecular disulfide bonds. A new amino acid, His, is also significant because of its role at the active center of enzymes where it binds to metal ions through the imidazole group. In the course of protein evolution, the recruitment of putative new amino acids may have generated new catalytic activities and more complicated and stable structures. This would be a reason for proteins to have employed an expanded set of amino acids rather than limiting themselves to primitive amino acids. Thus, not only putative primitive amino acids but also certain new amino acids, such as Cys and His as described above, may have been needed in the design of artificial proteins, depending on the target function.

6. Conclusion

Soluble, functional proteins tend to occur more frequently in libraries based on limited sets of primitive amino acids than in libraries based on limited sets of new amino acids and library based on the full set of 20 amino acids. Thus, the evolutionary engineering of proteins using limited sets of primitive amino acids may be an effective tool for the creation of artificial proteins, such as industrial enzymes and monoclonal antibodies that are used in the pharmaceutical industry.

7. Acknowledgments

We thank members of our laboratory at Keio University for helpful comments and discussions. We also thank Drs Hideaki Takashima, Kenichi Horisawa, Seiji Tateyama and Etsuko Miyamoto-Sato in particular for their help with mRNA display and Dr Toru Tsuji for help with characterization of the SH3 domain variants. This work was supported in part by a Grant-in-Aid for Scientific Research (19657073) from the JSPS (Japan Society for the Promotion of Science) and a Grant-in-Aid from the Keio University Global Center of Excellence (G-COE) Program entitled 'Center of Human Metabolomic Systems Biology' from MEXT (the Ministry of Education, Culture, Sports, Science and Technology) of Japan.

8. References

Akanuma, S.; Kigawa, T. & Yokoyama, S. (2002). Combinatorial mutagenesis to restrict amino acid usage in an enzyme to a reduced set. *Proceedings of the National Academy of Sciences of the United States of America*, Vol. 99, No. 21, pp. 13549-13553.

Altman, S. (1981). Transfer RNA processing enzymes. *Cell*, Vol. 23, No. 1, pp. 3-4.

Babajide, A.; Hofacker, I.L.; Sippl, M.J. & Stadler, P.F. (1997). Neutral networks in protein space: a computational study based on knowledge-based potentials of mean force. *Folding & Design*, Vol. 2, No. 5, pp. 261-269.

Bloom, J.D.; Meyer, M.M.; Meinhold, P.; Otey, C.R.; MacMillan, D. & Arnold F.H. (2005). Evolving strategies for enzyme engineering. *Current Opinion in Structural Biology*, Vol. 15, No. 4, pp. 447-452.

Cech, T.R.; Zaug, A.J. & Grabowski, P.J. (1981). *In vitro* splicing of the ribosomal RNA precursor of Tetrahymena: involvement of a guanosine nucleotide in the excision of the intervening sequence. *Cell*, Vol. 27, No. 3, pp. 487-496.

Chaput, J.C. & Szostak, J.W. (2004). Evolutionary optimization of a nonbiological ATP binding protein for improved folding stability. *Chemistry & Biology*, Vol. 11, No. 6, pp. 865-874.

Davidson, A.R. & Sauer, R.T. (1994). Folded proteins occur frequently in libraries of random amino acid sequences. *Proceedings of the National Academy of Sciences of the United States of America*, Vol. 91, No. 6, pp. 2146-2150.

Davidson, A.R.; Lumb, K.J. & Sauer, R.T. (1995). Cooperatively folded proteins in random sequence libraries. *Nature Structural Biology*, Vol. 2, No. 10, pp. 856-864.

Doi, N. & Yanagawa, H. (1999). STABLE: protein-DNA fusion system for screening of combinatorial protein libraries in vitro. *FEBS Letters*, Vol. 457, No. 2, pp. 227-230.

Doi, N. & Yanagawa, H. (2001). Genotype-phenotype linkage for directed evolution and screening of combinatorial protein libraries. *Combinatorial Chemistry & High Throughput Screening*, Vol. 4, No. 6, pp. 497-509.

Doi, N.; Kakukawa, K.; Oishi, Y. & Yanagawa, H. (2005). High solubility of random-sequence proteins consisting of five kinds of primitive amino acids. *Protein Engineering, Design & Selection*, Vol. 18, No. 6, pp. 279-284.

Dryden, D.T.F.; Thomson, A.R. & White, J.H. (2008). How much of protein sequence space has been explored by life on Earth? *Journal of The Royal Society Interface*, Vol. 5, No. 25, pp. 953-956.

Eigen, M. & Schuster, P. (1978). The hypercycle. A principle of natural self-organization. Part C: The realistic hypercycle. *Naturwissenschaften*, Vol. 65, No. 7, pp. 341-369.

Fellouse, F.A.; Li, B.; Compaan, D.M.; Peden, A.A.; Hymowitz, S.G. & Sidhu, S.S. (2005). Molecular recognition by a binary code. *Journal of Molecular Biology*, Vol. 348, No. 5, pp. 1153-1162.

Fellouse, F.A.; Wiesmann, C. & Sidhu, S.S. (2004). Synthetic antibodies from a four-amino-acid code: a dominant role for tyrosine in antigen recognition. *Proceedings of the National Academy of Sciences of the United States of America*, Vol. 101, No. 34, pp. 12467-12472.

Feng, S.; Kasahara, C.; Rickles, R.J. & Schreiber, S.L. (1995). Specific interactions outside the proline-rich core of two classes of Src homology 3 ligands. *Proceedings of the National Academy of Sciences of the United States of America*, Vol. 92, No. 26, pp. 12408-12415.

Georgiou, G.; Stathopoulos, C.; Daugherty, P.S.; Nayak, A.R.; Iverson, B.L. & Curtiss, R. 3rd. (1997). Display of heterologous proteins on the surface of microorganisms: from the screening of combinatorial libraries to live recombinant vaccines. *Nature Biotechnology*, Vol. 15, No. 1, pp. 29-34.

Go, A.; Kim, S.; Baum, J. & Hecht, M.H. (2008). Structure and dynamics of de novo proteins from a designed superfamily of 4-helix bundles. *Protein Science*, Vol. 17, No. 5, pp. 821-832.

Hanes, J. & Plückthun, A. (1997). *In vitro* selection and evolution of functional proteins by using ribosome display. *Proceedings of the National Academy of Sciences of the United States of America*, Vol. 94, No. 10, pp. 4937-4942.

Hoogenboom, H.R. (2005). Selecting and screening recombinant antibody libraries. *Nature Biotechnology*, Vol. 23, No. 9, pp. 1105-1116.

Huang, J.; Koide, A.; Makabe, K. & Koide, S. (2008). Design of protein function leaps by directed domain interface evolution. *Proceedings of the National Academy of Sciences of the United States of America*, Vol. 105, No. 18, pp. 6578-6583.

Jordan, I.K.; Kondrashov, F.A.; Adzhubei, I.A.; Wolf, Y.I.; Koonin, E.V.; Kondrashov, A.S. & Sunyaev, S. (2005). A universal trend of amino acid gain and loss in protein evolution. *Nature*, Vol. 433, No. 7026, pp. 633-638.

Jumawid, M.T.; Takahashi, T.; Yamazaki, T.; Ashigai, H. & Mihara, H. (2009). Selection and structural analysis of de novo proteins from an α3β3 genetic library. *Protein Science*, Vol. 18, No. 2, pp. 384-398.

Kamtekar, S.; Schiffer, J.M.; Xiong, H.; Babik, J.M. & Hecht, M.H. (1993). Protein design by binary patterning of polar and nonpolar amino acids. *Science*, Vol. 262, No. 5140, pp. 1680-1685.

Keefe, A.D. & Szostak, J.W. (2001). Functional proteins from a random-sequence library. *Nature*, Vol. 410, No. 6829, pp. 715-718.

Kyte, J. & Doolittle, R.F. (1982). A simple method for displaying the hydropathic character of a protein. *Journal of Molecular Biology*, Vol. 157, No. 1, pp.105-132.

Larson, S.M. & Davidson, A.R. (2000). The identification of conserved interactions within the SH3 domain by alignment of sequences and structures. *Protein Science*, Vol. 9, No. 11, pp. 2170-2180.

Leemhuis, H.; Stein, V.; Griffiths, A.D. & Hollfelder, F. (2005). New genotype-phenotype linkages for directed evolution of functional proteins. *Current Opinion in Structural Biology*, Vol. 15, No. 4, pp. 472-478.

Mandecki, W. (1990). A method for construction of long randomized open reading frames and polypeptides. *Protein Engineering*, Vol. 3, No. 3, pp. 221-226.

Mandecki, W. (1998). The game of chess and searches in protein sequence space. *Trends in Biotechnology*, Vol. 16, No. 5, pp. 200-202.

Matsumura, N.; Doi, N. & Yanagawa, H. (2006). Recent progress and future prospects in protein display technologies as tools for proteomics. *Current Proteomics*, Vol. 3, No. 3, pp. 199-215.

Miller, S.L. (1987). Which organic compounds could have occurred on the prebiotic earth? *Cold Spring Harbor Symposia on Quantitative Biology*, Vol. 52, pp. 17-27.

Miyamoto-Sato, E.; Nemoto, N.; Kobayashi, K. & Yanagawa, H. (2000). Specific bonding of puromycin to full-length protein at the C-terminus. *Nucleic Acids Research*, Vol. 28, No. 5, pp. 1176-1182.

Murai, T.; Ueda, M.; Atomi, H.; Shibasaki, Y.; Kamasawa, N.; Osumi, M.; Kawaguchi, T.; Arai, M. & Tanaka, A. Genetic immobilization of cellulase on the cell surface of Saccharomyces cerevisiae. (1997). *Applied Microbiology and Biotechnology*, Vol. 48, No. 4, pp. 499-503.

Nemoto, N.; Miyamoto-Sato, E.; Husimi, Y. & Yanagawa, H. (1997). In vitro virus: Bonding of mRNA bearing puromycin at the 3'-terminal end to the C-terminal end of its encoded protein on the ribosome in vitro. *FEBS Letters*, Vol. 414, No. 2, pp. 405-408.

Ohgushi, M. & Wada, A. (1983). 'Molten-globule state': a compact form of globular proteins with mobile side-chains. *FEBS Letters*, Vol. 164, No. 1, pp. 21-24.

Oparin, A.I. (1961). *Life, its nature, origin and development*, Academic Press, New York.

Patel, S.C.; Bradley, L.H.; Jinadasa, S.P. & Hecht, M.H. (2009). Cofactor binding and enzymatic activity in an unevolved superfamily of *de novo* designed 4-helix bundle proteins. *Protein Science*, Vol. 18, No. 7, pp. 1388-1400.

Peisajovich, S.G.; Rockah, L. & Tawfik, D.S. (2006). Evolution of new protein topologies through multistep gene rearrangements. *Nature Genetics*, Vol. 38, No. 2, pp. 168-174.

Prijambada, I.D.; Yomo, T.; Tanaka, F.; Kawama, T.; Yamamoto, K.; Hasegawa, A.; Shima, Y.; Negoro, S. & Urabe, I. (1996). Solubility of artificial proteins with random sequences. *FEBS Letters*, Vol. 382, No. 1-2, pp. 21-25.

Reetz, M.T.; Kahakeaw, D. & Lohmer, R. (2008). Addressing the numbers problem in directed evolution. *ChemBioChem*, Vol. 9, No. 11, pp. 1797-1804.

Riddle, D.S.; Santiago, J.V.; Bray-Hall, S.T.; Doshi, N.; Grantcharova, V.P.; Yi, Q. & Baker, D. (1997). Functional rapidly folding proteins from simplified amino acid sequences. *Nature Structural Biology*, Vol. 4, No. 10, pp. 805-809.

Roberts, R.W. & Szostak, J.W. (1997). RNA-peptide fusions for the *in vitro* selection of peptides and proteins. *Proceedings of the National Academy of Sciences of the United States of America*, Vol. 94, No. 23, pp. 12297-12302.

Romero, P.A. & Arnold, F.H. (2009). Exploring protein fitness landscapes by directed evolution. *Nature Reviews Molecular Cell Biology*, Vol. 10, No. 12, pp. 866-876.

Silverman, J.A.; Balakrishnan, R. & Harbury, P.B. (2001). Reverse engineering the $(\beta/\alpha)_8$ barrel fold. *Proceedings of the National Academy of Sciences of the United States of America*, Vol. 98, No. 6, pp. 3092-3097.

Smith, G.P. (1985). Filamentous fusion phage: novel expression vectors that display cloned antigens on the virion surface. *Science*, Vol. 228, No. 4705, pp. 1315-1317.

Tanaka, J.; Doi, N.; Takashima, H. & Yanagawa, H. (2010). Comparative characterization of random-sequence proteins consisting of 5, 12, and 20 kinds of amino acids. *Protein Science*, Vol. 19, No. 4, pp. 786-795.

Tanaka, J.; Yanagawa, H. & Doi, N. (2011). Comparison of the frequency of functional SH3 domains with different limited sets of amino acids using mRNA display. *PLoS ONE*, Vol. 6, No. 3, e18034.

Tokuriki, N. & Tawfik, D.S. (2009a). Chaperonin overexpression promotes genetic variation and enzyme evolution. *Nature*, Vol. 459, No. 7247, pp. 668-673.

Tokuriki, N. & Tawfik, D.S. (2009b). Protein dynamism and evolvability. *Science*, Vol. 324, No. 5924, pp. 203-207.

Trifonov, E.N. (2004). The triplet code from first principles. *Journal of Biomolecular Structure & Dynamics*, Vol. 22, No. 1, pp. 1-11.

Vamvaca, K.; Vögeli, B.; Kast, P.; Pervushin, K. & Hilvert, D. (2004). An enzymatic molten globule: efficient coupling of folding and catalysis. *Proceedings of the National Academy of Sciences of the United States of America*, Vol. 101, No. 35, pp. 12860-12864.

Walter, K.U.; Vamvaca, K. & Hilvert, D. (2005). An active enzyme constructed from a 9-amino acid alphabet. *The Journal of Biological Chemistry*, Vol. 280, No. 45, pp. 37742-37746.

Watters, A.L. & Baker, D. (2004). Searching for folded proteins *in vitro* and *in silico*. *European Journal of Biochemistry*, Vol. 271, No. 9, pp. 1615-1622.

Wilkinson, D.L. & Harrison, R.G. (1991). Predicting the solubility of recombinant proteins in *Escherichia coli*. *Biotechnology*, Vol. 9, No. 5, pp. 443-448.

Wolkowicz, R.; Jager, G.C. & Nolan, G.P. (2005). A random peptide library fused to CCR5 for selection of mimetopes expressed on the mammalian cell surface via retroviral vectors. *The Journal of Biological Chemistry*, Vol. 280, No. 15, pp. 15195-15201.

Wong, J.T. (1975). A co-evolution theory of the genetic code. *Proceedings of the National Academy of Sciences of the United States of America*, Vol. 72, No. 5, pp. 1909-1912.

Wong, J.T. (1988). Evolution of the genetic code. *Microbiological Sciences*, Vol. 5, No. 6, pp. 174-181.

Wong, J.T. (2005). Coevolution theory of the genetic code at age thirty, *BioEssays*, Vol. 27, No. 4, pp. 416-425.

Wright, P.E. & Dyson, H.J. (1999). Intrinsically unstructured proteins: re-assessing the protein structure-function paradigm. *Journal of Molecular Biology*, Vol. 293, No. 2, pp. 321-331.

Wu, S.; Acevedo, J.P. & Reetz, M.T. (2010). Induced allostery in the directed evolution of an enantioselective Baeyer-Villiger monooxygenase. *Proceedings of the National Academy of Sciences of the United States of America*, Vol. 107, No. 7, pp. 2775-2780.

Yamauchi, A.; Yomo, T.; Tanaka, F.; Prijambada, I.D.; Ohhashi, S.; Yamamoto, K.; Shima, Y.; Ogasahara, K.; Yutani, K.; Kataoka, M. & Urabe, I. (1998). Characterization of soluble artificial proteins with random sequences. *FEBS Letters*, Vol. 421, No. 2, pp. 147-151.

Zheng, H. & Reetz, M.T. (2010). Manipulating the stereoselectivity of limonene epoxide hydrolase by directed evolution based on iterative saturation mutagenesis. *Journal of the American Chemical Society*, Vol. 132, No. 44, pp. 15744-15751.

5

Engineering High-Affinity T Cell Receptor/Cytokine Fusions for Therapeutic Targeting

Jennifer D. Stone[1], Yiyuan Yin[2], Min Mo[2], K. Scott Weber[3], David L. Donermeyer[3], Paul M. Allen[3], Roy A. Mariuzza[2,4] and David M. Kranz [1]
[1]*Department of Biochemistry, University of Illinois at Urbana-Champaign, Urbana,*
[2]*Institute for Bioscience and Biotechnology Research, University of Maryland, Rockville,*
[3]*Department of Pathology and Immunology,*
Washington University School of Medicine, St. Louis,
[4]*Department of Cell Biology and Molecular Genetics,*
University of Maryland, College Park,
USA

1. Introduction

1.1 Background: Inflammatory diseases

Inflammatory diseases or conditions, including autoimmune diseases, allergies, and transplant/graft rejection, constitute a significant health issue. Among these diseases, current treatments include various global immunosuppression strategies such as steroids for autoimmune diseases and asthma, antithymocyte globulin (ATG) for lymphocyte depletion or cyclosporine A for transplant tolerance, and other treatments (reviewed in (Fort & Narayanan 2010)); however, these can leave the patient with increased susceptibility to infection, tumor development, or other immune challenge (reviewed in (Belkaid 2007; Sabat et al. 2010)). Therefore, more targeted, antigen-specific immune tolerance is desirable.

While excessive inflammation is a complex process including many immune effector cells, CD4+ helper T cells are thought to be crucial to disease progression for many conditions (Davidson et al. 1996; Gebe et al. 2008; Reijonen et al. 2002; Sospedra & Martin 2005; Tesmer et al. 2008). Pathogenic CD4+ T cells in these diseases are thought to express T cell receptors (TCRs) that recognize specific peptides bound to class II major histocompatibility complex (MHC) proteins on the surface of antigen-presenting cells (APCs), where the presented peptide is derived not from a pathogen, but from an allergen, transplant antigen, or self protein. Specific peptides derived from allergens or "self" proteins have been identified as linked to the pathological responses in many cases (Haselden et al. 1999; Hemmer et al. 1998; Jahn-Schmid et al. 2002; Muraro et al. 1997; Reijonen et al. 2002; Vergelli et al. 1997), and links between particular class II MHC alleles and disease susceptibility have been identified (Ebers et al. 1996; Oksenberg et al. 1996; Wucherpfennig & Sethi 2011). Hence, one attractive option would be to selectively inhibit only those CD4+ T cells that initiate or sustain pathological inflammation. Some efforts at more targeted induction of tolerance have been explored, including natalizumab, an antibody against the α4 integrin, which selectively inhibits leukocyte attachment to and extravasation through inflamed vascular endothelium (Rice et al. 2005), but has no epitope or tissue specificity. In another example, a pooled, randomized peptide vaccine called glatiramer acetate (GA) has been approved to induce tolerance in multiple sclerosis (Johnson et al. 1995; Teitelbaum et al. 1988; Teitelbaum et al. 1996), but acts across a variety of epitopes, and may affect the patient's ability to respond normally to novel infectious challenges. Truly selective immunosuppression, therefore, remains an outstanding challenge.

1.2 Immunosuppressive cytokine IL-10

Interleukin-10 (IL-10), first identified as a cytokine synthesis inhibitory factor secreted by Th2 cells (Fiorentino et al. 1989), is a member of the Class II cytokine family largely associated with immune suppression. While IL-10 can be produced broadly by many different T cell subsets, along with B cells, NK cells, dendritic cells, macrophages, and other immune cells (reviewed in (Asadullah et al. 2003; O'Garra et al. 2008; Ouyang et al. 2011; Saraiva & O'Garra 2010)), it is strongly associated with regulatory T cells (Tregs). Tregs both produce IL-10 (and sometimes TGF-β) in the absence of other effector cytokines, and require IL-10 for their function. The ability of IL-10 to support Treg differentiation as well as suppress inflammatory responses in antigen-specific T cells makes it an attractive potential therapeutic.

1.2.1 Biological / immunological effects

IL-10 functions *in vivo* to protect against excessive inflammation and tissue damage. It suppresses the expression of proinflammatory cytokines by adaptive and innate immune cells (reviewed in (Asadullah et al. 2003; Moore et al. 2001)). In the absence of IL-10, exaggerated inflammatory responses to infection occur, often resulting in tissue damage. For example, IL-10-deficient mice develop chronic inflammatory bowel disease after exposure to enteric bacteria (Kuhn et al. 1993; Sellon et al. 1998). Many general immunosuppressants currently used to treat inflammatory diseases are thought to be effective in part by increasing the ratio of IL-10 to other cytokines (O'Garra et al. 2008). Despite its predominant function to suppress inflammatory responses, IL-10 can also enhance certain functions of the immune system, including humoral immunity, NK cell activity, or CD8+ T cell responses (reviewed in (Asadullah et al. 2003; O'Garra et al. 2008)). These various effects depend on the specific conditions and concentrations of IL-10.

Defects in IL-10 levels have been linked to steroid-insensitive asthma (Hawrylowicz et al. 2002), and can increase the pathology of autoimmune diabetes. Clinical trials of systemic administration of IL-10 for treatment of the autoimmune conditions psoriasis, Crohn's disease, and rheumatoid arthritis have been designed to take advantage of the broad immunosuppressive action of this cytokine (Asadullah et al. 2003). While no particular benefit was seen in the treatment of rheumatoid arthritis, some promising results were seen in psoriasis and Crohn's disease; however, the effects were modest, and they depended heavily on finding the appropriate dosing (Asadullah et al. 2003; O'Garra et al. 2008). Studies of experimental autoimmune encephalomyelitis (EAE), an animal model of multiple sclerosis, suggested that IL-10 is crucial to protect against inflammation and damage in the central nervous system, but localization of the cytokine within the target organ is required for effectiveness (Bettelli et al. 1998; Kennedy et al. 1992; Samoilova et al. 1998). As a soluble therapeutic, IL-10 seems most effective when combined with targeted delivery methods such as T cells transduced to express IL-10, or genetically engineered *Lactococcus lactis* that produce IL-10, both having shown promise in treatment of murine inflammatory bowel disease (Lindsay & Hodgson 2001; Steidler et al. 2000; Wirtz et al. 1999).

1.2.2 Structure

IL-10, like other Class II cytokine family members, is composed of alpha-helical bundles (six helices designated A-F), similar to common cytokine-receptor γ-chain (γ_c) family cytokines IL-2 and IL-15. However, unlike IL-2, but like interferon gamma, IL-10 exists as an intercalated homodimer, where the E and F helices from one polypeptide chain fold into a bundle with the A-D helices from a second polypeptide chain (Zdanov et al. 1995). As a dimer, its affinity for the high-affinity IL-10 receptor is in the picomolar range (Tan et al. 1995). However, introduction of an extended flexible linker between N116 and K117 allowed folding into a soluble, bioactive, monomeric form with a K_D of 30 nM for IL-10R1 (Josephson et al. 2000).

Some viruses have evolved an ability to take advantage of the suppressive functions of IL-10 to down modulate the immune response by producing viral IL-10 homologues (Fleming et al. 1997; Hsu et al. 1990; Kotenko et al. 2000; Moore et al. 1990; Spencer et al. 2002). Study of these homologues have led to identification of a point mutation, I87A, which, when included in murine or human IL-10, leads to a cytokine with the suppressive, but not stimulatory functions of IL-10 (Ding et al. 2000).

1.3 Specific tissue targeting of immunoactive proteins

Systemic immunomodulation carries potential risks associated with the disruption of the balance of inflammation and tolerance in the immune system (Belkaid 2007; Fort & Narayanan 2010). In contrast, targeted delivery of a cytokine like IL-10 to particular sites of interest could greatly increases its desired immunosuppressive effect (Kennedy et al. 1992; Kuhn et al. 1993; Lindsay & Hodgson 2001). In order to deliver immunomodulatory proteins selectively to the tissues where their actions are desired, strategies involving covalent linkages of soluble targeting elements to immunoactive proteins have been explored with some success. For example, these targeted approaches have been studied for cancer treatments with the hope of directing immunostimulatory cytokines to the tumor microenvironment (Gillies 2009; Kaspar et al. 2007; Kaspar et al. 2006; Schanzer et al. 2006; Sommavilla et al. 2010). By analogy, immunosuppressive treatments for inflammatory disease would be desired locally, rather than systemically. In addition, the tissue-specific, graft-specific, or allergen-specific nature of most inflammatory diseases make them strong candidates for targeted delivery of immunosuppression. Discussion of several specific targeting strategies for the delivery of soluble effector molecules follows.

1.3.1 Antibodies and scFv

The use of monoclonal antibodies that bind specific target epitopes has been a common strategy for many purposes, including diagnostics and treatments. Some antibodies have been used directly for targeting, taking advantage of Fc-mediated effector functions, such as with rituximab (anti-CD20 for B cell carcinoma) or with infliximab (anti-TNF-alpha for rheumatoid arthritis or Crohn's disease). Although beyond the scope of this article, antibodies have also been linked to bioactive payloads such as toxins, radioisotopes, or chemotherapeutic agents for the treatment of cancer (Ma et al. 2004; Pohlman et al. 2006; Schanzer et al. 2006).

Antibody fragments termed single-chain variable fragments (scFv) that consist of the variable domains of the heavy and light chains of an antibody, connected by a flexible linker into a single polypeptide chain, have been employed to target immunoactive molecules locally to particular tissues. Fusions of scFv specific for tumor-associated epitopes with cytokines including IL-15, IL-12, GM-CSF, TNF-α, and IFN-α have been studied for localized immune stimulation in the case of cancer with some success (Gillies 2009; Kaspar et al. 2007; Kaspar et al. 2006; Schanzer et al. 2006; Sommavilla et al. 2010). In one study, targeting of IL-10 with the scFv L19, specific for an angiogenesis marker, led to significant improvement in arthritis symptoms over irrelevantly-targeted IL-10 (Trachsel et al. 2007). However, ideal cell surface tissue antigens that can be targeted by monoclonal antibodies may not be available for every application, especially if they are unrelated to the progression of the disease being addressed.

1.3.2 Benefits of targeting MHC-restricted epitopes using TCRs

Since many inflammatory diseases are thought to be mediated by T cells that recognize specific, sometimes known, peptide-MHC complexes (Haselden et al. 1999; Hemmer et al. 1998; Jahn-Schmid et al. 2002; Muraro et al. 1997; Reijonen et al. 2002; Vergelli et al. 1997), it is reasonable to target immunosuppressive treatments directly to those complexes. As indicated above, the natural receptor for peptide-MHC is the $\alpha\beta$ heterodimeric TCR (reviewed in (Davis et al. 1998)). TCRs have diversity that is similar to antibodies, and they have evolved to recognize peptide-MHC ligands with a high degree of peptide specificity, and thus the ability to distinguish among small variations in these complexes. However, TCRs, unlike antibodies, are not expressed in soluble form, but only as transmembrane proteins, and only as part of a larger TCR complex with additional transmembrane subunits called CD3. The wild-type affinities of TCR:peptide-MHC interactions are typically low — on the order of 1-500 µM (reviewed in (Stone et al. 2009)). While this affinity is sufficient to trigger cellular responses in the T cell, it is likely too low to efficiently target a soluble therapeutic. Nevertheless, there is some precedence for use of a wild-type affinity TCR, expressed as a three-domain soluble fusion with IL-2 or IL-15 (Belmont et al. 2006; Card et al. 2004; Wong et al. 2011). Naturally occurring autoimmune TCRs are thought to be even weaker binders than TCRs against foreign peptide-MHC ligands; in fact, the affinity of one autoimmune-specific TCR was impossible to measure for a soluble monomer by surface plasmon resonance (Li et al. 2005a). However, advances in protein engineering have allowed the generation of soluble TCRs with affinities that are appropriate for targeting.

1.3.3 Affinity considerations / engineering

The goal of using a TCR as a targeting moiety for a soluble immunotherapeutic has become plausible due to advances in engineering and directed evolution (reviewed in (Richman & Kranz 2007)) using techniques such as yeast surface display, phage display, or more recently, T cell display (Chervin et al. 2008). Engineered receptors with 1000-fold or more increased affinity compared to their original, wild-type TCRs have been isolated, and the soluble forms of these TCRs bind to the targeted peptide-MHC with a high level of specificity (Chervin et al. 2008; Holler et al. 2003; Holler et al. 2000; Li et al. 2005b; Weber et al. 2005). The higher affinities achieved for TCR binding to peptide-MHC (nanomolar and picomolar) are in the range of affinity-matured antibodies, and are thus sufficient to evaluate for their ability to serve as soluble therapeutics.

Obtaining sufficient quantities of purified, soluble, recombinant TCR posed a significant challenge. One solution was to produce in E. coli soluble TCRs that contain the full extracellular alpha and beta chains with an additional disulfide bond between the $C\alpha$ and $C\beta$ domains (Boulter et al. 2003). By analogy with scFv (V_{heavy}-linker-V_{light}) constructs from antibodies, single-chain TCR variable domain, or scTv, constructs (V_{alpha}-linker-V_{beta}) have been generated for several TCRs, and this form would be well suited for covalent fusion to immune effector molecules. Unlike scFvs, scTvs generally require additional engineering to be soluble and stable, a process that has been carried out through yeast display (Kieke et al. 1999; Richman et al. 2009). Also, increased experience with TCR engineering has led to identifying a particular human variable domain called $V\alpha2$ that is best suited for engineering, facilitating future efforts at scTv construction (Aggen et al. 2011).

1.4 Model system 3.L2 and high-affinity TCR M15 engineered by yeast surface display

Inflammatory disease is critically linked to specific TCR interactions with class II MHC that present particular peptides on the surface of APCs. Our goal to target the immunosuppressive effects of IL-10 locally to particular class II MHC epitopes may be modeled by targeting the murine class II MHC allele I-E^k presenting a peptide derived from the beta chain of the minor d allele of hemoglobin (Hbβ^d [64-76]). This Hb/I-E^k complex is recognized by a wild-type TCR called 3.L2 with an affinity of 20 μM for the peptide-MHC (Weber et al. 2005). Engineering efforts to improve this interaction are detailed below.

1.4.1 Engineered for stability as a soluble, single-chain TCR variable fragment (scTv)

The variable genes for the 3.L2 TCR were used to generate a scTv construct in a $V\beta8.3$-linker-$V\alpha18$ orientation (Weber et al 2005). This construct was cloned as a fusion to the yeast mating protein Aga-2 for engineering by yeast surface display. The scTv with wild-type variable sequences was not expressed as a folded protein on the surface of yeast, as has been observed for most scTvs. To identify key mutations that might enable yeast surface display and expression in E. coli, random mutations were introduced into the scTv gene by error-prone PCR, and a library of scTv mutants was transformed into yeast. From yeast expressing the library of mutants, a 3.L2 clonotypic antibody was used to select yeast with the most stable 3.L2 scTv mutants. The most stable and highest expressed mutant, called M2, contained six point mutations (Weber et al. 2005).

1.4.2 Engineered for high affinity to Hb/I-E^k complex

Using the M2 gene as a scaffold, directed hypervariable libraries were generated with mutations in the complementarity determining regions 3 (CDR3) of the scTv (Weber et al. 2005). Successive rounds of selection of the yeast libraries for binding to the specific ligand,

Hb/I-E^k, led to isolation of a high-affinity mutant named M15. The M15 mutant and an M1 mutant containing completely wild-type CDR sequences (identical to 3.L2) were expressed in *E. coli* and refolded as soluble scTv proteins. Binding to Hb/I-E^k was measured by surface plasmon resonance for both scTvs, and an 800-fold increase in affinity was found for the high-affinity M15 (K_D=25 nM) as compared to 3.L2 (K_D=20 µM for M1) (Weber et al. 2005). The affinity of M15 for Hb/I-E^k is thus in the range seen for monoclonal antibodies.

1.5 M15-targeted IL-10 fusion molecules

To evaluate the possibility of targeting the immunosuppressive cytokine IL-10 to cells expressing a particular class II MHC-peptide complex, we engineered a series of fusion molecules linking the M15 scTv to various IL-10 proteins; these IL-10 proteins included human, murine, dimeric, monomeric, and I87A mutants. The proteins were produced and purified from *E. coli* or HighFive insect cells, and were found to be stable in solution. The various constructs were evaluated for proper folding of the IL-10 cytokine as detected by ELISA, and the M15 targeting element was shown to bind specifically to cells presenting the proper peptide-MHC. Importantly, the constructs functioned in various IL-10 cytokine-dependent assays, indicating that the approach has the potential to deliver targeted immunosuppressive molecules, and providing a valuable model for further evaluation of this strategy.

2. Methods

2.1 Cloning, production, and purification of scTv M15 and M15:IL-10 fusions

For *in vitro* folding, full-length human IL-10 (hIL-10) was attached to the C-terminus of scTv M15 via either a 18-mer linker (VNAKTTAPSVYPLAPVSG) (Card et al. 2004) or a 15-mer linker (SSSSG)$_3$ (Trachsel et al. 2007). In addition to these wild-type M15:hIL-10 constructs, constructs containing modified hIL-10 sequences were also made. One modification involved inserting six amino acids (GGGSGG) between hIL-10 residues Asn115 and Lys116 (Josephson et al. 2000), in order to prevent formation of IL-10 homodimers that could complicate production of M15:hIL-10 conjugates. The other modification involved a point mutation (I87A) to reduce the affinity of hIL-10 for hIL-10 receptor (hIL-10R) by ~100-fold, while preserving its biological activity (Ding et al. 2000). The rationale for reducing the affinity of hIL-10 for hIL-10R, whose K_D is in the picomolar range (Tan et al. 1995), was to prevent the hIL-10–hIL-10R interaction from potentially dominating the binding of scTv M15 to its peptide–MHC ligand (Hb–I-E^k) on target cells (nanomolar K_D).

Each M15:hIL-10 construct was cloned into the bacterial expression vector pET-26b(+) and produced as inclusion bodies in BL21(DE3) *E. coli* cells. The inclusion bodies were washed with 50 mM Tris-HCl (pH 8.0) containing 5% (v/v) Triton X-100, then dissolved in 6 M guanidine, 50 mM Tris-HCl (pH 8.0), and 10 mM DTT. For *in vitro* folding, the inclusion bodies were diluted into ice-cold folding buffer containing 0.4 M L-arginine-HCl, 50 mM Tris-HCl (pH 8.0), 1 mM EDTA, 3 mM reduced glutathione, and 0.9 mM oxidized glutathione to a final protein concentration of ~40 mg/l. After 72 h at 4 °C, the mixtures were concentrated and dialyzed against PBS, prior to purification using sequential Superdex S-200 and Mono Q columns.

Wild-type human and mouse IL-10 (mIL-10) conjugated to scTv M15 were also produced using a baculovirus expression system. The homologous IL-10 genes were fused to scTv M15 in the same way as for *E. coli* expression, using both 18-mer and 15-mer linkers. Accordingly, we cloned the following constructs into the pAcGP67B baculovirus transfer vector: 1) M15:mIL-10 containing the 18-mer linker; 2) M15:SG15:hIL-10 containing the 15-mer linker; and 3) M15:SG15:mIL-10. In addition, the constructs included a C-terminal FLAG tag (DYKDDDDK) for affinity purification.

The pAcGP67B vector uses a baculovirus signal sequence (gp67) to direct secretion of expressed proteins into the culture supernatant. Each construct was co-transfected into Sf9 insect cells with linearized baculovirus DNA to generate a recombinant baculovirus. Secretion of all three constructs was confirmed by Western blotting using the anti-FLAG monoclonal antibody M2. For protein production, 1–3 liters of HighFive insect cells at $\sim 10^6$ cells/ml in serum-free medium were incubated with recombinant baculovirus at a multiplicity of infection of ~3 pfu/cell. Culture supernatants were harvested 3–4 days after infection and concentrated ~20-fold. Following dialysis against TBS (pH 7.3), supernatants were loaded on an anti-FLAG M2 agarose column and bound proteins were eluted using 100 µg/ml FLAG peptide. Recombinant M15:mIL-10, M15:SG15:mIL-10, and M15:SG15:hIL-10 were further purified with sequential Superdex S-75 and Mono Q columns.

2.2 IL-10 ELISA

Properly folded IL-10 was detected in preparations of free IL-10 or M15:IL-10 fusions using the Ready-Set-Go sandwich ELISA kit from eBioscience (San Diego, CA). Briefly, a monoclonal antibody against IL-10, murine or human, was diluted in a coating buffer of phosphate-buffered saline (PBS) and coated onto the wells of high-binding, clear, flat-bottomed 96-well plates by incubating overnight at 4°C. The solution was removed, and wells were then blocked by filling with a solution of 1% bovine serum albumin in phosphate-buffered saline (PBS-BSA), pH 7.4, and incubating for at least 2 hours at room temperature or overnight at 4°C. After blocking, known concentrations of standard IL-10 or various dilutions of the M15 fusion protein of interest in the provided assay diluent buffer were added to the wells and incubated for 2 hours at room temperature. Wells were washed three times with PBS containing 0.1% Tween-20 detergent (PBST), and then a biotinylated polyclonal detection antibody solution was added to each well and incubated for 2 hours at room temperature. Wells were then washed three times with PBST, and a solution of Avidin conjugated to horseradish peroxidase was added to each well and incubated for 45 minutes at room temperature. After the avidin solution was removed, wells were again washed three times with PBST, and 50 microliters per well of tetramethylbenzidine (TMB) solution was added, which, when acted upon by peroxidase, results in a colored precipitate. After development, the reaction was stopped by adding 50 microliters per well of a 1N sulfuric acid solution. Absorbance in each well at 450nm was measured using an ELx800 Universal Microplate Reader (BioTek Instruments, Inc, Winooski, VT).

2.3 Specific binding to peptide-loaded cells

The I-E^{k+} CH27 immortalized B cell line (Pennell et al. 1985) was loaded incubated with various concentrations of a hemoglobin-derived peptide, Hb^d[64-76] in RPMI 1640 media supplemented with overnight at 37°C penicillin, streptomycin, hepes, glutamine, beta mercaptoethanol, and 10% fetal calf serum (complete RPMI media). Cells were then washed in ice-cold PBS-BSA, and incubated with 20 ng/mL of scTv M15:IL-10 fusion constructs or 10 ng/mL biotinylated scTv M15 on ice for 2 hours. Cells were then washed in excess PBS-BSA, and cells stained with M15:IL-10 fusions were incubated for 2 hours on ice with 1:100 diluted biotinylated anti-IL-10 polyclonal antibody from the Ready-Set-Go ELISA kit (eBioscience, San Diego, CA). Cells were again washed with PBS-BSA, and then all the cells incubated for one hour on ice with streptavidin fluorescently labeled with phycoerythrin (1:100, BD Pharmingen, San Diego, CA). Cells were washed twice with PBS-BSA, and then resuspended in PBS-BSA for analysis by analytical flow cytometry using a BD FACSCanto (BD Biosciences, San Jose, CA).

2.4 MC/9 proliferation functional assay

MC/9 mouse liver mast cells (American Type Culture Collection, ATCC, Manassas, VA) were maintained in complete RPMI media supplemented with 5% Rat ConA supernatant (TStim, BD Biosciences, San Jose, CA). In the absence of ConA supernatant, MC/9 cells selectively proliferate in response to functional human or murine IL-10 cytokine. A standard number of MC/9 cells per well were cultured with 5 pg/mL murine IL-4 (insufficient to drive proliferation on its own, but can potentiate responses to IL-10) and various concentrations of IL-10 standard concentrations or scTv M15:IL-10 fusion proteins for 48 hours at 37°C. After this time, 10 microliters per well of 3-(4, 5-dimethylthiazolyl-2)-2, 5-diphenyltetrazolium bromide (MTT kit, ATCC, Manassas, VA) solution was added, and the cells were cultured an additional 2 hours at 37°C. Metabolically active cells reduced this to a purple formazan precipitate, which was then dissolved by adding detergent to each well and incubating at least two hours in the dark at room temperature. Absorbance was measured at 570nm using an ELx800 Universal Microplate Reader (BioTek Instruments, Inc, Winooski, VT).

2.5 Inhibition of T cell activation by soluble scTv

CH27 antigen presenting cells ($3x10^4$) were cultured at 37°C for 30 minutes with soluble scTv M15 at the indicated concentrations (0, 10, or 50 µg/ml) and a range of Hb peptide concentrations. 3.L2 T cell hybridoma cells ($1x10^5$) were then added to the antigen presenting cells for 24 hours and activation was determined by measurement of IL-2 in the tissue culture supernatant. Supernatants (100 µl) were added to the IL-2-dependent CTLL-2 line ($5x10^3$ cells), followed by culture for 48 hours. CTLL-2 cells were pulsed with 0.4 µCi [³H]thymidine for the final 18–24 h. IL-2 production was quantified by measurement of [³H]thymidine incorporation.

3. Results

3.1 Cloning, expression, and purification of M15:IL-10 fusions

To conjugate high-affinity TCR M15 to the anti-inflammatory cytokine IL-10, we fused a single-chain version of M15 (scTv M15) to the N-terminus of IL-10. The scTv M15 construct, which was engineered by linking Vβ to Vα via a 17-mer peptide (GSADDAKEDAAKKDGES), was then connected to different versions of hIL-10, including wild-type hIL-10, hIL-10 (I87A) which contains the I87A point mutation, monomeric hIL-10 (mono) which has the GGGSGG insertion, and hIL-10 (I87A, mono) with both the point mutation and insertion. Wild-type IL-10 exists as an intercalated (domain-swapped) homodimer in solution with picomolar affinity for IL-10R (Zdanov et al. 1995). Insertion of GGGSGG into the loop connecting the swapped secondary structural elements of dimeric IL-10 generates a monomeric version of this cytokine with a K_D for IL-10R of 30 nM, without significantly affecting immunosuppressive activity (Josephson et al. 2000).

All M15–IL-10 constructs were tested for *in vitro* folding. Interestingly, only the monomeric versions could be folded successfully, whereas conjugates containing wild-type (i.e. dimeric) IL-10 showed poor folding behavior under a wide range of conditions. As expected, M15–hIL-10 (mono) and M15–hIL-10 (I87A, mono) behaved as monomers in gel filtration (Figure 1A).

The linker used for attaching scTv M15 to IL-10 was either an 18-mer or 15-mer peptide. The 18-mer linker (VNAKTTAPSVYPLAPVSG) (Card et al. 2004) was originally adopted from the hinge region between the V_H and C_H1 domains of an antibody, whereas the 15-mer linker (SSSSG)$_3$ (Trachsel et al. 2007) was designed to maximize flexibility and solubility. Proteins with the two different linkers showed similar properties, except that extensive hydrolysis was detected for conjugates using the 18-mer linker after two weeks storage at 4 °C (Figure 1C).

To express conjugates with wild type IL-10, we adopted the baculovirus expression system. Constructs containing scTv M15 fused to wild type mouse or human IL-10 genes were inserted into the pAcGP67B baculovirus transfer vector, which uses the baculovirus polyhedrin promoter to drive expression. Secretion of M15–IL-10 conjugates was directed by the gp67 signal sequence. Importantly, all three wild-type IL-10 conjugates (M15:mIL-10, 18aa linker, M15:hIL-10, 15aa linker, and M15:mIL-10, 15aa linker) eluted as dimers in gel filtration (Figure 1B), indicating that IL-10 retained its native state when conjugated to scTCR M15. The yield of the conjugates from infected High Five cells was typically 1 mg per liter of culture.

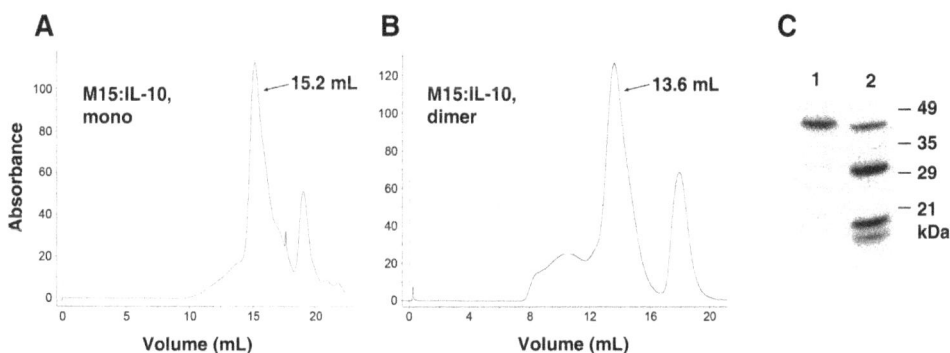

Fig. 1. Purification and characterization of soluble, scTv M15:IL-10 fusion proteins.

Size exclusion chromatography of *(A) in vitro* folded human M15:IL-10, mono, and *(B)* insect cell-expressed human M15:IL-10, dimer. Gel filtration was performed with a Superdex S-200 column in phosphate-buffered saline at a flow rate of 0.5 ml/min. M15:IL-10, mono eluted at

15.2 mL and M15:IL-10, dimer eluted at 13.6 mL. *(C)* SDS-PAGE analysis of M15:IL-10 conjugates with two different linkers. Purified samples of M15:IL-10, I87A, mono, 15aa linker (Lane 1) and M15: IL-10, I87A, mono, 18aa linker (Lane 2), both from *in vitro* folding and containing human IL-10, were stored at 4 °C for two weeks before being analyzed.

3.2 Detection by IL-10 ELISA

To examine initially whether the IL-10 domain of the M15 fusion constructs were folded properly, they were assayed with IL-10-specific antibodies in a sandwich ELISA, as shown in Figure 2. M15:IL-10 fusions expressed in *E. coli* containing human IL-10 with an additional linker to allow for monovalent IL-10 folding (M15:IL-10, mono and M15:IL-10, mono,I87A) were detected equivalently with recombinant, commercially available human IL-10 (Figure 2A). M15:IL-10, I87A, which is designed for IL-10 folding into an intercalated dimer, was detected less efficiently. This corresponds to difficulty folding the dimeric IL-10 construct *in vitro* (See Section 3.1). Additional dimeric constructs expressed in insect cells were more stable, and were detected by ELISA with identical efficiency with commercially available free IL-10 (human M15:IL-10 proteins: Figure 2B; murine M15:IL-10 proteins: Figure 2C). These results show that the M15:IL-10 fusion proteins contain properly folded IL-10.

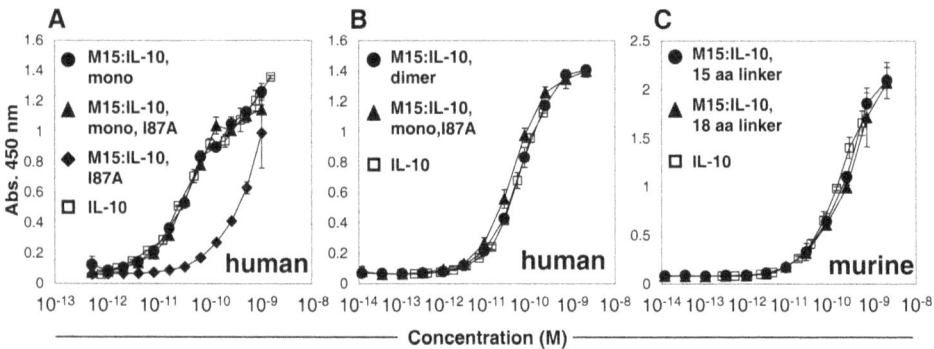

Fig. 2. Properly folded IL-10 was detected in ELISAs of the various soluble M15:IL-10 fusions.

(A-C) Absorbance of the ELISA-linked TMB substrate at 450 nm is shown for various concentrations of IL-10 and M15:IL-10 fusion proteins. *(A)* Human IL-10 was detected by sandwich ELISA for M15:IL-10 fusions produced in *E. coli*: M15:IL-10, mono (closed circles), M15:IL-10, mono, I87A (closed triangles), M15:IL-10, I87A (dimer construct, closed diamonds), and human IL-10 alone (open squares). *(B)* Human IL-10 was detected by sandwich ELISA for M15:IL-10 fusion produced in HighFive insect cells as well as *E. coli*: M15:IL-10, dimer (HighFive protein, closed circles), M15:IL-10, mono, I87A (*E. coli* protein, closed triangles), and human IL-10 alone (open squares). *(C)* Murine IL-10 was detected by sandwich ELISA for dimeric M15:IL-10 fusions produced in HighFive: M15:IL-10, 15 amino acid linker (closed circles), M15:IL-10, 18 amino acid linker (closed triangles), and free murine IL-10 (open squares).

3.2.1 Constructs with human IL-10, produced in *E. coli*

Preliminary tests were carried out with human IL-10 fused to scTv M15 in several configurations, including IL-10 as a monomer with an additional linker inserted between residues 116 and 117 (M15:IL-10, mono), IL-10 as a dimer containing the point mutation I87A (M15:IL-10, I87A), purported to eliminate stimulatory effects of the cytokine while maintaining suppressive effects, and IL-10 as a monomer plus the I87A mutation (M15:IL-10, mono,I87A). Initial protein expression was carried out in *E. coli* as inclusion bodies, followed by solubilization and re-folding *in vitro*. Analysis of these purified proteins by ELISA confirmed that monomeric IL-10 constructs were detected equivalently on a molar basis as free IL-10, with or without the I87A mutation (Figure 2A, closed circles and triangles). The fusion designed to allow IL-10 to form the native intercalated dimer (M15:IL-10, I87A) was detected at much lower levels, corresponding to less stable protein (See Section 2). In order to be able to obtain more stable IL-10 dimer fusions to accurately compare monomer and dimer constructs for biological function, a second, eukaryotic expression system was employed.

3.2.2 Constructs with human IL-10, produced in insect cells

A dimeric M15:IL-10 fusion protein (M15:IL-10, dimer) was designed and expressed as a stable dimer (See Section 2) by HighFive insect cells. This purified protein was compared directly with a similar, monomeric protein from *E. coli* (M15:IL-10, mono,I87A). Both proteins contained a 15-amino acid linker between the scTv and the cytokine. As can be seen in Figure 2B, the dimeric human IL-10 fusion (closed circles) was detected equivalently on a per-molar basis as both the *E. coli* monomer fusion (closed triangles) and free human IL-10 alone (open squares).

3.2.3 Constructs with murine IL-10

Dimeric M15:IL-10 fusions containing murine IL-10 were also expressed in insect cells. Two linkers were analyzed between the scTv and the cytokine: a 15-amino acid linker similar to the human constructs in Figure 2B (M15:IL-10, 15aa linker), and an 18-amino acid linker (M15:IL-10, 18aa linker) similar to the original E. coli constructs in Figure 2A. The 18-amino acid linker had proven to be susceptible to degradation over time. Side-by-side analysis by murine ELISA (Figure 2C) shows that fusion proteins with both linkers are detected equivalently on a per molar basis by ELISA (closed circles and closed triangles), and they are, in turn, detected equivalently with free murine IL-10 (open squares).

3.3 Specific binding/targeting to cells displaying target epitope

Having confirmed proper folding of the IL-10 portion of the fusions by ELISA, specific binding of the scTv portion was tested by detecting binding specifically to APCs presenting Hb/I-E^k. Flow cytometry experiments verified that M15:IL-10 fusion proteins bound specifically, and carried with them covalently-linked IL-10.

3.3.1 Detection of binding through scTv via fused IL-10

CH27 cells (I-E^{k+}) were loaded with peptide by overnight incubation with 1 or 5 μM Hb peptide in culture medium, and were then incubated with soluble, biotinylated scTv M15 or with various M15:IL-10 fusion constructs on ice. Bound fusion was detected using biotinylated, polyclonal anti-IL-10 followed by fluorescently-labeled streptavidin, while bound scTv was detected by fluorescently-labeled streptavidin only. Staining results are shown in Figure 3.

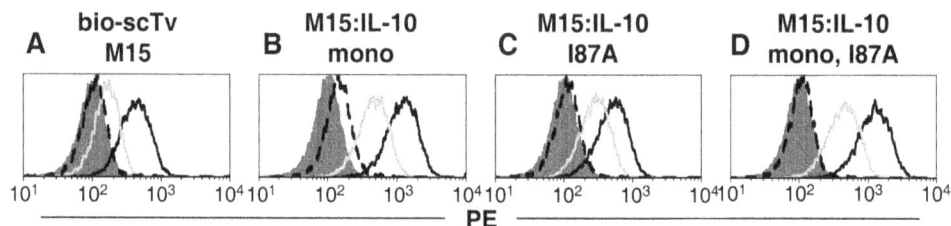

Fig. 3. M15 scTv and M15:IL-10 fusions bind specifically to cells presenting Hb/I-E^k.

(A-D) Flow cytometry histograms are shown in each panel for CH27 cells loaded with 5 μM Hb peptide (solid black trace), 1 μM Hb peptide (solid gray trace), no added peptide (dashed black trace), or secondary detection only (dark gray shaded trace). *(A)* Cells stained with biotinylated scTv M15 without any IL-10, detected by streptavidin phycoerythrin (PE) alone. *(B-C)* Binding of M15:IL-10 fusions refolded from *E. coli* was detected with biotinylated polyclonal anti-IL-10 antibody, followed by streptavidin-PE for *(B)* M15:IL-10 containing monovalent human IL-10, *(C)* M15:IL-10 containing dimeric human IL-10 and the I87A point mutation, and *(D)* M15:IL-10 containing monovalent human IL-10 with the I87A mutation.

Figure 3A shows staining of CH27 cells with free M15 scTv. At 5 μM Hb peptide, the cells stain four-fold brighter than without added peptide (MFI of 46.5 and 11.3, respectively). As can be seen in Figure 3B-D, all of the M15:IL-10 constructs bound specifically to cells which had been loaded with Hb peptide. The magnitude of staining increase seen with the dimeric M15:IL-10, I87A protein was lower (roughly 5-fold) than that seen for the two monomeric constructs containing the additional linker between N116 and K117 (8-fold for M15:IL-10, mono, and 12–fold for M15:IL-10, mono,I87A, respectively). This is consistent with evidence that the dimer construct made in *E. coli* was less stable than the monomer constructs (See Sections 3.1 and 3.2, and Figure 2A).

3.4 Functional assays of M15 scTv and IL-10 fusions

M15 and its fusions can act by direct inhibition of T cell recognition, by preventing the binding of the cellular TCRs on T cells to the peptide-MHC on APCs, or of course could act to immunosuppress resident T cells through the action of the cytokine IL-10. The direct T cell inhibitory potential was examined with soluble M15 scTv (Figure 4), and the biological function of the cytokine portion of the M15-IL-10 fusions was evaluated using a murine mast cell line that selectively proliferates in the presence of IL-10 (Figure 5). This cell line can respond to both human and murine IL-10.

3.4.1 Direct inhibition of T cell activation by soluble M15 scTv

The high-affinity scTv M15 binds to its antigen, a Hb peptide loaded onto I-E^{k+}. For purposes of using this scTv as a model for inhibiting pathogenic T cells that are specific for a target antigen, a soluble scTv could not only serve as a specific IL-10 targeting agent, but could directly inhibit activation of the T cells by preventing recognition of APCs. To explore this possibility, soluble high-affinity M15 was added in a T cell activation assay (Figure 4). IL-2 secretion from the specific T cells was induced at various concentrations of Hb peptide loaded onto I-E^{k+} APCs and this activity was inhibited by the soluble scTv M15. This suggests that the binding of scTv alone can reduce the magnitude of an immune response against a self- or allegen-associated peptide-MHC complex.

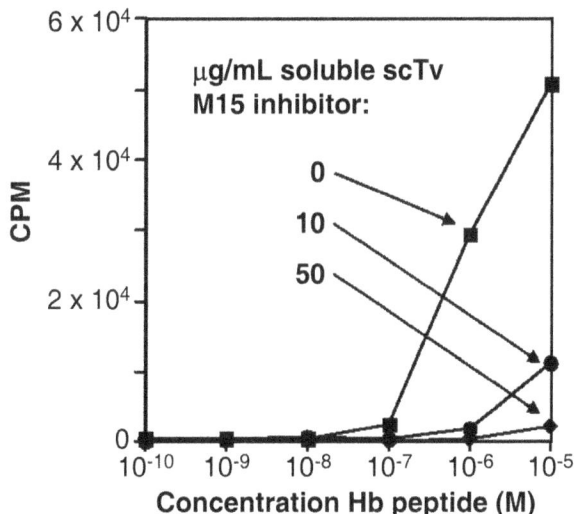

Fig. 4. Activation of 3.L2 T cells is inhibited by binding of scTv M15.

Activation of T cells carrying the 3.L2 TCR (wild-type affinity) by I-E^{k+} APCs incubated with various concentration of stimulatory Hb peptide alone (closed squares), or with 10 μg/mL (closed circles) or 50 μg/mL (closed diamonds) soluble, high-affinity scTv M15. IL 2 secretion by the 3.L2 T cells was detected by proliferation of an IL-2-dependent cell line, measured by incorporation of tritiated thymidine.

3.4.2 Induced proliferation of a cytokine-dependent cell line (MC/9)

To determine whether the M15:IL-10 fusions exhibited functional activity, we determined if they induced proliferation of the IL-10-dependent cell line, MC/9. Although less potent than commercial recombinant IL-10, the M15:IL-10 dimer protein (closed diamonds) was functional in this biological assay, whereas the M15:IL-10, mono-I87A (closed circles) was less potent than the dimer (Figure 5A). The murine M15:IL-10 proteins (closed triangles and closed squares) were equally capable of stimulating proliferation of MC/9 cells (Figure 5B).

(A,B) Proliferation of the cytokine-dependent MC/9 cell line cultured with free IL-10 or different M15:IL-10 constructs was evaluated by MTT reduction by measuring absorbance of solubilized formazan precipitate at 570 nM (See Section 2.4). (A) Proliferation induced by human IL-10 (open squares) or the human IL-10 fusion proteins M15:IL-10, dimer (closed diamonds) or M15:IL-10, mono,I87A (closed circles). (B) Proliferation induced by murine IL-

10 (open squares) or the murine IL-10 fusions M15:IL-10, 15aa linker (closed triangles) or M15:IL-10, 18aa linker (closed squares).

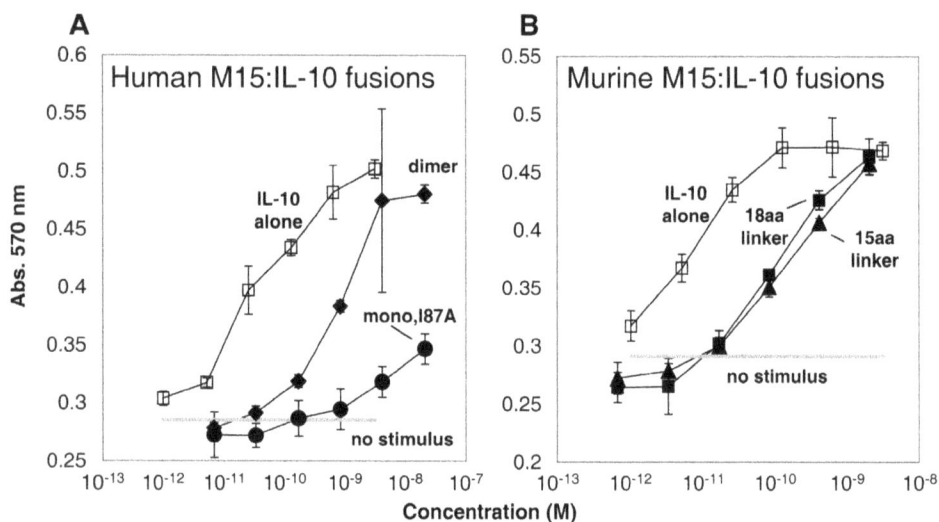

Fig. 5. Functional activity of IL-10 and M15:IL-10 fusion proteins.

4. Conclusions

We have described here the first soluble IL-10 protein fusions targeted by a high-affinity single-chain variable TCR (scTv) protein, M15, directed to a class II MHC-peptide complex (Hb/I-E^k). Many inflammatory diseases are driven by recognition of specific class II MHC-peptide complexes, providing an opportunity to use such scTv molecules to target soluble therapeutics, or as direct inhibitors of pathogenic T cells. To our knowledge, the M15 TCR is the only class II TCR with high affinity for its target peptide-MHC (an 800-fold increased affinity compared to wild type 3.L2). The fusion proteins we generated contained N-terminal scTv linked by a flexible 15- or 18-amino acid peptide to IL-10 at the C-terminus. Proper folding of the cytokine portion of the fusions was verified (Figure 2), as was specific binding of the scTv to the target ligand, Hb/I-E^k (Figure 3).

Functional assays confirmed that the IL-10 fusion proteins could stimulate biological activity in a cytokine-dependent cell line (Figure 5); however, the fusion proteins were less efficient than free murine or human IL-10 at stimulating that activity. It is possible that some steric hindrance from the N-terminal scTv prevents fully efficient signaling that is not ameliorated by extending the linker between the two proteins from 15 to 18 residues (Figure 5B). The ELISA data shown in Figure 2 suggests that most of the epitopes are, indeed, accessible to antibodies. The published structure of IL-10 bound to its high-affinity receptor, IL-10R1 (Josephson et al. 2001) suggests that an N-terminal extension would not interfere with that interaction. Based on modeling and mutagenesis studies, the low-affinity receptor, IL-10R2, has been suggested to dock much more closely to the N-terminus (Yoon et al. 2010). Hence, this lower affinity receptor interaction may be interrupted by the N-terminal scTv in the fusion protein. Switching the domain order may result in more potent function from the scTv fusions. However, this level of reduced potency may not, in fact, be a disadvantage for these targeted therapeutics, as it may reduce potentially problematic systemic effects of IL-10 (Belkaid 2007; Fort & Narayanan 2010), while allowing the biodistribution to be dominated by the high-affinity scTv targeting element (Weber et al. 2005).

Despite data that monomeric IL-10 created by inserting a 6-amino acid linker between residues 116 and 117 of the protein is still functional (Josephson et al. 2000), the fusions reported here with monomeric IL-10 displayed low activity, even when the I87A mutation, which abrogates stimulatory activities of IL-10 (Ding et al. 2000), was not included (Figure 4A, and data not shown). Perhaps signaling through an IL-10 monomer is particularly sensitive to any steric interference reducing the ability to co-localize IL-10R1 with IL-10R2 resulting from the N-terminal attachment of the scTv domain.

Further characterization of scTv:IL-10 fusions will include examining the effects (including suppressive capacity) of the IL-10 fusions at different doses, ideally for different cell types, and particularly various helper T cell lineages including Th1, Th2, and Th17 cells. Biodistribution and pharmacokinetic studies will also be required to verify appropriate *in vivo* targeting by the scTv, and sufficient serum half-life for effectiveness. Results of these studies may suggest further redesign of either the delivery system for the protein or, possibly, of the fusion constructs themselves.

The use of engineered scTv TCR fragments as targeting elements for soluble molecules is an increasingly attractive and viable option (Aggen et al. 2011; Richman & Kranz 2007). T cells play a central role in many inflammatory diseases, and effective T cell responses can combat serious infections or cancer. Selective modulation of the relevant T cell populations by downregulation or upregulation, respectively, may represent an advance in treatment with reduced side effects compared to systemic treatments. Additionally, in many cases relevant T cell peptide-MHC epitopes related to these diseases are known, and immunodominant or specific TCRs have been isolated. While the wild-type receptors are not suited for incorporation into soluble therapeutics, advances in TCR engineering for affinity and stability as a soluble scTv (Aggen et al. 2011; Kieke et al. 1999; Li et al. 2005b; Richman et al. 2009; Weber et al. 2005) transform this into a more practical strategy.

As has been seen, the location of delivered IL-10 has a dramatic impact on its effect as a therapeutic (Asadullah et al. 2003; Kennedy et al. 1992; Lindsay & Hodgson 2001; O'Garra et al. 2008). Maintenance of an appropriate dose at the site of action is also important, as high doses of IL-10 can be less effective than lower doses at controlling inflammation in patients (Asadullah et al. 2003). Targeting by inclusion of a specific scTv in IL-10 fusions may be effective alone, and may synergize when combined with promising delivery methods used for free IL-10, such as local secretion by virally-transduced cells or by genetically modified bacteria (Lindsay & Hodgson 2001; Steidler et al. 2000; Wirtz et al. 1999). Sustained activity at the desired site, combined with facile, continuous delivery may make targeted IL-10 therapy a valuable addition to current treatments against various excessive inflammatory conditions.

5. References

Aggen, D. H., A. S. Chervin, et al. (2011). "Identification and engineering of human variable regions that allow expression of stable single-chain T cell receptors." Protein Eng Des Sel.

Asadullah, K., W. Sterry, et al. (2003). "Interleukin-10 therapy--review of a new approach." Pharmacol Rev 55(2): 241-69.

Belkaid, Y. (2007). "Regulatory T cells and infection: a dangerous necessity." Nat Rev Immunol 7(11): 875-88.

Belmont, H. J., S. Price-Schiavi, et al. (2006). "Potent antitumor activity of a tumor-specific soluble TCR/IL-2 fusion protein." Clin Immunol 121(1): 29-39.

Bettelli, E., M. P. Das, et al. (1998). "IL-10 is critical in the regulation of autoimmune encephalomyelitis as demonstrated by studies of IL-10- and IL-4-deficient and transgenic mice." J Immunol 161(7): 3299-306.

Boulter, J. M., M. Glick, et al. (2003). "Stable, soluble T-cell receptor molecules for crystallization and therapeutics." Protein Eng 16(9): 707-11.

Card, K. F., S. A. Price-Schiavi, et al. (2004). "A soluble single-chain T-cell receptor IL-2 fusion protein retains MHC-restricted peptide specificity and IL-2 bioactivity." Cancer Immunol Immunother 53(4): 345-57.

Chervin, A. S., D. H. Aggen, et al. (2008). "Engineering higher affinity T cell receptors using a T cell display system." J Immunol Methods 339(2): 175-84.

Davidson, N. J., M. W. Leach, et al. (1996). "T helper cell 1-type CD4+ T cells, but not B cells, mediate colitis in interleukin 10-deficient mice." J Exp Med 184(1): 241-51.

Davis, M. M., J. J. Boniface, et al. (1998). "Ligand recognition by alpha beta T cell receptors." Annu Rev Immunol 16: 523-44.

Ding, Y., L. Qin, et al. (2000). "A single amino acid determines the immunostimulatory activity of interleukin 10." J Exp Med 191(2): 213-24.

Ebers, G. C., K. Kukay, et al. (1996). "A full genome search in multiple sclerosis." Nat Genet 13(4): 472-6.

Fiorentino, D. F., M. W. Bond, et al. (1989). "Two types of mouse T helper cell. IV. Th2 clones secrete a factor that inhibits cytokine production by Th1 clones." J Exp Med 170(6): 2081-95.

Fleming, S. B., C. A. McCaughan, et al. (1997). "A homolog of interleukin-10 is encoded by the poxvirus orf virus." J Virol 71(6): 4857-61.

Fort, M. M. and P. K. Narayanan (2010). "Manipulation of regulatory T-cell function by immunomodulators: a boon or a curse?" Toxicol Sci 117(2): 253-62.

Gebe, J. A., K. A. Unrath, et al. (2008). "Autoreactive human T-cell receptor initiates insulitis and impaired glucose tolerance in HLA DR4 transgenic mice." J Autoimmun 30(4): 197-206.

Gillies, S. D. (2009). Immunocytokines: A Novel Approach to Cancer Immune Therapy. Targeted Cancer Immune Therapy. J. L. e. al. Waltham, MA, Springer Science+Business Media, LLC: 241-256.

Haselden, B. M., A. B. Kay, et al. (1999). "Immunoglobulin E-independent major histocompatibility complex-restricted T cell peptide epitope-induced late asthmatic reactions." J Exp Med 189(12): 1885-94.

Hawrylowicz, C., D. Richards, et al. (2002). "A defect in corticosteroid-induced IL-10 production in T lymphocytes from corticosteroid-resistant asthmatic patients." J Allergy Clin Immunol 109(2): 369-70.

Hemmer, B., M. Vergelli, et al. (1998). "Predictable TCR antigen recognition based on peptide scans leads to the identification of agonist ligands with no sequence homology." J Immunol 160(8): 3631-6.

Holler, P. D., L. K. Chlewicki, et al. (2003). "TCRs with high affinity for foreign pMHC show self-reactivity." Nat Immunol 4(1): 55-62.

Holler, P. D., P. O. Holman, et al. (2000). "In vitro evolution of a T cell receptor with high affinity for peptide/MHC." Proc Natl Acad Sci U S A 97(10): 5387-92.

Hsu, D. H., R. de Waal Malefyt, et al. (1990). "Expression of interleukin-10 activity by Epstein-Barr virus protein BCRF1." Science 250(4982): 830-2.

Jahn-Schmid, B., P. Kelemen, et al. (2002). "The T cell response to Art v 1, the major mugwort pollen allergen, is dominated by one epitope." J Immunol 169(10): 6005-11.

Johnson, K. P., B. R. Brooks, et al. (1995). "Copolymer 1 reduces relapse rate and improves disability in relapsing-remitting multiple sclerosis: results of a phase III multicenter, double-blind placebo-controlled trial. The Copolymer 1 Multiple Sclerosis Study Group." Neurology 45(7): 1268-76.

Josephson, K., R. DiGiacomo, et al. (2000). "Design and analysis of an engineered human interleukin-10 monomer." J Biol Chem 275(18): 13552-7.

Josephson, K., N. J. Logsdon, et al. (2001). "Crystal structure of the IL-10/IL-10R1 complex reveals a shared receptor binding site." Immunity 15(1): 35-46.

Kaspar, M., E. Trachsel, et al. (2007). "The antibody-mediated targeted delivery of interleukin-15 and GM-CSF to the tumor neovasculature inhibits tumor growth and metastasis." Cancer Res 67(10): 4940-8.

Kaspar, M., L. Zardi, et al. (2006). "Fibronectin as target for tumor therapy." Int J Cancer 118(6): 1331-9.

Kennedy, M. K., D. S. Torrance, et al. (1992). "Analysis of cytokine mRNA expression in the central nervous system of mice with experimental autoimmune encephalomyelitis reveals that IL-10 mRNA expression correlates with recovery." J Immunol 149(7): 2496-505.

Kieke, M. C., E. V. Shusta, et al. (1999). "Selection of functional T cell receptor mutants from a yeast surface-display library." Proc Natl Acad Sci U S A 96(10): 5651-6.

Kotenko, S. V., S. Saccani, et al. (2000). "Human cytomegalovirus harbors its own unique IL-10 homolog (cmvIL-10)." Proc Natl Acad Sci U S A 97(4): 1695-700.

Kuhn, R., J. Lohler, et al. (1993). "Interleukin-10-deficient mice develop chronic enterocolitis." Cell 75(2): 263-74.

Li, Y., Y. Huang, et al. (2005). "Structure of a human autoimmune TCR bound to a myelin basic protein self-peptide and a multiple sclerosis-associated MHC class II molecule." Embo J 24(17): 2968-79.

Li, Y., R. Moysey, et al. (2005). "Directed evolution of human T-cell receptors with picomolar affinities by phage display." Nat Biotechnol 23(3): 349-54.

Lindsay, J. O. and H. J. Hodgson (2001). "Review article: the immunoregulatory cytokine interleukin-10--a therapy for Crohn's disease?" Aliment Pharmacol Ther 15(11): 1709-16.

Ma, W., H. Yu, et al. (2004). "A novel approach for cancer immunotherapy: tumor cells with anchored superantigen SEA generate effective antitumor immunity." J Clin Immunol 24(3): 294-301.

Moore, K. W., R. de Waal Malefyt, et al. (2001). "Interleukin-10 and the interleukin-10 receptor." Annu Rev Immunol 19: 683-765.

Moore, K. W., P. Vieira, et al. (1990). "Homology of cytokine synthesis inhibitory factor (IL-10) to the Epstein-Barr virus gene BCRFI." Science 248(4960): 1230-4.

Muraro, P. A., M. Vergelli, et al. (1997). "Immunodominance of a low-affinity major histocompatibility complex-binding myelin basic protein epitope (residues 111-129) in HLA-DR4 (B1*0401) subjects is associated with a restricted T cell receptor repertoire." J Clin Invest 100(2): 339-49.

O'Garra, A., F. J. Barrat, et al. (2008). "Strategies for use of IL-10 or its antagonists in human disease." Immunol Rev 223: 114-31.

Oksenberg, J. R., E. Seboun, et al. (1996). "Genetics of demyelinating diseases." Brain Pathol 6(3): 289-302.

Ouyang, W., S. Rutz, et al. (2011). "Regulation and functions of the IL-10 family of cytokines in inflammation and disease." Annu Rev Immunol 29: 71-109.

Pennell, C. A., L. W. Arnold, et al. (1985). "Cross-reactive idiotypes and common antigen binding specificities expressed by a series of murine B-cell lymphomas: etiological implications." Proc Natl Acad Sci U S A 82(11): 3799-803.

Pohlman, B., J. Sweetenham, et al. (2006). "Review of clinical radioimmunotherapy." Expert Rev Anticancer Ther 6(3): 445-61.

Reijonen, H., E. J. Novak, et al. (2002). "Detection of GAD65-specific T-cells by major histocompatibility complex class II tetramers in type 1 diabetic patients and at-risk subjects." Diabetes 51(5): 1375-82.

Rice, G. P., H. P. Hartung, et al. (2005). "Anti-alpha4 integrin therapy for multiple sclerosis: mechanisms and rationale." Neurology 64(8): 1336-42.

Richman, S. A., D. H. Aggen, et al. (2009). "Structural features of T cell receptor variable regions that enhance domain stability and enable expression as single-chain ValphaVbeta fragments." Mol Immunol 46(5): 902-16.

Richman, S. A. and D. M. Kranz (2007). "Display, engineering, and applications of antigen-specific T cell receptors." Biomol Eng 24(4): 361-73.

Sabat, R., G. Grutz, et al. (2010) "Biology of interleukin-10." Cytokine Growth Factor Rev 21(5): 331-44.

Samoilova, E. B., J. L. Horton, et al. (1998). "Acceleration of experimental autoimmune encephalomyelitis in interleukin-10-deficient mice: roles of interleukin-10 in disease progression and recovery." Cell Immunol 188(2): 118-24.

Saraiva, M. and A. O'Garra (2010). "The regulation of IL-10 production by immune cells." Nat Rev Immunol 10(3): 170-81.

Schanzer, J. M., P. A. Baeuerle, et al. (2006). "A human cytokine/single-chain antibody fusion protein for simultaneous delivery of GM-CSF and IL-2 to Ep-CAM overexpressing tumor cells." Cancer Immun 6: 4.

Sellon, R. K., S. Tonkonogy, et al. (1998). "Resident enteric bacteria are necessary for development of spontaneous colitis and immune system activation in interleukin-10-deficient mice." Infect Immun 66(11): 5224-31.

Sommavilla, R., N. Pasche, et al. (2010). "Expression, engineering and characterization of the tumor-targeting heterodimeric immunocytokine F8-IL12." Protein Eng Des Sel 23(8): 653-61.

Sospedra, M. and R. Martin (2005). "Immunology of multiple sclerosis." Annu Rev Immunol 23: 683-747.

Spencer, J. V., K. M. Lockridge, et al. (2002). "Potent immunosuppressive activities of cytomegalovirus-encoded interleukin-10." J Virol 76(3): 1285-92.

Steidler, L., W. Hans, et al. (2000). "Treatment of murine colitis by Lactococcus lactis secreting interleukin-10." Science 289(5483): 1352-5.

Stone, J. D., A. S. Chervin, et al. (2009). "T-cell receptor binding affinities and kinetics: impact on T-cell activity and specificity." Immunology 126(2): 165-76.

Tan, J. C., S. Braun, et al. (1995). "Characterization of recombinant extracellular domain of human interleukin-10 receptor." J Biol Chem 270(21): 12906-11.

Teitelbaum, D., R. Aharoni, et al. (1988). "Specific inhibition of the T-cell response to myelin basic protein by the synthetic copolymer Cop 1." Proc Natl Acad Sci U S A 85(24): 9724-8.

Teitelbaum, D., M. Fridkis-Hareli, et al. (1996). "Copolymer 1 inhibits chronic relapsing experimental allergic encephalomyelitis induced by proteolipid protein (PLP) peptides in mice and interferes with PLP-specific T cell responses." J Neuroimmunol 64(2): 209-17.

Tesmer, L. A., S. K. Lundy, et al. (2008). "Th17 cells in human disease." Immunol Rev 223: 87-113.

Trachsel, E., F. Bootz, et al. (2007). "Antibody-mediated delivery of IL-10 inhibits the progression of established collagen-induced arthritis." Arthritis Res Ther 9(1): R9.

Vergelli, M., M. Kalbus, et al. (1997). "T cell response to myelin basic protein in the context of the multiple sclerosis-associated HLA-DR15 haplotype: peptide binding, immunodominance and effector functions of T cells." J Neuroimmunol 77(2): 195-203.

Weber, K. S., D. L. Donermeyer, et al. (2005). "Class II-restricted T cell receptor engineered in vitro for higher affinity retains peptide specificity and function." Proc Natl Acad Sci U S A 102(52): 19033-8.

Wirtz, S., P. R. Galle, et al. (1999). "Efficient gene delivery to the inflamed colon by local administration of recombinant adenoviruses with normal or modified fibre structure." Gut 44(6): 800-7.

Wong, R. L., B. Liu, et al. (2011). "Interleukin-15:Interleukin-15 receptor {alpha} scaffold for creation of multivalent targeted immune molecules." Protein Eng Des Sel.

Wucherpfennig, K. W. and D. Sethi (2011). "T cell receptor recognition of self and foreign antigens in the induction of autoimmunity." Semin Immunol 23(2): 84-91.

Yoon, S. I., B. C. Jones, et al. (2010). "Structure and mechanism of receptor sharing by the IL-10R2 common chain." Structure 18(5): 638-48.

Zdanov, A., C. Schalk-Hihi, et al. (1995). "Crystal structure of interleukin-10 reveals the functional dimer with an unexpected topological similarity to interferon gamma." Structure 3(6): 591-601.

6

FN3 Domain Engineering

Steven Jacobs and Karyn O'Neil
Centyrex Venture, Johnson & Johnson
USA

1. Introduction

The Fibronectin Type III domain (FN3) is a small globular protein domain of 90-100 amino acids found in thousands of proteins, over all of protein sequence space from bacteria to humans. In fact, sequence analysis estimates that the FN3 domain can be found in approximately 2% of all animal proteins (Bork & Doolittle, 1992). The frequency that these domains are found is even greater when considering the larger family, the Ig-like fold, of which FN3 is a specific member. Despite weak to no sequence identity and greatly divergent functions, FN3 domains fold into a common structure. Over 230 structures of FN3 domains have been deposited to the PDB database as of 2011 (> 990 when considering all Ig-like domains), showing that the fold consists of 7 antiparallel β-strands that pack into two β-sheets (Main et al., 1992) (Figure 1). These β-strands are designated with the nomenclature A-F with the loops connecting each strand designated by the strands it connects (For example, the AB loop connects strands A and B). With a few exceptions, structural variability of these domains is found exclusively in the loops connecting the β-strands while the strands remain structurally conserved.

The small size, simple structure, and robust prokaryotic expression levels of many FN3 domains has led to the development of these proteins as important tools for the understanding of protein folding and stability. To date, two members of the FN3 family, the tenth FN3 domain from human Fibronectin (FnFN10) and the 3rd FN3 domain from human Tenascin-C (TnFN3) have been the most widely studied. Despite low levels of sequence identity (23%), these two FN3 domains fold into very similar structures with the majority of structural deviations occurring in the loops connecting the β-strands (Leahy et al., 1992, Main, et al., 1992). Of great importance to the use of FN3 domains in novel applications is the ability of proteins with very diverse sequences to fold into a common structure, thus allowing for these domains to be heavily mutated in order to incur a desired function. This property has been demonstrated in a series of experiments published from the Clarke lab in which a "fold approach" was taken to characterize the folding and stability of a number of FN3 and Ig domains. The resulting work has lead to the proposal that proteins of the FN3 and related families fold by similar mechanisms involving nucleation from the B, C, E and F strands, even when no sequence similarity is detectable (Clarke et al., 1999, Hamill et al., 2000a, Hamill et al., 2000b). Despite such similar structures and folding pathways, FN3 domains can exhibit markedly different conformational stabilities. For example, the FnFN10 and TnFN3 domains have conformational stabilities of 9.4 and 6.7 kcal mol^{-1} respectively and FnFN9, which resides directly adjacent to FnFN10 in the Fibronectin protein is approximately 5-fold less stable than FnFN10 (Clarke et al., 1997, Cota & Clarke, 2000, Hamill et al., 1998, Plaxco et al., 1997). Understanding the causes of these differences is important for producing FN3 domains for pharmaceutical or industrial applications. One explanation for the high stability of FnFn10 may be the rapid rate at which this isolated domain folds, despite the presence of 8 proline residues that might be expected to slow down folding due to cis-trans isomerization (Plaxco et al., 1996, Plaxco, et al., 1997). Indeed the data available so far for FN3 domains supports the hypothesis that the refolding rates of β-sheet proteins reflect overall stability (Finkelstein, 1991). Another contribution to the differences in conformational stability of these proteins may be differences in structural plasticity. NMR spectroscopy was used to determine that the side-chains in the hydrophobic core of TnFN3 had a higher degree of mobility than those of FnFN10, while mutational analysis has shown that FnFN10 is more amenable to surface mutation due to greater flexibility in peripheral regions of the protein (Best et al., 2004, Cota et al., 2000). In support of this model, Billings and colleagues produced a chimera of FnFN10 and TnFN3 which was composed of the hydrophobic core of TnFN3 and the surface and loops of FnFN10. This chimeric FN3 domain showed greater stability than that of TnFN3, suggesting that the surface composition and loops can significantly contribute to stability (Billings et al., 2008).

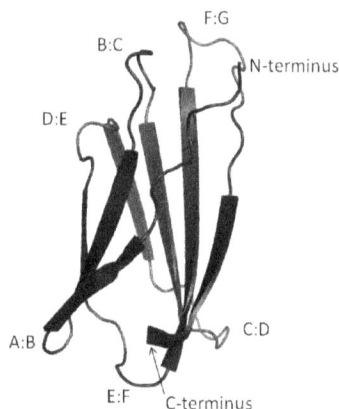

Fig. 1. Solution structure of FN3 domain FnFN10 (PDB 1TTF). Labels indicate the identity of loop regions as well as the N and C-termini.

The contributions of surface residues to FN3 stability and the ability of such residues to tolerate mutation are keys to the development of these domains for biotechnology applications in which various surfaces or loops are mutated to employ a new function (see below). This property was first exploited by Koide et al. who demonstrated that the surface exposed loops of FnFN10 could be randomized and used to select FN3 domains that interact with a novel ligand, ubiquitin, thereby imparting a non-natural function to the FN3 domain (Koide et al., 1998). Importantly, the authors were able to confirm that the FN3 domains produced with such foreign loops were able to maintain the overall FN3 structure, although the resulting binders had reduced conformational stability compared to that of FnFN10. Numerous subsequent biochemical and biotechnological applications of FN3 domains are described in this chapter. The success of these techniques is strongly dependent upon the ability of highly diverse FN3 domains to fold into a common structure while tolerating a high level of sequence mutation. In the next section, selection systems which have been used to select FN3 domains of novel function are described, followed by descriptions of protein engineering studies designed to increase the overall stability of the FN3 scaffold and to evolve FN3 scaffolds with high affinity and selectivity for a particular target. Finally, a report of novel applications of engineered FN3 domains is provided.

2. Selection technologies

Over the last twenty years, a variety of display technologies have evolved as tools for protein engineering and selection of binding ligands. Such techniques can be broadly separated into phage display, *in vitro* display (including ribosome display, mRNA display, and DNA display) and yeast display strategies (Figure 2). Phage display requires propogation and secretion from bacteria (ie. *E. coli*) using prokaryotic secretion machinery while yeast display relies on the growth and secretion properties of eukaryotic yeast. In contrast, the *in vitro* display technologies are accomplished completely *in vitro* which allows for significant manipulation of the selection conditions and much larger libraries.

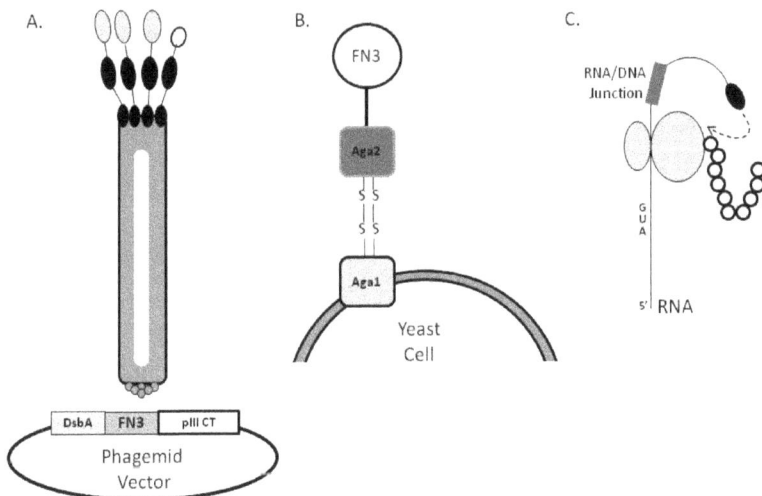

Fig. 2. Display platforms used for engineering and selection of FN3 domains. A) Phage display using a phagemid vector where the FN3 domain is fused to the C-terminal domain of M13 pIII. The signal sequence used to direct secretion is the DsbA sequence. B) Yeast display platform involving fusion of the FN3 domain to the yeast cell membrane protein Aga2p. Aga2p associates with membrane bound Aga1 via disulfide linkages. C) mRNA display involving an RNA/DNA hybrid wherein the RNA encodes the FN3 domain; the ribosome is depicted with the 2 shaded ovals with the transcribed protein illustrated by the hollow circles. Puromycin is depicted with a solid oval with the linkage site to the nascent protein strand shown by a dotted arrow.

Koide and coworkers first demonstrated the utility of the FN3 domain as a scaffold suitable for creation of libraries of variants (Koide, et al., 1998). Based on analysis of FN3 domains involved in protein interaction motifs that suggested loop residues were frequently involved in binding, a library containing randomized sequences in the FG and BC loops was designed. The FN3 scaffold library was fused to the C-terminal fragment of pIII using a phagemid vector and pIII display was achieved by infection with helper phage. Following five rounds of selection, a ubiquitin specific binder with approximately µM affinity for ubiquitin was isolated. Despite reduced stability and solubility compared to the native parental FN3 domain from fibronectin, this study set the groundwork for future studies expanding the applications of FN3 domains to therapeutic proteins. Richards and coworkers in the Koide lab subsequently demonstrated display and selection of FN3 domains on pVIII using a phagemid display vector (Julie Richards et al., 2003). Because of the relatively low affinity of the parent FnFN10 domain for integrin binding, multivalent pVIII display was used to identify a high affinity, selective binder to $\alpha v \beta 3$. The selected FN3 domain inhibits $\alpha v \beta 3$ dependent processes *in vitro* in a manner similar to that described for antibodies thereby providing additional support for the utility of FN3 scaffolds as antibody-mimetics.

In 2006, Steiner and coworkers reported that highly stable, highly expressed proteins often demonstrate compromised display from standard phage display vectors (Steiner et al., 2006). For filamentous phages, translated proteins need to traverse the bacterial wall to the periplasm before assembling into the mature phage particle. Transolocation is triggered by the signal sequences that precede the mature gene sequence and most standard phage display vectors rely on Sec-dependent translocation sequences (ie. OmpA). Steiner et al. showed that exchange of the phage-fusion protein Sec-dependent signal sequence to the cotranslational signal recognition particle (SRP) translocation pathway enabled display improvements of up to 1000-fold. As illustrated in Figure 2A, Koide and coworkers applied this strategy to develop an improved phage display vector employing the DsbA signal sequence instead of the OmpA sequence used in earlier work. (Koide et al., 2007)

In vitro display strategies have proved to be a valuable tool for selection of high affinity scFvs, protease resistant peptides, and other scaffold protein libraries (Binz et al., 2005, Eldridge et al., 2009, Hanes & Pluckthun, 1997). Xu and coworkers applied a novel mRNA display strategy to construct a library based on the structural analogy between FN3 domains and antibody V-domains (Xu et al., 2002). The library encoded 21 random residues in the BC, DE, and FG loops of the FnFN10 domain with a theoretical diversity of 20^{21}. While one advantage of *in vitro* display libraries is the size of libraries it is possible to generate, the practical size of such libraries is $\sim 10^{12}$. The Xu library was constructed in multiple pieces using PCR and random oligonucleotides to introduce diversity in to the loop regions. To remove non-productive sequences from the library, each segment was fused in frame to a T7 promoter and amino and carboxy terminal tags and the translated product subjected to affinity purification on the appropriate affinity tag ligands. Productive library members were recovered by PCR and joined to prepare the full length library. For mRNA display, a library of dsDNA is transcribed to generate mRNA which is ligated to a puromycin containing oligonucleotide (Takahashi, 2003)(Figure 2B). An *in vitro* translation reaction is performed to yield the RNA/DNA/protein complex that can be selected for binding. The final step in the selection cycle is to recover the mRNA-protein complexes by PCR to regenerate an enriched dsDNA library. Xu et al. selected high affinity, highly selective inhibitors to TNFα thereby confirming that the FnFN10 scaffold had broad utility for protein therapeutics and diagnostics applications. Reports from the Roberts lab have described mRNA display selections for FN3 binders in a variety of applications involving both intracellular and extracellular targets (Liao et al., 2009, Olson & Roberts, 2007).

Odegrip and coworkers have described the optimization of an *in vitro* display system that addresses the RNA instability challenges seen with mRNA and ribosome display (Odegrip et al., 2004). The method, termed CIS display, makes use of the property of the DNA replication initiator protein (RepA) to bind exclusively to the DNA from which it has been expressed (ie. in cis). Recently, Jacobs and coworkers have applied CIS display for identification of stable, high affinity binders from a consensus sequence based on FN3 domains from human tenascin C (manuscript in preparation).

One of the limitations of phage display systems is their reliance on bacterial expression. Wittrup and coworkers have exploited the eukaryotic secretion machinery in yeast to develop a display platform that can be screened to discriminate between domains of different stability and different affinity (Shusta et al., 1999) (Figure 2C). One advantage of yeast display is the ability to use fluorescence activated cell sorting (FACS) to sort specific clones on the basis of scaffold display level and binding activity. Yeast display has the disadvantage of limiting library diversity to ~10^8 due to the much lower transformation efficiency of yeast compared to bacteria. Despite the limitations on library size dictated by yeast, the studies by Lipovsek et al. demonstrate the utility of yeast display for maturation of high affinity binders (Lipovsek et al., 2007). Randomizing only two loops (14 residues) of the ^{10}Fn3 domain and using libraries several orders of magnitude smaller than those possible with mRNA display, Lipovsek was able to select binders to TNFα with sub-nanomolar affinity, a 340-fold improvement from the parental clone. Hackel and coworkers have extended the application of yeast display with more complex library designs building on the basic principle of improved display correlating with improved stability. Using FACS to sort clones on the basis of display level and binding activity together with a step-wise affinity maturation scheme that relies on the yeast homologous recombination machinery for *in vivo* shuffling, stable, high affinity binders to a variety of targets were identified (Hackel et al., 2010, Hackel et al., 2008a).

3. Engineering for stability and solubility

One of the significant advantages of alternative scaffolds compared to antibodies is the improved biophysical properties for scaffold domains. Indeed, the stability and solubility properties of FN3 domains are an important point of differentiation that may allow for increased tissue penetration, reduced immunogenicity, and development of high concentration formulations. Often, sequence modifications to very small stable protein domains can have a significant impact on protein stability and solubility. Thus, researchers working on protein scaffolds have developed strategies to maximize and screen for domain stability as part of the selection process.

As the originators of the FN3 scaffold design work, Koide and coworkers were early leaders in exploring domain stability. In contrast to the generally accepted premise that surface electrostatic interactions have little role in overall protein stability, the studies by Koide demonstrate that a negatively charged patch on the surface of FnFN10 plays an important role in limiting the stability of the domain (Koide et al., 2001). Mutation of one of the negatively charged residues to a neutral asparagine residue or to a positively charged lysine residue increased the T_m for the proteins by 7° or 9°C compared to the native domain. These studies suggest the potential for improved stability of selected FnFN10 domains by engineering surface residues.

As researchers began to consider introduction of diversity into a variety of alternative scaffold domains to enable selection of novel therapeutic and diagnostic candidates, Batori and coworkers systematically explored the effects of loop elongation on each of the loops in FnFN10 (Batori et al., 2002). In order to determine which loops of the domain might be best utilized for target binding, they assessed the impact of elongation on the conformational stability of the domain. While it was possible to introduce up to four additional glycine residues in all six loops of FnFN10 while retaining the global fold, EF loop elongation was highly destabilizing. Mutations in the other five loops had only modest destabilizing effects suggesting the potential to use the loops for engineering binding affinity.

Dutta et al. explored the potential to apply yeast two-hybrid fragment complementation to select FN3 domains with improved stability (Dutta et al., 2005). To assess complementation, FN3 fragment pairs were designed wherein the C-terminal fragment of each pair was fused to the LexA DNA binding domain and the N-terminal fragment was fused to the B42 activation domain. A yeast two-hybrid β-galactosidase assay was used to evaluate each of the combinations for the ability to regenerate a highly stable domain. The scientists hypothesized that an increase in the affinity of fragment complementation would increase the stability of the uncut parental protein. Using three previously identified destabilizing mutations, a library of mutants was designed with the intent to select for compensating mutations in the opposite fragment. Mutations obtained from the selection re-introduced into the wild type and mutant domains resulted in ~2 kcal/mol increase in stability and demonstrated the utility of the fragment complementation method for identification of stabilizing mutations.

Stability and solubility properties of candidate therapeutics are of key importance for successful development. From a chemistry, manufacturing and controls (CMC) perspective, the stability properties of a biotherapeutic candidate often play a major role in the design of purification strategies, protease sensitivity and storage conditions. For example, an early step in the purification of bacterially expressed FN3 domains with a high melting temperature can involve heating the bacterial lysate to > 60°C where most host proteins unfold irreversibly. The FN3 domain remains in solution and significant purification is achieved. For storage of final product, high stability products offer the potential advantage of room temperature storage. Parker and coworkers (Parker et al., 2005) reported on a strategy to improve the thermal stability of selected FN3 domains. In previous experiments, two anti-VEGF-R2 binding clones were identified with only six amino acid differences between sequences. Nevertheless, there were significant differences in the biophysical properties of the clones. Clone VR28 had a relatively low affinity of 13 nM for VEGR2 and relatively higher thermal stability of 62°C while clone 159 was higher affinity with a Kd of 0.34 nM for VEGF-R2 and a thermal stability of 32-52°C. In addition, the two clones exhibit significant differences in solubility and aggregation state. Examination of the sequences for the wt and two FnFN3 derivatives enabled design of mutants with high affinity (0.59 nM) and high thermal stability (59°C).

Olson and Roberts took a novel approach to identifying FN3 domain variants with good stability and solubility properties using a green fluorescent protein (GFP) reporter screen

(Garcia-Ibilcieta et al., 2008), SDS-PAGE analysis and chemical denaturation (Olson & Roberts, 2007). A library comprising 7 random residues in the BC loop and 10 random residues in the FG loop was synthesized by overlap PCR and inserted in frame between a T7 promoter and a tobacco mosaic virus translation enhancer for mRNA display. The library was also cloned into a GFP fusion vector for assessment of library quality (ie. in frame sequences, no stop codons) and folding. Approximately 45% of the library sequences appeared to be in frame with no stop codons. Given the large size of the library (> 10^{13} sequences), the functional size remains significantly larger than is possible with other display methods. Overall fluorescence intensities of "functional" clones populated a continuum of fluorescence values from <1%-130% of the parental domain standard. Approximately 20% of the library had fluorescence values between 80%-130%, suggesting that these variants are well folded with an additional 30% of the library exhibiting fluorescence values consistent with potential utility. This study demonstrates the value of GFP screening to evaluate complex FN3 domain libraries for library quality and maintenance of proper folding.

Olson and coworkers applied the GFP screening strategy for identification of FN3 scaffold IkBa binders with improved solubility (Olson et al., 2008). Three rounds of evolution using error-prone PCR on a low solubility binder were performed. Following each round of evolution, approximately 2000 colonies were screened for increased fluorescence using the GFP fusion technology described above. By combining two point mutations correlated with improved solubility, Olson was able identify an FN3 clone with expression improved to ~ 1mg/L culture while maintaining affinity. The improvements in solubility and binding suggest that the overall stability of the domain could be improved with only modest changes to the overall sequence.

Most recently, a report by Jacobs et al. describes the use of consensus design to prepare a highly stable FN3 domain (Jacobs, et al. 2012). Following an alignment of the 15 FN3 domains found in the human protein tenascin C or 15 FN3 domains from human fibronectin, the researchers designed an alignment based on the most conserved amino acid at each position. The resulting consensus domains (Tencon, Tenascin consensus, or Fibcon, Fibronectin consensus) demonstrate excellent stability with T_ms of 78°C and 90°C respectively and adopt the designed fold as confirmed by x-ray crystallography. A series of mutations were designed to increase the stability of Tencon even further by consideration of core packing and secondary structure preferences for amino acids. Several of the mutations which were shown to improve domain stability individually were combined to yield scaffold with significantly improved stability and a T_m ~ 93°C. The authors note that preliminary data suggests that high affinity, high stability Tencon FN3 domain binders have been identified from libraries based on the engineered consensus domain and that the improved stability and biophysical properties of library members can be traced to the increase stability of the parent scaffold.

4. Engineering to improve binding affinity

The compact, simple structure and exceptional stability of the FN3 domain has enabled the development of numerous methods for the engineering of high-affinity binders. The engineering of an FN3 domain to specifically bind to a non-natural target protein was first described by Koide and colleagues, who engineered the FnFN10 to bind specifically to ubiquitin (Koide, et al., 1998). In this seminal paper, the authors introduced random amino acids into the BC and FG loops of FN3 and selected binders via phage display. The authors chose to randomize these particular loops due to the sequence variations of these loops among FN3 family members as well as the fact that FG and/or BC loops have been shown to be responsible for interactions with integrins and the human growth hormone receptor (de Vos et al., 1992, Main, et al., 1992). One variation found in FN3 structure is the length of the loops connecting the β-strands. Koide et al. randomized replaced the BC loop with 5 random amino acids, the same length as that of the native FnFN10, while the randomized FG loop was shortened from 8 residues to 5. The resulting library was estimated to be 10^8 in complexity. After selection by phage display, one dominant clone was identified that bound specifically to ubiquitin, albeit with low affinity (~ 5 μM). Alanine scanning experiments demonstrated that residues of both the BC and FG loop contribute to ubiquitin binding.

A similar FnFN10 phage display library randomizing 5 residues in each of the BC and FG loops was displayed on M13 phage, this time with a complexity of 2×10^9 (Karatan et al., 2004). After three rounds of selection against the SH3 domain of human c-Src, 6 variants were found that bound to the target specifically, with affinities ranging from 0.25 to 1.3 μM. Most of the clones discovered contained proline rich loops, reminiscent of the binding of natural proteins to SH3 domains (Olson, et al., 2008). As with the previous example, both the BC and FG loops were found to contribute to SH3 binding.

A more complex library consisting of a mixture of three separately made libraries in which the FG loop, the BC and FG loops, and the BC, FG, and DE loops were randomized was produced for selection using the mRNA display system (Dineen et al., 2008, Xu, et al., 2002). This library design randomized 21 FN3 residues simultaneously, 10 residues from the FG loop, 7 residues from the BC loop, and 4 residues from the DE loop. Because of the use of an *in vitro* display system, a library of very high complexity (10^{12}) was produced. The preselected library was panned against TNFα which was coupled to sepharose beads and screening for binding completed after 9 and 10 rounds of selection. After 10 rounds, a diverse set of TNFα binding sequences were identified with K_D values ranging from 1-24 nM. In order to drive the selection process to higher affinity, 4 additional rounds of selection were completed either with or without the incorporation of error-prone PCR to increase diversity. The tightest binder obtained (K_D 20 pM) was found from the selection strategy incorporating error-prone PCR, indicating that adding diversity to a pool of enriched binders can lead to higher affinity binding. The effect on binding of framework mutations outside of the randomized loops that might arise from error-prone mutagenesis was not investigated in this study. Interestingly the tightest binding clone found had an FG loop that was truncated to 4 residues, highlighting the importance of minimizing entropic factors when producing high affinity interactions.

The same FnFN10 mRNA Display library used to isolate TNFα binders was also utilized to select FN3 domains that bind to the extracellular domain of VEGFR2 (Getmanova et al., 2006). After 6 rounds of selection, clones binding to VEGFR2 that could compete with the binding of VEGF to this receptor were identified. As the affinities of clones from the original rounds of selection showed only modest affinity (~ 10 nM), an affinity maturation strategy employing hypermutagenic PCR to randomize the loops of the best binder from round 6 was employed. Although this technique led to only slight improvements in affinity, 2 serendipitous mutations in the N-terminal A-strand were found to improve the binding to VEGFR2, leading to a second affinity maturation strategy where the FG loop was further diversified in combination with truncation of the N-terminus. This alternative strategy led to the discovery of anti-VEGFR2 FN3 molecules with affinity as high as 320 pM.

The previous examples utilizing mRNA display demonstrate the power of large libraries to generate high affinity binding molecules. However, several studies have shown that sub-nanomolar binding affinities can be obtained even with smaller functional libraries. Lipovsek et al. used yeast surface display (Boder & Wittrup, 1997) to select lysozyme binders from FnFN10 libraries in which 7 BC loop and 7 FG loop residues were randomized simultaneously or individually (Lipovsek, et al., 2007). In this case, triplet codons were used to generate DNA diversity, thus incorporating all 20 amino acids and no stop codons into the functional library. 1 round of magnetic-activated cell sorting and 3 rounds of fluorescence-activated cell sorting were used to select yeast displaying FN3 molecules that bound to lysozyme. No enrichment was obtained from a library with only the BC loop randomized and specific binders were selected from all other libraries. Affinity maturation libraries in which 1 loop of a binder was fixed and the other randomized were produced and subjected to 4 more rounds of FACS sorting. FN3 domains binding to lysozyme with affinities as high as 350 pM were obtained after maturation. Interestingly, the majority of FN3 molecules found after affinity maturation, and all of the tightest binders, were found to have conserved cysteine residues in each of the BC and FG loops, which presumably stabilize the confirmations of the loops by forming a disulfide bond leading to a decrease in the entropic penalty of binding and resulting in higher affinities. Similarly, disulfides bonds between CDR1 and CDR3 have been found to stabilize camelid single domain antibodies (Dong et al., 2011, Sweeney et al., 2008, Tanha et al., 2001).

A series of studies from the Wittrup lab investigated the effect of amino acid composition and loop length on the ability to select high affinity binders by yeast surface display (Hackel, et al., 2010, Hackel et al., 2008b, Hackel & Wittrup, 2010). A library, produced by PCR and recombination in yeast, randomized the BC, DE, and FG loops of FnFN10 simultaneously. Diversity was achieved not only by incorporating random amino acids via NNB diversity, but also by varying the loop lengths within the library from 4-7 residues (DE), 6-9 residues (BC), and 5-10 residues (FG) (Hackel, et al., 2008b). As this library was displayed on yeast and could only be made to a diversity of 2.3×10^7 clones, additional diversity was incorporated between rounds of sorting using error-prone PCR of the loops in combination with loop shuffling. 3 rounds of selection against lysozyme followed by 4 additional rounds with affinity maturation and decreasing concentrations of target protein were employed to produce FN3 molecules with binding affinities ranging from 1 uM to 2 pM. Sequence analysis of the selected clones suggests that both loop shuffling and error-prone PCR can contribute to increased affinities. Mutagenesis of the highest affinity binder indicated that binding is dominated by the BC and FG loop and revealed that framework mutations derived from error-prone PCR can have a large effect on affinity. The highest affinity binders were all found to have length deviations compared to wild-type FnFN10 in at least one of the loops, highlighting the potential benefits of incorporating loop length diversity into a library. In support of this idea, loop length diversity was also found to be important in selecting FnFN10 variants against MBP, hSUMO4, and ySUMO4 using a minimalist library composed of only Tyr and Ser at diversified positions (Koide, et al., 2007).

It will be interesting to examine if dominant loop lengths emerge as these libraries are used to select binders to additional targets.

A variation of this yeast display library was produced in which the diversity of certain FnFN10 loop positions thought to play a role in protein stability were biased towards that of the WT FnFN10 protein in order to maximize the stability and folding of library members (Hackel, et al., 2010). Three library compositions were compared, the first in which the BC, DE, and FG loops were completely randomized, a second in which certain positions in the DE loop were biased towards WT sequences and the remaining loop sequences randomized as 50% Ser, 50% Tyr, and finally a third design in which structurally important residues in the BC and FG loop were biased towards WT sequences and the diversity of the remaining residues was tailored to the diversity found in CDR3 of antibodies (Zemlin et al., 2003). Yeast surface display levels of random clones from these libraries confirmed that molecules from the 3rd library were generally more stably folded, demonstrating the benefit of incorporating WT sequences into certain loop positions. For analysis, these three individual libraries were pooled and selections carried out against 7 different targets: human A33, mouse A33, epidermal growth factor receptor (EGFR), mouse IgG, human serum albumin, Fcγ receptor IIA and Fcγ receptor IIIA. After selections by yeast surface display, binders were selected against each target, although only one binding sequence was found for several of the targets. 90% of the recovered sequences were derived from the 3rd library design described above, evidence that the improved stability of this library compared to the others leads to easier selection of binders. Due to the complex nature of the designs and the mixing of the libraries for selections, it is not possible to determine if the increased binding efficiency of this library is due to the stabilizing mutations or to the use of diversity tailored to resemble the CDRs of antibodies.

An alternative approach to achieve high affinity relies not on the design of FN3 libraries, but solely on an increase in avidity for target binding (Duan et al., 2007). A FnFN10 variant with an enhanced RGD sequence in the FG loop that has improved affinity for αvβ3 integrin (J. Richards et al., 2003) was produced as a fusion protein to the COMP assembly domain, which forms a pentamer via a five-stranded a-helical bundle (Malashkevich et al., 1996). In this construct, the FN3 domain was extended from the COMP domain with a 20 amino acid helical linker followed by a 25 residue flexible linker. The majority of the fusion protein produced in E. coli was found to be pentameric, however monomers, dimers, trimers, and tetramers could also be detected by SDS-PAGE. Biacore analysis demonstrated a > 100-fold decrease in k_{off} for the pentameric FN3 form in comparison to the monomeric form, indicating an avidity effect in binding.

The studies reviewed here collectively show that a wide range of FN3 library designs can lead to the selection of high-affinity binding molecules. The choice of loops to be randomized, amino acid composition of the randomizations, length of randomized loops, specific positions randomized within the loop, and choice of affinity maturation strategy can all have a large effect on the quality and affinity of binders isolated. It has been demonstrated that large in vitro display libraries or comparatively smaller phage display and smaller yet yeast surface display libraries can all be utilized effectively. No consensus has yet emerged as to which loop design strategy yields the best binders, although current studies indicate that the FG and BC loops are most important for target binding, while the DE loop may help to stabilize the confirmations of the other loops. Strict comparisons of the effectiveness of the different library designs described are difficult as the target molecules as well as the display systems and selection methods utilized are often different between studies. However, the pace of published studies describing the selection of FN3 binding reagents has quickened considerably since the first description by Koide *et al*. We anticipate that as more detailed characterizations for FN3 binders are reported, rules governing library design for this scaffold class will begin to emerge.

Targets	Display System	Library	Maturation Strategy	Highest Affinity	Reference
Ubiquitin	P	BC5 + FG5	none	~ 5 µM	Koide, Bailey et al. 1998
Src SH3	P	BC5 + FG5	none	250 nM	Karatan, Merguerian et al. 2004
TNF-α	R	BC7 + DE4 + FG10	error prone PCR	20 pM	Xu, Aha et al. 2002
VEGFR2	R	BC7 + DE4 + FG10	hypermutagenic PCR, N-terminal truncation	320 pM	Getmanova, Chen et al. 2006
MBP, hSUMO4, ySUMO4	P + Y	Tyr/Ser BC(6-10) + DE(4-10) + FG(9-13)	none	5 nM	Koide, Gilbreth et al. 2007
lysozyme	Y	BC7, FG7, BC7 + FG7	re-randomize BC loop	350 pM	Lipovsek, Lippow et al. 2007
lysozyme	Y	BC(6-9) + DE(4-7) + FG(5-10)	error prone PCR + loop shuffling	1.1 pM	Hackel, Kapila et al. 2008
Human A33, mouse A33, EGFR, mIgG, HSA, FcγRIIA and FcγRIIIA.	Y	various loop compositions	error prone PCR + loop shuffling	250 pM	Hackel, Ackerman et al. 2010
rabbit IgG, goat IgG	Y	various loop compositions and lengths	error prone PCR + loop shuffling	51 pM	Hackel and Wittrup 2010
Abl kinase SH2 domain	P		none		Wojcik, Hantschel et al. 2010
Phosphorylated IκBα	R	BC7 + FG10	none	18 nM	Olson, Liao et al. 2008
N Protein of SARS	R	BC7 + FG10	none	1.7 nM	Liao, Olson et al. 2009
Estrogen Receptor α	H	FG7, AB7	none	not reported	Koide, Abbatiello et al. 2002

Table 1. Selected binders from naïve FN3 libraries. Display systems are abbreviated as P for phage display, Y for yeast surface display, R for mRNA display, and H for yeast 2-hybrid. Numbers in parenthesis after loop descriptions refer to the length of randomized loops. Tyr/Ser refers to a loop composition made entirely of tyrosine and serine amino acids.

5. Applications

Over the last decade since the first reports on the utility of FN3 domains as antibody mimics or alternative scaffold proteins, this class of proteins has been engineered for a variety of applications. Affinities of FN3 binders have now been described with values and specificity similar to those obtained for antibody antigen interactions. The inherent biophysical properties of FN3 domains and the development of strategies for stability and affinity evolution have extended the utility of the scaffold. On the basis of promising preclinical tumor xenograft studies for an anti VEGFR2- FN3 domain (Dineen, et al., 2008), the first FN3-based therapeutic has been evaluated in human patients in a Phase 1 study (Molckovsky & Siu, 2008, Sweeney, et al., 2008). Initial pharmacokinetic studies for a PEGylated anti-VEGFR2 FN3 domain dosed intravenously have demonstrated that the scaffold has a terminal half-life of ~69 hours with a maximum tolerated dose of 2 mg/kg weekly. No clinically significant immunogenicity was observed and a biomarker for VEGFR2 demonstrated target engagement. These studies provide the foundation for advancement of additional FN3 domain scaffolds into the clinic.

One of the many attractive features of FN3 domain scaffolds is the ability to link domains with different specificities together to form multi-specific therapeutics. As many inflammatory and oncologic diseases have been demonstrated to be dependent on activities in multiple pathways, the ability to create multi-specific inhibitors has significant potential for biotherapeutics. Emanuel and coworkers recently provided the first report of a bispecific FN3-domain inhibitor (Emanuel et al., 2011). Adnectins™ with high affinity to epidermal growth factor receptor (EGFR) and insulin-like growth factor-1 receptor (IGF-1R) were selected using mRNA display. Given that both receptors mediate proliferative and survival cell signaling in cancer, bispecific constructs with Adnectins™ fused in tandem via a flexible glycine-serine linker were evaluated *in vitro*. In *in vivo* studies, pegylated bispecific constructs inhibited growth of EGFR and IGF-1R driven tumor xenografts, induced degradation of EGFR and reduced EGFR phosphorylation. These studies demonstrate that engineering of multi-specific FN3 domains as anti-tumor agents may have the potential to improve efficacy over mono-specific biologic therapies.

Active sites of proteins are often located at domain interfaces in multi-domain proteins. Huang et al. explored the potential to engineer a novel interdomain "active site" by linking a low-affinity peptide binding domain (the primary domain) with an unrelated FN3 domain (the enhancer domain) and evolving for high affinity, high specificity binding (Huang et al., 2008). Evolution was initiated from the Erbin PDZ domain that has micromolar affinity for the C-termini of p-120-related catenins. An FN3 library was linked to the PDZ domain to create a two-domain protein interface not naturally observed. The strategy, termed "directed domain interface evolution" was used to select a novel protein that bound to the target peptide with an affinity enhancement of nearly 500 fold significantly greater than that attainable by evolution of the PDZ domain alone. The x-ray crystal structure showed a clamshell structure quite similar to the intended design. The novel "affinity-clamp" protein demonstrates the utility of FN3 domains as affinity reagents that may be useful in a variety of applications. Huang and Koide (Huang & Koide, 210) demonstrated the utility of the affinity-clamp strategy to monitor a peptide biomarker using FRET-optimized fluorescent proteins. Fusion of one fluorescent protein to the clamp PDZ domain and a second fluorescent protein to the FN3 domain allowed for strong FRET activity for the unbound affinity clamp. In the presence of target peptide, the affinity clamp undergoes a significant conformational change and a large decrease in FRET. The researchers demonstrated that peptide concentration in crude cell lysates could be readily quantified over 3 orders of magnitude using FRET measurements thus providing a general strategy for monitoring peptide motifs via such label-free sensors. Olson and coworkers took a similar approach using mRNA display to select high-affinity phosphorylation-specific IκBα-binding FN3 domains (Olson, et al., 2008). The selected domain specifically recognized endogenous phosphorylated IκBα from mammalian cell extracts. The utililty of such specific reagents for monitoring kinase activity was demonstrated by creation of n IκB kinase (IKK) FRET biosensor. To engineer the biosensor, cyan fluorescent protein (CFP) was fused at one end of the FN3 domain and yellow fluorescent protein (YFP) at the other end. The IκBα peptide sequence, an IKK substrate, was inserted between the YFP and the FN3 domain where it is readily phosphorylated in the presence of IKK. Binding of the phospho-peptide to the FN3 domain induces a conformational change bringing the two fluorescent domains close enough to enable FRET.

The absence of disulfide bonds in FN3 domains is an attractive feature that suggests the potential for these domains to function intracellularly. Liao and coworkers demonstrated this potential with selection of FN3 domains that bind to the nucleocapsid protein (N) from several acute respiratory syndrome (SARS) coronavirus (Liao, et al., 2009). The researchers demonstrated that intrabodies are well expressed in mammalian cells and that they co-localize with the N-protein. Most significantly, they were able to demonstrate that selected intrabodies inhibited virus replication and that intrabodies binding two unique sites on N-protein can synergize to inhibit virus replication. This work represents the first demonstration of intracellularly functional FN3 domains and provides the framework for expanding the utility of FN3 scaffolds beyond extracellular targets. Ishikawa and coworkers from the same lab used the selected SARS N-protein binders to prepare nanowire/nanotube biosensors (Ishikawa et al., 2009). In2O3 nanowire based biosensors were configured with the SARS N-protein FN3 domain binder to enable detection of subnanolar concentrations of N-protein in the presence of high concentrations of background protein. The concentration dependent binding of N-protein demonstrates the utility of biosensors modified with engineered FN3 domains for selective and sensitive detection of biomarker proteins.

A particularly elegant application of FN3 domain binders as fluorescent biosensors was recently reported by Gulyani et al. (Gulyani et al., 2011) A Src-family kinase (SFK) biosensor was prepared using a FN3 domain specific for activated SFKs. Binding of the FN3 domain to its target was monitored by attachment of a bright, environmentally sensitive fluorescent dye to the scaffold protein where target protein binding was shown to increase fluorescence. Here again, the lack of disulfide bonds in FN3 domains provides the ability for the biosensor to function in living cells where SRC activation dynamics can be monitored. Using automated image analysis, the biosensor showed specific activation of SFKs during protrusion with a level of activity proportional to the velocity of the extending edge. Such studies may be extended to other cellular activities via the combination of high throughput FN3 domain selections, intracellular functionality, optimization of fluorescent dye attachment and biosensor engineering.

6. Conclusion

The applications described above represent an evolving landscape for FN3 scaffold domains with strategies designed to take advantage of the novel features of such domains. The advent of multiple display platforms together with the development of strategies for improving stability, solubility and affinity of FN3 domain binders has laid the foundation for novel applications of the platform. We anticipate that bispecific FN3 domain scaffolds will become an important therapeutic modality with the potential for increasing specificity and efficacy. In addition, the chemical properties of FN3 domains will likely extend their utility to intracellular targets. Increasingly, diagnostics and biosensors based on FN3 domains are expected to play a role in monitoring biomarkers, intracellular signaling networks and providing high sensitivity images of intracellular activities. Finally, we expect that the biophysical properties of the FN3 scaffolds will play an important role in novel targeting and delivery technologies.

7. References

Batori, V., Koide, A. & Koide, S. (2002). Exploring the potential of the monobody scaffold: effects of loop elongation on the stability of a fibronectin type III domain. *Protein Eng*, Vol.15, No.12, (December 2002), pp. 1015-1020.

Best, R.B., Rutherford, T.J., Freund, S.M. & Clarke, J. (2004). Hydrophobic core fluidity of homologous protein domains: relation of side-chain dynamics to core composition and packing. *Biochemistry*, Vol.43, No.5, (February 2004), pp. 1145-1155.

Billings, K.S., Best, R.B., Rutherford, T.J. & Clarke, J. (2008). Crosstalk between the protein surface and hydrophobic core in a core-swapped fibronectin type III domain. *J Mol Biol*, Vol.375, No.2, (January 2008), pp. 560-571.

Binz, H.K., Amstutz, P. & Pluckthun, A. (2005). Engineering novel binding proteins from nonimmunoglobulin domains. *Nat Biotechnol*, Vol.23, No.10, (October 2005), pp. 1257-1268.

Boder, E.T. & Wittrup, K.D. (1997). Yeast surface display for screening combinatorial polypeptide libraries. *Nat Biotechnol*, Vol.15, No.6, (June 1997), pp. 553-557.

Bork, P. & Doolittle, R.F. (1992). Proposed acquisition of an animal protein domain by bacteria. *Proc Natl Acad Sci U S A*, Vol.89, No.19, (October 1992), pp. 8990-8994. ISSN 0027-8424

Clarke, J., Cota, E., Fowler, S.B. & Hamill, S.J. (1999). Folding studies of immunoglobulin-like beta-sandwich proteins suggest that they share a common folding pathway. *Structure*, Vol.7, No.9, (September 1999), pp. 1145-1153.

Clarke, J., Hamill, S.J. & Johnson, C.M. (1997). Folding and stability of a fibronectin type III domain of human tenascin. *J Mol Biol*, Vol.270, No.5, (August 1997), pp. 771-778.

Cota, E. & Clarke, J. (2000). Folding of beta-sandwich proteins: three-state transition of a fibronectin type III module. *Protein Sci*, Vol.9, No.1, (January 2000), pp. 112-120. ISSN 0961-8368

Cota, E., Hamill, S.J., Fowler, S.B. & Clarke, J. (2000). Two proteins with the same structure respond very differently to mutation: the role of plasticity in protein stability. *J Mol Biol*, Vol.302, No.3, (September 2000), pp. 713-725.

de Vos, A.M., Ultsch, M. & Kossiakoff, A.A. (1992). Human growth hormone and extracellular domain of its receptor: crystal structure of the complex. *Science*, Vol.255, No.5042, (January 1992), pp. 306-312. ISSN 0036-8075

Dineen, S.P., Sullivan, L.A., Beck, A.W., Miller, A.F., Carbon, J.G., Mamluk, R., Wong, H. & Brekken, R.A. (2008). The Adnectin CT-322 is a novel VEGF receptor 2 inhibitor that decreases tumor burden in an orthotopic mouse model of pancreatic cancer. *BMC cancer*, Vol.8, 2008), pp. 352. ISSN 1471-2407

Dong, J., Sereno, A., Aivazian, D., Langley, E., Miller, B.R., Snyder, W.B., Chan, E., Cantele, M., Morena, R., Joseph, I.B.J.K., Boccia, A., Virata, C., Gamez, J., Yco, G., Favis, M., Wu, X., Graff, C.P., Wang, Q., Rohde, E., Rennard, R., Berquist, L., Huang, F., Zhang, Y., Gao, S.X., Ho, S.N., Demarest, S.J., Reff, M.E., Hariharan, K. & Glaser, S.M. (2011). A stable IgG-like bispecific antibody targeting the epidermal growth factor receptor and the type I insulin-like growth factor receptor demonstrates superior anti-tumor activity. *mAbs*, Vol.3, No.3, 2011), pp. 273-288. ISSN 1942-0862

Duan, J., Wu, J., Valencia, C.A. & Liu, R. (2007). Fibronectin type III domain based monobody with high avidity. *Biochemistry*, Vol.46, No.44, (November 2007), pp. 12656-12664. ISSN 0006-2960

Dutta, S., Batori, V., Koide, A. & Koide, S. (2005). High-affinity fragment complementation of a fibronectin type III domain and its application to stability enhancement. *Protein Sci*, Vol.14, No.11, (November 2005), pp. 2838-2848.

Eldridge, B., Cooley, R.N., Odegrip, R., McGregor, D.P., Fitzgerald, K.J. & Ullman, C.G. (2009). An in vitro selection strategy for conferring protease resistance to ligand binding peptides. *Protein engineering, design & selection : PEDS*, Vol.22, No.11, (November 2009), pp. 691-698. ISSN 1741-0134

Emanuel, S.L., Engle, L.J., Chao, G., Zhu, R.-R., Cao, C., Lin, Z., Yamniuk, A.P., Hosbach, J., Brown, J., Fitzpatrick, E., Gokemeijer, J., Morin, P., Morse, B.A., Carvajal, I.M., Fabrizio, D., Wright, M.C., Das Gupta, R., Gosselin, M., Cataldo, D., Ryseck, R.P., Doyle, M.L., Wong, T.W., Camphausen, R.T., Cload, S.T., Marsh, H.N., Gottardis, M.M. & Furfine, E.S. (2011). A fibronectin scaffold approach to bispecific inhibitors of epidermal growth factor receptor and insulin-like growth factor-I receptor. *mAbs*, Vol.3, No.1, 2011), pp. 38-48. 1942-0862

Finkelstein, A.V. (1991). Rate of beta-structure formation in polypeptides. *Proteins*, Vol.9, No.1, 1991), pp. 23-27.

Garcia-Ibilcieta, D., Bokov, M., Cherkasov, V., Sveshnikov, P. & Hanson, S.F. (2008). Simple method for production of randomized human tenth fibronectin domain III libraries for use in combinatorial screening procedures. *BioTechniques*, Vol.44, No.4, (April 2008), pp. 559-562. ISSN 0736-6205

Getmanova, E.V., Chen, Y., Bloom, L., Gokemeijer, J., Shamah, S., Warikoo, V., Wang, J., Ling, V. & Sun, L. (2006). Antagonists to human and mouse vascular endothelial growth factor receptor 2 generated by directed protein evolution in vitro. *Chem Biol*, Vol.13, No.5, (May 2006), pp. 549-556.

Gulyani, A., Vitriol, E., Allen, R., Wu, J., Gremyachinskiy, D., Lewis, S., Dewar, B., Graves, L.M., Kay, B.K., Kuhlman, B., Elston, T. & Hahn, K.M. (2011). A biosensor generated via high-throughput screening quantifies cell edge Src dynamics. *Nature Chemical Biology* Vol.7, 2011), pp. 437-444.

Hackel, B.J., Ackerman, M.E., Howland, S.W. & Wittrup, K.D. (2010). Stability and CDR composition biases enrich binder functionality landscapes. *Journal of molecular biology*, Vol.401, No.1, (August 2010), pp. 84-96. ISSN 1089-8638

Hackel, B.J., Kapila, A. & Wittrup, K.D. (2008a). Picomolar affinity fibronectin domains engineered utilizing loop length diversity, recursive mutagenesis, and loop shuffling. *J Mol Biol*, Vol.381, No.5, (September 2008), pp. 1238-1252.

Hackel, B.J. & Wittrup, K.D. (2010). The full amino acid repertoire is superior to serine/tyrosine for selection of high affinity immunoglobulin G binders from the fibronectin scaffold. *Protein engineering, design & selection : PEDS*, Vol.23, No.4, (April 2010), pp. 211-219. ISSN 1741-0134

Hamill, S.J., Cota, E., Chothia, C. & Clarke, J. (2000a). Conservation of folding and stability within a protein family: the tyrosine corner as an evolutionary cul-de-sac. *J Mol Biol*, Vol.295, No.3, (January 2000), pp. 641-649.

Hamill, S.J., Meekhof, A.E. & Clarke, J. (1998). The effect of boundary selection on the stability and folding of the third fibronectin type III domain from human tenascin. *Biochemistry*, Vol.37, No.22, (June 1998), pp. 8071-8079.

Hamill, S.J., Steward, A. & Clarke, J. (2000b). The folding of an immunoglobulin-like Greek key protein is defined by a common-core nucleus and regions constrained by topology. *J Mol Biol*, Vol.297, No.1, (March 2000), pp. 165-178.

Hanes, J. & Pluckthun, A. (1997). *In vitro* selection and evolution of funcitonal proteins by using ribosome display. *Proc Natl Acad Sci U S A*, Vol.94, 1997), pp. 4937-4942.

Huang, J., Koide, A., Makabe, K. & Koide, S. (2008). Design of protein function leaps by directed domain interface evolution. *Proceedings of the National Academy of Sciences of the United States of America*, Vol.105, No.18, (May 2008), pp. 6578-6583. ISSN 1091-6490

Huang, J. & Koide, S. (210). Rational Conversion of Affinity Reagents into Label-Free Sensors for Peptide Motifs by Designed Allostery. *ACS Chemical Biology*, Vol.5, No.3, 210), pp. 273-277.

Ishikawa, F.N., Chang, H.-K., Curreli, M., Liao, H.I., Olson, C.A., Chen, P.-C., Zhang, R., Roberts, R.W., Sun, R., Cote, R.J., Thompson, M.E. & Zhou, C. (2009). Label-free, Electrical Detection of the SARS Verus N-Protein with Nanowire Biosensors Utilizing Antibody Mimics as Capture Probes *ACS Nano* Vol.3, No.5, 2009), pp. 1219-1224.

Jacobs, Steven; Diem, Michael; Luo, Jinquan; Teplyakov, Alexey; Obmolova, Galina; Malia, Thomas; Gilliland, Gary; O'Neil, Karyn (2012) Design of Novel FN3 Domains with High Stability by a Consensus Sequence Approach PEDS *in press.*

Karatan, E., Merguerian, M., Han, Z., Scholle, M.D., Koide, S. & Kay, B.K. (2004). Molecular recognition properties of FN3 monobodies that bind the Src SH3 domain. *Chem Biol*, Vol.11, No.6, (June 2004), pp. 835-844.

Koide, A., Bailey, C.W., Huang, X. & Koide, S. (1998). The fibronectin type III domain as a scaffold for novel binding proteins. *J Mol Biol*, Vol.284, No.4, (December 1998), pp. 1141-1151.

Koide, A., Gilbreth, R.N., Esaki, K., Tereshko, V. & Koide, S. (2007). High-affinity single-domain binding proteins with a binary-code interface. *Proc Natl Acad Sci U S A*, Vol.104, No.16, (April 2007), pp. 6632-6637.

Koide, A., Jordan, M.R., Horner, S.R., Batori, V. & Koide, S. (2001). Stabilization of a fibronectin type III domain by the removal of unfavorable electrostatic interactions on the protein surface. *Biochemistry*, Vol.40, No.34, (August 2001), pp. 10326-10333.

Leahy, D.J., Hendrickson, W.A., Aukhil, I. & Erickson, H.P. (1992). Structure of a fibronectin type III domain from tenascin phased by MAD analysis of the selenomethionyl protein. *Science*, Vol.258, No.5084, (November 1992), pp. 987-991.

Liao, H.I., Olson, C.A., Hwang, S., Deng, H., Wong, E., Baric, R.S., Roberts, R.W. & Sun, R. (2009). mRNA display design of fibronectin-based intrabodies that detect and inhibit severe acute respiratory syndrome coronavirus nucleocapsid protein. *The Journal of biological chemistry*, Vol.284, No.26, (June 2009), pp. 17512-17520. ISSN 0021-9258

Lipovsek, D., Lippow, S.M., Hackel, B.J., Gregson, M.W., Cheng, P., Kapila, A. & Wittrup, K.D. (2007). Evolution of an interloop disulfide bond in high-affinity antibody mimics based on fibronectin type III domain and selected by yeast surface display: molecular convergence with single-domain camelid and shark antibodies. *J Mol Biol*, Vol.368, No.4, (May 2007), pp. 1024-1041.

Main, A.L., Harvey, T.S., Baron, M., Boyd, J. & Campbell, I.D. (1992). The three-dimensional structure of the tenth type III module of fibronectin: an insight into RGD-mediated interactions. *Cell*, Vol.71, No.4, (November 1992), pp. 671-678.

Malashkevich, V.N., Kammerer, R.A., Efimov, V.P., Schulthess, T. & Engel, J. (1996). The crystal structure of a five-stranded coiled coil in COMP: a prototype ion channel? *Science*, Vol.274, No.5288, (November 1996), pp. 761-765. ISSN 0036-8075

Molckovsky, A. Siu, L.L. (2008). First-in-class, first -in-human phase I results of targeted agents: highlights of the 2008 American society of clinical oncology meeting. *Journal of hematology & oncology, Vol.1, 2008), pp 20. ISSN 1756-8722*

Odegrip, R., Coomber, D., Eldridge, B., Hederer, R., Kuhlman, P.A., Ullman, C., FitzGerald, K. & McGregor, D. (2004). CIS display: In vitro selection of peptides from libraries of protein-DNA complexes. *Proc Natl Acad Sci U S A*, Vol.101, No.9, (March 2004), pp. 2806-2810.

Olson, C.A., Liao, H.I., Sun, R. & Roberts, R.W. (2008). mRNA Display Selection of a High-Affinity Modification-Specific Phospho-IkBa-Binding Fibronectin. *ACS Chemical Biology*, Vol.3, November 2008), pp. 480-485.

Olson, C.A. & Roberts, R.W. (2007). Design, expression, and stability of a diverse protein library based on the human fibronectin type III domain. *Protein Sci*, Vol.16, No.3, (March 2007), pp. 476-484. ISSN 0961-8368

Parker, M.H., Chen, Y., Danehy, F., Dufu, K., Ekstrom, J., Getmanova, E., Gokemeijer, J., Xu, L. & Lipovsek, D. (2005). Antibody mimics based on human fibronectin type three domain engineered for thermostability and high-affinity binding to vascular endothelial growth factor receptor two. *Protein Eng Des Sel*, Vol.18, No.9, (September 2005), pp. 435-444.

Plaxco, K.W., Spitzfaden, C., Campbell, I.D. & Dobson, C.M. (1996). Rapid refolding of a proline-rich all-beta-sheet fibronectin type III module. *Proc Natl Acad Sci U S A*, Vol.93, No.20, (October 1996), pp. 10703-10706. ISSN 0027-8424

Plaxco, K.W., Spitzfaden, C., Campbell, I.D. & Dobson, C.M. (1997). A comparison of the folding kinetics and thermodynamics of two homologous fibronectin type III modules. *J Mol Biol*, Vol.270, No.5, (August 1997), pp. 763-770. ISSN 0022-2836

Richards, J., Miller, M., Abend, J., Koide, A., Koide, S. & Dewhurst, S. (2003). Engineered fibronectin type III domain with a RGDWXE sequence binds with enhanced affinity and specificity to human alphavbeta3 integrin. *Journal of molecular biology*, Vol.326, No.5, (March 2003), pp. 1475-1488. ISSN 0022-2836

Richards, J., Miller, M., Abend, J., Koide, A., Koide, S. & Dewhurst, S. (2003). Engineered Fibronectin Type III Domain with a RGDWXE Sequence Binds with Enhanced Affinity and Specificity to Human $\alpha v\beta 3$ Integrin. *Journal of Molecular Biology*, Vol.326, No.5, 2003), pp. 1475-1488. ISSN 00222836

Shusta, E.V., Kieke, M.C., Parke, E., Kranz, D.M. & Wittrup, K.D. (1999). Yeast polypeptide fusion surface display levels predict thermal stability and soluble secretion efficiency. *J Mol Biol*, Vol.292, No.5, (October 1999), pp. 949-956.

Steiner, D., Forrer, P., Stumpp, M.T. & Pluckthun, A. (2006). Signal sequences directing cotranslational translocation expand the range of proteins amenable to phage display. *Nat Biotechnol*, Vol.24, No.7, (July 2006), pp. 823-831. ISSN 1087-0156

Sweeney, C.J., Chrioean, E.G., Mita, M.M., Papadopoulos, K.P., Silver, B., Freed, M., Gokemeijer, J., Eaton, C., Furfine, E. & Tolcher, A.W. (2008). Phase I study of CT-322, first Adnectin protein therapeutic and potent inhibitor of VEGFR-2, in patients with advanced solid tumors. *journal of Clinical Oncology*, Vol.26, No.15S, 2008), pp. 3523.

Takahashi, T. (2003). mRNA display: ligand discovery, interaction analysis and beyond. *Trends in Biochemical Sciences*, Vol.28, No.3, 2003), pp. 159-165. ISSN 09680004

Tanha, J., Xu, P., Chen, Z., Ni, F., Kaplan, H., Narang, S.A. & MacKenzie, C.R. (2001). Optimal design features of camelized human single-domain antibody libraries. *The Journal of biological chemistry*, Vol.276, No.27, (July 2001), pp. 24774-24780. ISSN 0021-9258

Xu, L., Aha, P., Gu, K., Kuimelis, R.G., Kurz, M., Lam, T., Lim, A.C., Liu, H., Lohse, P.A., Sun, L., Weng, S., Wagner, R.W. & Lipovsek, D. (2002). Directed evolution of high-affinity antibody mimics using mRNA display. *Chem Biol*, Vol.9, No.8, (August 2002), pp. 933-942.

Zemlin, M., Klinger, M., Link, J., Zemlin, C., Bauer, K., Engler, J.A., Schroeder, H.W., Jr. & Kirkham, P.M. (2003). Expressed murine and human CDR-H3 intervals of equal length exhibit distinct repertoires that differ in their amino acid composition and predicted range of structures. *Journal of molecular biology*, Vol.334, No.4, (December 2003), pp. 733-749. ISSN 0022-2836

7

Molecular Evolution Within Protease Family C2, or Calpains

Liudmila Lysenko and Nina Nemova
Institute of Biology, KarRC of Russian Academy of Science
Russian Federation

1. Introduction

Calpains, or the intracellular Ca^{2+}-dependent proteases (EC 3.4.22.17, family C2, cysteine protease clan CA) are one of the most important proteolytic systems in cytosol of eukaryotic and some prokaryotic cells (reviews: Croall & DeMartino, 1991; Goll et al., 2003; Sorimachi et al., 2011a). Calpain family is ancient and diverse since highly variable modules flanking conservative proteolytic core are found in the structure of these proteins. Linking of protease catalytic domain with ancillary domains, i.e. specialized functional modules with their own spectrum of non-proteolytic activities and binding partners, expands calpain functions on multiple cellular processes. Calpains are processing proteases cleaving their specific substrates at one or a limited number of sites to modulate their structure and activity rather than degrade it. Calpains are versatile proteases that have been implicated in diverse cellular signaling pathways mediated by calcium, such as cytoskeleton remodeling, cell-cycle regulation, differentiation, and death (Croall & DeMartino, 1991; Carafoli & Molinari, 1998; Goll et al., 2003; Nemova et al., 2010). Calpains were also considered to participate in chromosome rearrangements during mitosis (Schollmeyer, 1988), microtubule assembly and disassembly (Billger et al., 1988), intracellular signaling, motility, and vesicle traffic (Choi et al., 1997; Huttenlocher et al., 1997; Lu et al., 2002; Li & Iyengar, 2002). In mammals and plants it is clear that calpains are of critical importance for development (Wang et al., 2003; Dutt et al., 2006; Sorimachi et al., 2010). Furthermore, impaired calpain activity due to mutations or misregulation of the calpains has been implicated in a variety of pathological conditions including muscular dystrophy, ischemia, diabetes mellitus, cancer, and neurodegenerative disease (Tidball & Spencer, 2000; Huang & Wang, 2001; Crocker et al., 2003; Mamoune et al., 2003; Suzuki et al., 2004; Zatz & Starling, 2005). Calpains have attracted much attention because of the recent discovery of correlations between calpain gene mutations and human diseases, together with elucidation of its three-dimensional structure (Hosfield et al., 1999; Strobl et al., 2000) and Ca^{2+}-induced activation mechanisms (Moldoveanu et al., 2002; 2004). Because the enzyme participates not only in normal intracellular signal transduction cascades but also in various pathological states, calpain research has attracted tremendous interest in wide areas of life sciences in both basic and clinical terms.

2. Distribution and diversity of calpains

A variety of identified sequences of calpains and calpain-like polypeptides was found in all eukaryotes and a few bacteria since 1964 when the first protein of this family (later named m-calpain) was isolated by Gordon Guroff from the rat brain (Guroff, 1964). m-Calpain and its close homologue μ-calpain (60% amino acid identity) were discovered prior to other calpains and serve as typical family members though possess very distinct features compared to other calpains. Firstly, μ- and m-calpains form heterodimers consisting of distinct catalytic subunit (calpain 1 or calpain 2 encoded by *CAPN1* or *CAPN2*, respectively) with approximate molecular weight 80 kDa and common regulator polypeptide CSS (calpain small subunit) with approximate molecular weight 28 kDa. Isolated calpains 1 and 2 possess full protease activity of original heterodimers if properly folded (Barrett et al., 1998). Secondly, μ- and m-calpains are highly susceptible to endogenic proteinaceous inhibitor, calpastatin, whereas the majority of calpains exist and function as monomeric calpastatin-insensitive proteins. The terms μ-calpain and m-calpain were first used in 1989 (Cong et al., 1989) to refer to the micromolar Ca^{2+}-requiring (μ-calpain) and millimolar Ca^{2+}-requiring (m-calpain) proteases, respectively.

Advances of the last few decades in the field of DNA cloning and sequencing led to identification of 895 calpain-coding sequences in wide range of organisms – from bacteria to human (according to *MEROPS* database that provides a comprehensive catalogue and structure-based classification of proteases and inhibitors) (Rawlings et al., 2010). Fifteen molecules in mammals, 4 in *Drosophila melanogaster*, 7 in *Anopheles gambiae*, 14 in *Caenorhabditis elegans*, 7 in *Shistosoma mansoni*, up to 24 in protozoan kinetoplastids, infusoria, and only one in fungi, plants, and some bacteria share more than 25% identity with the catalytic domain of m-calpain. The biochemical, immunological, and functional characteristics of the possible members of the calpain family significantly vary.

2.1 Functional diversity of calpains

2.1.1 Information from transgenic organisms

Calpain function has been investigated by both genetic and cell-biological routes (review: Sorimachi et al., 2010). To understand the physiological function of calpain, targeted deletion, and more recently the conditional deletion of calpain genes have been performed. Individual knockout of two calpain genes in mouse, *Capn2* and *Css1*, appeared to be preimplantation lethal, demonstrating that the calpain system is essential for early embryogenesis (Arthur et al., 2000; Zimmerman et al., 2000; Dutt et al., 2006; Tan et al., 2006); the function of calpains in development is not yet known, however. The targeted deletion of only one of the two "ubiquitous" calpains, *Capn1*, does not result in embryonic death (Azam et al., 2001). Loss of *Capn9* in NIH3T3 cells results in a more transformed phenotype (Liu et al., 2000), but this gene has not yet been targeted in whole organism yet. Targeted deletion of *Capn3* in mice produces a model for assessing its specific role in muscle function and repair (Duguez et al., 2006; Cohen et al., 2006) as multiple genetic loss-of-function defects in calpain 3a seem to be a cause of limb-girdle muscular dystrophy type IIa in human (Zatz & Starling, 2005; Duguez et al., 2006).

The importance of calpains in physiology of lower organisms is demonstrated by a variety of defects caused by compromised calpain function including deficient sex determination in *C. elegance*, defects in aleurone cell development in plants and alkaline stress susceptibility in ascomycete *Aspergillus nidulans*, lethality of calpain-like polypeptide depletion by RNA interference in procyclic cells of kinetoplastid parasites (reviews: Croall & Ersfeld, 2007; Sorimachi et al., 2011b).

2.1.2 Physiological substrates of calpains

The calpain superfamily includes a diverse proteases that vary greatly in terms of domain structure, expression (*see below*), and substrate specificity. As such, calpains have been implicated in wide range biological processes. Despite wealth of new information, the precise physiological function(s) of the calpain system remain poorly defined, although many potential substrate proteins have been identified *in vivo* and *in vitro*. A large number of proteins (over 100 have been reported) are cleaved by the calpains in *in vitro* assays (reviews: Goll et al., 1999; 2003). Many of these protein substrates can be placed in one of four general categories:

a. *Cytoskeletal proteins especially those involved in cytoskeletal/plasma membrane interactions (actin-binding proteins, microtubule-associated proteins, etc.):* adducing, ankyrin, caldesmon, cadherin, calponin, catenin, C-protein, desmin, dystrophin, gelsolin, filamin/actin-binding protein, MAP1, MAP2, myosin, nebulin, neurofilament proteins, NFH, NFM, NFL, protein 4.1a, 4.1b, αII-spectrin, β-spectrin, synemin, talin, tau, titin, tropomyosin, troponin I, troponin T, tubulin, vimentin, vinculin.
b. *Membrane-associated proteins:* growth factor receptors (EGF receptors), adhesion molecules (integrin, cadherin, N-CAM), ion transporters (Ca^{2+}-ATPase), etc.
c. *Enzymes, cytosolic or membrane-associated:* kinases (protein kinase C, myosin light-chain kinase, calcium/calmodulin-dependent protein kinase, doublecortin-like kinase, EGF receptor kinase, pp60src, ppFAK[125]); phosphatases (calcineurin, inositol polyphosphate 4-phosphatase, protein tyrosine phosphatase 1B); phospholipases (phospholipase C), etc.
d. *Others:* cytokines (interleukin Iα), transcription factors (Fos, Jun), lens proteins (crystallins), etc.

Calpains are not specific for certain amino acid residues or sequences but recognize bonds between domains. As a consequence, calpains hydrolyze substrate proteins in a limited manner, and large fragments retaining intact domains are produced by hydrolysis. Thus calpains are regarded as the biomodulators of its substrate proteins (Saido et al., 1994; Sorimachi et al., 1997; Carafoli & Molinari, 1998; Sorimachi & Suzuki, 2001; Suzuki et al., 2004). Not all *in vitro* calpain substrates are cleaved by calpains *in vivo*. Certain extracellular proteins, such as fibronectin or factor V, are unlikely cleaved by calpains *in vivo* as the enzymes are localized exclusively intracellularly except in diseased or injured tissues.

2.1.3 Some well-documented and proposed calpain functions

Wide distribution of the protease family among living organisms indicates evolutionary conservation of the essential calpain functions. The physiological cell-based techniques show that the calpains system are involved in functions as diverse as cytoskeletal/plasma membrane attachments, cell motility, signal transduction by the activation of some signaling molecules and assembly of focal adhesions, progression through the cell cycle and regulation of gene expression, some apoptotic pathways, and long-term potentiation (review: Goll et al., 2003).

Modifying integrins, cadherins, and cytoskeletal proteins calpains may contribute to the developmental regulation of cell adhesion (Franco & Huttenlocher, 2005; Mellgren et al., 2009). Some cases of membrane fusion, as in myoblast differentiation (Kwak et al., 1993), have also been reported to involve calpain activity. In *S. mansoni*, calpain may regulate surface membrane biogenesis (Sorimachi et al., 1993). A controversial but attractive topic is the possible involvement of calpain in cell division (Schollmeyer, 1988) and in long-term potentiation (del Cerro et al., 1990; Suzuki et al., 1992).

Two non-classical calpains, Tra3 and PalB (human orthologs of calpain 5 and calpain 7), mediate signal transduction pathways for sex determination in nematodes (Sokol & Kuwabara, 2000) and alkaline pH adaptation in fungi (Nozawa et al., 2003), respectively.

Non-proteolytic calpain homologues with substituted amino acids in classical Cys/His/Asn triad or truncated catalytic core domain supposed to be involved in calpain system regulation by non-productive substrate binding as well as inhibitor binding. The existence of calpains with this structural feature indicates that they have specific non-proteolytic functions the nature of that has not been established yet. The non-proteolytic function(s) may be extended on proteolytically active homologues as well.

2.1.4 Pathological implications of calpains

A number of pathologic conditions have been associated with disturbances of the calpain system. Calpain-associated pathologies, so called "calpainopathies", are related with genetic defects in calpain polypeptides as well as with calpain misregulation due to loss of Ca^{2+} homeostasis in cell.

Mutations in some calpain genes are also linked with diseases such as senile diabetes and muscular dystrophy. Disease-related disruptions in *CAPN3* gene are linked to loss of proteolytic activity of muscle-specific calpain 3a (Zatz & Starling, 2005; Dugues et al., 2006). Specific splice variants of *CAPN3* occur in the lens of the eye and are linked to the formation of cataracts (Shih et al., 2006). Limited proteolysis of crystallins by abnormally activated calpain is likely to induce lens protein insolubilization during aging. Variations in the *CAPN10* gene may be a factor contributing to increased susceptibility to a multifactorial disease – type 2 diabetes (Horikawa et al., 2000; Goll et al., 2003; Turner et al., 2005). Calpain 10 may contribute to stimulated secretion and(or) pancreatic cell death (Johnson et al., 2004; Marshall et al., 2005), and thereby be relevant to this disease.

Calpainopathy appears to be primarily caused by compromised protease activity, rather than by damaged structural properties. Calpains are believed to be strongly related to certain conditions accompanied by calcium misregulation including neurodegeneration. Calpains are involved in long-term potentiation, Alzheimer's disease (Suzuki et al., 1992; Saito et al., 1993; Muller et al., 1995) and ischemia (Blomgren et al., 1995; review: Siman et al., 1996). Taking into account the absence of brain-specific calpain species ubiquitously expressed μ- and(or) m-calpains are involved in brain-specific functions. Neuronal degeneration in postischemic brain has been shown to involve the activation of various signal transduction-related elements such as excitatory amino acid receptors, protein kinase C, μ-calpain, and other Ca^{2+}-dependent enzymes (Schmidt-Kastner & Freund, 1991).

Besides specific role of tissue-specific calpains in target organs, conventional calpains tend to be overactivated in muscular dystrophies, cardiomyopathies, traumatic ischemia, and lissencephaly, probably due to compromised intracellular Ca^{2+} homeostasis caused by these diseases. These observations suggest that calpains function as the mediators of the pathological process, not necessarily as the ultimate destroyers of cells. The involvement of calpain in pathological states is of particular clinical interest because it is expected that further research may eventually lead to therapeutic applications.

2.2 Domain architecture of calpains

2.2.1 Catalytic core

Calpain *domain II* is the proteolytic papain-like domain with typical catalytic triade Cys/His/Asn (clan CA) (Arthur et al., 1995) ('CysPc' or cd00044 in the *NCBI* conserved domain database) that shares however little sequence homology with other cysteine proteases, and it is likely that it evolved from a different ancestral gene. Predicted subdomain IIa contains active site Cys105, whereas the remainder amino acid residues of catalytic triade (His262 and Asn286) are located in subdomain IIb. The crystallographic data indicate that the catalytic residues of subdomains IIa and IIb are uncoupled in the absence of Ca^{2+} (distance 10.5 Å) (Hosfield et al., 1999; Strobl et al., 2000), and Ca^{2+} binding induces conformational changes in the full-length molecule, allowing assembly of active site. Domain II has affinity to Ca^{2+} essential for enzyme activity since contain two Ca^{2+}-binding sites located in peptide loops (Moldoveanu et al., 2004).

Due to complex domain structure of calpains with highly divergent *N*- and *C*-terminal regions flanking conservative catalytic core only catalytic domain II sequence but not complete amino acid alignment is used to establish protein homology. The degree of similarity that allows assigning a polypeptide to the calpain family has not been determined yet. Depending on the molecule, the amino acid identity in domain II region varies from less than 30% to more than 75%. For example, recently assigned to family protease from prokaryote *Porphyromonas gingivalis* contains catalytic domain almost equally similar to μ-calpain on the one hand and papain sequence from the other hand and the enzymatic properties of the protease drastically differ from the characteristics of μ- and m-calpains (Bourgeau et al., 1992).

Some family members (calpain 6 and demi-calpain of vertebrates, CALPC of *Drosophila*, some of *C. elegance* homologues, and numerous calpains of kinetoplastid parasites), seem to be catalytically inactive proteases owing to substitutions in specific amino acid residues (one or more) that are located in critical active-site regions or constitutively truncated catalytic domain.

2.2.2 Other structural and functional modules

Several functional domains are found on *N*- or *C*-terminus of protease core. Calpain genes are the products of evolutionary combinations of several ancestral unit genes, that is, genes for the C2-like, penta-EF-hand, C2-containing T, SOH, PBH, calpastatin-like, Zn-finger, and transmembrane domains.

A typical member of C2 family, calpain 2 (or the catalytic 80-kDa subunit of m-calpain), is a prototype used for designation other polypeptides as calpains. Calpain 2 has classical four-domain architecture, where domain II is proteolytic (Fig. 1), domain III is C2-like domain and domain IV is Ca^{2+}-binding domain of PEF family. Other calpain homologues have a combination of the "classical" domains and the others listed above.

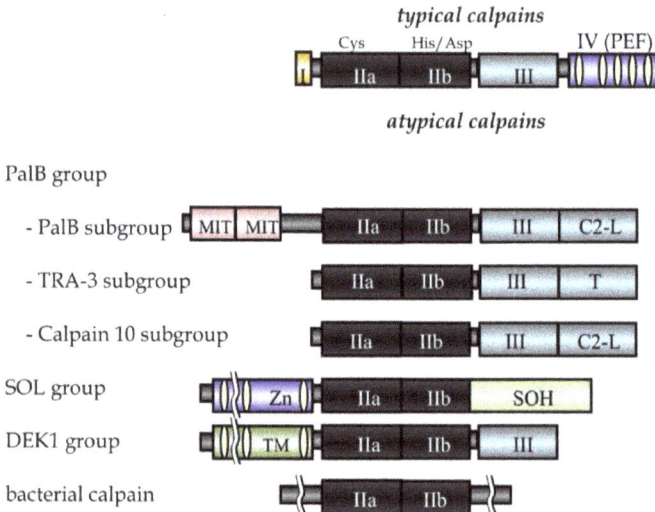

Fig. 1. Calpain classification based on protein domain structure. Typical calpains are composed of four domains (I–IV): I – highly divergent *N*-terminal prodomain; II – protease core domain subdivided on IIa and IIb subdomains; III – C2 domain-like; IV (PEF) – penta-EFhand-containing domain (*see details in the text*). Atypical calpains have no classical Ca^{2+}-binding PEF domain and contain additional functional modules either on *C*-terminus: C2-L – C2 domain-like; T – C2 domain; SOH – SOL-homology domain; or on *N*-terminus: MIT – microtubule-interacting-and-transport motif; Zn – Zn-finger-like domain; TM – transmembrane domain.

Based on the amino acid sequence, calpain 2 consists of four domains (Suzuki, 1990); according to the X-ray diffraction analysis data (Hosfield et al., 1999; Strobl et al., 2000), in the absence of Ca^{2+} domain II is subdivided on two subdomains carrying different components of catalytic triade (Cys or His/Asn, respectively), and there is a linker region between domains III and IV.

N-terminal *domain I* is only 18–20 amino acid residues in length and has no sequence homology with any known polypeptide. This propeptide region is autolyzed upon activation. According to crystallography data, the propeptide does not block sterically the active site of the enzyme (Hosfield et al., 1999; Strobl et al., 2000) as it is shown for a most of protease zymogens; the activation mechanism for calpains is recognized as calcium-induced interdomain autoactivation (Goll et al., 2003; Benyamin, 2006).

Domain III provides the coupling of the catalytic domain II and the Ca^{2+}-binding domain IV and accelerates Ca^{2+}-induced conformational changes (Tompa et al., 2001). The spatial organization of domain III is determined by eight antiparallel β-strands. This domain shows a functional (but low sequence) resemblance to other Ca^{2+}-regulated proteins such as the conserved domain 2 of protein kinase C, or C2; owing to this fact, calpain domain III is assigned as C2-like domain (or C2-L) (Fig. 1). Functionally C2 and the C2-L domains have an affinity to Ca^{2+} and enable to associate with internal membrane phospholipids in Ca^{2+}-dependent manner (Kawasaki & Kawashima, 1996; Hosfield et al., 1999; Tompa et al., 2001). Analysis of domain III amino acid sequence indicates that it includes two potential Ca^{2+}-binding EF-hand motifs functionally active in the calpain from flat worm *S. mansoni* but apparently not binding Ca^{2+} in mammalian calpains (Goll et al., 2003).

The sequence of calpain 2 *domain IV* represents a distinct module in the protein; it has a similarity with calmodulin (24–44% identical residues for human molecules) and contains five putative Ca^{2+}-binding EF-hand motifs (Sorimachi & Suzuki, 2001) (Fig. 1) thus classical calpains belong to penta-EF-hand (PEF) protein family. In addition several alternative mechanisms for binding calcium and associating with membrane phospholipids are found throughout the family. Since several substrates have been shown to bind to domain IV (Noguchi et al., 1997; Shinozaki et al., 1998), it is likely that this domain is important for substrate recognition by calpain. In this case, the function can be assumed by other protein-interactive domains, such as the C2-containing domain T, and, possibly, the SOH and PBH domains (*see below*).

SOH domain – conservative C-terminal module with putative substrate-recognition function in protein SOL (small optic lobes) of *Drosophila* or in calpain 15, vertebrate homologue of SOL.

Zn-fingers motifs localized into N-terminal *Zn-finger domain* of some calpains, such as SOL (Kamei et al., 1998), function as intermolecular interaction modules enabling to bind DNA, RNA, proteins or small molecules.

T-module, or classic C2-domain, shares both structural and functional similarity with conserved domain 2 of protein kinase C, or C2. This Ca^{2+}-binding module is associated with membranes and involved in signal transduction and transmembrane traffic (Barnes & Hodgkin, 1996).

PBH domain, or MIT (microtubule-interacting and transport) domain, is usually found in AAA-ATPases and some proteins lacking ATPase domains, such as PalB from *Aspergillus* (Mugita et al., 1997); it is responsible for the association with microtubules and intracellular traffic of molecules.

Domain IQ found only in calpain 16, or demi-calpain, is responsible for the interaction with Ca^{2+}-binding molecule – calmodulin in the absence of their own Ca^{2+}-binding motifs.

Some calpains from protists contain *domain CSTN* weakly similar to calpastatin; their role is not elucidated yet.

Transmembrane domain (usually in tandem) was found on N-terminus of plant calpain, or phytocalpain, and some ciliate calpains ascribed to constitutive membrane anchoring proteins. Although membrane binding is not well substantiated for classical calpains, predicted transmembrane segments in phytocalpain and some ciliate calpains suggest an evolutionary link between calpain function and membranes.

Domain I^K (K for kinetoplastids) of calpain-like proteins – possesses a high degree of sequence conservation in one of two kinetoplastid phyletic branches and contains mainly three consensus motifs: GLLF/Y toward the N-terminus, WAFYNDT in the center, and VYPxETE toward the C-terminus; its function is not elucidated yet. Only in I^K-containing kinetoplastid proteins the acylation motifs are localized. The other kinetoplastid calpain-like proteins have *domain I^H* (H for heterogeneous) – N-terminal sequence highly variable in both intra- and interspecies comparisons (Ersfeld et al., 2005).

Critical role in calpain activation and calpain response to Ca^{2+} signaling play temporary interaction or co-localization with cellular membranes (Cong et al., 1989; Gil-Parrado et al., 2003). Several ancillary domains or even auxiliary non-calpain polypeptides are responsible for such interaction: (1) acylated N-terminal region in calpain-like proteins of the kinetoplastids *L. major* and *T. brucei* allowing association with the cytoplasmic face of membranes and lipid rafts and contribution to signal transduction (Tull et al., 2004; Croall & Ersfeld, 2007); (2) Gly-clustering (conserved peptide GTAMRILGGVI) located in small subunit N-terminal domain and formed a membrane-penetrating α-helical structure (Dennison et al., 2005), providing a mechanism for calpains 1 and 2 binding to membranes; (3) for many calpains, the C2-L domain III provides an additional or alternative mechanism for membrane association via its phospholipid-binding properties. None of the conventional calpains is acylated, however, that some of the established *in vivo* substrates of mammalian calpains, the cytoskeletal proteins vinculin, spectrin, ankyrin, and band 4.1, are themselves acylated and membrane-associated (Maretzki et al., 1990; Bhatt et al., 2002).

2.3 Classification of calpains

Conventionally calpains are subdivided on the basis of domain composition onto two classes – typical and atypical and on the basis of polypeptide composition onto monomeric and oligomeric proteins. Functionally it is rationally to separate calpains on ubiquitous and tissue-specific as well as on constitutively proteolytically active and inactive enzymes.

2.3.1 Structural classification

Based on the similarity of the domain organization to that of calpains 1 or 2 and the presence of penta-EF-hand motif in domain IV, it is proposed to divide calpains into two general classes – typical and atypical. Typical calpains have a C-terminal calmodulin-like domain carrying EF-hand motifs lacking in atypical calpains. Atypical calpains further subdivided into six groups based on homology in the region of ancillary domains (Fig. 1).

Typical calpains include fifteen known molecules: (1) ten vertebrate proteins such as calpains 1-3, 8, 9, 11–14, and μ/m-calpain from chicken and *Xenopus*, (2) three calpains from *Drosophila* such as CALPA, CALPB, and CALPC, and (3) two trematode proteins Sm-calpain from *Shistosoma mansoni* and Sj-calpain from *S. japonicum* (reviews: Goll et al., 2003; Sorimachi et al., 2011a). The complete amino acid sequences of these proteins are highly conservative and folding is described as "classical" four-domain architecture. Some data suggest that calpains 12 and 14 have domain IV with degenerative EF-hand motifs that are unlikely to bind calcium (Croall & Ersfeld, 2007); thus these molecules could be functionally assigned to atypical group.

The products of vertebrate genes *CAPN(5–7)*, *CAPN10*, *CAPN15*, and *CAPN16*, all calpains from *C. elegans*, plants, fungi, trypanosomes, and *S. cerevisiae* are *atypical calpains* lacking domain IV with EF-motifs. Despite this, almost all atypical calpains are Ca^{2+}-regulated; most probably the interaction of these enzymes with Ca^{2+} may involve structures distinct from EF-hand motifs such as located in domain III (Hood et al., 2004; Shao et al., 2006; Samanta et al., 2007) or some possible cofactors (Goll et al., 2003). Conserved domains found on C-terminus instead of the calmodulin-like domain IV permits one to divide atypical calpains on SOL and PalB subfamilies (Fig. 1); and the latter includes calpains 5, 6, 7 and 10 because of structural similarity and evolutionary relation of domains III (C2-L) and T (C2).

Other criterion for structural classification of calpains is their polypeptide composition. While a most of identified calpains are monomeric proteins there are some calpains which potentially form hetero- or homooligomers in the native form. So, μ- and m-calpains *in vivo* and *in vitro* forms heterodimers with 28 kDa polypeptide (calpain small subunit, or CSS1, formerly referred as calpain 4) through fifth EF-hand motif in domain IV and similar EF-hand motif in 28 kDa subunit. It means that this EF-hand fragment does not bind Ca^{2+} but is involved in dimerization instead. Full proteolytic activity of digestive tract-specific calpain 9 (nCL-4) also requires dimerization with CSS1 (Lee et al., 1999). No other calpains identified to date bind small subunit. Presence of CSS1 is of critical importance for calpastatin regulation as it provides one of three binding site for effective calpain/inhibitor interaction. Stomach-specific calpain 8 (nCL-2) exists in both monomeric and homooligomeric forms, but not as a heterodimer with CSS1. The oligomerization occurs through domains other than the 5EF-hand domain IV, most probably through domain III, suggesting a novel regulatory system for nCL-2 (Hata et al., 2007). Calpain 8 sensitivity to calpastatin suggests effective enzyme/inhibitor binding through motifs distinct from those localized in small subunit.

2.3.2 Classification on the basis of tissue distribution

In addition to the structural features, vertebrate calpains are independently classified into two categories according to their tissue (or organ) distribution: (1) calpains with wide tissue distribution, so called "ubiquitous" (human μ-calpain, m-calpain, calpains 5, 7, 10, 13, 15, 16, and μ/m-calpain from chicken and *Xenopus*), and (2) tissue-specific calpains (calpains 3, 6, 8, 9, 11, 12). The products of genes *CAPN1* and *CAPN2* are found almost in all tissues of higher animals with the exception of erythrocytes possessing only one of two forms: μ-calpain in human, μ/m-calpain in avian, and m-calpain in fish (Murakami et al., 1981; Nemova et al., 2000). Calpains 5, 7, 10, 13, 15, and 16 are widely expressed in tissues. Calpains 3a, 3b, 6, 8, 9, 11, and 12 are tissue-specific; they are expressed in skeletal muscles, the crystalline lens, placenta, smooth muscles, alimentary tract, testis, and hair follicle, respectively (review: Goll et al., 2003). In mammalian tissues, mRNA transcripts of calpain 14 have not been found yet.

2.4 Distributions of calpains in living organisms

Calpain genes are distributed among all kingdoms of life except archea and viruses. There are 895 calpain sequences in studied organisms identified to date according to protease database MEROPS (Rawlings et al., 2010). Now 15 calpains are known in human (and in almost all vertebrates). Other organisms have far fewer genes coding for calpain-like proteins. With few exceptions, most organisms outside the animal kingdom (fungi, plants, and bacteria) and a most of protists have only a single calpain gene if any (Sorimachi & Suzuki, 2001; Goll et al., 2003) while other cysteine proteases are abundant in lower eukaryotes and parasitic protozoa (Sajid & McKerrow, 2002; Mottram et al., 2003).

Unexpected variety of calpain genes were shown in a number of parasitic kinetoplastids and infusoria. The discovery of this surprisingly large family of calpain-like proteins in lower eukaryotes contributes to our understanding of the molecular evolution of this abundant protein family.

2.4.1 Calpains in bacteria

Proteins homologous to calpains have been detected only in few prokaryotes, whereas 96% of bacteria, including *Escherichia coli* and all archea, have no calpain genes. Trp protease from *Porphyromonas gingivalis* illustrates some of the difficulties encountered when assigning molecules to the calpain family on the basis of cDNA-derived amino acid sequence alone. Although the predicted amino acid sequence of the *P. gingivalis* enzyme has 53.1% similarity (23.7% identity) to domain IIa/IIb of human µ-calpain, the *P. gingivalis* enzyme has nearly the same homology (22.5% identity) with the amino acid sequence of papain; the expressed protease is not inhibited by leupeptin, an inhibitor of the calpains, reveal distinct substrate specificity and is inhibited, not activated, by Ca^{2+} (Bourgeau et al., 1992).

2.4.2 Calpains in fungi

Most fungal genomes sequenced have only a single gene for calpain-like proteins, the exception being *Neurospora crassa*, where three genes have been identified (data from *MEROPS* database). Highly modular structure of calpains and calpain-like proteins in lower eukaryotes that combines novel and conserved sequence modules suggests that they are involved in diverse cellular activities. PalB protein from *Aspergillus* is an ancestor of the most evolutionarily conserved subfamily of calpains with known members in vertebrates (human calpain 7, or PalBH), yeasts, fungi, protists, nematodes, and insects (except *Drosophila*), but not in plants. PalBH homologues commonly contain two C-terminal C2-L domains in tandem and conserved microtubule-interacting and transport (MIT) motifs at the N-terminus.

2.4.3 Calpains in plants

Phytocalpain genes encoding a highly conserved, unique plant-specific calpain-like molecule (phytocalpain) derive their names from the *Zea mays* Defective Kernel-1 (DEK1) gene, which was the first to be phenotypically characterized and cloned (Becraft et al., 2002; Lid et al., 2002). The polypeptide encoded by gene *DEK1* consists of C-terminal intracellular domain with significant sequence homology to domains IIa/IIb and III of calpains 1 or 2 and extended N-terminal region that is predicted to contain 21 transmembrane domains interrupted by an extracellular loop and an extended cytoplasmic juxtamembrane region showing little homology to other proteins (Lid et al., 2002). Highly conserved homologues of DEK1 have been described across the plant kingdom, including basal plants such as *Physcomitrella*. Interestingly, these genes occur in one copy in plant genomes sequenced thus far, including that of the model plant *Arabidopsis thaliana*. Phytocalpain has been shown to be essential for the correct development of both an embryonic epidermal cell layer and the specialized outer layer of the endosperm (aleurone) during seed development in maize and *Arabidopsis* (Lid et al., 2002; Ahn et al., 2004; Johnson et al., 2005).

2.4.4 Calpains in invertebrates

Based on screening of sequenced or partially sequenced genomes a most of unicellular protozoan (*Plasmodium falciparum, Theileria annulata, Cryptosporidium parvum, Entamoeba histolytica*, etc.) has a single copy of calpain-coding gene whereas no calpain-like sequences were identified in others (for instance, *Giardia lamblia*) (review: Croall & Ersfeld, 2007).

Uniquely within protozoa, some kinetoplastid parasites and the ciliate *Tetrahymena thermophila* display expansion of calpain genes. Fourteen genes encoding calpain-related proteins have been identified in *Trypanosoma brucei*, 17 in *Leishmania major*, 15 in *T. cruzi* (Ersfeld et al., 2005), and 26 in macronuclear genome of the ciliate *T. thermophila* (Ersfeld et al., 2005; review: Croall & Ersfeld, 2007). The presence of numerous calpain-like proteins, exceeding the numbers found in most other organisms including vertebrates, and their unique protein architecture point to important trypanosomatid-specific functions of these proteins that have not been elucidated yet; however, a number of calpain-like proteins are differentially expressed during the life cycle of the parasites (Matthews & Gull, 1994).

Only atypical calpains are found in lower eukaryotes, including protozoan and fungi, as well as in *C. elegans*, plants, and *S. cerevisiae* (Sorimachi et al., 2011a), whereas three of four *Drosophila* calpains, four *S. mansoni* calpains and three *A. gambiae* calpains are typical. All calpains in trypanosome and one typical calpain in *Drosophila* lack one or more of the Cys/His/Asn residues at their catalytic site, so they probably are not proteolytic enzymes. It is predicted that proteomes of obligate parasites is enriched by the genes of longer conserved proteins of fundamental function (Brocchieri & Karlin, 2005) owing to elimination low expressed and poorly conserved genes under no or weak selection (Haigh, 1978). Calpain diversity in kinetoplastid parasite genomes emphasizes house-keeping function of calpains including non-proteolytic ones.

Only a few calpain-like molecules isolated from *Schistosoma mansoni, Drosophila*, and crustacean have been enzymatically characterized (reviews: Mykles, 1998; Goll et al., 2003; Cantserova et al., 2010; Lysenko et al., n.d.). Some of them such as SOL calpain from *Drosophila* are closely related to mammalian calpains and the others are highly distinct from that ones.

Interestingly, neither calpastatin activity nor a calpastatin-like DNA sequence has yet been detected in invertebrates; a search of the *Drosophila* genome failed to detect a gene homologous to calpastatin (Laval & Pascal, 2002). CSS polypeptide is also has not been found in invertebrates (Pintér et al., 1992).

2.4.5 Calpain-calpastatin system in vertebrates

Today, fifteen calpain genes (*CAPN1-3, CAPN5*-16) (Table 1) and calpain-related genes such as calpastatin (*CAST*) and calpain small subunite (*CSS1* and *CSS2*) have been identified in human genome. Two or more of the three well-characterized members of the calpain system, μ-calpain, m-calpain, and calpastatin, has been detected in any vertebrate cell that has been carefully examined for their presence. Different tissues (or cells), however, differ widely in the ratios of the three proteins (Thompson & Goll, 2000). Other calpains are expressed in more or lesser extent in tissue-specific manner.

Gene	Chromosome	Gene product	Other names	Protease activity	Domains			Expression
					C2-L	C2(T)	PEF	
CAPN1	11q13	calpain 1	μ-calpain large subunite	+	+	–	+	ubiquitous
CAPN2	1q41-q42	calpain 2	m-calpain large subunite	+	+	–	+	ubiquitous (except for erythrocyte)
CAPN3	15q15.1-q21.1	calpain 3a	p94	+	+	–	+	skeletal muscle
		calpain 3b	Lp82, Lp85	+	+	–	+	lens, retina
		calpain 3c		+	+	–	+	ubiquitous
CAPN5	11q14	calpain 5	hTRA-3	+	+	+	–	ubiquitous (abundant in testis, brain)
CAPN6	Xq23	calpain 6	calpamodulin	–	+	+	–	embrionic muscles, placenta
CAPN7	3p24	calpain 7	PalBH	+	++	–	–	ubiquitous
CAPN8	1q41	calpain 8	nCL-2	+	+	–	+	gastrointestinal tract
		calpain 8b	nCL-2'	+	–	–	–	
		calpain 8c		+	+	–	+/–	
CAPN9	1q42.11-q42.3	calpain 9	nCL-4	+	+	–	+	gastrointestinal tract
		calpain 9b		+	+	–	+	
CAPN10	2q37.3	calpain 10		+	++	–	–	ubiquitous
		calpain 10b-h		n.d.	+/–	–	–	
CAPN11	6p12	calpain 11		n.d.	+	–	–	testis
CAPN12	19q13.2	calpain 12		n.d.	+	–	+	hair follicle
		calpain 12b, 12c		n.d.	+	–	+/–	
CAPN13	2p22-p21	calpain 13		n.d.	+	–	+	ubiquitous
CAPN14	2p23.1-p21	calpain 14		n.d.	+	–	+	ubiquitous
CAPN15	16p13.3	calpain 15	SOLH	n.d.	–	–	–	ubiquitous
CAPN16	6q24.3	calpain 16	demi-calpain	–	–	–	–	ubiquitous

Table 1. Some characteristics of human calpains

Furthermore, several calpain genes from vertebrates (*Capn3*, *Capn8-10*, and *Capn12*) and a few calpains from invertebrates (*CalpA* from *Drosophila*) possess specific sites for alternative splicing. For example, at least eight splice variants of calpain 10 (calpains 10a through 10h) have been identified, including three variants that lack intact domain II and thus have no protease activity (Horikawa et al., 2000).

Chickens and *Xenopus laevis* have been shown to contain μ/m-calpain with intermediate structural and enzymatic characteristics between common mammalian calpains and was recently shown to correspond to mammalian calpain 11 from an evolutionary viewpoint (Sorimachi et al., 1995; Macqueen et al., 2010). Fish due to polyploidization events in nature history have a duplicate set of most of the fifteen mammalian calpains and calpain-related molecules (Lepage & Bruce, 2008; Macqueen et al., 2010).

Only oligomeric molecules of calpains are inhibited by calpastatin; μ- and m-calpains have similar susceptibility to calpastatin. Among other homologues only calpains 8 and 9 are also inhibited by calpastatin (Lee et al., 1999; Hata et al., 2001; 2007).

Significantly that higher organisms along with enzymes having "advanced" characteristics (dimere structure, tissue-specific patterns of expression, alternative promotors, regulation by calpastatin, etc.), contain almost whole set of non-classical calpain homologues conservative from lower eukaryotes and invertebrates (Bondareva & Nemova, 2008).

3. Phylogenetic analysis of calpains

Phylogenetic trees have been constructed for isolated domains (Sorimachi & Suzuki, 2001; Jékely & Friedrich, 1999) and for the defining catalytic core domain II in conjunction with the most common auxiliary domain III of selected species (Jékely & Friedrich, 1999; Wang et al., 2003). The phylogenetic approach confirms clear segregation of two main groups of calpains: the first clade includes EF-hand-containing *CAPN* genes (*Schistosoma*, *Drosophila* A/B and the classic vertebrate calpains) and also EF-hand-free *C. elegans* CLP-1, and the second cluster contains capn5(TRA-3) and capn6 typical and atypical (Jékely and Friedrich, 1999). This segregation clearly corresponds to common groups of typical and atypical calpains except for CLP-1 protease.

Heterodimeric vertebrate calpains (μ- and m-calpains and calpain 9) are grouped together with a reliability of 96%, and it is highly probable that calpains of *S. mansoni* and *Drosophila* also belong to this group. The results of the phylogenetic approach to the study of calpains indicate that the scenario of their evolution is similar to that proposed for the evolution of other multigene families (actin, troponin, Hox and others) in vertebrates (Holland et al., 1994; Ohta, 1991).

3.1 Evolutionary roots of calpains

For the first time, Koichi Suzuki has suggested in 1987 the hypothesis that calpains are chimeric proteins originated by the fusion of two genes coding for proteins with completely different functions and origins – cysteine protease and calmodulin (Suzuki et al., 1987). Initially it was predicted that the catalytic domain of calpains and other cysteine proteases, such as papain and cathepsins B, H, and L, arose from a single archetypical protease. Based on this presumption, these enzymes were grouped into a unified family of papain peptidases C1. According to the refined structural information calpains were included in the database MEROPS (Rawlings et al., 2010) as an individual family C2 of the papain-like clan CA of cysteine peptidases. Although common catalytic triade Cys/His/Asn and papain-like catalytic mechanism (Arthur et al., 1995) calpains share little sequence homology with other cysteine proteases, and it is likely evolved from a different ancestral gene.

3.2 Possible scenario of molecular evolution in calpain family

Structural features, together with the organization of mammalian calpain genes, strongly suggest that calpains evolved by the arrangement and restructuring of genes of the ancestral calpain-type cysteine protease with other functional units. Evolutionary history of calpains is a classical example of a multigene protein family evolution.

A small-scale or large-scale genome duplication occurred in different lineages generates additional new genes, the majority of which (about 70-90%) have since been degraded and(or) lost ("nonfunctionalization"). The functions of the ancestral gene may be distributed among the duplicates (so called "subfunctionalization") or one of the duplicates retains the ancestral functions while the other acquires a completely new function (neofunctionalization) (Braasch & Salzburger, 2009). The gene duplication with subsequent divergent resolution, subfunctionalization and neofunctionalization ("duplication-diversification hypothesis") is applicable to calpain evolution as well.

3.2.1 Gene duplication

Any gene family is the product of gene duplication. The importance of gene duplication in molecular evolution is well established (Nei, 1969; Ohno, 1970). Gene copies can be generated through one of two main mechanisms, namely small-scale or large-scale duplication events, with the most extreme large-scale event being duplication of the entire genome (Hakes et al., 2007).

It has been hypothesized that the highly conserved sequence and protein structures of typical calpain family members arose from multiple tandem duplications of a single ancestral calpain gene with subsequent functional diversification of copies (Jekely & Friedrich, 1999; Hata et al., 2001). The topology of the phylogenetic tree of calpains together with the analysis of chromosomal localization of calpain genes shed light on the time of two episodes of major calpain gene duplications coinciding with early chordate whole-genome duplications (Holland et al., 1994; Jékely & Friedrich, 1999).

One episode of gene duplication occurred later than the separation of lines plants–animals and fungi–animals (Jékely & Friedrich, 1999). This conclusion was made after the detection of atypical T-domain-containing calpains 5 and 6 in vertebrates (Mugita et al., 1997; Dear et al., 1997), homologous to *C. elegans* TRA-3, probably, their direct ancestor. Since plants and fungi lack TRA-3 homologues and typical EF-hand Ca^{2+}-binding calpains (Wolfe et al., 1989; Mewes et al., 1997) it was supposed that the the first round of duplication that led to the separation of TRA-3 and typical calpains occurred earlier than the divergence of the lines of protostomatic and deuterostomatic animals. The acquisition of conventional domain IV is considered a late event in calpain evolution that has occurred during animal evolution but not in other eukaryotic branches (Ohno et al., 1984).

The origination of calpains specific to vertebrates is refered to the second episode of gene duplication occurred at early stages of the evolution of chordates. The mapping of the human genome showed that 15 calpain genes as well as the gene of the 28 kDa calpain subunit are grouped on nine chromosomes: 1, 2, 3, 6, 11, 15, 16, 19, and X (Hata et al., 2001; Dear & Boehm, 2001) (Table 1). Tandem chromosomal localization of calpain genes may also suggest their coordinated regulation (e.g., through the Ca^{2+} sensitivity, the involvement of common enhancer elements or locus control regions). Proposed gene duplication events may explain the closer evolutionary relationships between the pairs *CAPN2* and *CAPN8*, *CAPN3* and *CAPN9*, and *CAPN1* and chicken μ/m calpain gene (Jékely & Friedrich, 1999).

The third predicted round of whole genome duplication (also known as the 3R duplication) is specific for teleost fish lineage (Hoegg et al., 2004; Hurley et al., 2007; Braasch & Salzburger, 2009) and endowed teleosts with additional new genes. Thus, both *capn1a* and *capn1b* in fish appear to be orthologous to the single *CAPN1* gene and *capn2a* and *capn2b* – orthologous to calpain 2 present in other vertebrate species. Furthermore, both *Css1a* and *Css1b* identified in zebrafish exhibit conserved synteny and significant sequence identity to human *CSS1*, whereas only a single *CAST* orthologue that produces multiple transcript variants through alternative splicing has been found in teleost (Lepage & Bruce, 2009). Fish-specific proteins of calpain system most probably arose from the 3R duplication during teleost evolution (Hoegg et al., 2004; Hurley et al., 2007).

The presence of five calpain-like proteins with three copies of domain II identified in kinetoplastids has to be a relatively late gene segment duplication in an ancestral species (Ersfeld et al., 2005).

3.2.2 Gene fusion

Protein domains are fundamental and largely independent units of protein structure and function which occur in a number of different combinations or domain architectures. Interestingly, more complex organisms have more complex domain architectures, as well as a greater variety of domain combinations (Chothia et al., 2003; Babushok et al., 2007). It has been proposed that the novel combinations of preexisting domains had a major role in the evolution of protein networks and more complex cellular activities (Pawson & Nash, 2003; Peisajovich et al., 2010).

Fusion of interacting single-function proteins into oligodomain units may facilitate the interaction between functional units and diminish the need to produce proteins in greater amounts to achieve appropriate concentrations of their complexes in the cell. It seems plausible that the acquisition of longer, polyfunctional proteins in eukaryote organisms may have evolved concomitant with the acquisition of multi-exon proteins (Brocchieri & Karlin, 2005).

Gene duplication has preceded domain gain in at least 80% of the gain events (Buljan et al., 2010). Genesis of multiple gene variants as a result of small- and large-scale gene duplication followed by their fusion, exon deletion, insertion of extended sequences, amino acid substitution, etc. are considered as the basis of calpain diversity. Novel domain combinations have a major role in evolutionary innovation. Newly formed structures underwent functional differentiation and caused the appearance of new morphological and cellular characteristics of vertebrates (Ohta, 1991). The major mechanism for gains of new domains in metazoan proteins is likely to be gene fusion through joining of exons from adjacent genes, possibly mediated by non-allelic homologous recombination (Buljan et al., 2010).

Acquisition of the penta-EF-hand module involved in calcium binding (and the formation of heterodimers for some calpains) seems to be a relatively late event in calpain evolution. Calpain homologues from protozoan, plant, *C. elegans*, and fungi lack calmodulin-like domain, and thus it seems likely that the proposed cysteine protease-calmodulin gene fusion leading to the classical calpain structure (Croall & DeMartino, 1991; Goll et al., 2003; Sorimachi et al., 2010) occurred exclusively within the animal lineage. Even similar functional units such as EF-hand module evolutionary gained by different ways. A phylogenetic tree rooted to the calpain-related sequence of the prokaryote *Porphyromonas gingivalis* and based only on the catalytic core suggests that the EF-hand-containing calpains from animals (*C*-terminal EF-hands) and *Tetrahymena* (*N*-terminal EF-hands) are phylogenetically well separated (Croall & Ersfeld, 2007). This raises the possibility that the acquisition of these motifs occurred through independent gene-fusion events in these groups.

Transmembrane motif present in calpains from different sources (ciliate *Tetrahymena* and plants). The phylogenetic analysis reveals a close relationship of indicated transmembrane motif-containing calpains, thus raising the possibility of a common origin for these unusual calpains (Croall & Ersfeld, 2007). Lateral gene transfer from a green alga-type endosymbiont of ciliates is one possible mechanism.

Apart from episodes of gene fusion there are elucidated some episodes of sequence partial deletion. Calpain 8 of vertebrates lacks the functional exon 10'. EF-hands module has been lost during evolution by some calpain genes. Calmodulin-like domain is absent in the structure of calpain CLP-1 from *C. elegans* but on the basis of phylogenetic analysis it occures to belong to EF-hand Ca^{2+}-binding clade (Jékely & Friedrich, 1999); that may suggest that this sequence might have lost it calmoduline-like domain secondarily.

Small subunit of calpains shares homology only with EF-hand-containing domain IV of typical calpains. Mamimum-likehood tree resolved the phylogeny of calpain small subunit which derived from *CAPN3* gene; chromosomal localization of corresponding calpain genes strongly supports this hypothesis (Jékely & Friedrich, 1999).

3.2.3 Divergent evolution of duplicated calpain genes

The domain structure of calpains was realized during several acts of fusion and duplication of genes with subsequent loss of functional exons or residues. It is highly probable that the variable C-terminal domains such as the C2-L domain (domain III), the EF-hand-containing calmodulin-like domain (domain IV), the TRA-3-homologous domain (or C2 domain, T-module), PalB-homologous domain (PBH), and SOL-homologous domain (SOH) originated from independent sources.

In splice-variant *CAPN3* (muscle-specific calpain 3a) and *Drosophila* CALPA there are unique inserted regions (Pintér et al., 1992). Insertion of extended sequences led to the formation of new functional associations and substantionaly changes the characteristics of the protein. For example, due to three unique fragments (NS, novel sequence; IS1, insert 1; and IS2, insert 2) in the structure of calpain 3a are responsible for high susceptibility to autoproteolysis, capability to binding skeletal muscle myofibrils, and nuclear localization (Kinbara et al., 1998; Goll et al., 2003).

Along with mechanisms of protease invention that are based on gene duplication and divergence, the evolution of calpains has also been driven by exon shuffling and the duplication of protein modules in protease genes to form new architectures. By these mechanisms protease substrate and binding specificities can be altered in an evolutionarily rapid and selectable way that leads to gene-family diversity and results in substrate specificity or diversity and new kinetic, inhibitory, and cell or tissue localization properties (Puente et al., 2003).

3.2.4 Functional diversification of calpains

Following gene duplication the newly emerging paralogues descendants may undergo functional differentiation (Ohta, 1991). Genetic variants of calpains were subjected to further functional diversification.

Acquired ancillary domains in calpain structure can interact with many subcellular organelles and molecules including phospholipids, calmodulin, calpastatin and even with each other (that was shown for calpains 8 and 9 and some invertebrate calpains). Structural diversity of calpains due to composition of highly conserved catalytic domain with other functional modules indicates that calpains are involved in a variety of fundamental processes at cellular (for example, motility) and organism (for example, embryogenesis) levels. Despite of overlapping or unique substrate specificities and inhibitor sensitivities of calpains the studies of their individual functions in cellular pathways have to be designed to acheieve their distinguishing. Nevertheless high selectivity and processing mode of their action allow considering calpains as intracellular modulator proteases.

Specific calpain-dependent processes in invertebrates vary from alkaline adaptation in *Aspergillus* (calpain PalB) (Nozawa et al., 2003) to sex determination in nematodes (TRA-3 from *C. elegans*) (Sokol & Kuwabara, 2000).

Both in genetic and cell-biological models on targeted and conditional deletion of *CAPN1*, *CAPN2*, and *CSS1* genes (review: Sorimachi et al., 2010) vital essentiality of calpain 2 in embryogenesis of mammals was shown (Dutt et al., 2006). Variable in structure and functionality splice variants of *CAPN3* gene play tissue-specific roles: calpain 3a in the skeletal muscles; calpains Lp82 (or calpain 3b), Lp85, and Lp74 in the crystalline lens; the others – in the retina (Rt88, Rt88', and Rt90) and in the cornea (Cn94) of the eye (Fukiage et al., 2002). Calpain 3a is essential for skeletal muscle integrity therefore multiple inactivated mutations in *CAPN3* seem to be a cause of limb-girdle muscular dystrophy type IIa (review: Zatz & Starling, 2005). Single nucleotide polymorphism in *CAPN10* gene is a hereditary factor of high susceptibility to type 2 diabetes (Turner et al., 2005). For calpain 5 (human orthologue of TRA-3 from *C. elegans*) it was proposed analogous role of in mammalian sex determination, since the increased expression of the enzyme have been found in testis in relation to colon, kidney, liver and trachea and other tissues.

For the family members lacking key catalytic residues, alternative functions await discovery. Inactive homologues found to be abundant in some protease families and might have important roles as regulatory or inhibitory molecules, acting as dominant negatives by binding substrates through the inactive catalytic or exosite ancillary domains in nonproductive complexes, or by titrating inhibitors from the milieu to increase the net proteolytic activity (López-Otín & Overall, 2002; Puente et al., 2003; Pils & Schultz, 2004). A recent report describes a role for the non-catalytic calpain 6 in the stabilization of microtubules (Tonami et al., 2007). The expression of non-protease homologues is interesting with an evolutionary viewpoint as the discovery of their physiological functions may elucidate calpain functions distinct from proteolysis. It has been argued that non-proteolytic calpains are derived from catalytic precursors, because the majority of family members are active (Todd et al., 2002); however, in kinetoplastids the number of calpain-like proteins with a nonstandard catalytic domain far outnumbers the few proteins with the classical Cys/His/Asn triad (Ersfeld et al., 2005).

3.2.5 Evolution of calpain-related genes

Evolutionary achievements in regulation of calpains implicate the proteins of other families: calpastatin, endogenic inhibitor of calpain activity (protease inhibitor family I27), and regulatory small subunit (PEF protein family). In addition to *CAPN1* and *CAPN2* gene products, *CSS1*, and *CAST* are regarded as a classic/ubiquitous components of the calpain system (Nakamura et al., 1988).

Calpain small subunit is not essential for protease activity (Yoshizawa et al., 1995; Pal et al., 2001) but indispensable for correct folding of catalytic subinit (Yoshizawa et al., 1995; Moldveanu et al., 2002) primarily acting as a chaperon. CSS1 also plays a role of adaptor protein allowing interaction of heterodimer calpains with membranes and consequently facilitates their activation. Dissociated CSS1 might have a function different from proteolysis after forming a homodimer (Pal et al., 2001); thus, it is required for the induction of senescence (Demarchi et al., 2010), Ca^{2+}-dependent repair of wounded plasma membranes (Mellgren et al., 2009), and macroautophagy (Demarchi et al., 2006).

Calpastatin functions as a major inhibitor with high affinity and strict specificity for calpain (Suzuki et al., 1987; Maki et al., 1990). Monomeric calpains including calpains 1 and 2 dissociated from native dimer are not inhibited and thus escape from the regulatory actions of calpastatin. Calpastatin displays molecular polymorphism, the biological significance of which is not yet understood though it seems to be associated somewhat with cellular differentiation (Maki et al., 1990; Lee et al., 1992).

Described proteins are found only in vertebrates. Thus, both calpain small subunit gene *CSS1* and calpastatin gene *CAST* are the "vertebrate genes" (Sorimachi & Suzuki, 2001; Cantserova et al., 2010; Lysenko et al., n.d.).

4. Conclusion

On the basis of molecular evolution within the calpain protease family there was demonstrated how acquisition of structural features facilitates spatial and temporal control of the protease activity. Evolutionary achievements developed in the course of calpain molecular evolution concern some aspects: (1) regulation of calpain synthesis (alternative splicing, tissue-specific patterns of expression), (2) structural characteristics (oligomeric structure, specialized functional domains and insertion sequences), (3) enzymatic characteristics (limited substrate specificity, increased Ca^{2+} sensitivity), (4) intracellular behavior (additional Ca^{2+}-binding sites, membrane-interacting modules, wide range of binding partners), and (5) mechanisms of regulation of their activity (endogenous specific inhibitor, calpastatin, activating proteins).

However calpain study is far for complete and future efforts are needed to determine how the modules associated with proteolytic core influence its function. There is likely to be interplay between protein-protein interactions, membrane binding, Ca^{2+} binding and, potentially, posttranslational modifications in the modulation of calpain function (Croall & Ersfeld, 2007). Many calpain proteins remain to be purified and characterized biochemically, so the challenge of identifying their relevant binding partners as well as specific functional activity remains. The increased knowledge of the structure, function and regulation of proteases will provide excellent opportunities to design new generations of therapeutic inhibitors, including those based on endogenous protease inhibitors.

5. Acknowledgment

This work was supported by project No 14.740.11.1034 (Ministry of Education & Science, Russian Federation), Russian Foundation for Basic Research (grant 11-04-00167), President RF grant "Leader Scientific Schools" (3731.2010.4), and RAS Program "Bioresources".

6. References

Ahn, J.W., Kim, M., Lim, J.H., Kim, G.T. & Pai, H.S. (2004). Phytocalpain controls the proliferation and differentiation fates of cells in plant organ development. *Plant J.*, Vol.38, No.6, pp. 969-981.

Arthur, J.S., Elce, J.S., Hegadorn, C., Williams, K. & Greer, PA. (2000). Disruption of the murine calpain small subunit gene, capn4: calpain is essential for embryonic development but not for cell growth and division. *Mol. Cell Biol.*, Vol.20, pp. 4474-4481.

Arthur, J.S., Gauthier, S. & Elce, J.S. (1995). Active site residues in m-calpain: identification by site-directed mutagenesis. *FEBS Lett.*, Vol.368, No.3, pp. 397-400.

Azam, M., Andarabi, S., Sahr, K., Kamath, L., Kuliopulos, A. & Chisti, A. (2001). Disruption of the mouse μ-calpain gene reveals an essential role in platelet function. *Mol. Cell Biol.*, Vol.21, pp.2213-2220.

Babushok, D.V., Ostertag, E.M. & Kazazian, H.H.Jr. (2007). Current topics in genome evolution: molecular mechanisms of new gene formation. *Cell Mol. Life Sci.*, Vol.64, No.5, pp. 542-554.

Barnes, T.M. & Hodgkin, J. (1996). The tra-3 sex determination gene of *Caenorhabditis elegans* encodes a member of the calpain regulatory protease family. *EMBO J.*, Vol.15, No.17, pp. 4477-4484.

Barrett, A.J., Rawlings, N.D. & Woessner, J.F. (1998). *Handbook of Proteolytic Enzymes.* San Diego: Academic Press, 1998.

Becraft, P.W., Li, K., Dey, N. & Asuncion-Crabb, Y. (2002). The maize dek1 gene functions in embryonic pattern formation and cell fate specification. *Development*, Vol.129, No.22, pp. 5217-5225.

Benyamin, Y. (2006). The structural basis of calpain behavior. *FEBS J.*, Vol.273, No.15, pp. 3413-3414.

Billger, M., Wallin, M. & Karlsson, J.O. (1988). Proteolysis of tubulin and microtubule-associated proteins 1 and 2 by calpain I and II. Difference in sensitivity of assembled and disassembled microtubules. *Cell. Calcium*, Vol.9, pp.33-44.

Blomgren, K., Kawashima, S., Saido, T. C., Karlsson, J.-O., Elmered, A. & Hagberg, H. (1995). Fodrin degradation and subcellular distribution of calpains after neonatal rat cerebral hypoxic-ischemia. *Brain Res.*, Vol.684, No.2, pp.143-149.

Bondareva, L.A. & Nemova, N.N. (2008). Molecular evolution of intracellular Ca^{2+}-dependent proteases. *Rus. J. Bioorg. Chem.* Vol.34, No.3, pp. 266-273.

Bourgeau, G., Lapointe, H., Péloquin, P. & Mayrand, D. (1992). Cloning, expression, and sequencing of a protease gene *(tpr)* from *Porphyromonas gingivalis*, W83, in *Escherichia coli. Infect. Immun.*, Vol.60, No.8, pp. 3186-3192.

Braasch, I. & Salzburger, W. (2009). *In ovo omnia*: diversification by duplication in fish and other vertebrates. *J. Biol.*, Vol.8, No.3, art. 25.

Brocchieri, L. & Karlin S. (2005). Protein length in eukaryotic and prokaryotic proteomes. *Nucleic Acids Res.*, Vol.33, No.10, pp. 3390-3400.

Buljan, M., Frankish, A. & Bateman, A. (2010). Quantifying the mechanisms of domain gain in animal proteins. *Genome Biology*, Vol.11, No.7, pp. R74.

Cantserova, N.P., Ushakova, N.V., Lysenko, L.A. & Nemova, N.N. (2010). Calcium-dependent proteinases of some invertebrates and fishes. *J. Evol. Biochem. Physiol.* Vol.46, No.6, pp. 481-494.

Carafoli, E. & Molinari, M. (1998). Calpain: a protease in search of a function? *Biochem. Biophys. Res. Commun.*, Vol.247, No.2, pp.193-203.

Choi, Y.H., Lee, S.J., Nguyen, P., Jang, J.S., Lee, J., Wu, M.L., et al. (1997). Regulation of cyclin D1 by calpain protease. *J. Biol. Chem.*, Vol.272, pp.28479–28484.

Chothia, C., Gough, J., Vogel, C. & Teichmann, S.A. (2003). Evolution of the protein repertoire. *Science*, Vol.300, No. 5626, pp. 1701-1703.

Cohen, N., Kudryashova, E., Kramerova, I., Anderson, L.V., Beckmann, J.S., Bushby, K. & Spencer, M.J. (2006). Identification of putative *in vivo* substrates of calpain 3 by comparative proteomics of overexpressing transgenic and nontransgenic mice. *Proteomics*, Vol.6, pp. 6075-6084.

Cong, J.Y., Goll, D.E., Peterson, A.M. & Kapprell, H.P. (1989). The role of autolysis in activity of the Ca^{2+}-dependent proteinases (μ-calpain and m-calpain). *J. Biol. Chem.*, Vol.264, No.17, pp. 10096–10103.

Croall, D.E. & DeMartino, G.N. (1991). Calcium-activated neutral protease (calpain) system: structure, function, and regulation. *Physiol. Rev.*, Vol.71, No.3, pp. 813–847.

Croall, D.E. & Ersfeld, K. (2007). The calpains: modular designs and functional diversity. *Genome Biology*, Vol.8, No.6, art. 218.

Crocker, S.J., Smith, P.D., Jackson-Lewis, V., Lamba, W.R., Hayley, S.P., Grimm, E., et al. (2003). Inhibition of calpains prevents neuronal and behavioral deficits in an MPTP mouse model of Parkinson's disease. *J. Neurosci.*, Vol.23, pp. 4081-4091.

Dear, T.N. & Boehm, T. (2001). Identification and characterization of two novel calpain large subunit genes. *Gene*, Vol.274, No.1-2, pp. 245-252.

Dear, T.N., Matena, K., Vingron, M. & Boehm, T. (1997). A new subfamily of vertebrate calpains lacking a calmodulin-like domain: implications for calpain regulation and evolution. *Genomics*, Vol.45, No.1, pp. 175–184.

del Cerro, S., Larson, J., Oliver, M. W. & Lynch, G. (1990). Development of hippocampal long-term potentiation is reduced by recently introduced calpain inhibitors. *Brain Res.*, Vol.530, pp. 91-95

Demarchi, F., Cataldo, F., Bertoli, C. & Schneider, C. (2010). DNA damage response links calpain to cellular senescence. *Cell Cycle*, Vol.9, pp. 755-760.

Demarchi, F., Bertoli, C., Copetti, T., Tanida, I., Brancolini, C., Eskelinen, E.L. & Schneider, C. (2006). Calpain is required for macroautophagy in mammalian cells. *J. Cell Biol.*, Vol.175, pp. 595-605.

Dennison, S.R., Dante, S., Hauss, T., Brandenburg, K., Harris, F. & Phoenix, D.A. (2005). Investigations into the membrane interactions of m-calpain domain V. *Biophys. J.*, Vol.88, No.4, pp. 3008-3017.

Duguez, S., Bartoli, M. & Richard, I. (2006). Calpain 3: a key regulator of the sarcomere? *FEBS J.*, Vol.273, pp. 3427-3436.

Dutt, P., Croall, D.E., Arthur, J.S.C., DeVeyra, T., Williams, K., Elce, J.S. & Greer, P.A. (2006). m-Calpain is required for preimplantation embryonic development in mice. *BMC Dev. Biol.*, Vol.6, art. 3.

Ersfeld, K., Barraclough, H. & Gull, K. (2005). Evolutionary relationships and protein domain architecture in an expanded calpain superfamily in kinetoplastid parasites. *J. Mol. Evol.*, Vol.61, No.6, pp. 742-757.

Franco, S.J. & Huttenlocher, A. (2005). Regulating cell migration: calpains make the cut. *J. Cell Sci.*, Vol.118, Pt.17, pp. 3829-3838.

Fukiage, C., Nakajima, E., Ma, H., Azuma, M. & Shearer, T.R. (2002). Characterization and regulation of lens-specific calpain Lp82. *J. Biol. Chem.*, Vol.277, No.23, pp. 20678-20685.

Gil-Parrado, S., Popp, O., Knoch, T.A., Zahler, T.A., Bestvater, F., Felgenträger, M., et al. (2003). Subcellular localization and *in vivo* subunit interactions of ubiquitous µ-calpain. *J. Biol. Chem.*, Vol.278, No.18, pp. 16336-16346.

Goll, D.E., Thompson, V.F., Taylor, R.G., Ouali, A. & Chou R.-G.R. (1999). The calpain system in muscle tissue. In: *Calpain: Pharmacology and Toxicology of Calcium-Dependent Protease*, Wang K.K.W. & Yuen P.-W. (eds), Philadelphia, PA, Taylor & Francis, 1999, p. 127-160.

Goll, D.E., Thompson, V.F., Li, H., Wei, W. & Cong, J. (2003) The calpain system. *Physiol. Rev.*, Vol.83, No.3, pp. 731-801.

Guroff, G. (1964). A neutral calcium-activated proteinase from the soluble fraction of rat brain. *J. Biol. Chem.*, Vol.239, pp. 149-155.

Haigh, J. (1978). The accumulation of deleterious genes in a population – Muller's Ratchet. *Theor. Popul. Biol.*, Vol.14, No.2, pp. 251-267.

Hakes, L., Pinney, J.W., Lovell, S.C., Oliver, S.G. & Robertson, D.L. (2007). All duplicates are not equal: the difference between small-scale and genome duplication. *Genome Biology*, Vol.8, No.10, art. R209.

Hata, S., Doi, N., Kitamura, F. & Sorimachi, H. (2007). Stomach-specific calpain, nCL-2/calpain 8, is active without calpain regulatory subunit and oligomerizes through C2-like domains. *J. Biol. Chem.*, Vol.282, No.38, pp. 27847-27856.

Hata, S., Nishi, K., Kawamoto, T., Lee, H.-J., Kawahara, H., Maeda, T., et al. (2001). Both the conserved and the unique gene structure of stomach-specific calpains reveal processes of calpain gene evolution. *J. Mol. Evol.*, Vol.53, No.3, pp. 191–203.

Hoegg, S., Brinkmann, H., Taylor, J.S. & Meyer, A. (2004). Phylogenetic timing of the fishspecific genome duplication correlates with the diversification of teleost fish. *J. Mol. Evol.*, Vol.59, pp. 190-203.

Holland, P.W., Garcia-Fernandez, J., Williams, N. & Sidow, A. (1994). Gene duplications and other the origins of vertebrate development. *Development Suppl.*, pp. 125–133.

Hood, J.L., Brooks, W.H. & Roszman, T.L. (2004). Differential compartmentalization of the calpain/calpastatin network with the endoplasmic reticulum and Golgi apparatus. *J. Biol. Chem.*, Vol.279, No. 41, pp. 43126–43135.

Horikawa, Y., Oda, N., Cox, N.J., Li, X., Orho-Melander, M., Hara, M., et al. (2000). Genetic variation in the gene encoding calpain-10 is associated with type 2 diabetes mellitus. *Nat. Genet.*, Vol.26, pp. 163-175.

Hosfield, C.M., Elce, J.S., Davies, P.L. & Jia, Z. (1999). Crystal structure of calpain reveals the structural basis for Ca^{2+}-dependent protease activity and a novel mode of enzyme activation. *EMBO J.*, Vol.18, No.24, pp. 6880-6889.

Huang, Y. & Wang, K.K.W. (2001). The calpain family and human disease. *Trend. Mol. Med.* Vol.7, pp.355-362.

Hurley, I.A., Mueller, R.L., Dunn, K.A., Schmidt, E.J., Friedman, M., Ho, R.K., et al. (2007). A new time-scale for ray-finned fish evolution. *Proc. Biol. Sci.* Vol.274, pp. 489-498.

Huttenlocher, A., Palecek, S.P., Lu, Q., Zhang, W., Mellgren, R.L., Lauffenburger, D.A., et al. (1997). Regulation of cell migration by the calcium-dependent protease calpain. *J. Biol. Chem.*, Vol.272, pp.32719-32722.

Jékely, G. & Friedrich, P. (1999). The evolution of the calpain family as reflected in paralogous chromosome regions. *J. Mol. Evol.*, Vol.49, pp. 272-281.

Johnson, K.L., Degnan, K.A., Ross Walker, J. & Ingram, G.C. (2005). AtDEK1 is essential for specification of embryonic epidermal cell fate. *Plant J.*, Vol.44, No.1, pp. 114-127.

Johnson, J.D., Han, Z., Otani, K., Ye, H., Zhang, Y., Wu, H., et al. (2004). RyR2 and calpain-10 delineate a novel apoptosis pathway in pancreatic islets. *J. Biol. Chem.*, Vol.279, pp. 24794-24802.

Kamei, M., Webb, G.C., Young, I.G. & Campbell, H.D. (1998). SOLH, a human homologue of the *Drosophila melanogaster* small optic lobes gene is a member of the calpain and zinc-finger gene families and maps to human chromosome 16p13.3 near CATM (cataract with microphthalmia). *Genomics*, Vol.51, No.2, pp. 197-206.

Kawasaki, H. & Kawashima, S. (1996). Regulation of the calpain-calpastatin system by membranes. *Mol. Membr. Biol.*, Vol.13, No.4, pp. 217-224.

Kinbara, K., Ishiura, S., Tomioka, S., Sorimachi, H., Jeong, S.-Y., Amano, S., et al. (1998). Purification of native p94, a muscle-specific calpain, and characterization of its autolysis. *Biochem. J.*, Vol.335, Pt.3, pp. 589-596.

Kwak, K.B., Kambayashi, J., Kang, M.S., Ha, D.B. & Chung, C. H.(1993). Cell-penetrating inhibitors of calpain block both membrane fusion and filamin cleavage in chick embryonic myoblasts. *FEBS Lett.*, Vol.323, pp. 151-154

Laval, M. & Pascal, M. (2002). A calpain-like activity insensitive to calpastatin in *Drosophila melanogaster*. *Biochim. Biophys. Acta*, Vol.1570, No.2, pp. 121-128.

Lee, W.J., Ma, H., Takano, E., Yang, H.Q., Hatanaka, L, & Maki, M. (1992). Molecular diversity in amino-terminal domains of human calpastatin by exon skipping. *J. Biol. Chem.*, Vol.267, pp. 8437-8442.

Lee, H.J., Tomioka, S., Kinbara, K., Masumoto, H., Jeong, S., Sorimachi, H., et al. (1999). Characterization of a human digestive tract-specific calpain, nCL-4, expressed in the baculovirus system. *Arch. Biochem. Biophys.*, Vol.362, No.1, pp. 22-31.

Lepage, S.E. & Bruce, A.E. (2008). Characterization and comparative expression of zebrafish calpain system genes during early development. *Dev. Dynam.*, Vol.237, pp. 819-829.

Li, G. & Iyengar, R. (2002). Calpain as an effector of the Gq signaling pathway for inhibition of Wnt/beta-catenin-regulated cell proliferation. *Proc. Natl. Acad. Sci. USA*, Vol.99, pp. 13254-13259.

Lid, S.E., Gruis, D., Jung, R., Lorentzen, J.A., Ananiev, E., Chamberlin, M., et al. (2002). The defective kernel 1 (dek1) gene required for aleurone cell development in the endosperm of maize grains encodes a membrane protein of the calpain gene superfamily. *Proc. Natl. Acad. Sci. USA*, Vol.99, No.8, pp. 5460-5465.

Liu, K., Li, L. & Cohen, S.N. (2000). Antisense RNA-mediated deficiency of the calpain protease, nCL-4, in NIH3T3 cells is associated with neoplastic transformation and tumorigenesis. *J. Biol. Chem.*, Vol.275, pp. 31093-31098.

López-Otín, C. & Overall, C.M. (2002). Protease degradomics: a new challenge for proteomics. *Nature Rev. Mol. Cell Biol.*, Vol.3, No.7, pp. 509–519.

Lysenko, L.A., Kantserova, N.P., Ushakova, N.V. & Nemova, N.N. Proteases of calpain family in water invertebrates and fish. *Rus. J. Bioorg. Chem.* n.d.

Macqueen, D.J., Delbridge, M.L., Manthri, S. & Johnston, I.A. (2010). A newly classified vertebrate calpain protease, directly ancestral to *CAPN1* and *2*, episodically evolved a restricted physiological function in placental mammals. *Mol. Biol. Evol.*, Vol.27, No.8, pp. 1886-1902.

Maki, M., Hatanaka, M., Takano, E., and Murachi, T. (1990). Structure-function relationship of calpastatins. In *Intracellular Calcium-dependent Proteolysis* (Mellgren, R.L. & Murachi, T., eds.), pp. 37-54, CRC Press, Boca Raton.

Mamoune, A., Luo, J.H., Lauffenburger, D.A. & Wells, A. (2003). Calpain-2 as a target for limiting prostate cancer invasion. *Cancer Res.*, Vol.63, pp. 4632-4640.

Maretzki, D., Mariani, M. & Lutz, H.U. (1990). Fatty acid acylation of membrane skeletal proteins in human erythrocytes. *FEBS Lett.*, Vol.259, pp. 305-310.

Marshall, C., Hitman, G.A., Partridge, C.J., Clark. A., Ma, H., Shearer, T.R., Turner, M.D. (2005). Evidence that an isoform of calpain-10 is a regulator of exocytosis in pancreatic β-cells. *Mol. Endocrinol.*, Vol.19, pp. 213-224.

Matthews, K.R. & Gull, K. (1994). Evidence for an interplay between cell cycle progression and the initiation of differentiation between life cycle forms of African trypanosomes. *J. Cell. Biol.*, Vol.125, pp. 1147-1156.

Mellgren, R.L., Miyake, K., Kramerova, I., Spencer, M.J., Bourg, N., Bartoli, M., et al. (2009). Calcium-dependent plasma membrane repair requires m- or mu-calpain, but not calpain-3, the proteasome, or caspases. *Biochim. Biophys. Acta,* Vol.1793, pp. 1886-1893.

Mewes, H.W., Albermann, K., Bahr, M., Frishman, D., Gleissner, A., Hani, J., et al. (1997). Overview of the yeast genome. *Nature*, Vol.387, No.6632 (Suppl.), pp. 7-65.

Moldoveanu, T., Jia, Z. & Davies, P.L. (2004). Calpain activation by cooperative Ca^{2+} binding at two non-EF-hand sites. *J. Biol. Chem.*, Vol.279, No.7, pp. 6106-6114.

Moldoveanu, T., Hosfield, C.M., Lim, D., Elce, J.S., Jia, Z. & Davies, P.L. (2002). A Ca^{2+} switch aligns the active site of calpain. *Cell,* Vol.108, pp. 649-660.

Mottram, J.C., Helms, M.J., Coombs, G.H. & Sajid, M. (2003). Clan CD cysteine peptidases of parasitic protozoa. *Trends Parasitol.*, Vol.19, pp. 182-187.

Mugita, N., Kimura, Y., Ogawa, M., Saya, H. & Nakao, M. (1997). Identification of a novel, tissue-specific calpain *htra-3*, a human homologue of the *Caenorhabditis elegans* sex determination gene. *Biochem. Biophys. Res. Commun.*, Vol.239, No.3, pp. 845-850.

Muller, D., Molinari, I., Soldati, L. & Bianchi, G. (1995). A genetic deficiency in calpastatin and isovalerylcarnitine treatment is associated with enhanced hippocampal long-term potentiation. *Synapse*, Vol.19, No.1, pp. 37-45.

Murakami, T., Hatanaka, M. & Murachi, T. (1981). The cytosol of human erythrocytes contains a highly Ca^{2+}-sensitive thiol protease (calpain I) and its specific inhibitor protein (calpastatin). *J. Biochem.*, Vol.90, No.6, pp. 1809-1816.

Mykles, D.L. (1998). Intracellular proteinases of invertebrates: calcium-dependent and proteasome/ubiquitin-dependent systems. *Int. Rev. Cytol.*, Vol.184, pp. 157-289.

Nakamura, M., Imahori, K. & Kawashima, S. (1988). Tissue distribution of an endogenous inhibitor of calcium-activated neutral protease and age-related changes in its activity in rats. *Comp. Biochem. Physiol. B*, Vol.89, No.2, pp. 381-384.

Nei, M. (1969). Gene duplication and nucleotide substitution in evolution. *Nature*, Vol.221, No.5175, pp. 40-42.

Nemova, N.N., Kaivarainen, E.I. & Bondareva, L.A. (2000). Ca^{2+}-activated neutral proteinase in some fish erythrocytes. *Vestn. Mosk. Univ. Ser. 2 Khimiya*, Vol.41, No.6, pp. 106-108.

Nemova, N.N., Lysenko, L.A. & Kantserova, N.P. (2010). Proteases of the calpain family: structure and functions. *Rus. J. Dev. Biol.*, Vol.41, No.5, pp. 318–325.

Noguchi, M., Sarin, A., Aman, M.J., Nakajima, H., Shores, E.W., Henkart, P.A. & Leonard, W.J. (1997). Functional cleavage of the common cytokine receptor γ chain (c) by calpain. *Proc. Natl. Acad. Sci. USA*, Vol.94, pp. 11534-11539.

Nozawa, S.R., May, G.S., Martinez-Rossi, N.M., Ferreira-Nozawa, M.S., Coutinho-Netto, J., Maccheroni, W.Jr. & Rossi, A. (2003). Mutation in a calpain-like protease affects the posttranslational mannosylation of phosphatases in *Aspergillus nidulans*. *Fungal Genet. Biol.*, Vol.38, No.2, pp. 220-227.

Ohno, S. (1970). *Evolution by Gene Duplication*, Springer-Verlag, ISBN 0-04-575015-7, Berlin.

Ohta, T. (1991). Multigene families and the evolution of complexity. *J. Mol. Evol.*, Vol.33, No.1, pp. 8201-8206.

Pal, G.P., Elce, J.S. & Jia, Z. (2001). Dissociation and aggregation of calpain in the presence of calcium. *J. Biol. Chem.*, Vol.276, pp. 47233-47238.

Pawson, T. & Nash, P (2003). Assembly of cell regulatory systems through protein interaction domains. *Science*, Vol.300, No.5618, pp. 445-452.

Peisajovich, S.G., Garbarino, J.E., Wei, P. & Lim, A.W. (2010). Rapid diversification of cell signaling phenotypes by modular domain recombination. *Science*, Vol.328, No.5976, pp. 368-372.

Pils, B. & Schultz J. (2004). Inactive enzyme-homologues find new function in regulatory processes. *J. Mol. Biol.*, Vol.340, No.3, pp. 399-404.

Pintér, M., Stierandova, A. & Friedrich, P. (1992). Purification and characterization of a Ca^{2+}-activated thiol protease from *Drosophila melanogaster*. *Biochemistry*, Vol.31, No.35, pp. 8201-8206.

Puente, X.S., Sánchez, L.M., Overall, C.M. & López-Otín, C. (2003). Human and mouse proteases: a comparative genomic approach. *Nature Rev. Genet.*, Vol.4, No.7, pp. 544-558.

Rawlings, N.D., Barrett, A.J. & Bateman, A. (2010). MEROPS: the peptidase database. *Nucleic Acids Res.*, Vol.38, Database issue, pp. D227–D233.

Saido, T.C., Sorimachi, H. & Suzuki, K. (1994). Calpain: new perspectives in molecular diversity and physiological, pathological involvement. *FASEB J.*, Vol.8, pp. 814-822.

Saito, K.-I., Elce, J.S., Hamos, J.E. & Nixon, R.A. (1993). Widespread activation of calcium-activated neutral proteinase (calpain) in the brain in Alzheimer disease: a potential molecular basis for neuronal degeneration. *Proc. Natl. Acad. Sci. USA*, Vol.90, No.7, pp. 2628-2632.

Sajid, M. & McKerrow, J.H. (2002). Cysteine proteases of parasitic organisms. *Mol. Biochem. Parasitol.*, Vol.120, pp. 1-21.

Samanta, K., Kar, P., Ghosh, B., Chakraborti, T. & Chakraborti, S. (2007). Localization of m-calpain and calpastatin and studies of their association in pulmonary smooth muscle endoplasmic reticulum. *Biochim. Biophys. Acta*, Vol.1770, No.9, pp. 1297-1307.

Schollmeyer, J.E. (1988). Calpain II involvement in mitosis. *Science*, Vol.240, pp.911-913.

Schmidt-Kastner, R. & Freund, T.F. (1991). Selective vulnerability of the hippocampus in brain ischemia. *Neuroscience*, Vol.40, pp. 599-636.

Shao, H., Chou, J., Baty, C.J., Burke, N.A., Watkins, S.C., Stolz, D.B. & Wells, A. (2006). Spatial localization of m-calpain to the plasma membrane by phosphoinositide biphosphate binding during epidermal growth factor receptor-mediatedactivation. *Mol. Cell. Biol.*, Vol.26, No.14, pp. 5481-5496.

Shih, M., Ma, H., Nakajima, E., David, L.L., Azuma, M. & Shearer, T.R. (2006). Biochemical properties of lens-specific calpain Lp85. *Exp. Eye Res.*, Vol.82, pp. 146-152.

Shinozaki, K, Maruyama, K., Kume, H., Tomita, T., Saido, T.C., Iwatsubo, T. & Obata, K. (1998). The presenilin 2 loop domain interacts with the μ-calpain C-terminal region. *Int. J. Mol. Med.*, Vol.1, pp. 797-799.

Siman, R., Bozyczko-Coyne, D., Savage, M.J. & Roberts-Lewis, J.M. (1996). The calcium-activated protease calpain I and ischemia-induced neurodegeneration. *Adv. Neurol.* Vol.71, pp. 167-174.

Sokol, S.B. & Kuwabara P.E. (2000). Proteolysis in *Caenorhabditis elegans* sex determination: cleavage of TRA-2A by TRA-3. *Genes Dev.*, Vol.14, pp. 901-906.

Sorimachi, H., Tsukahara, T, Okada-Ban, M., Sugita, H., Ishiura, S. & Suzuki, K. (1995). Identification of a third ubiquitous calpain species – Chicken muscle expresses four distinct calpains. *Biochim. Biophys. Acta*, Vol.1261, pp. 381-393.

Sorimachi, H., Ishiura, S. & Suzuki, K. (1997). Structure and physiological function of calpains. *Biochem. J.*, Vol.328, pp. 721-732.

Sorimachi, H. & Suzuki, K. (2001). The structure of calpain. *J. Biochem.*, Vol.129, No.5, pp. 653-664.

Sorimachi, H., Hata, S. & Ono, Y. (2010). Expanding members and roles of the calpain superfamily and their genetically modified animals. *Exp. Anim.*, Vol.59, No.5, pp. 549-566.

Sorimachi, H., Hata, S. & Ono, Y. (2011a). Calpain chronicle – an enzyme family under multidisciplinary characterization. *Proc. Jpn. Acad., Ser. B*, Vol. 87, No.6, pp. 287-327.

Sorimachi, H., Hata, S. & Ono, Y. (2011b). Impact of genetic insights into calpain biology. *J. Biochem.*, Vol.150, No.1, pp.23-37.

Strobl, S., Fernandez-Catalan, C., Braun, M., Huber, R., Masumoto, H., Nakagawa, K., et al. (2000). The crystal structure of calcium-free human m-calpain suggests an electrostatic switch mechanism for activation by calcium. *Proc. Natl. Acad. Sci. USA*, Vol.97, No.2, pp. 588-592.

Suzuki, K. (1990). The structure of calpains and the calpain gene. In: *Intracellular Calcium-Dependent Proteolysis*, R.L. Mellgren & T. Murachi, (eds.), pp. 25-35, CRC Press, Boca Raton, FL.

Suzuki, T., Okumura-Noji, K., Ogura, A., Tanaka, R., Nakamura, K. & Kudo, Y. (1992). Calpain may produce a Ca^{2+}-independent form of protein kinase C in long-term potentiation. *Biochim. Biophys. Res. Commun.* Vol.189, pp. 1515-1520.

Suzuki, K., Ohno, S., Emori, Y., Imajoh, S. & Kawasaki, H. (1987). Primary structure and evolution of calcium-activated neutral protease (CANP). *J. Prot. Chem.*, Vol.6, No.1, pp. 7-15.

Suzuki, K., Hata, S., Kawabata, Y. & Sorimachi, H. (2004). Structure, activation, and biology of calpain. *Diabetes*, Vol.53 (Suppl 1), pp. S12–S18.

Tan, Y., Dourdin, N., Wu, C., De Veyra, T., Elce, J.S. & Greer, P.A. (2006). Conditional disruption of ubiquitous calpains in the mouse. *Genesis*, Vol.44, pp. 297-303.

Thompson, V.F. & Goll, D.E. (2000). Purification of μ-calpain, m-calpain, and calpastatin from animal tissues. *Meth. Mol. Biol.*, Vol.144, pp. 3-16.

Tidball, J.G. & Spenser, M.J. (2000). Calpains and muscular dystrophies. *Int. J. Biochem. Cell Biol.*, Vol.32, No.1, pp. 1-5.

Todd, A.E., Orengo, C.A. & Thornton, J.M. (2002). Sequence and structural differences between enzyme and nonenzyme homologs. *Structure (Cambr)*, Vol.10, pp. 1435-1451.

Tompa, P., Emori, Y., Sorimachi, H., Suzuki, K. & Friedrich, P. (2001). Domain III of calpain is a Ca^{2+}-regulated phospholipid-binding domain. *Biochem. Biophys. Res. Commun.*, Vol.280, No.5, pp. 1333-1339.

Tonami, K., Kurihara, Y., Aburatani, H., Uchijima, Y., Asano, T., Kurihara, H. (2007). Calpain 6 is involved in microtubule stabilization and cytoskeletal organization. *Mol. Cell Biol.*, Vol.27, No.7, pp. 2548-2561.

Tull, D., Vince, J.E., Callaghan, J.M., Naderer, T., Spurck, T., McFadden, G.I., et al. (2004). SMP-1, a member of a new family of small myristoylated proteins in kinetoplastid parasites, is targeted to the flagellum membrane in *Leishmania*. *Mol. Biol. Cell*, Vol.15, No.11, pp. 4775-4786.

Turner, M.D., Cassell, P.G. & Hitman, G.A. (2005). Calpain-10: from genome search to function. *Diabetes Metab. Res. Rev.*, Vol.21, No.6, pp. 505-514.

Wang, C., Barry, J.K., Min, Z., Tordsen, G., Rao, A.G. & Olsen, O.A. (2003). The calpain domain of the maize DEK1 protein contains the conserved catalytic triad and functions as a cysteine proteinase. *J. Biol. Chem.*, Vol.278, No. 36, pp. 34467-34474.

Wolfe, F.H., Szpacenko, A., McGee, K. & Goll, D.E. (1989). Failure to find Ca^{2+}-dependent proteinase (calpain) activity in a plant species, *Elodea densa*. *Life Sci.*, Vol.45, No.22, pp. 2093-2101.

Yoshizawa, T., Sorimachi, H., Tomioka, S., Ishiura, S. & Suzuki, K. (1995). Calpain dissociates into subunits in the presence calcium ions. *Biochem. Biophys. Res. Commun.*, Vol.208, pp. 376-383.

Zatz, M. & Starling, A. (2005). Calpains and disease. *New Engl. J. Med.*, Vol.352, No.23, pp. 2413-2423.

Zimmerman, U.-J.P., Boring, L., Park, J.H., Mukerjee, N. & Wang, K.K.W. (2000). The calpain small subunit gene is essential: its inactivation results in embryonic lethality. *Life,* Vol.50, pp. 63-68.

8

Engineered Derivatives of Maltose-Binding Protein

Paul D. Riggs
New England Biolabs
U.S.A.

1. Introduction

Maltose-binding protein (MBP), a member of the periplasmic binding protein family of Gram negative bacteria, is a versatile substrate for protein engineering. In common with other periplasmic proteins, it is extremely protease resistant, and it can fold properly in both the cytoplasmic and periplasmic compartments. It binds a variety of glucose-α1→4-glucose polysaccharides, from maltose and longer chain maltodextrins to β-cyclodextrin. Upon binding its ligand, it undergoes a large conformational change. These properties have made MBP attractive for a number of engineering studies that have elucidated its role in maltodextrin transport, tuned its properties as an affinity and solubility tag, and transformed it into an allosteric effector or a biosensor for both its natural ligand and for compounds as varied as zinc and TNT (Marvin & Hellinga, 2001, Naal et al., 2002, Wu et al., 1997).

1.1 MBP's function in *Escherichia coli*

MBP's native role in *E. coli* is to shepard maltodextrins from 2 to 7 glucose units in length through the periplasm to the transport apparatus in the cytoplasmic membrane. *E. coli* can grow on high concentrations of maltose in the absence of MBP, but at low concentrations it requires MBP for growth. In the present model of maltodextrin transport (Oldham & Chen, 2011), maltodextrins enter the periplasm by facilitated diffusion through the outer membrane porin LamB, where they are bound by MBP. In binding maltodextrin, MBP shifts from the open, unliganded conformation to the closed conformation. The liganded MBP diffuses to the inner membrane, where it binds to MalFG. Upon binding to MalFG, MBP shifts to the open conformation and releases the maltodextrin to its binding site in MalFG. This transmits a signal to the MalK ATPase subunit bound to MalFG on the cytoplasmic side, promoting ATP hydrolysis. Upon hydrolysis, MalFGK changes conformation and releases the maltodextrin on the cytoplasmic side of the membrane and MBP on the periplasmic side.

1.2 X-ray structures of MBP

The foundation of both understanding the function of MBP and protein engineering using MBP is a series of exquisite crystal structures from the Quiocho lab (Duan et al., 2001, Duan

& Quiocho, 2002, Quiocho et al., 1997, Sharff et al., 1992, Spurlino et al., 1991, Spurlino et al., 1992)(Fig.1). MBP consists of two globular domains, named domains I and II. The binding site is positioned in the cleft between the two domains. The binding site consists of regions that interact with the glucose residues, via nonpolar interactions with the sugar rings (primarily donated by domain II), and a large number of hydrogen bonds that interact with sugar hydroxyls (largely donated by residues in domain I). The structures of the liganded and unliganded forms show that it undergoes a large hinge-twist motion upon binding most of its ligands (Fig.2). Structures of MBP complexed with ligands that support the growth of *E. coli* show a fully closed conformation (Quiocho et al., 1997, Spurlino et al., 1991); the structure with β-cyclodextrin and one of the maltotetraitol structures, which will not support the growth of *E. coli*, show MBP in a open form (Sharff et al., 1993).

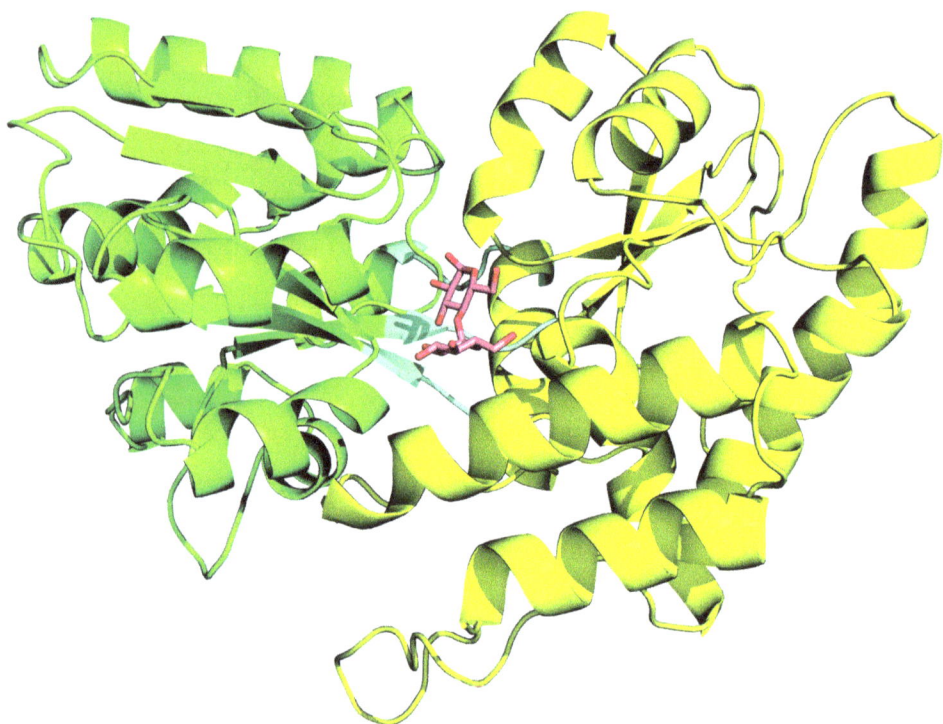

Fig. 1. Cartoon of the structure of MBP with bound maltose. Structure from PDB file 1anf rendered using PyMOL. Domain I is in green, domain II is in yellow, hinge regions are in cyan, and maltose carbons are in salmon.

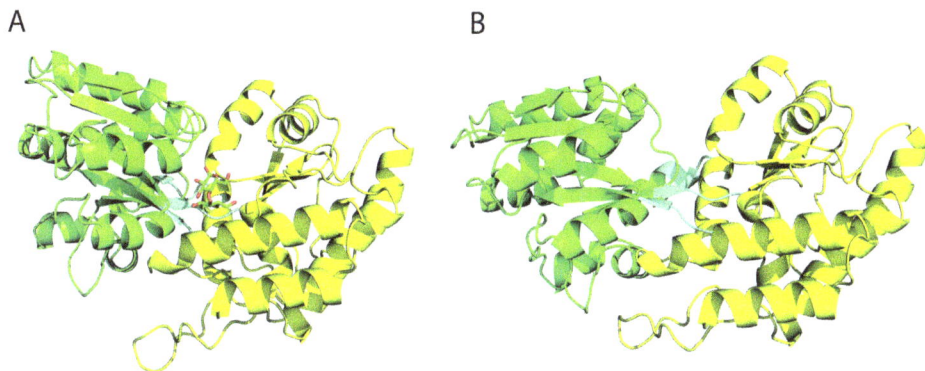

Fig. 2. Ribbon diagram of the two conformations of MBP. Colors are as in Fig 1
A. Closed, maltose-bound form. B. Open, unliganded form (PDB 1jw4)

2 Engineering to understand function

2.1.1 Examination of the open-closed equilibrium

The demonstration of the open and closed conformation of MBP by structural studies led to an examination of the role of this conformational change in the function of MBP. The Nikaido lab first showed that physical techniques such as fluorescence spectroscopy and electron paramagnetic resonance (EPR) spectroscopy could be used to probe the conformation of MBP (Hall et al., 1997a, Hall et al., 1997b). For EPR, they used an engineered MBP that contained an Asp41 to Cys (D41C) mutation in domain I and a Ser211 to Cys (S211C) mutation in domain II. This allowed them to attach spin labels to the cysteines via disulfide linkages. They showed that the double mutant, with and without spin labels, behaved normally in binding maltose. They were then able to measure the distance between the spin labels in the presence of maltose, maltotetraose, maltotrietol and β-cyclodextrin, and showed that upon binding ligands that allow the closed conformation in X-ray structures, the spin-labelled residues are closer together. The Clore lab extended this line of study using paramagnetic relaxation enhancement (PRE) to examine the conformational state of the ligand-bound and unliganded MBP (Tang et al., 2007). Using a different spin-label on the same two mutated cysteine residues, they measured the percentage of MBP in the closed and open form in both states. They found that while the PRE measurements of MBP-maltotriose were consistent with it being in the closed form, the unliganded MBP was a mixture of about 95% open and 5% in a modified closed form. They examined the structure of the modified closed form and found it to be slightly different from the ligand-bound closed form, with the two domains not completely closed and the sugar-binding site of domain II accessible. A third study examined the change in conformation mechanically. Choi et al. attached a single-stranded DNA linker to the two lobes of MBP via a N-terminal His-tag and an L202C mutation in domain II (Choi et al., 2005). Upon adding the complement of the DNA, the stiffness of the double strand exerts force on MBP to push it toward the open conformation. By varying the length of the DNA linker, they were able to measure a change in the Kd for maltose that could be explained by the physical properties of the DNA.

2.1.2 Use of mutant derivatives in structure-function studies

An understanding of the open-closed conformational shift has also led to an engineered derivative of MBP that has been used in structural studies to elucidate maltodextrin transport. Zhang et al. engineered a double mutant of MBP which locks it in the closed conformation. They substituted cysteines for G69 and S337, which are located on separate domains but adjacent to each other in the closed conformation. Upon purification, about 80% of this MBP mutant forms an intramolecular disulfide bond that locks it in the closed conformation. Two labs have taken advantage of this mutant MBP in their stuctural studies of the MalFGK transport apparatus (Oldham & Chen, 2011, Orelle et al., 2010). Orelle et al. used the locked MBP in EPR studies of the transport complex to show that the conformational changes that lead to ATP hydrolysis by MalK only take place after MBP in its closed conformation binds to MalFG. Oldham and Chen then followed this up with a crystal structure of MalFGK with the closed MBP bound, which along with their earlier structure of the open MBP-MalFGK complex (Oldham et al., 2007) gives a nearly complete picture of the steps in maltodextrin transport.

2.1.3 Probing folding and unfolding by mutation

In common with other periplasmic proteins, MBP is remarkably stable and resistant to proteolysis. In spite of this, it has been shown to fold relatively slowly, and becomes incompetent for export to the periplasm if it folds in the cytoplasm. Aspects of its folding/unfolding have been studied by mutating residues that disrupt the folding pathway or lower the energy barrier to unfolding (Betton, J. & Hofnung, 1996, Chang & Park, 2009). Betton and Hofnung found the MBP double mutant G32D I33P among random mutants that were unable to grow on maltose. The mutant MBP was expressed and secreted, but formed inclusion bodies in the periplasm; a derivative without a signal sequence formed inclusion bodies in the cytoplasm. If the inclusion bodies were purified, denatured and refolded they showed near normal affinity for maltose, indicating that the defect was in the folding pathway and not in the structure of the mutant protein. Chang and Park studied the unfolding pathway of MBP by examining its suceptibility to protease digestion during partial denaturation. Many proteins, when treated with a protease under partial denaturation conditions, give proteolytic fragments that indicate the domain or subdomain structure. However, some proteins, including MBP, show an all-or-nothing response to this treatment, where the first proteolytic cleavage leads to rapid unfolding and proteolysis of the entire protein. These researchers used a two step mutagenesis approach to identify the region of MBP that unfolds to allow the initial cleavage. They first surveyed the protein by making mutations in buried residues throughout its sequence to find mutations that destabilized the structure. Upon identifying the susceptible region, they made additional mutants that defined the final two C-terminal α-helices as a subdomain that unfolds to allow the first cleavage.

3 Engineered binding

3.1 High affinity derivatives

3.1.1 Site-directed based on open-closed conformations

The equilibrium between the open and closed form of MBP provides a route to altering its affinity for maltodextrins without changing the sugar binding site on the protein. Since MBP is predominantly in the closed form when bound to maltodextrin, it is possible to alter its affinity for ligand by biasing the equilibrium of the unliganded MBP towards the closed form. In the open form an interface between the two domains forms in the area behind the hinge (Fig. 3). An examination of the interface shows close packing of the side chains. As the conformation shifts to the closed form this interface opens and becomes solvent accessible, and the binding energy of the contact surface is lost. Hellinga and coworkers were the first to take advantage of this by mutating a residue in the interface, I329, to residues with smaller or larger volume side chains, as well as to cysteine to allow attachment of bulkier substituents via the sulfhydryl. They found that larger groups at position 329 yielded proteins with higher affinity for maltose, while the one example with a smaller group (I329A) gave a lower affinity. The largest improvement with a natural substitution was found with I329Y, which gave a 23-fold tighter Kd that wild type. A cysteine at position 329 derivatized with thio-nitropyridine gave a protein with a Kd more than 100-fold tighter than wild type.

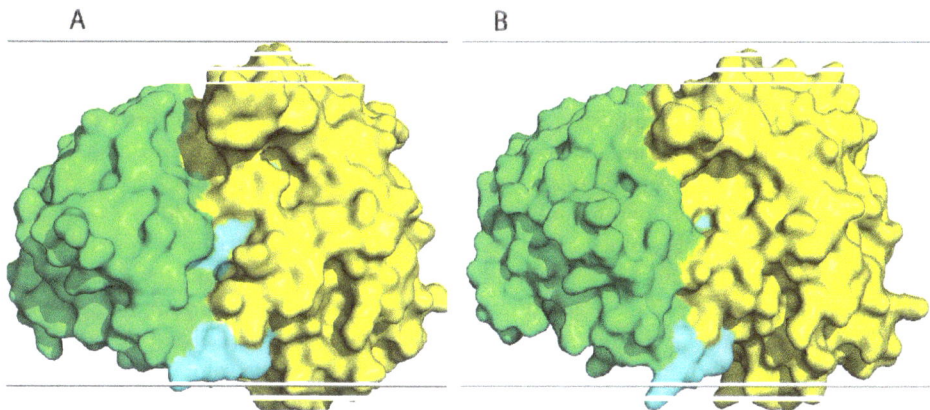

Fig. 3. Surface representation of MBP viewed from the opposite the binding cleft, showing the interface that forms behind the hinge region. A. Closed, maltose-bound form (PDB file 1anf). B. Open, unliganded form (PDB file 1jw4).

Telmer and Shilton took a slightly different approach, examining the interface for alterations that would disrupt the interface by removal of important contacts which stabilize the interface behind the hinge (Telmer & Shilton, 2003). Their analysis considered residues that had higher temperature factors in the closed conformation, indicating higher mobility, as well as the structural contacts of those residues that formed in the open conformation. They identified the side chain of M321 as fitting into a pocket formed by hydrophobic side chains on the opposite domain, and Q325 as an important sidechain that shields the M321 contact from solvent in the open conformation. An M321A Q325A double mutant increased the affinity of MBP for maltotriose about 6-fold. Another interaction their analysis identified was a loop consisting of residues 171-177, which makes contact with residues in domain II in the open conformation. Because G174 is part of a β-turn within the loop, they chose to preserve this residue and delete residues 172-173 and 175-176 on either side to shorten the loop. This deletion mutant also showed about a 6-fold increase in affinity for maltotriose. They combined the M321A/Q325A with the deletions on domain I to obtain a mutant they called MBP-DM. This mutant could be produced *in vivo,* and had a 100-fold higher affinity for maltotriose than wild type.

3.1.2 Random mutagenesis for higher affinity

MBP, in addition to its attractiveness as a substrate for protein engineering for study of its function, is a useful affinity and solubility tag for recombinant protein expression. While MBP is one of the best tags for its ability to give high expression of soluble protein in *E. coli,* it affinity for maltodextrins is sometimes compromised when fused to another protein. The elegant studies that demonstrated higher affinity MBP's by manipulating the open/closed conformational equilibrium led experimenters to try two of these derivatives in the context of a fusion protein, to see if it could improve the yield during purification. Nallamsetty and Waugh fused MBP-DM and MBP (I329Y) to three proteins whose solubilty they had previously shown to be greatly enhanced by expression as an MBP fusion construct (Nallamsetty & Waugh, 2007). They found these derivatives of MBP, while soluble when expressed by themselves, completely loose the ability to enhance the solubility of the fusion partners.

The possibility that there might exist other mutations in MBP that would increase its affinity without damaging its solubility enhancement properties led Walker et al. to do a random mutagenesis and then screen for higher yield and solubility enhancement (Walker et al., 2010). A screen of 4000 random mutants yielded 19 that had increased yield in a small-scale purification, and five that retained solubility enhancement for two fusion proteins that had been demonstrated to tend toward insolubility. Mutations were found in residues that previous studies had identified as important in the open-closed equilibrium, namely M321 and Q325, as well as a number of others that could be rationalized to affect the conformational equilibrium similarly. Mutations that preserved the solubility enhancement of MBP showed modest improvements in affinity, in the range of 2- to 4-fold in both yield and Kd measurements. By combining two of these mutations, a mutant with a 10-fold tighter affinity was obtained that still functioned well as a solubility tag.

3.2 Altered affinity: Zn binding

Hellinga's lab has also done extensive work in altering the affinity of MBP from maltodextrins to zinc by computational design (Benson et al., 2001, Benson et al., 2002, Marvin & Hellinga, 2001). Using the design program *DEZYMER,* they searched the region in and around the maltose binding site in the cleft between the two domains for potential sites, modelled on tetrahedral zinc binding by three histidine residues and a water molecule. They started with the closed structure of MBP, and imposed the constraint that one of the three His residues be in the opposite domain from the other two. This biases the proposed sites towards those where zinc binding would cause the conformational shift from the open to closed form, similar to maltose binding for wild type. Twenty potential sites were identified, and they constructed four of these by site directed mutagenesis; two that replaced residues that make up part of the maltose-binding site and two that were located on the rim of the maltose binding site and might be compatible with retained affinity for maltose. By mutating a surface aspartate residue at position 95 to cysteine and attaching a fluorescent label, they were able to monitor binding by fluorescence intensity. As predicted, the first class bound zinc but not maltose. The second class bound zinc only in the presence of maltose, suggesting that the zinc binding site depended on maltose- induced shift to the closed form for assembly of the zinc binding site. A closer examination of models of the first class indicated that some wild type residues were involved in binding, making fortuitous contacts to the zinc. This was confirmed by mutagenesis to improve the geometry of binding, with a site consisting of two His residues in domain I and two Glu residues in domain II giving the highest affinity for zinc. Telmer and Shilton examined this $His_2 Glu_2$ zinc-binding MBP by low angle X-ray scattering and crystallography, and to their surpise found that the MBP derivative bound zinc in the open conformation (Telmer & Shilton, 2005). All the zinc contacts were donated by sidechains from domain I. They confirmed the requirement of the Glu residues in domain II for zinc binding, but found them 8 A away from the zinc in the crystal structure, and concluded they must contribute to the electronegative environment of the binding site rather than providing direct zinc contacts.

4 MBP as a biosensensor

A number of labs have exploited MBP's specificity and the large conformational shift between the liganded and unliganded forms to develop biosensors. Both sensors that can be used in solution, *in vitro* or *in vivo*, and sensors immobilized on a surface have been explored. The most common readout for these sensors is a change in fluorescence, but alternatives that offer electronic and enzymatic read-outs have also been constructed. Besides solubility and the type of readout, important parameters include the strength of the signal, the dynamic range of ligand that can be detected, and whether the system is designed to use reagents or not.

4.1 Fluorescent biosensors with a single fluorescent reporter

The simplest form of MBP biosensor carries a single flourescent dye, such that a change of fluorescence intensity occurs when the protein undergoes the open/closed conformational shift. These biosensors use the fact that the local environment of the fluorescent group changes when the ligand is bound and the conformation of MBP changes. Gilardi et al. followed this approach with an S337C derivative of MBP by attaching a nitrobenzoxadiazole (NBD) group to the substituted cysteine (Gilardi et al., 1994, Gilardi et al., 1997). They found a increase in fluorescence intensity of 1.8-fold with their sensor. Hellinga and coworkers studied this kind of biosensor in detail, both with MBP and with their derivative that binds zinc (de Lorimier et al., 2002, Hellinga & Marvin, 1998, Marvin & Hellinga, 2001). They mutated a number of MBP residues to cysteine, then tested a number of fluorescent dyes to find the best combination of position and dye. The residue to attach the dye was either near the binding site of maltodextrin, e.g. S233, or one that contacts the opposite domain in one or the other conformation, e.g. D95, F92 and I329. A ratio ΔR was defined, which describes the difference in fluorescent intensity between ligated and unliganded sensor at two wavelength bands. With some experimentation as to fluorescent dye used and placement on MBP, ΔR's of 3 to 4 could be obtained. By mutating the MBP, they obtained a set of derivatives that could sense maltose at concentrations from 0.1 µM to 10 mM (Marvin et al., 1997). Sherman et al.

took a similar approach, attaching fluorescent dyes via different linker arms to an S237C mutant of MBP, and getting a maximum difference in fluorescence intensity of 3-fold between unliganded and liganded forms (Sherman et al., 2006). Jeong et al. produced a sensor that relies on a split green fluorescent protein, which they fused to the N- and C- termini of MBP. Upon maltose binding, the termini move closer together, allowing the split GFP to assemble and leading to a 5-fold increase in fluorescence (Jeong et al., 2006).

One way to make a biosensor convenient and reusable is to immobilize it on a surface. Topoglidis et al. used a nanocrystallin TiO_2 surface to immobilize a fluorophore-labeled MBP and showed a change in fluorescence in response to maltose (Topoglidis et al., 1998). Dattelbaum et al. incorporated a fluorophore-labeled MBP into a sol-gel silica matrix, and demonstrated fluorescence change in response to maltose at close to the same μmolar sensitivity as MBP in solution (Dattelbaum et al., 2009). They subsequently pegylated the MBP to help maintain it in an aqueous environment, and increased the intensity of the fluorescent signal. As we will see below, other forms of MBP-based biosensors can also be adapted to work as immobilized biosensors.

4.2 FRET sensors

Numerous MBP-based biosensors have been develped that use Förstner resonance energy transfer (FRET) to capture information about the binding state of MBP and its derivatives. FRET energy transfer is sensitive to both the distance between the two fluorophores and their relative orientations, making it attractive way to capture confomational information that changes upon ligand binding. Measures of FRET efficiency are quantified by measuring the change in the ratio of fluorescence intensity at the respective wavelengths of the two fluorophores.

4.2.1 MBP-GFP fusion FRET sensors

One way to arrange the FRET donor and acceptor is to fuse green fluorescent protein variants to the N- and C-terminus of MBP. Frommer and coworkers constructed enhanced cyan fluorescent protein-MBP-enhanced yellow fluorescent protein fusion (CFP-MBP-YFP) biosensors that they characterized *in vitro* and *in vivo* in yeast (Fehr et al., 2002, Fehr et al., 2005). The change in fluorescence ratio they observed for these sensors was around 0.1. They constructed CFP-MBP-YFP variants that had mutations in or near the maltose binding site of MBP to reduce its affinity and widen the dynamic range of the biosensor. In later work they improved their biosensor by shortening the linkers between the GFP variants and MBP, which improved its ratio to about 0.2 (Kaper et al., 2008). They used this improved sensor to detect sugar concentrations in *E. coli* upon the addition of maltose to the medium. Ha et al. took a similar approach with their CFP-MBP-YFP sensor, making systematic changes to the linkers to get a derivative with a fluorescence ratio of 0.5 (Ha et al., 2007). They then mutated Trp62 to decrease its affinity for maltose, once again to increase the sensor's dynamic range, and as an unexpected by-product got variants that showed fluorescence ratios of 0.7 and 1.0. They expressed their sensors in yeast and demonstrated the appropriate response to the addition of maltose to the medium. Park et al. also studied a CFP-MBP-YFP FRET sensor, using the characteristics of the energy transfer to measure the distance between the lobes of MBP (Park, K. et al., 2009b).

4.2.2 Dye-labeled FRET sensors

In principle, one should be able to construct a FRET biosensor by attaching fluorescent dyes to the lobes of MBP in place of the fluorescent proteins in the fusions described above. The difficulty lies in devising two site-specific labeling strategies so that each MBP gets labeled on domain I with one member of the donor/acceptor pair and on domain II with its partner. The most convenient and widely used method of site-specific labeleing, used in all studies described above, uses the sulfhydryl on cysteine substitutions as attachment sites on MBP. Hellinga and coworkers devised an elegant protocol to reversibly blocks a labeling site, by fusing a BZif or ZifQNK zinc finger domain to an MBP with a A141C substitution (Smith et al., 2005). Binding of zinc (BZif) or disulfide bond formation (ZifQNK) blocks the cysteines from reacting. The zinc finger domains were fused to the N- or C-terminus of MBP A141C, and in blocked form allowed specific labeling of the cysteine substitution at position 141. The block was then reversed and a second dye was conjugated to the cyteines in the zinc finger domain. A number of combinations of donor/acceptor at the three positions were constructed, including a triple-labelled MBP. With the dye tetramethylrhodamine-5-maleimide at position 141 and Cy5 at the C-terminus, a threefold change in the ratio of the donor:acceptor emission intensities was observed upon the addition of maltose.

4.2.3 Quantum dot FRET sensors

Quantum dots (QD) are colloidal nanocrystal fluorophores that have broad absorption spectra and tunable emmission spectra, making them particularly interesting for the development of FRET sensors. Medintz and coworkers have explored the properties of 555-nm emitting CdSe-ZnS quantum dots with MBP attached via a C-terminal His-tag (Medintz et al., 2003b, Medintz et al., 2004b, Medintz et al., 2005). In their initial experiments, they pre-bound the QD-MBP with β-cyclodextrin conjugated to Cy5 or the quencher QSY9, and measured the increase in fluorescence when the β-cyclodextrin was displaced by maltose (Medintz et al., 2003a). They followed this work with development of a reagentless biosensor, by attaching a Cy3 at a cystein substitution H41C, near the maltose binding site (Medintz et al., 2005), where conformational change upon maltose binding reduces the dyes efficiency as a FRET acceptor. Pons et al. extended this work by developing a single particle QD biosensor using the same MBP H41C labelled with Cy3 (Pons et al., 2006). Multiple MBP Cy3 complexes were immobilized on the QD, and the response to maltose was compared in ensemble and in single particles.

4.2.4 Immobilized FRET sensors

As mentioned above, immobilizing a sensor on a surface has advantages in methodology and reusability, and could allow the fabrication of integrated microfluidic biosensing devices. One FRET biosensor has been adapted to work as a surface-tethered assembly by coating a glass slide with neutravidin and tethering a His-tagged MBP labelled with a quenching dye to the surface via a biotin-Ni-NTA linker (Fig. 4)(Medintz et al., 2004a). A signalling dye linked to β-cyclodextrin is tethered to the same surface via a biotinylated DNA linker. Upon addition of maltose, the β-cyclodextrin is displaced from the MBP binding site and thus removed from the vicinity of the quencher, and fluorescence intensity increases. Sapsford et al. used a similar tethering strategy to link a MBP-Cy3.5-quantum dot sensor to a glass slide, extending the advantages of quantum dot sensors to this format (Sapsford et al., 2004).

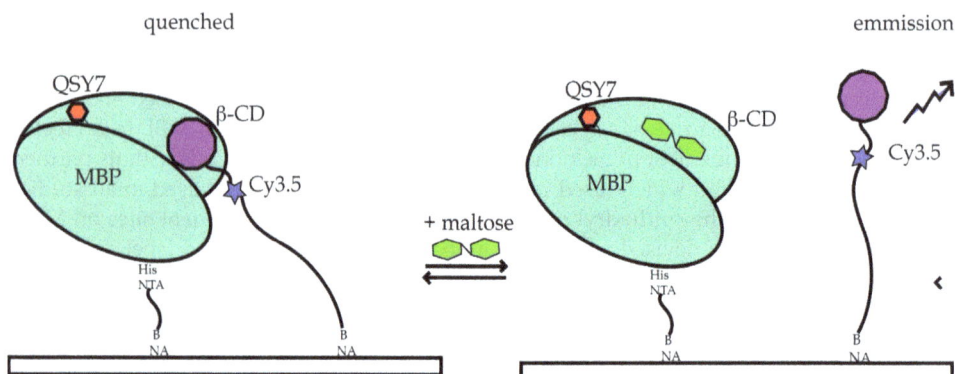

Fig 4. Reagentless surface tethered FRET sensor for maltose. When β-cyclodextrin (β–CD) is bound, the QSY7 on MBP 95C quenches the Cy3.5. Maltose displaces the β –CD and allows fluorescence.

4.3 Enzymatic sensors

Linking a biosensor to an enzyme allows a biochemical read-out that can greatly amplify the signal. Hofnung's lab has mapped certain sites, termed permissive sites, in MBP that can accept peptides without disturbing the folding and function of MBP (Clement et al., 1991). An enzyme encoding ampicillin resistance, TEM β-lactamase, can be inserted into two of these permissive sites, and the fusion protein retains activity for both maltose binding and ampicillin cleavage (Betton, J. M. et al., 1997). In order to make a bifunctional protein where enzymatic activity responds to ligand-binding, Guntas et al. isolated, circularized, then linearized the gene for β-lactamase, then inserted this circularly-permuted collection into a plasmid containing the gene for MBP (Guntas et al., 2004). They identified isolates from this library that exhibited both ampicillin resistance and conferred growth on maltose for a strain deficient in MBP, then screened in a microplate format for an isolate where β-lactamase activity depended on maltose. They obtained an isolate, called RG13, where enzymatic activity in the presence of maltose was 25-fold greater than in its absence. In later work it was demonstrated that mutations that effect MBP's conformational change also effect the β-lactamase activity of the bifunctional protein, strengthening the conclusion that β-lactamase activity was dependent on formation of the closed conformation of MBP (Kim & Ostermeier, 2006). This MBP-β-lactamase biosensor was subsequently immobilized via a C-terminal His-Tag on a gold surface derivatized with Ni-NTA, and maltose dependent β-lactamase activity was confirmed (Zayats et al., 2011).

4.4 Electrochemical sensors

Fluorescent and enzymatic sensors may suffer from the background of natural fluorescence and enzymatic activity that can be present in biological samples, making the development of biosensors with other forms of read-out attractive. Two MBP biosensors with that produce electrical signals have been prototyped, one that produces a electrochemical signal through a redox reaction and one that uses the conformational change in MBP to effect electrical current directly. Hellinga and coworkers attached an MBP derivatized with a Ru(II) reporter to a Ni-NTA derivatized gold electrode via a C-terminal His tag (Benson et al., 2001). A change in redox potential dependent on maltose could be measured by monitoring current as a function of voltage. This general idea was extended to MBP-Ru(II) tethered to ZnS coated CdSe nanoparticles, where the photoluminescense of the nanoparticles responds to the MBP conformational shift (Sandros et al., 2005, Sandros et al., 2006). Park et al. took a different approach, fabricating an ion sensitive field effect transistor by coating a standard CMOS transistor with nickel and assembling MBP-His on the surface (Park, H. J. et al., 2009a). Upon addition of maltose, the charge on MBP effected the gate capacitance of the transistor differentially as the MBP changed from the open to close conformation, which led to a drop in current.

4.5 AFM sensor

While MBP biosensors with electrical read-out have advantages for fabrication of microfluidic devices, the relatively weak signal poses a challenge for detection of very low amounts of ligand. Staii et al. addressed this problem by placing and sensing an MBP derivative on a gold surface using an atomic force microscope (AFM) (Staii et al., 2008). A gold electrode was derivatized with a thiol compound, then activated by scanning with the AFM at a relatively high force. This allowed MBP fused to a Cys-Cys dipeptide at the C-terminus to immobilize at the activated spot by formation of a disulfide bond(s). The placement of the MBP-Cys-Cys could then be detected by the AFM probe scanning at low force. It was found that unliganded MBP, for unknown reasons, produced greater friction interacting with the probe than MBP liganded to maltose. These researchers used the difference in friction to measure the K_D of MBP for maltose at about 1 μM, in good agreement with measurements done on the wild type protein in solution. Their sensor could detect about 10^4 maltose molecules (10 nM concentration), a much higher sensitivity than can be obtained by electrochemical sensors, and uses an MBP derivative that can be purified and incorporated without *in vitro* modification with a dye or quantum dot.

4.6 Biosensors using MBP as a scaffold

MBP has been used by a number of labs as a scaffold in biosensors, not as a conformational switch but simply taking advantage of its robustness and ease of modification. Vardar Schara et al. modified a CFP-MBP-YFP fusion protein by cysteine substitution at several residues centrally located on domain I of MBP (Vardar-Schara et al., 2007). This allowed a binding domain to be attached between the GFP variants, and binding of its ligand could be detected by FRET. Another method of immobilizing a biosensor takes advantage of MBP itself as a linker. MBP has a natural affinity for a pyrolyl-propyl bipyridine surface, allowing an MBP-nitrate reductase fusion protein to be immobilized on the surface (Naal et al., 2002, Takada et al., 2002). This immobilization strategy preserves the enzymatic activity of the nitrate reductase, and allows sensing of TNT by electrochemical detection in a potentiostat: the nitrate reductase reduces the NO_2 groups on TNT, with the electrons ultimately donated by the PBB layer. While these methods do not involve the conformational shift caused by binding maltose, the tools developed for MBP modification made them much easier to fabricate.

5 MBP fusion engineering

MBP has a twenty-year history as an expression, affinity and solublility tag for production of recombinant fusion proteins, and continues to be one of the best tags for producing soluble protein in E. coli. This has led to a number of variations on the basic fusion protein scheme that have facilitated research in diverse areas. An exploration of orthologs and mutants of MBP has extended its utility, and the foundation of so much research using MBP has made it an attractive tool for production of novel peptides and proteins.

5.1 Peptide production

At first glance, producing peptides as fusions to MBP appears to be unattractive, since for a 4 kDa peptide one needs to produce 10 mgs of fusion for every milligram of peptide produced. But the problems in producing synthetic or partially synthetic sequences in a soluble and stable form have led to a number of applications where MBP can be a useful scaffold. As related above, Hofnung and coworkers mapped regions in MBP that could accept insertions of foreign sequences (Clement et al., 1991). They made extensive use of this method to insert epitopes and study the immune response to poliovirus (LeClerc et al., 1990, Leclerc et al., 1991, Lo-Man et al., 1994, Martineau et al., 1996), and binding of HIV to its CD4 receptor (Clement et al., 1996, Lo-Man et al., 1994, Szmelcman et al., 1990), among other studies. Another way in which MBP has been used to study peptides is as a carrier for peptides identified from phage display libraries (Zwick et al., 1998). Restriction sites that allow subcloning from phage identified in commercially-available libraries simplify the transfer of DNA encoding the peptide to the N-terminus of the gene for MBP, and the peptide-MBP fusion can be affinity purified. In both these examples, fusion to MBP not only made purification of the peptide simpler, but most likely avoided problems of stability and solubility.

5.2 Exploration of solubility enhancement

The ability of MBP to enhance the solubility of recombinant fusion proteins in E. coli is one of its most attractive features, but the basis of this enhancement is not well understood. Waugh and coworkers examined the hypothesis that the somewhat hydrophobic binding cleft between the domains of MBP is responsible for solubility enhancement (Fox et al., 2001). They mutated hydrophobic residues that are exposed on the surface of MBP, and tested the mutants for their ability to enhance the solubility of three proteins that tend to insolubility when expressed in E. coli. They identified a region, not in but near one end of the binding cleft, that seemed to be important for this quality, but found the mutations also effected the global stability of MBP and thus were unable to distinguish between a chaperone effect of the folded protein and an effect on the folding pathway of the fusion proteins. Given the wider distribution of mutations that reduce or destroy solubility enhancement in the study of Walker et al. (Walker et al., 2010), it is difficult to imagine that the effect arises from a patch on the folded protein. It remains to be examined whether the effect stems from interactions during folding. In a later study, Waugh and coworkers cloned orthologs of MBP from five bacteria and archae and tested them as fusions with eight proteins that tend to be insoluble when expressed unfused in E. coli (Fox et al., 2003). All of the orthologs could confer solubility on the test proteins, although to different extents, so it seems that this is a common property of these periplasmic binding proteins. The availability of thermostable MBP's may turn out to be attractive for the production of thermostable fusion proteins, as heat treatment is a useful purification step for production of these proteins in mesophiles such as E. coli.

5.3 Determining the structure of MBP fusions

The idea of crystallizing and determining the X-ray structure of a protein fused to MBP, similar to the production of peptides, may at first seem counterintuitive and unnecessarily complicated. The relatively large size of MBP (~40kDa) added to the protein whose structure is to be determined would result in a more complex diffraction pattern. However, the fact that MBP enhances solubility, the possibility that MBP might enhance the formation of crystal contacts, and the fact that the MBP structure is solved, all combine to make this approach one of the ways in which difficult protein structures are solved (Derewenda, 2010, Smyth et al., 2003). In most cases, conformational flexibility between the MBP and the target protein works against crystallization, so fusion proteins expressed in the most common commercial MBP vectors are often modified to shorten the spacer between MBP and the target. This approach has been used to solve the structures of a number of proteins, among them human T cell leukemia virus type 1 gp21 (Kobe et al., 1999), the SarR protein from Staphylococcus aureus (Liu et al., 2001), yeast MATa1 (Ke & Wolberger, 2003), and RACK1 from Arabidopsis thaliana (Ullah et al., 2008).

6 Conclusion

Maltose binding protein has seen wide and extensive use in protein engineering, and continues to be used in ways that could not be forseen when it was first discovered forty-odd years ago. Some of the properties that have made it so attractive for these studies are byproducts of its nature as a periplasmic binding protein in E. coli: it expresses well, it is naturally exported to the periplasm, it is very stable, and it undergoes a large conformational change upon binding maltodextrins. Other properties are somewhat fortuitous: it expresses and folds well when expressed in the cytoplasm, it has no cysteines (and thus no disulfide bonds), and it enhances the solubility of proteins to which it is fused. In addition, early studies that determined its structure and function in such detail made subsequent experiments much easier. All these characteristics have made MBP a prime component of the molecular biologist's toolkit, and will continue to keep it there in the foreseable future.

7 References

Benson, D.E., Conrad, D.W., de Lorimier, R.M., Trammell, S.A. & Hellinga, H.W. (2001). Design of bioelectronic interfaces by exploiting hinge-bending motions in proteins. Science 293, 5535, (Aug 31), pp. 1641-1644.

Benson, D.E., Haddy, A.E. & Hellinga, H.W. (2002). Converting a maltose receptor into a nascent binuclear copper oxygenase by computational design. Biochemistry 41, 9, (Mar 5), pp. 3262-3269.

Betton, J. & Hofnung, M. (1996). Folding of a mutant maltose-binding protein of Escherichia coli which forms inclusion bodies. J Biol Chem 271, 14, pp. 8046-8052.

Betton, J.M., Jacob, J.P., Hofnung, M. & Broome-Smith, J.K. (1997). Creating a bifunctional protein by insertion of beta-lactamase into the maltodextrin-binding protein. *Nat Biotechnol* 15, 12, (Nov), pp. 1276-1279, 1087-0156.

Chang, Y. & Park, C. (2009). Mapping transient partial unfolding by protein engineering and native-state proteolysis. *J Mol Biol* 393, 2, (Oct 23), pp. 543-556, 1089-8638.

Choi, B., Zocchi, G., Canale, S., Wu, Y., Chan, S. & Perry, L.J. (2005). Artificial allosteric control of maltose binding protein. *Phys Rev Lett* 94, 3, (Jan 28), pp. 038103.

Clement, J.M., Charbit, A., Martineau, P., O'Callaghan, D., Szmelcman, S., Leclerc, C. & Hofnung, M. (1991). Bacterial vectors to target and/or purify polypeptides: their use in immunological studies. *Ann Biol Clin (Paris)* 49, 4, pp. 249-254.

Clement, J.M., Jehanno, M., Popescu, O., Saurin, W. & Hofnung, M. (1996). Expression and biological activity of genetic fusions between MalE, the maltose binding protein from Escherichia coli and portions of CD4, the T-cell receptor of the AIDS virus. *Protein Expr Purif* 8, 3, pp. 319-331.

Dattelbaum, A.M., Baker, G.A., Fox, J.M., Iyer, S. & Dattelbaum, J.D. (2009). PEGylation of a maltose biosensor promotes enhanced signal response when immobilized in a silica sol-gel. *Bioconjug Chem* 20, 12, (Dec), pp. 2381-2384, 1520-4812.

de Lorimier, R.M., Smith, J.J., Dwyer, M.A., Looger, L.L., Sali, K.M., Paavola, C.D., Rizk, S.S., Sadigov, S., Conrad, D.W., Loew, L. & Hellinga, H.W. (2002). Construction of a fluorescent biosensor family. *Protein Sci* 11, 11, (Nov), pp. 2655-2675.

Derewenda, Z.S. (2010). Application of protein engineering to enhance crystallizability and improve crystal properties. *Acta Crystallogr D Biol Crystallogr* 66, Pt 5, (May), pp. 604-615, 1399-0047.

Duan, X., Hall, J.A., Nikaido, H. & Quiocho, F.A. (2001). Crystal structures of the maltodextrin/maltose-binding protein complexed with reduced oligosaccharides: flexibility of tertiary structure and ligand binding. *J Mol Biol* 306, 5, (Mar 9), pp. 1115-1126, 0022-2836.

Duan, X. & Quiocho, F.A. (2002). Structural evidence for a dominant role of nonpolar interactions in the binding of a transport/chemosensory receptor to its highly polar ligands. *Biochemistry* 41, 3, (Jan 22), pp. 706-712, 0006-2960.

Fehr, M., Frommer, W.B. & Lalonde, S. (2002). Visualization of maltose uptake in living yeast cells by fluorescent nanosensors. *Proc Natl Acad Sci U S A* 99, 15, (Jul 23), pp. 9846-9851, 0027-8424.

Fehr, M., Okumoto, S., Deuschle, K., Lager, I., Looger, L.L., Persson, J., Kozhukh, L., Lalonde, S. & Frommer, W.B. (2005). Development and use of fluorescent nanosensors for metabolite imaging in living cells. *Biochem Soc Trans* 33, Pt 1, (Feb), pp. 287-290, 0300-5127.

Fox, J.D., Kapust, R.B. & Waugh, D.S. (2001). Single amino acid substitutions on the surface of Escherichia maltose-binding protein can have a profound impact on the fusion proteins. *Protein Sci* 10, 3, (Mar), pp. 622-630.

Fox, J.D., Routzahn, K.M., Bucher, M.H. & Waugh, D.S. (2003). Maltodextrin-binding proteins from diverse bacteria and archaea are potent solubility enhancers. *FEBS Lett* 537, 1-3, (Feb 27), pp. 53-57.

Gilardi, G., Zhou, L.Q., Hibbert, L. & Cass, A.E. (1994). Engineering the maltose binding protein for reagentless fluorescence sensing. *Anal Chem* 66, 21, (Nov 1), pp. 3840-3847, 0003-2700.

Gilardi, G., Mei, G., Rosato, N., Agro, A.F. & Cass, A.E. (1997). Spectroscopic properties of an engineered maltose binding protein. *Protein Eng* 10, 5, (May), pp. 479-486, 0269-2139.

Guntas, G., Mitchell, S.F. & Ostermeier, M. (2004). A molecular switch created by in vitro recombination of nonhomologous genes. *Chem Biol* 11, 11, (Nov), pp. 1483-1487.

Ha, J.S., Song, J.J., Lee, Y.M., Kim, S.J., Sohn, J.H., Shin, C.S. & Lee, S.G. (2007). Design and application of highly responsive fluorescence resonance energy transfer biosensors for detection of sugar in living Saccharomyces cerevisiae cells. *Appl Environ Microbiol* 73, 22, (Nov), pp. 7408-7414, 0099-2240.

Hall, J.A., Gehring, K. & Nikaido, H. (1997a). Two modes of ligand binding in maltose-binding protein of Escherichia coli. Correlation with the structure of ligands and the structure of binding protein. *J Biol Chem* 272, 28, (Jul 11), pp. 17605-17609, 0021-9258.

Hall, J.A., Thorgeirsson, T.E., Liu, J., Shin, Y.K. & Nikaido, H. (1997b). Two modes of ligand binding in maltose-binding protein of Escherichia coli. Electron paramagnetic resonance study of ligand-induced global conformational changes by site-directed spin labeling. *J Biol Chem* 272, 28, (Jul 11), pp. 17610-17614, 0021-9258.

Hellinga, H.W. & Marvin, J.S. (1998). Protein engineering and the development of generic biosensors. *Trends Biotechnol* 16, 4, (Apr), pp. 183-189.

Jeong, J., Kim, S.K., Ahn, J., Park, K., Jeong, E.J., Kim, M. & Chung, B.H. (2006). Monitoring of conformational change in maltose binding protein using split green fluorescent protein. *Biochem Biophys Res Commun* 339, 2, (Jan 13), pp. 647-651, 0006-291X.

Kaper, T., Lager, I., Looger, L.L., Chermak, D. & Frommer, W.B. (2008). Fluorescence resonance energy transfer sensors for quantitative monitoring of pentose and disaccharide accumulation in bacteria. *Biotechnol Biofuels* 1, 1, pp. 11, 1754-6834.

Ke, A. & Wolberger, C. (2003). Insights into binding cooperativity of MATa1/MATalpha2 from the crystal structure of a MATa1 homeodomain-maltose binding protein chimera. *Protein Sci* 12, 2, (Feb), pp. 306-312, 0961-8368.

Kim, J.R. & Ostermeier, M. (2006). Modulation of effector affinity by hinge region mutations also modulates switching activity in an engineered allosteric TEM1 beta-lactamase switch. *Arch Biochem Biophys* 446, 1, (Feb 1), pp. 44-51, 0003-9861.

Kobe, B., Center, R.J., Kemp, B.E. & Poumbourios, P. (1999). Crystal structure of human T cell leukemia virus type 1 gp21 ectodomain crystallized as a maltose-binding protein chimera reveals structural evolution of retroviral transmembrane proteins. *Proc Natl Acad Sci U S A* 96, 8, (Apr 13), pp. 4319-4324, 0027-8424.

LeClerc, C., Martineau, P., Van der Werf, S., Deriaud, E., Duplay, P. & Hofnung, M. (1990). Induction of virus-neutralizing antibodies by bacteria expressing poliovirus epitope in the periplasm. The route of immunization the isotypic distribution and the biologic activity of the antibodies. *J Immunol* 144, 8, (Apr 15), pp. 3174-3182.

Leclerc, C., Charbit, A., Martineau, P., Deriaud, E. & Hofnung, M. (1991). The cellular location of a foreign B cell epitope expressed by recombinant bacteria determines its T cell-independent or T cell-dependent characteristics. *J Immunol* 147, 10, pp. 3545-3552.

Liu, Y., Manna, A., Li, R., Martin, W.E., Murphy, R.C., Cheung, A.L. & Zhang, G. (2001). Crystal structure of the SarR protein from Staphylococcus aureus. *Proc Natl Acad Sci U S A* 98, 12, (Jun 5), pp. 6877-6882, 0027-8424.

Lo-Man, R., Martineau, P., Betton, J.M., Hofnung, M. & Leclerc, C. (1994). Molecular context of a viral T cell determinant within a chimeric bacterial protein alters the diversity of its T cell recognition. *J Immunol* 152, 12, (Jun 15), pp. 5660-5669.

Martineau, P., Leclerc, C. & Hofnung, M. (1996). Modulating the immunological properties of a linear B-cell insertion into permissive sites of the MalE protein. *Mol Immunol* 33, 17-18, (Dec), pp. 1345-1358.

Marvin, J.S., Corcoran, E.E., Hattangadi, N.A., Zhang, J.V., Gere, S.A. & Hellinga, H.W. (1997). The rational design of allosteric interactions in a monomeric protein and its applications to the construction of biosensors. *Proc Natl Acad Sci U S A* 94, 9, (Apr 29), pp. 4366-4371.

Marvin, J.S. & Hellinga, H.W. (2001). Conversion of a maltose receptor into a zinc biosensor by computational design. *Proc Natl Acad Sci U S A* 98, 9, (Apr 24), pp. 4955-4960.

Medintz, I.L., Clapp, A.R., Mattoussi, H., Goldman, E.R., Fisher, B. & Mauro, J.M. (2003a). Self-assembled nanoscale biosensors based on quantum dot FRET donors. *Nat Mater* 2, 9, (Sep), pp. 630-638, 1476-1122.

Medintz, I.L., Goldman, E.R., Lassman, M.E. & Mauro, J.M. (2003b). A fluorescence resonance energy transfer sensor based on maltose binding protein. *Bioconjug Chem* 14, 5, (Sep-Oct), pp. 909-918, 1043-1802.

Medintz, I.L., Anderson, G.P., Lassman, M.E., Goldman, E.R., Bettencourt, L.A. & Mauro, J.M. (2004a). General strategy for biosensor design and construction employing multifunctional surface-tethered components. *Anal Chem* 76, 19, (Oct 1), pp. 5620-5629, 0003-2700.

Medintz, I.L., Konnert, J.H., Clapp, A.R., Stanish, I., Twigg, M.E., Mattoussi, H., Mauro, J.M. & Deschamps, J.R. (2004b). A fluorescence resonance energy transfer-derived structure of a quantum dot-protein bioconjugate nanoassembly. *Proc Natl Acad Sci U S A* 101, 26, (Jun 29), pp. 9612-9617, 0027-8424.

Medintz, I.L., Clapp, A.R., Melinger, J.S., Deschamps, J.R. & Mattoussi, H. (2005). A reagentless biosensing assembly based on quantum dot donor Förstner resonance energy transfer. *Adv Mater* 17, 20, pp. 2450-2455.

Naal, Z., Park, J.H., Bernhard, S., Shapleigh, J.P., Batt, C.A. & Abruna, H.D. (2002). Amperometric TNT biosensor based on the oriented immobilization of a nitroreductase maltose binding protein fusion. *Anal Chem* 74, 1, (Jan 1), pp. 140-148, 0003-2700.

Nallamsetty, S. & Waugh, D.S. (2007). Mutations that alter the equilibrium between open and closed conformations of Escherichia coli maltose-binding protein impede its ability to enhance the solubility of passenger proteins. *Biochem Biophys Res Commun* 364, 3, (Dec 21), pp. 639-644.

Oldham, M.L., Khare, D., Quiocho, F.A., Davidson, A.L. & Chen, J. (2007). Crystal structure of a catalytic intermediate of the maltose transporter. *Nature* 450, 7169, (Nov 22), pp. 515-521, 1476-4687.

Oldham, M.L. & Chen, J. (2011). Crystal structure of the maltose transporter in a pretranslocation intermediate state. *Science* 332, 6034, (Jun 3), pp. 1202-1205, 1095-9203.

Orelle, C., Alvarez, F.J., Oldham, M.L., Orelle, A., Wiley, T.E., Chen, J. & Davidson, A.L. (2010). Dynamics of alpha-helical subdomain rotation in the intact maltose ATP-binding cassette transporter. *Proc Natl Acad Sci U S A* 107, 47, (Nov 23), pp. 20293-20298, 1091-6490.

Park, H.J., Kim, S.K., Park, K., Lyu, H.K., Lee, C.S., Chung, S.J., Yun, W.S., Kim, M. & Chung, B.H. (2009a). An ISFET biosensor for the monitoring of maltose-induced conformational changes in MBP. *FEBS Lett* 583, 1, (Jan 5), pp. 157-162, 1873-3468.

Park, K., Lee, L.H., Shin, Y.B., Yi, S.Y., Kang, Y.W., Sok, D.E., Chung, J.W., Chung, B.H. & Kim, M. (2009b). Detection of conformationally changed MBP using intramolecular FRET. *Biochem Biophys Res Commun* 388, 3, (Oct 23), pp. 560-564, 1090-2104.

Pons, T., Medintz, I.L., Wang, X., English, D.S. & Mattoussi, H. (2006). Solution-phase single quantum dot fluorescence resonance energy transfer. *J Am Chem Soc* 128, 47, (Nov 29), pp. 15324-15331, 0002-7863.

Quiocho, F.A., Spurlino, J.C. & Rodseth, L.E. (1997). Extensive features of tight oligosaccharide binding revealed in high-resolution structures of the maltodextrin transport/chemosensory receptor. *Structure* 5, 8, (Aug 15), pp. 997-1015.

Sandros, M.G., Gao, D. & Benson, D.E. (2005). A modular nanoparticle-based system for reagentless small molecule biosensing. *J Am Chem Soc* 127, 35, (Sep 7), pp. 12198-12199, 0002-7863.

Sandros, M.G., Shete, V. & Benson, D.E. (2006). Selective, reversible, reagentless maltose biosensing with core-shell semiconducting nanoparticles. *Analyst* 131, 2, (Feb), pp. 229-235, 0003-2654.

Sapsford, K.E., Medintz, I.L., Golden, J.P., Deschamps, J.R., Uyeda, H.T. & Mattoussi, H. (2004). Surface-immobilized self-assembled protein-based quantum dot nanoassemblies. *Langmuir* 20, 18, (Aug 31), pp. 7720-7728, 0743-7463.

Sharff, A.J., Rodseth, L.E., Spurlino, J.C. & Quiocho, F.A. (1992). Crystallographic evidence of a large ligand-induced hinge-twist motion between the two domains of the maltodextrin binding protein involved in active transport and chemotaxis. *Biochemistry* 31, 44, (Nov 10), pp. 10657-10663.

Sharff, A.J., Rodseth, L.E. & Quiocho, F.A. (1993). Refined 1.8-A structure reveals the mode of binding of beta-cyclodextrin to the maltodextrin binding protein. *Biochemistry* 32, 40, (Oct 12), pp. 10553-10559, 0006 2960 (Print) 0006-2960 (Linking).

Sherman, D.B., Pitner, J.B., Ambroise, A. & Thomas, K.J. (2006). Synthesis of thiol-reactive, long-wavelength fluorescent phenoxazine derivatives for biosensor applications. *Bioconjug Chem* 17, 2, (Mar-Apr), pp. 387-392, 1043-1802.

Smith, J.J., Conrad, D.W., Cuneo, M.J. & Hellinga, H.W. (2005). Orthogonal site-specific protein modification by engineering reversible thiol protection mechanisms. *Protein Sci* 14, 1, (Jan), pp. 64-73.

Smyth, D.R., Mrozkiewicz, M.K., McGrath, W.J., Listwan, P. & Kobe, B. (2003). Crystal structures of fusion proteins with large-affinity tags. *Protein Sci* 12, 7, (Jul), pp. 1313-1322, 0961-8368.

Spurlino, J.C., Lu, G.Y. & Quiocho, F.A. (1991). The 2.3-A resolution structure of the maltose- or maltodextrin-binding protein, a primary receptor of bacterial active transport and chemotaxis. *J Biol Chem* 266, 8, (Mar 15), pp. 5202-5219.

Spurlino, J.C., Rodseth, L.E. & Quiocho, F.A. (1992). Atomic interactions in protein-carbohydrate complexes. Tryptophan residues in the periplasmic maltodextrin receptor for active transport and chemotaxis. *J Mol Biol* 226, 1, (Jul 5), pp. 15-22.

Staii, C., Wood, D.W. & Scoles, G. (2008). Verification of biochemical activity for proteins nanografted on gold surfaces. *J Am Chem Soc* 130, 2, (Jan 16), pp. 640-646, 1520-5126.

Szmelcman, S., Clement, J.M., Jehanno, M., Schwartz, O., Montagnier, L. & Hofnung, M. (1990). Export and one-step purification from Escherichia coli of a MalE-CD4 hybrid protein that neutralizes HIV in vitro. *J Acquir Immune Defic Syndr* 3, 9, pp. 859-872.

Takada, K., Naal, Z., Park, J.-H., Shapleigh, J.P., Bernhard, S., Batt, C.A. & Abruña, H.D. (2002). Study of Specific Binding of Maltose Binding Protein to Pyrrole-Derived Bipyridinium Film by Quartz Crystal Microbalance. *Langmuir* 18, pp. 4892-4897.

Tang, C., Schwieters, C.D. & Clore, G.M. (2007). Open-to-closed transition in apo maltose-binding protein observed by paramagnetic NMR. *Nature* 449, 7165, (Oct 25), pp. 1078-1082, 1476-4687.

Telmer, P.G. & Shilton, B.H. (2003). Insights into the conformational equilibria of maltose-binding protein by analysis of high affinity mutants. *J Biol Chem* 278, 36, (Sep 5), pp. 34555-34567.

Telmer, P.G. & Shilton, B.H. (2005). Structural studies of an engineered zinc biosensor reveal an unanticipated mode of zinc binding. *J Mol Biol* 354, 4, (Dec 9), pp. 829-840.

Topoglidis, E., Cass, A.E., Gilardi, G., Sadeghi, S., Beaumont, N. & Durrant, J.R. (1998). Protein Adsorption on Nanocrystalline TiO(2) Films: An Immobilization Strategy for Bioanalytical Devices. *Anal Chem* 70, 23, (Dec 1), pp. 5111-5113, 0003-2700.

Ullah, H., Scappini, E.L., Moon, A.F., Williams, L.V., Armstrong, D.L. & Pedersen, L.C. (2008). Structure of a signal transduction regulator, RACK1, from Arabidopsis thaliana. *Protein Sci* 17, 10, (Oct), pp. 1771-1780, 1469-896X.

Vardar-Schara, G., Krab, I.M., Yi, G. & Su, W.W. (2007). A homogeneous fluorometric assay platform based on novel synthetic proteins. *Biochem Biophys Res Commun* 361, 1, (Sep 14), pp. 103-108, 0006-291X.

Walker, I.H., Hsieh, P.C. & Riggs, P.D. (2010). Mutations in maltose-binding protein that alter affinity and solubility properties. *Appl Microbiol Biotechnol* 88, 1, (Sep), pp. 187-197, 1432-0614.

Wu, Q., Storrier, G.D., Pariente, F., Wang, Y., Shapleigh, J.P. & Abruna, H.D. (1997). A nitrite biosensor based on a maltose binding protein nitrite reductase fusion immobilized on an electropolymerized film of a pyrrole-derived bipyridinium. *Anal Chem* 69, 23, (Dec 1), pp. 4856-4863, 0003-2700.

Zayats, M., Kanwar, M., Ostermeier, M. & Searson, P.C. (2011). Surface-tethered protein switches. *Chem Commun (Camb)* 47, 12, (Mar 28), pp. 3398-3400, 1364-548X.

Zwick, M.B., Bonnycastle, L.L., Noren, K.A., Venturini, S., Leong, E., Barbas, C.F., 3rd, Noren, C.J. & Scott, J.K. (1998). The maltose-binding protein as a scaffold for monovalent display of peptides derived from phage libraries. *Anal Biochem* 264, 1, (Nov 1), pp. 87-97, 0003-2697.

9

Protein Folding, Binding and Energy Landscape: A Synthesis

Shu-Qun Liu[1,2], Yue-Hui Xie[3], Xing-Lai Ji[1,2], Yan Tao[1], De-Yong Tan[4],
Ke-Qin Zhang[1] and Yun-Xin Fu[1,5]

[1]*Laboratory for Conservation and Utilization of Bio-Resources & Key Laboratory for
Southwest Biodiversity, Yunnan University, Kunming,*
[2] *Sino-Dutch Biomedial and Information Engineering School,
Northeastern University, Shenyang,*
[3]*Teaching and Research Section of computer,
Department of Basic Medical, Kunming Medical College, Kunming*
[4]*School of Life Sciences, Yunnan University, Kunming,*
[5]*Human Genetics Center, School of Public Health,
The University of Texas Health Science Center, Houston, Texas,*
[1,2,3,4]*P. R. China*
[5]*USA*

1. Introduction

Protein folding and molecular recognition and binding provide the basis for life on earth. The native 3D structure of a protein is necessary for its biological function; and the protein-associated molecular recognition and binding is the fundamental principle of all biological processes, thereby unraveling the mechanisms of protein folding and binding is fundamental to describing life at molecular level (Perozzo et al., 2004).

Of particular interest is that protein folding and binding are similar processes with the only difference between them being the presence and absence of the chain connectivity (Ma et al., 1999). Among many models such as diffusion-collision (Karplus & Weaver, 1994), nucleation-condensation (Itzhaki et al., 1995), jigsaw puzzle (Harrison & Durbin, 1985), hydrophobic collapse (Agashe et al., 1995) and stoichiometry models (Mittal et al., 2010) proposed to describe the mechanism of protein folding or, to some extent, protein binding, the "folding funnel" model (Leopold et al., 1992) based on the free energy landscape theory (Bryngelson et al., 1995; Wales, 2003) has now been most widely accepted. In this model, protein folding can be regarded as going down the funnel-like free energy landscape through multiple parallel pathways towards the bottom of the funnel (Noe et al., 2009; Yon, 2002); and protein binding occurring along the rugged free energy surface around the funnel bottom can be viewed as microfunnel fusion (Ma et al., 1999; Tsai et al., 2001) or downward extension of free energy well, which provides the thermodynamic interpretation for the binding models such as lock-and-key (Fischer, 1894), induced fit (Koshland, 1958) and conformational selection (Foote & Milstein, 1994; Monod et al., 1965). Thus, the energy landscape provides a basis to synthesize and explain the mechanisms of protein folding, binding and dynamics. Under the energy landscape theory, the protein folding and binding are essentially thermodynamically controlled processes involving various types of driving forces, including the enthalpic contribution of noncovalent bond formations, entropic effects such as uptake and release of solvent molecules, burial of apolar protein surface area (hydrophobic effect), restrictions of degrees of freedom of protein and ligand, and loss of the rotational and translational freedom of the interacting partners (Chaires, 2008; Perozzo et al., 2004). Briefly, these two processes, which are driven by a decrease in total Gibbs free energy, are dictated by the mechanism of a delicate balance of the opposing effects involving enthalpic and entropic contributions. It should be noted that it is the thermodynamically driven subtle enthalpy-entropy compensation that leads to the global free energy minimum of the protein-solvent system, but that the specific inter-atomic interactions observed in the folded 3D protein structure are to a large extent the consequence of thermodynamic equilibrium and can not fully define the driving forces for the folding process (Ji & Liu, 2011a). Such a subtle enthalpy-entropy compensation also underlies the mechanisms of both the induced fit and conformational selection models of protein-ligand binding since most of the binding processes occur through both mechanisms: the conformational selection, which is under entropy control, generally plays a role in the initial contact between the protein and ligand and determines the binding specificity; the induced fit, which is mainly driven by the negative enthalpy of bond formation, plays a role in the subsequent conformational adjustment and determines the binding affinity.

Interestingly, we speculate that many other processes can be explained by thermodynamic enthalpy-entropy compensation (Ji & Liu, 2011b), i.e., the Yin and Yang balance in traditional Chinese medicine theory could correspond to the enthalpy and entropy compensation of the second law of thermodynamics; global warming can be considered as the consequence of excessive production of positive entropy (carbon dioxide) from chemically ordered fossil fuel (enthalpy storage), urging people to slow resource consumption to delay the inevitable death by entropy; the cancer occurrence can be considered as the consequence of accumulation of nucleotide mutations (which could be seen as increase in entropy) caused by gradual loss of DNA repair ability, leading to disordered gene expressions and uncontrolled cell proliferation.

In this chapter the problems of protein folding are first introduced and discussed. In order to address these problems, several protein folding models or hypotheses are discussed and compared. Subsequently, the free energy landscape theory, including the concept of the energy landscape, the funnel-like folding landscape and the detailed folding process occurring in such a folding funnel, the roughness/ruggedness of the energy landscape, the amplitude and timescale of the protein dynamics and their associated functions, and the dynamic feature of the energy landscape, are introduced and discussed. The protein-ligand binding models such as the lock-and-key, induced fit and conformational selection are introduced and the driving forces and thermodynamic mechanism underlying these models are discussed in depth based on the free energy landscape theory. Finally, the mechanisms of protein folding and binding are synthesized to elicit a general funnel model of the energy landscape, under which the folding and binding are commonly driven by the decrease in total Gibbs free energy through the entropy-enthalpy compensation, with different steps being dominated by either the entropic or enthalpic contribution. We highlight that it is the free energy landscape theory that synthesize the protein folding and binding mechanisms together to bring about the common funnel model. In addition, the energy landscape-associated concepts such as the thermodynamics, kinetics, free energy, enthalpy, entropy, and frustration/ruggedness of the landscape are also introduced. This chapter will facilitate not only the protein engineering studies aimed at modifying protein structure and function but also the rational drug design and understanding of life in the post-genomic era.

2. The problems of protein folding

Protein folding is the process of transformation of one-dimensional linear information encoded in the amino acid sequence into a functional 3D structure. This is essentially a physical chemistry process for which the mechanism remains elusive due to the following three incompletely resolved problems (Dill et al., 2007): i) the thermodynamic question of how a native structure results from the inter-atomic forces acting on an amino acid sequence — the folding code; ii) the kinetic problem of how a native structure can fold so fast — the folding rate; iii) the computational problem of how to predict the native structure of a protein from its amino acid sequence — the protein structure prediction.

2.1 The folding code and thermodynamics

The apparent complexity of folded structures and the abundant diversity of conformational states of the denatured proteins make it difficult to understand and differentiate various types of forces that drive protein folding. A popular opinion, which was first proposed by (Anfinsen & Scheraga, 1975), considers that the protein folding code is the sum of many different small interactions such as electrostatic, van der Waals and hydrogen bonding interactions, which are embodied in the static secondary structures and are mainly distributed locally in the sequence. However, from a statistical mechanics point of view, the folding code can be viewed as the thermodynamic profile of interaction process that represents various types of forces that drive folding, including the changes in folding free energy (ΔG), enthalpy (ΔH) and entropy (ΔS). The predominant force that drives folding is considered the hydrophobic interaction or desolvation, which is essentially the entropic effect and is distributed both locally and non-locally in the sequence. The native secondary structures and the finally observed interactions (such as electrostatic, van der Waals and hydrogen bonding interactions) are the consequence rather than the cause of folding forces (Dill, 1999; Dill et al., 2007), which drive the search for the global free energy minimum of the protein-solvent system through entropy-enthalpy compensation (Ji & Liu, 2011a).

2.2 The folding rate and kinetics

In 1961, Anfinsen's classic experiments on renaturation of denatured ribonuclease showed that small globular proteins can fold spontaneously into their native 3D structures in the absence of any catalytic biomolecules, thus leading to an assumption that the folded proteins exist in the global-minimum free energy state (Anfinsen et al., 1961). Soon afterward, Levinthal (Levinthal, 1968, 1969) recognized that proteins have far too vast conformational spaces to permit a thorough search for the folded native structure in a biologically relevant time. To resolve this "paradox", Levinthal proposed the notion of protein folding pathway through which proteins can search and converge quickly to the native states. Using various experimental techniques including fast laser temperature-jump methods (Yang & Gruebele, 2003), protein engineering technique such as site-directed mutagenesis that identify those amino acids that control fold speed (Matouschek et al., 1989; Sosnick et al., 2004), fluorescence resonance energy transfer (FRET) methods that can watch the formation of particular contacts (Magg et al., 2006; Schuler et al., 2002), and hydrogen exchange methods that can see the structural folding events (Maity et al., 2005), the timescales of folding events occurring on several model proteins (such as cytochrome c, chymotrypsin inhibitor 2, barnase, src, fyn SH3 domains, Trp-cage, etc.) were extensively studied. The results show that the protein folding rates correlate mainly with the topology of the native proteins, exhibiting variable timescales with a range of more than eight orders of magnitude (Dill et al., 2007; Plaxco et al., 1998). The proteins that fold relatively fast usually have more local structural elements such as helices and tight turns, whereas the proteins that fold relatively slowly have more non-local structural elements such as β sheets. However, is there a single pathway or multiple parallel routes by which a protein folds up? How does one characterize the protein folding pathways connecting the denatured and native states? These are two aspects of the kinetic problem of protein folding. Full answers to these questions require the introduction of concept of the "protein folding funnel" (Dill, 1985; Dill & Chan, 1997; Dobson, 2000; Leopold et al., 1992) based on the free energy landscape theory (Bryngelson et al., 1995; Wales, 2003). Briefly, protein folding can be viewed as going down the funnel-like energy landscape via multiple parallel pathways from the vast majority of individual non-native conformations to the native states around the bottom of the funnel. At any stage the protein exists as an ensemble of conformations and can be trapped transiently in many local energy minimum wells.

It must keep in mind that the folding time and pathways are concepts related to the kinetics, which defines the energy barriers between different conformational states; while that the free energy corresponds to the conformational states in relation to thermodynamics, which defines the relative probabilities/populations/lifetimes of the different conformational substates (Henzler-Wildman & Kern, 2007). Both of these concepts belong to the field of statistical mechanics and are of crucial importance in understanding the protein folding mechanism.

2.3 The protein structure prediction

The major motivations for protein structure prediction arise from i) the fundamental desire of human beings to explore natural laws and the needs to test or verify the putative protein folding mechanism; ii) the requirements to determine the functions of encoded protein sequences in the post-genomic era and to make drug discovery faster and more efficient through substitution fast cheap computer simulation for slow and expensive structural biology experiments (Dill et al., 2007). The protein structure prediction methods can be classified as physics-based and bioinformatics-based ones (Dill et al., 2007). The basis for developing the purely physics-based approaches is the thorough recognition and understanding of the protein folding mechanism. An ideal aim of such methods is that they are able to direct an amino acid sequence of a protein to fold into its native 3D structure through mimicking various possible driving forces without knowledge derived from databases. On the contrary, the bioinformatics-based methods rely mainly on various priori information derived from databases such as statistical energy functions, secondary structure and structure classification information of proteins, in addition to the information of the primary amino acid sequence.

2.3.1 Bioinformatics-based methods and CASP

Critical Assessment of Techniques for Protein Structure Prediction (CASP) is a community-wide, worldwide experiment for computer-based protein structure prediction taking place every two years since 1994 (Moult et al., 1995). The primary goal of CASP is to facilitate the development of the approaches for predicting protein 3D structure from its amino acid sequence. To date, there have been nine previous CASP experiments: from CASP1 in 1994 to CASP9 in 2010. Description of these experiments and the full data (such as targets, predictions, interactive tables with numerical evaluation results, dynamic graphs and prediction visualization tools) can be accessed by following the link at: http://predictioncenter.org/index.cgi.

Currently, the ranking results show that the most efficient and accurate bioinformatics-based method is the homology modeling (Sali & Blundell, 1993), which has not only the speed to compute approximate folds for large fractions of whole genomes (Pieper et al., 2006), but also sufficient accuracy for the predicted structures generally with errors better than 3 Å if the sequence identity between target and template is higher than 30% (Baker & Sali, 2001). For single-domain globular proteins having < ~ 90 amino acids in length, several web server tools can predict the structures often having C_α RMSD (root mean square deviation) values of about 2–6 Å with respect to their experimental structures (Baker, 2006; Bradley et al., 2005; Dill et al., 2007; Zhang et al., 2005). Very recently, a newly developed homology-based structure prediction method achieved 1-2 Å (mean 1.6 Å) C_α RMSD to the reference crystal structures from the full spectrum of test domains in recent CASP experiments (Soundararajan et al., 2010). This approach is mainly based on defining and extracting the fold-conserved PCAIN (protein core atomic interaction network) from distinct protein domain families, which is found to be significantly distinguished among different domain families and as such, can be considered as the "signature" of a domain's native fold. It should be noted that the bioinformatics methods are mainly limited to structural prediction of the single domains with relatively small size. Predicting the structures of large multidomain or domain-swapped proteins (Dill et al., 2007) and membrane proteins (Bowie, 2005) is the major challenge in the near future.

2.3.2 Physics-based methods

Due to the objective existence of some limitations such as the incomplete understanding of the folding mechanism, inaccurate description of the force-fields and insufficient computational power, there have been only a few successes of physics-based methods in computing structures with very small size in the past decade. For example, in 1988, Duan and Kollman performed nearly one microsecond simulation on the unfolded 36-residue villin headpiece in explicit solvent, yielding a collapsed state with a 4.5 Å C_α RMSD relative to its NMR structure (Duan & Kollman, 1998). In 2003, the IBM blue gene team achieved the nearly native folded structure of the 20-residue Trp-cage peptide with C_α RMSD value of < 1 Å relative to its experimental structure using 92 ns of replica-exchange molecular dynamics in implicit solvent (Pitera & Swope, 2003). In 2010, Folding@home researchers Voelz et al. reported a computer simulation of *ab initio* protein folding on the millisecond timescale for the 39-residue protein NTL9(1−39) using distributed molecular dynamics simulations in implicit solvent on GPU processors, and observed a small number of productive folding events at a temperature lower than the melting point of the force field (Voelz et al., 2010).

Despite these limited successes in only small simple proteins, we believe that, as was stated by (Dill et al., 2007), "*once physics-based approaches succeed, the advantages would be the ability for us to predict conformational changes, such as induced fit, a common and important unsolved problem in computational drug discovery; the ability to understand protein mechanisms, motions, folding processes, conformational transitions and other situations in which protein behavior requires more than just knowledge of the static native structure; the ability to design synthetic proteins for new applications or to design foldable polymers from nonbiological backbones; and the ability to systematically improve protein modeling based on the laws of physics*". We also believe that the combination of physics-based with bioinformatics-based structural prediction methods will incubate more powerful approaches that can not only predict accurately and quickly the protein structures but also facilitate the solution of the two key problems relevant to protein folding: the folding code and the folding rate.

3. Protein folding models and hypotheses

Due to the complexity of the protein folding problems, several simplified models have been proposed to explore the conformational and sequence spaces, describe the sequence of events in the folding process, and further, to probe, examine and uncover the protein folding mechanism. These include diffusion collision model, framework model, nucleation-condensation model, Zipping-and-assembly (ZA), hydrophobic collapse model, folding funnel model as well as a stoichiometry-driven protein folding hypothesis.

3.1 Diffusion collision model

In order to discuss the dynamics of protein folding and introduce a possible mechanism by which a protein can make use of a more sophisticated procedure than a simple random search of all conformational possibilities (i.e., the Levinthal paradox), (Karplus & Weaver, 1976, 1994) proposed the diffusion collision model (Figure 1A), which is based on dividing the folding of a protein molecule into parts such that the informationn stored in the sequence of each part can be used independently. These divided parts, called microdomains, can be portions of incipient secondary structures (such as α-helices, β-strands and connecting turns) or hydrophobic clusters (i.e., hydrophobic amino acids close to each other in the amino acid sequence). In this model, the folding process could start with a polypeptide chain in an extended random coiled state: the microdomains are formed first; then they move diffusively with collisions taking place between them, leading to the coalescence into the multi-microdomain intermediates or higher aggregates; and finally, these multi-microdomain intermediates form the exact tertiary structure through mutual rearrangements. The formation of the microdomains or secondary structures (caused by local interactions) and their interactions rather than interactions between individual amino acids would greatly reduce folding times to reasonable values. However, such a simplified model can neither characterize detailed motions involved in the folding process at the atomic level nor provide complete information on energy barriers and populations of individual conformational states. Therefore, the diffusion collision model is only useful for a semiquantitative analysis of protein folding dynamics through concentrating on the microdomains.

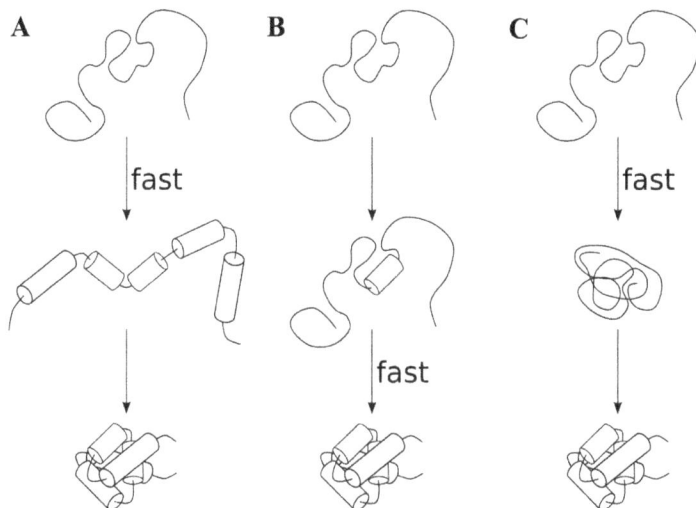

Fig. 1. Schematic representation of several protein folding models. (A) Diffusion collision model. (B) Nucleation condensation model. (C) Hydrophobic collapse model. The fast step of folding is indicated. For details, see the text.

In 1982, Kim and Baldwin (Kim & Baldwin, 1982) proposed the "framework" model for protein folding in which the native secondary structures are formed before the tertiary structures. The framework model is essentially similar to the diffusion collision model since both consider that folding is a hierarchical process in which simple local structures are formed first and this is followed by the formation of more complex structures through interactions or collisions among these simple structures.

3.2 Nucleation-condensation model

The nucleation-condensation mechanism (Figure 1B) suggests that the reaction is initiated by the formation of a marginally stable nucleus that acts as a template or seed from which the native structure would propagate through rapid condensation of further structure around it (Nolting & Agard, 2008). In this way, the number of conformations required to be sampled in the folding reaction is dramatically reduced. An essential feature of the nucleation–condensation model is that the nucleus occurrence requires the simultaneous formation of both secondary and tertiary interactions. This contrasts the diffusion collision model, which, as described above, involves a hierarchical process of structural assembly where the secondary structure elements, guided by local contacts, are initially formed independently of tertiary structure, followed by the packing of secondary structure elements to coalesce into the native tertiary structure. Therefore, the essential difference between these two folding models is whether the folding reaction can proceed in a concerted manner (such as in the nucleation-condensation model) or in a stepwise manner (such as in the diffusion collision and/or framework models).

3.3 Zipping-and-assembly model

ZA model (also termed hydrophobic zipper model) was proposed by (Dill et al., 1993) to explain how a protein finds its native state without a globally exhaustive search. In this model, protein folding reaction starts with the independent formations of local structural pieces along the chain, then those pieces either grow (zip) or coalesce (assemble) with other structures to form the native fold. At first glance, such a folding process is very similar to that of the collision-diffusion model since both describe the hierarchical process of structural assembly. However, these two models differ from each other in some aspects. In the ZA model there is a fast collapse that is driven by nonlocal hydrophobic interactions leading to the concurrent formations of the hydrophobic clusters, helices, and sheets, and the hydrophobic zipping process further result in a broad ensemble of compact intermediate states that have much secondary but little tertiary structure; the subsequent slow process involves breaking/unzipping incorrect (nonnative) hydrophobic-hydrophobic contacts to proceed towards the native structure. The ZA therefore highlights that the folding is dominated by nonlocal interactions, i.e., with collapse as the driving force, and the secondary structures are a consequence of the collapse, rather than its cause (Dill et al., 1995). On the contrary, the diffusion collision model assumes that the local interactions are important factors in reducing conformational search, implying that the major driving forces for protein folding is the hydrogen bonding, electrostatic and van der Waals interactions, leading to the early formations of the secondary structures and the subsequent collapse and assembly into the tertiary structure. The ZA is therefore similar, to some extent, to the nucleation-condensation since in both models the nonlocal interactions dominate the formation of the collapse or the nucleus. An advantage of the ZA model is that it explicitly points out the cause of the collapse — the hydrophobic force. In this way, the ZA model is essentially similar to the hydrophobic collapse model because both models suggest that it is the solvent-mediated hydrophobic force that drives the first stage — the collapse, in the protein folding process.

The ZA mechanism has been recently applied to the folding of proteins through using the AMBER96 force field with a generalized Born/surface area implicit solvent model and replica exchange molecular dynamics sampling. The results show that such a physics-based model can find approximately correct folds for chain length up to ~ 100 amino acids (Ozkan et al., 2007).

3.4 Jigsaw puzzle model

The jigsaw puzzle model, a qualitative mechanism for explaining the folding of the compact domains of proteins, was proposed by (Harrison & Durbin, 1985). They argued that proteins fold by large numbers of different, parallel pathways rather than by a single definitive sequence of events, thus making folding more robust to mutations that do not affect adversely the native structure. Here the existence of multiple protein folding pathways towards the native structure resembles the multiple routes to reach a unique solution in the assembly of a jigsaw puzzle. According to this model, the identification of intermediates represents a kinetic description rather than a structural one since each intermediate consists of heterogeneous species in rapid equilibrium (Yon, 2002). This model presents some similarities with the diffusion-collision model because its kinetics could be obtained through the diffusions and collisions of the "microdomains" if all the elementary microdomains have similar properties and multi-microdomain intermediates of the same size have similar folding and unfolding rates. The computer simulations based on lattice models support the jigsaw puzzle model (Sali et al., 1994). The multiple pathways folding highlighted by the jigsaw puzzle model is actually consistent with a "new view" of protein folding funnel concept based on the energy landscape theory: the native state of a protein at its global free energy minimum is located at the bottom of funnel-like energy landscape; and each molecule may follow a different microscopic route from the top to the bottom. The folding funnel model and energy landscape theory will be discussed in more detail later.

3.5 Hydrophobic collapse model and folding funnel hypothesis

The hydrophobic collapse model (Figure 1C), first proposed by (Dill, 1985), is mainly based on the observation that the globular proteins' native states often contain a hydrophobic core of apolar amino acid side chains in the interior of proteins, while leaving most of the polar or charged residues on the solvent-exposed protein surface. Therefore this model considers that the folding of a globular protein starts with a rapid collapse of the chain driven by the hydrophobic force, resulting in the formation of the collapsed intermediates, i.e., the molten globule, within which the secondary and/or tertiary structure is subsequently formed. This model highlights that the "earliest event" of the folding is that the side chains of hydrophobic residues are driven together by the action of water, which is in much the same way as aggregation of oil droplets in water (Kauzmann, 1954, 1959; Lum et al., 1999; Stillinger, 1973; R. Zhou et al., 2004). Upon dripping oil into a water solvent or upon recovering the conditions for protein folding, the nonlocal and nonspecific hydrophobic force mediated by the water molecules forces the polypeptide chain or the oil to cluster together rapidly to minimize the surface area of the aggregates and to maximize the entropy of the solvent, thus lowering the total Gibbs free energy of the systems. In the case of the protein folding, the free energy is further lowered by favorable energetic contacts such as the isolation of electrostatically charged side chains on the protein surface and neutralization of salt bridges within the protein's interior. The subsequent step of the protein folding is a slow annealing to the native structure: mutual rearrangement and adjustment of the already formed building blocks through disruption and reformation of various noncovalent bonds, which further lowers the free energy of the system via enthalpy contribution. As mentioned above, the hydrophobic collapse model is similar to the hydrophobic zipper model, both point out that the folding is dominated by water-mediated hydrophobic interaction.

Compared to the hydrophobic collapse model, folding funnel hypothesis characterizes a more complete scenario of how a population of unfolded peptide chains goes down from the top of funnel-like energy landscape to the bottom of the native states via multiple routes. The hydrophobic collapse is just one of the most important events occurring in early-stage folding. Other events such as the formation of the molten globule, trapping of folding intermediates in the local minima, overcoming the local energy barriers, formation of the transition state ensemble, reconfiguration of the tertiary interactions and aspects of thermodynamics and kinetics of protein structure can also be characterized by the folding funnel (these will be introduced and discussed in the "Free energy landscape" section). In the canonical depiction of the folding funnel, the depth of the well represents the energetic stabilization of the native state versus the denatured state; the width of the well represents the conformational entropy of the system; and the surface outside the well is shown to be relatively flat to represent the heterogeneity of the random coil state.

3.6 Stoichiometry-driven protein folding hypothesis

The stoichiometry-driven protein folding hypothesis has been proposed very recently by Mittal et al., whose statistical analyses of 3718 folded protein structures reveal a surprisingly simple unifying principle of backbone organization, which is interpreted as Chargaff's Rules related to protein folding, i.e., a stoichiometry-driven protein folding (Mittal & Jayaram, 2011b, 2011a; Mittal et al., 2010). One of the interesting findings is that the total number of possible contacts for the C_α atom of a given amino acid correlates excellently with its occurrence percentage in the amino acid sequences, leading to a conclusion that protein folding is a direct consequence of a narrow band of the stoichiometric occurrences of amino acids in the primary sequences, regardless of the size and the fold of a protein. On the other hand, the statistical results reveal that there is no "preferential interactions" between amino acids, thus leading to a conclusion that the "preferential interactions" between amino-acids do not drive protein folding.

However, although interesting, the above two conclusions need to be interpreted carefully and more work need to be done to verify this hypothesis and to elucidate the mechanism underlying these statistical phenomena (Berendsen, 2011; Chan, 2011; Ji & Liu, 2011a). For example, how to elucidate the relationship between the stoichiometric occurrences of amino acids and the driving force of protein folding? Is it possible that the higher the occurrence frequency an amino acid has, the greater tendency for it to occupy the core of a folded protein, and the more it contributes to the hydrophobic interaction? Further work is required to examine the relationship between the burial extent of an amino acid and its occurrence percentage. The statistical result of the lack of preferential interactions between amino acids is based on a large number of already folded protein structures, and therefore it is safe to interpret it as the consequence rather than the cause of protein folding. However, it must be noted that the preferential interactions between amino acids are the basis of the development of knowledge-based potentials, which in turn form the underpinning of protein structure prediction by modeling and simulation (Ji & Liu, 2011a; Rackovsky & Schraga, 2011; Sarma, 2011) that are now routinely performed in many laboratories across the globe (Aman et al., 2010; Sklenovský & Otyepka, 2010; Tao et al., 2010). The lack of preferential interactions between amino acids in a large sample set suggests that the non-preferential or random inter-residue interactions might maintain the structural stability of the already folded proteins. However, using only the C_α atom as the statistical object may shield the effect of the preferential interactions occurring to a very large extent between the side chain groups of different amino acids (Galzitskaya et al., 2011; Matthews, 2011). It is reasonable to consider that the preferential interactions do not dominate protein folding when the folding process is thought to be under the thermodynamic control such as in the hydrophobic collapse and/or folding funnel models. The entropy effect is the dominant force that drives the early rapid collapses of the peptide chain, contributing the most to the decrease of the free energy of the protein-solvent system. The later slow rearrangement of the local building block elements is driven by enthalpic contribution, i.e., the loss and the formation of noncovalent bonds. We consider that the later slow process inevitably results in preferential or special interactions between amino acids and contributes, although to a lesser extent compared to the entropy effect, to the decrease of the free energy of the system. Therefore, the stoichiometry-driven folding hypothesis needs further analyses and verification using both theoretical and experimental methods, as has been suggested and discussed by dozens of protein structural chemists in "a conversation on protein folding", which was organized by professor Sarma (Mittal & Jayaram, 2011b; Sarma, 2011), the Editor-in-chief of the Journal of Biomolecular Structure and Dynamics.

3.7 Association between different protein folding models

The protein folding models or hypotheses described above are not independent and mutually exclusive but rather they are inextricably intertwined and commonly attempt to grasp different aspects of protein folding and to solve one or more of the three problems concerning protein folding: the folding code, folding rate and protein structure prediction. Experimental results have provided some support to each of the models. For example, the diffusion collision and framework models are essentially similar, commonly emphasizing a hierarchical folding process in which the local interactions drive the initial formations of simple structures (Figure 1A). Such a stepwise process is in line with our intuitive thinking for sequential events and facilitates the development of secondary structure prediction methods. In addition, if we consider the early formation of the secondary structures as individual nucleation processes, the diffusion collision model could be thought of as special case of the nucleation-condensation model. The hydrophobic collapse model highlights that the nonlocal hydrophobic force or the entropy effect mediated by the water is the predominant driving force for protein folding (Figure 1C). This model can be considered as another extreme of nucleation-condensation model because the collapse leads to one or more hydrophobic nucleuses composed of both the transient secondary and tertiary structures. Interestingly, the ZA model may be considered as the combined case of both the diffusion collision and the hydrophobic collapse models because of its hierarchical assembly feature and the dominant nonlocal hydrophobic force in collapse and zipping processes. In addition, the hydrophobic collapse is only a specific event occurring in the folding process described by the folding funnel model. The latter emphasizes the multiple parallel routes towards the global free energy minimum, which is also the essential feature of the jigsaw puzzle model. The advantage of the folding funnel model is the characterization of two essential elements of protein folding: the thermodynamics and kinetics, which play crucial roles not only in the protein folding but also in the binding and interaction between protein and its ligands.

4. Free energy landscape

As described above, protein folding is a complex problem and involves a series of complicated processes. Much of the complexity can be described and understood by resorting to a statistical mechanics approach to the energetics of protein conformation: the energy landscape (Bryngelson et al., 1995; Frauenfelder et al., 1991). It has been considered as the most realistic model of protein (Henzler-Wildman & Kern, 2007; Shea et al., 1999) because it provides a complete quantitative description of protein conformational space, including the folded native state (which is generally considered as the state of a global free energy minimum), ensembles of various conformational substates near or far way from the native state, various unfolded or denatured states, and a large ensemble of folding intermediates (such as molten globule, transition states, trapped nonnative states, etc.). In the case of protein folding, the energy landscape is embodied by the folding funnel model, which provides not only a simple way of understanding why the Levinthal paradox (Levinthal, 1969) is not a real problem, but also a conceptual framework for understanding the different scenarios of protein folding highlighted by different folding models described above. Therefore, the energy landscape and folding funnel views have progressively replaced the classical model of a hierarchical pathway of folding (Yon, 2002). In the case of protein binding and function, it has now been widely accepted that the dynamics of proteins govern ultimately their function (Henzler-Wildman & Kern, 2007). The protein dynamics are defined as any time-dependent change in atomic coordinates, including both equilibrium fluctuations and non-equilibrium effects (Henzler-Wildman & Kern, 2007). The equilibrium fluctuations result in distinct probability distributions of conformational substates over the rugged energy landscape. These conformational substates reside in different energy wells near or far away from the well of the global energy minimum, in which the dominant population of the native state resides. The dynamics at the equilibrium, therefore, are often thought to govern biological function in processes both near and far from equilibrium (Henzler-Wildman & Kern, 2007) such as the protein-ligand binding through conformational selection or induced fit; and most studies focus on these motions and their functional consequences (Liu et al., 2007c; Liu et al., 2008; Liu et al., 2010, 2011; Tao et al., 2010). The non-equilibrium fluctuations arise from the conformation transition between conformational substates, which needs to overcome the energy barrier between them and involves the transition conformational states with transient lifetime and therefore, is hard to detect experimentally. Taken together, the energy landscape has two key factors that can characterize its shape: i) the thermodynamics that defines the relative probabilities of the conformational states; ii) the kinetics that defines the energy barriers between the conformational states (Henzler-Wildman & Kern, 2007) and the transition between them.

4.1 The origin of energy landscape and its concept

The concept of the energy landscape, although being now most familiar in the field of protein folding, is first proposed to be applied to the folded protein more than 30 years ago by Frauenfelder and colleagues (Austin et al., 1975). Using the low-temperature flash photolysis technique, they investigated the kinetics of carbon monoxide (CO) and oxygen (O_2) rebinding to myoglobin (Mb) as a function of temperature and ligand concentration. Moreover, through computer-solving the differential equation for the motion of a ligand molecule over four barriers, the rates for all important steps were obtained. The temperature dependences of the rates yield enthalpy, entropy, and free-energy changes at all barriers (Austin et al., 1975). In order to explain the observations of non-exponential kinetics and the four successive barriers below 210 K, an energy landscape model was proposed, which was described by Frauenfelder et al.: *"The energy landscape describes the potential energy Ec of the protein as a function of conformational coordinates; it is a hypersurface in the high-dimensional space of the coordinates of all atoms in Mb"* (Frauenfelder et al., 1991). Accordingly, the relation of the energy landscape to the myoglobin function was elucidated; and the features of the landscape, such as the existence of a large number of nearly isoenergetic conformations (conformational substates) and the heights of the barriers between energy wells were characterized (Frauenfelder et al., 1979; Frauenfelder et al., 1991).

In the filed of physical chemistry, the energy landscape of a protein-solvent system is defined as an energy function $F(x) = F(x_1, x_2, \ldots, x_n)$, where $x_1, x_2, \ldots, x_n$ are variables specifying the protein microscopic states (Dill, 1999), which can be all the dihedral angles of the peptide chain, the eigenvector projections derived from essential dynamics analysis (Amadei et al., 1993; Tao et al., 2010), end-to-end distance of the peptide chain, the number of native contacts, and an order parameter that describes the similarity of the structure to the native one and others (Kapon et al., 2008) or any degree of freedom (Dill, 1999). $F(x)$ is then usually the free energy as a function of the conformation of the protein to describe the protein-solvent system, where the entropic part of the free energy comes from all possible solvent configurations and solute conformational states; and the enthalpic part comes from the noncovalent bond formation within the protein interior and formation and loss of bonds (such as the hydrogen bonds and van der Waals interactions) between the protein and solvent. Thus, the protein folding process can be seen as a search for the global free energy minimum value through solving the free energy function. For a protein with a rough energy surface at the bottom of the energy landscape, the locally stable conformations can be found by determining the set of values $x_1, x_2, \ldots, x_n$ that gives the local minimum values of the free energy function.

4.2 The funnel-shaped energy landscapes

Although the energy landscapes are highly multi-dimensional, they are generally pictured as a surface in the three dimensions due to the difficulty to draw a multiple-dimensional space. In the 3D energy landscape, the vertical axis represents the free energy and the horizontal axes represent the conformational degrees of freedom of a polypeptide chain (conformational entropy). Random heteropolymers, i.e., a polypeptide chain containing random amino acid sequences, may have either a very rugged energy landscape with too many local minima (Plotkin et al., 1996) or a very large flat energy landscape (Dill & Chan, 1997). Systems like this can either easily get trapped in one of the local minima or never find the minimum when walking randomly on the very flat surface and as such, usually do not have a unique, well-defined and stable conformational state. On the contrary, the real proteins have evolved to contain optimized sequences so that they can fold rapidly and efficiently into well-defined native conformational states (Onuchic & Wolynes, 2004). In order to fold quickly and efficiently, proteins must have the tilt, funnel-shaped energy landscape. Hydrophobic collapse driven by the entropy gain of the water solvent leads to compact chain conformations of the polypeptide and a reduced ensemble of peptide chain conformations, thus narrowing the energy landscape. There are many non-native states with high free energy but only one native state with a global free energy minimum. The ensemble of conformations is further reduced by enthalpic contribution such as the isolation of electrostatically charged or polar amino acid side chains on the protein surface and the neutralization of ion pairing within the interior of the protein, thus further narrowing the landscape. The final step occurring at a narrower part (near the bottom of the landscape) of the energy landscape is the reconfiguration of a very small set of near-native conformations through cooperative formation and mutual adjustment of tertiary interactions, resulting in the unique native conformation located at the narrowest part of the energy landscape.

Therefore, the progressive reduction in dimensionality of the accessible conformational space (or the number of conformational states) makes the protein folding landscape look like a funnel, meaning that many conformations have high free energy and few have low energy. A large number of the denatured states are located on the surface outside or the upper part of the funnel and have high conformational entropy, whereas few native states or near-native states are located at the narrowest bottom of the funnel and have low conformational entropy. Intuitively, "funnel" also implies that the landscapes are relatively smooth, meaning that the barriers are small so the folding process happens quickly. However, the "real" protein folding funnel should carry information about kinetics or barrier heights or smoothness or any landscape shape feature (Dill, 1999). Depending on the folding rate, properties, flexibility and the native structure of the proteins, folding energy landscapes are divided into the following types (Dill & Chan, 1997).

i.   The idealized protein folding funnel with smooth surface (Figure 2A). This funnel contains no traps or bumps, showing how the many denatured conformations can roll down along different routes to become the fewer compact conformations, and finally to one native conformation. Such a smooth funnel often results in fast folding and two-state (single-exponential) kinetics (Huang & Oas, 1995; Jackson & Fersht, 1991; Schindler et al., 1995; Sosnick et al., 1996).

ii. The rugged landscape with hills, traps and energy barriers (Figure 2B). Like the real mountain ranges, this rugged funnel can also have much broader array of shapes involving hills, valleys, ridges, channels, moguls, plains and valleys inside valleys, moats, varying slopes and ups and downs of all kinds. Such a rugged funnel often results in slow folding and the folding kinetics is likely multiple-exponential.

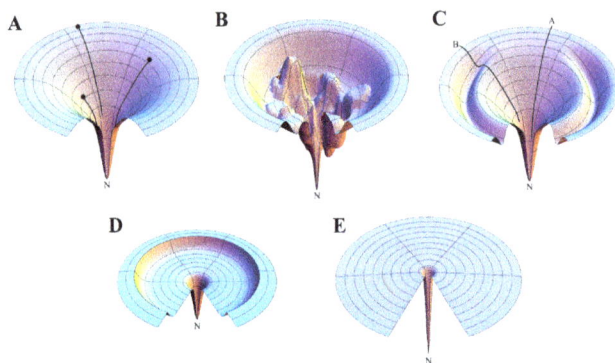

Fig. 2. Different types of free energy landscapes. (A) An idealized funnel-shaped landscape with smooth surface, which could guide the two-state folding kinetics at maximum attainable speed. (B) A rugged energy landscape with hills, traps and energy barriers. Folding kinetics is likely multiple-exponential. (C) The moat landscape illustrates how a protein could involve a fast (via path A) or slow (via path B) folding process. (D) The champagne glass landscape exhibits how conformational entropy can bring about "free-energy barriers" through random wandering on the flat plateau. (E) The Levinthal "golf course" landscape illustrates that a random search could not find the native state of the global energy minimum at the bottom of the "golf hole". N denotes the native state. Images were cited from (Dill & Chan, 1997).

iii.   The moat landscape on which the "moat" represents the kinetic trap (Figure 2C). As shown in Figure 2C, the A route is a funnel-like "throughway" path, while the folding molecules should have to pass through an obligatory folding intermediate trap when they follow the B route.
iv.   The champagne glass landscape (Figure 2D). This type of landscape is proposed to illustrate that the "bottleneck" or rate limit to folding is due to conformational entropy (Dill & Chan, 1997). The polypeptide chain is delayed en route to the bottom by aimless wandering on the flat plateau to find the remaining routes downhill, indicating how conformational entropy can cause "free-energy barriers" to folding. Here the "free-energy barriers" do not correspond to a process that goes "uphill in free energy" to overcome a free-energy barrier in the conventional sense, but just mean that the folding process is delayed compared to some reference rate.
v.   The Levinthal "golf course" landscape (Figure 2E). This type of landscape is proposed to illustrate Levinthal's argument (Levinthal, 1968, 1969) that a random search could not find the native state of the global energy minimum located at the bottom of the "golf hole". It is just like a ball rolling randomly on a very large flat course, taking a long time to find and fall in the hole.

4.3 Protein folding in the funneled energy landscape

Although there exist the smooth funnel-shaped energy landscapes without significant kinetic traps (Figure 2A), the most realistic folding funnel is rugged, guiding multi-exponential slow folding of most proteins (Figure 2B). Under inappropriate folding conditions, the funnel is shallow and therefore the polypeptide chains spend most of their time meandering around the upper part of the funnel, leading to conformational heterogeneity of the unfolded proteins and very slow folding due to the shallow slope (Dill & Chan, 1997). However, once restoration of the appropriate conditions, the funnel becomes stretched down leading to the steep slope, thereby initiating the downhill movements of molecules towards the native state (Figure 3).

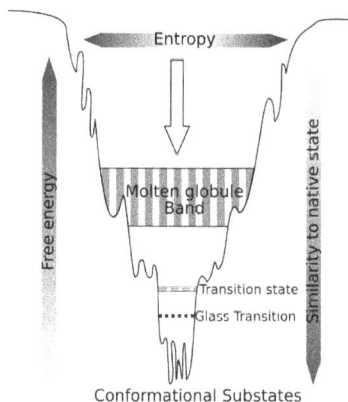

Fig. 3. A rugged funnel-like free energy landscape that describes the detailed processes and various intermediates in the folding reaction. The width of the funnel represents the conformational entropy; the depth of the funnel represents the change in total Gibbs free energy between the denatured and native states. This figure is modified from (Onuchic et al., 1997; Onuchic et al., 1995; Wolynes et al., 1995).

Numerous events occur during the downhill process and lead to the formation of numerous different conformational states, such as the unfolded state, collapse, molten globule, glass transition, transition state, intermediate, native state, and conformational substate. It must keep in mind that each of these conformations just represents a certain type of thermodynamic state, which exists in an ensemble of conformers and exhibits distinct population time. The conformational states occurring in the folding process should be seen as ensembles but not as specific structures. For example, the intermediate is a very large ensemble of conformational states because it contains all the states with only the exception of two states, the initial unfolded and the final native ones.

A funnel-shaped energy landscape illustrating the detailed processes and various intermediates in protein folding is shown in Figure 3. Just like an ensemble of skiers jumping and skiing down the steep slope (in this case the major driving force is the gravity), the first phase of protein folding is that an ensemble of unfolded polypeptide chains collapses rapidly to a compact ensemble. In this case the driving force is the chemical potential of the reduction in Gibbs free energy, to which the entropy gain via solvent sequestration makes a substantial contribution. Each skier skis along a different route, down the mountainside as shown in Figure 2B and Figure 3 and subsequently enters a rugged region composed of hills, valleys and traps. Correspondingly, the rapid hydrophobic collapse leads to the accumulation of "misfolded" compact intermediates, many of which are trapped in the kinetic traps or local free energy minima. Such an ensemble of compact folding intermediates, which is termed molten globule, contain some transient secondary structures and nonspecific tertiary interactions (Onuchic et al., 1995). The subsequent processes are slow steps, which arise from climbing an uphill slope, then reach a mountain pass and continue to the next downhill search (Dill & Chan, 1997). In these processes, the ensemble of the conformations located at the hilltop can be considered as "transition state", which, in the sense of the energy landscape, is defined as "*the bottleneck processes of flows down different mountainsides*" by (Dill & Chan, 1997) to highlight the aspect of the rates but not the specific structures. In order to overcome the bottleneck, the already collapsed conformations need to be pulled apart. This may be driven by the conformational entropy of the molten globules. In addition, uphill climbing does not necessarily mean a full opening of the compact collapses. This process usually is the breaking of a few non-covalent bonds here or there to resume progress towards the native state. The process of the transient trapping, uphill, downhill, transient trapping can repeat many times, thereby slowing the exploration of the routes towards the native state. Therefore such a repetitive process is a bottleneck for folding and can be viewed as the folding transition state as defined by Dill et al. (Figure 3) (Dill & Chan, 1997). Under thermodynamic conditions near the folding transition midpoint, entropy and energy do not completely compensate for each other; thus, intermediates are not present at equilibrium, i.e., a free thermodynamic energy barrier intercedes (Wolynes et al., 1995). The repetition of the uphills and downhills ultimately leads to the arrival of a so-called glass transition state (Bryngelson et al., 1995; Onuchic et al., 1995). Such a trapping resembles the way a liquid becomes a glass when cooled, remaining fixed in one of many structures and unable to reconfigure to the lowest energy crystal state (Wolynes et al., 1995). In the case of protein folding, therefore, the process of reconfiguring the glass transition to the native state is very slow (Onuchic et al., 1995), requiring a sufficient overall slope of the energy landscape so that the numerous valleys flow in a funnel towards the native structure. In addition, when the folding process enters into the glass transition, the intermediates have only a few paths to the native structure. The lifetimes of the discrete glass transition states are relatively longer than those of other intermediates but shorter than the average folding time. The formation of native contacts will further reduce both the entropy and enthalpy, with the larger negative enthalpy compensating for the entropy loss, putting the glass transition → native state under enthalpy control. For a more detailed and quantitative description of the funnel-like protein energy landscape, please refer to (Onuchic et al., 1997).

4.4 The roughness/ruggedness of energy landscapes

For most of proteins, the funnel-like energy landscapes, either on their tube part or at the bottom, are largely rugged or rough due to the kinetic traps, energy wells and barriers. Rough energy landscapes occur in problems in which there are many competing interactions which can not be satisfied at once (Bryngelson et al., 1995; Kapon et al., 2008). For example, in the entire folding process there are too many "mistaken interactions" that are caused by competition between different chain parts. The initial hydrophobic collapse expels the bound water molecules and leads to a compact ensemble which, although having relatively lower free energy, contains many non-native contacts, energetically costly cavities generated by imperfect fit of side-chains, and locally favorable secondary structures. These inappropriate interactions and orientations have to be disassembled and readjusted in order to form the final tertiary structure. Analogously, in the process of protein-ligand binding, the water network formed around the surface of the interacting partners need to be broken upon interaction and the changes in conformations of the protein or ligand may also be required, and sometimes a large scale conformational changes are required. Such competitive interactions, on the one hand, lead to relatively stable residence of conformational sates in wells of the local free energy minima, on the other hand, facilitate conformational motions that can help the molecules to jump out of the wells. In fact, for a bio-macromolecule such as the protein, its folding or binding processes will inevitably lead to a single structure or a complex that can not satisfy completely all of the constraints posed on the system and thus the reactions are said to be frustrated (Kapon et al., 2008; Shea et al., 1999).

Frustration, or ruggedness, a well-known trait of the complex systems such as glasses, solute-solvent systems and even social and economic networks, arises essentially from the fluctuation of free energy caused by non-complementary changes between entropy and enthalpy. For instance for the protein-solvent system, the change in enthalpy is the consequences of the competitive interactions between different parts of the protein and between the protein and solvent; while the change in entropy is caused by the nature of protein and solvent to increase their disorder or randomness. When the enthalpy change can not compensate for the entropy change, the fluctuation of the free energy occurs. In the energy landscapes, the local free energy fluctuations manifest themselves as a series of hills, valleys and traps of various heights and widths. In the case of protein folding, although the local free energy surface is rough or rugged, the entire free energy landscape is funnel-shaped (Figure 2A, 2B and 3), satisfying the need for the global decrease in free energy of the protein-solvent system. In the case of protein function, although most functional studies are concentrated on static X-ray crystallographic structures of proteins, the biologically relevant processes, such as receptor-acceptor binding, enzymatic catalysis, neuron activity and selective passing of ions through ion-channels are essentially dynamic. It is the ruggedness of the energy surface that brings about the dynamic behaviors and molecular motions of proteins. For example, the polar or charged residues residing inside the hydrophobic core can often bring about local instability and contribute to frustration. Nevertheless, it is such a frustration that serves to exchange ligands or degrade substrates (Tao et al., 2010). Loops are observed to exist universally in protein structures. Long flexible loops often assume a large number of conformations due to their inherent large conformational entropy. However, the large flexibility of loops makes them very useful for contact, recognition, binding and catching of the ligands, in which they are frequently found to play a pivotal role. Interestingly, for HIV gp120 glycoprotein, several long variable loop regions, i.e., V1-V5, located outside the surface of the molecule, shield the antigenic sites and make substantial contribution to the immune escape of HIV (Liu et al., 2007a; Liu et al., 2007b; Liu et al., 2007d, 2008).

4.4.1 Ruggedness, folding rate and protein function

As described in the section "The problems of protein folding", one of the protein folding problems is the folding rate. The funnel-shaped landscape seems to address this problem. It has been predicted that the speed limit for the folding of small globular proteins in a smooth folding funnel is ~ $100/n$ μs, where n denotes the number of amino acid residues in the protein (Kubelka et al., 2004). However, most natural proteins fold at least two or three orders of magnitude slower than the predicted empirical relationship due to the presence of the roughness of the energy landscape. Furthermore, the different degree of the ruggedness raises another question: what is a good speed for a protein to fold (Kapon et al., 2008)? One intuitive answer is that proteins should fold as fast as possible in order to avoid degradation and maintain maximum stability of their folded structure in the cellular environment. However, this raises the question of why roughness, with its detrimental effect on folding rates, has been preserved by evolution. It has been proposed that the answer is the functional requirement of the proteins because God has created proteins to perform a task but not to fold as fast as possible (Gruebele, 2005). It has been shown that some proteins which have smooth energy landscapes and fast folding rates tend to aggregation and proteolysis since they can also unfold very rapidly (Gruebele, 2005; Jacob et al., 1997; Kapon et al., 2008). On the other hand, if the energy landscape is too rough, the protein may also be subjected to similar risks due to the attenuation of diffusion along the energy surface (Kapon et al., 2008). Therefore, every protein must have evolved an optimum energy landscape roughness to satisfy simultaneously the two conflicting requirements: the survival and the function.

The ruggedness in the energy landscapes makes it possible for a protein to assume many states that may be quite distinct from each other in structure but have similar free energy. Furthermore, the transition between states or substates causes protein motions on different time scales and at different amplitude and directionality, which may involve different biological processes. The ability to adopt multiple conformational states also increases the repertoire of molecules with which a ligand can interact and provides robustness against deleterious mutations (Kapon et al., 2008).

4.4.2 Amplitude and timescale of protein motions

The protein dynamics contain two elementary components, i.e., the kinetic component and the structural component, which characterize the timescale and amplitude of the fluctuations, respectively (Ansari et al., 1985; Henzler-Wildman & Kern, 2007). Figure 4 shows a free energy landscape defining these two components. The protein dynamics are divided into three tiers, i.e., tier-0, tier-1 and tier-2 dynamics that occur on timescales ranging from the "slow" to "fast" timescale.

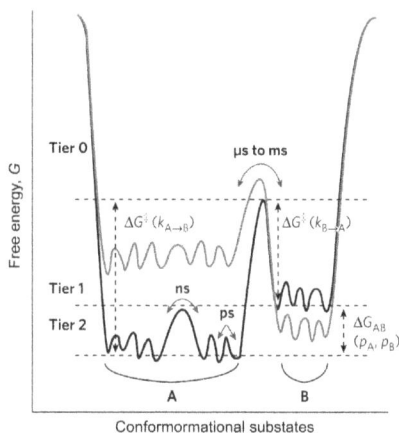

Fig. 4. The energy landscape that defines the amplitude and timescale of prtein dynamics. the conformational states are defined as minima in the energy surface and are located in the energy wells. The transition states can be seen as the free energy maxima between the wells and are located at the tops of hills. Tier 0 includes two conformational states, A and B. Conformational transition between them are rare because of the high energy barriers $\Delta G^{\ddagger}$ ($k_{A \rightarrow B}$) and $\Delta G^{\ddagger}$ ($k_{B \rightarrow A}$), resulting in a slow fluctuations between these two states with timescale of μs to ms, i.e., tier-0 dynamics. The poputlations of the tier-0 states, A and B (p_A, p_B), are defined as Boltzmann distributions based on their differences in free energy (ΔG_{AB}). Tier 1 includes several confomational substates located within tier 0 wells. Tier-1 dynamics are fast fluctuations on timescale of ns to allow fast transition between tier 1 substates. Similarly, Tier 2 describes faster fluctuations on timescale of ps between a large number of closely realted substates located within tier 1 wells. The difference between the two landscapes, depicted by the dark line and grey line, respectively, indicates that the energy lanscape is dynamic, with its shape being affected and changed by physical (such as temperature, pressure, pH, ionic strength, presence of denaturant, etc.) or functional (such as ligand binding and protein mutation) conditions. This figure is modified from (Ansari et al., 1985; Henzler-Wildman & Kern, 2007).

The tier-0 dynamics, which occur on a slow timescale, define fluctuations between kinetically distinct states that are separated by energy barriers of several kT (the product of the Boltzmann constant k and the absolute temperature T). Such slow timescales are evaluated to be microseconds (μs) or slower at physiological temperature; and the corresponding motions are typical larger-amplitude concerted motions that can propagate over the entire molecular structure, leading to a relatively small number of conformational states with relatively large conformational difference between them (Figure 4). The protein molecules within one of these tier-0 states (the largest wells in Figure 4) are also dynamic, fluctuating around the average structure on a faster timescale and exploring a large ensemble of closely related structures. This brings out the tier-1 or even tier-2 states located in smaller wells relative to the tier-0's large wells. An important feature of tier-0 states is that the conformational transitions between these states are rare because of i) high free energy barrier between tier-0 states; ii) long lifetimes of tier-0 states and stable equilibrium fluctuations around these states; iii) the low probability of the conformation that can initiate the transition. Such large-scale concerted motions occurring on the slow timescale are now receiving more and more attention since they govern many biological processes such as the enzymatic catalysis (i.e., the opening of substrate binding channel), signal transduction and protein-protein interactions (i.e., the conformational selection process). The relatively long lifetimes of individual states within the tier-0 wells make it possible for direct trap and observation of these states, and even for detection of the kinetics of inter-conversion between these states using experimental and computational methods. For the detailed description and discussion of these methods, refer to (Henzler-Wildman & Kern, 2007).

The tier-1 and tier-2 dynamics occur on "fast" timescale and define fluctuations within the wells of a tier-0 state. These many smaller wells within the tier-0 wells are occupied by a large ensemble of structurally similar substates. The energy barriers between them are less than $1\ kT$, resulting in more local, small-amplitude picosecond (ps)-to-nanosecond (ns) fluctuations at physiological temperature, with the tier-1 and tier-2 dynamics corresponding to the ns and ps fluctuations, respectively (Figure 4). The dynamics on the fast timescale, which are the consequent of occasional Brownian bombardments, can sample a large number of conformational substates. In other words, the conformational diversity can be considered as the subsequence of the entropy of the system. According to (Henzler-Wildman & Kern, 2007), the tier-1 dynamics can be used to describe ns-timescale fluctuations of a small group of atoms, such as loop motions, while the tier-2 dynamics correspond to ps-timescale fluctuations of local atoms. such as the rotations of the amino acid side chains (Figure 4) . The even higher tiers of dynamics also exist, such as bond vibrations, which occur on the femtosecond (fs) timescale.

4.5 The energy landscapes are dynamic

The concept of a dynamic landscape was first proposed by (Gulukota & Wolynes, 1994) in their statistical model of how chaperones can work to help to form the normal folding funnel of proteins to their native states. The shape of energy landscape can be affected by an individual set of physical environments, such as pH, ionic concentration, presence of denaturant, pressure and temperature, or functional environment such as binding to other molecules (Figure 4). For example, under the denaturing conditions of high temperature or presence of denaturant, the energy landscape of a protein is shallow and its surface is relatively flat, with an ensemble of unfolded polypeptide chains walking randomly around the flat surface. However, upon restoration of the native-like condition, the energy landscape is stretched down and the slope becomes steep. Despite the existence of the hills, valleys and traps, the emergence of the steep wall allows the chains to roll down the funnel. Therefore, switching the conditions between the denaturation and refolding is essentially to change the shape of the energy landscapes of proteins. Furthermore, changes in energy landscape can also be the consequence of the bound state of the protein, that is, whether it is free, or bound to one or more ligands, can change the depth and width of the wells around the bottom of the energy landscape. The works by (Freire, 1999; Todd & Freire, 1999) showed that residues far away from the active site can be stabilized by intermolecular association, which provide an example illustrating that the binding event shifts the energy landscape of a folding funnel (Tsai et al., 1999b). Changing or manipulating the physical or functional conditions is the most common way to change the relative populations of the conformational states or substates, and the kinetics of the conversion between them (Henzler-Wildman & Kern, 2007). In summary, although the protein function is often inferred from its single stable conformations derived from X-ray crystallographic technique, a deep and complete understanding of the protein function requires analysis of the changes in populations of its conformational substates caused by the dynamic energy landscape upon change in the physical or functional conditions.

5. Protein-ligand interactions and binding mechanisms

A fundamental principle of all biological processes is molecular organization (e.g., protein folding) and recognition, e.g., protein–ligand interactions and binding (Perozzo et al., 2004). Essentially, proteins perform their function through interaction with molecules such as proteins and peptides, nucleic acids, ligands and substrates, and other small molecules such as oxygen or metal ions. Thus, a detailed understanding of biological processes requires to investigate not only the protein folding mechanism but also the mechanism of protein-ligand binding.

Molecular recognition is a process by which biological macromolecules interact with each other or with small molecules to form a specific complex (Demchenko, 2001; Janin, 1995; Otlewski & Apostoluk, 1997). It has two important characteristics: specificity and affinity. The specificity is that biological macromolecules are able to distinguish the highly specific ligand from less specific ones. In order to make an interaction specific, the bonds between correct partners should be strong, while for other partners showing only minor differences in structure they should be weak or even repulsive. Therefore, once a correct ligand binds to its acceptor molecule, the affinity between them should be strong enough to prevent disassociation. The most important aspect of the molecular recognition is, as pointed out by (Demchenko, 2001), that the recognition is usually not a process in itself, but is an element of a more complex, functionally important mechanism such as allosteric regulation of enzyme activity, signal transduction, protein folding or the formation of multi-subunit and supramolecular structures. This requires important and sometimes dramatic changes in properties of the interacting partners. In this section, the protein-ligand binding mechanisms, the relationship between protein folding and binding, and the thermodynamics and kinetics of protein-ligand interaction will be described and discussed.

5.1 Protein-ligand binding mechanisms

The protein-ligand binding mechanisms have evolved from the early "lock-and-key" (Fischer, 1894) to the "induced fit" (Koshland, 1958) and to the now popular "conformational selection" models (Frauenfelder et al., 1991; Ma et al., 1999; Tsai et al., 1999a) (Figure 5). The lock-and-key model (Figure 5A) was first proposed by (Fischer, 1894) to explain binding of a single substrate to the enzyme. In this analogy, the lock is the enzyme and the key is the substrate. The prerequisite of this model is that the enzyme and substrate are both rigid and their surfaces should be complementary, and therefore only the correctly sized key (substrate) fits into the key hole (active site) of the lock (enzyme); the keys with incorrect size or incorrectly positioned teeth (incorrectly shaped or sized substrate molecules) do not fit into the lock (enzyme) because both the key and the lock can not change their shape and size. However, not all experimental evidence can be adequately explained by the lock-and-key model. Therefore, a modification called the induced fit model was proposed to compensate for the rigid, structurally invariable protein and ligand. The induced fit model (Figure 5B) assumes that the ligand plays a role in determining the final shape of the protein, i.e., the ligand binding induces a conformational change in the protein (Koshland, 1958). However, it seems that the induced fit mechanism is only suitable for the proteins showing minor or moderate conformational change after the ligand bindings. For the proteins undergoing substantially larger conformational changes upon ligand binding, Bosshard (Bosshard, 2001) suggested that *"induced fit is possible only if the match between the interacting sites is strong enough to provide the initial complex enough strength and longevity so that induced fit takes place within a reasonable time"*. Therefore, the induced fit mechanism alone can not explain well the association between ligand and proteins undergoing large conformational changes after binding (Tobi & Bahar, 2005).

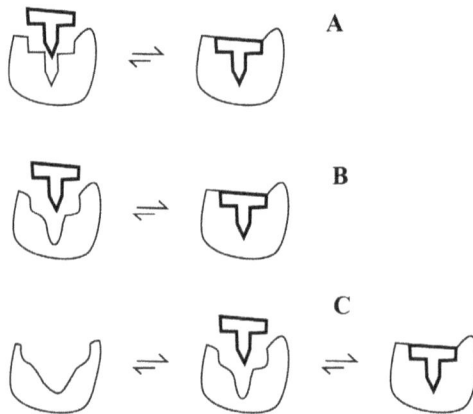

Fig. 5. Mechanisms for protein-ligand binding. (A) lock-and-key model. (B) Induced fit. (C) Conformational selection followed by induced fit. This figure is modified from (Tobi & Bahar, 2005).

In recent years, the conformational selection model (also termed preexisting equilibrium, population selection, fluctuation fit, or selected fit; see Figure 5C) has emerged as an alternative for induced fit and is becoming prevalent. Essentially, such a mechanism is the extension of the folding funnel model, postulating that the native state of a protein is not defined by a single rigid conformation but rather by an ensemble of closely related conformational substates coexisting in equilibrium, and the ligand can select to bind to the most suitable conformer among these conformational substates, shifting the equilibrium toward complex formation. As described above, at the bottom of the folding funnel, the rugged energy landscape near the native state contains several minima corresponding to these conformational substates. The more flexible the protein is, the lower the barrier height between the substates, and the larger the ensemble of acceptable near-native conformations may exist. Generally, the conformational changes in this model are beyond local side chains rearrangements near the binding site, but rather display large concerted motions or entire domain movements.

The early studies by (Berger et al., 1999) and (Foote & Milstein, 1994) demonstrate that conformational selection is responsible for antigen–antibody complex formation. For example, of the two isomeric conformations determined by X-ray crystallography for the SPE7 antibody (L.C. James et al., 2003), only one possesses a promiscuous, low-affinity binding site for haptens. The initial selective binding of haptens to this site is followed by an induced fit, resulting in a high-affinity SPE7-haptens complex. Recently, more and more data from the X-ray and cryo-electron microscope images, kinetics studies, extensive single molecule fluorescence and, in particular, NMR technique show that a unliganded protein can assume a repertoire of conformational states, including conformations corresponding to the bound form (Boehr et al., 2009; Lange et al., 2008; Tobi & Bahar, 2005; Wu et al., 2009).

5.2 The relationship between lock-and-key, induced fit and conformational selection

Interestingly, when considering the lock-and-key and conformational selection models under the background of the funnel-like energy landscape, the former can be viewed as an extremity of the later. The lock-and-key model is useful for explaining binding of a ligand to a very rigid protein, which has a smooth folding funnel with almost no ruggedness around the bottom of the energy landscape. Although a protein with relatively large structural flexibility has the folding funnel with massive ruggedness, the conformational selection model considers that, generally, only one ensemble of conformational substates residing in one energy well is appropriate for selective ligand binding. If we consider such an energy well as a small funnel and neglect the other wells as well as the kinetics of conversion between conformational substates, the binding process occurring in this well can be described by the lock-and-key mechanism.

Furthermore, induced fit and conformational selection are not two independent and exclusive processes but rather they both play a joint role in molecular binding (Perica & Chothia, 2010). What seems to be the general trend is that the conformational selection is more important for the initial recognition/contact between the interacting partners, while the induced fit is more important for the subsequent mutual conformational adjustment. Therefore, we consider that the large-scale concerted motions, i.e., tier-0 dynamics, may govern the conformational selection, while the tier-1 and tier-2 dynamics, which describe the fast-timescale motions (i.e., loop motions and side chain rotation), may play a role in fine-tuning the interactions at the later "induced-fit" stage. The induced fit model can also be viewed as an extreme version of the conformational selection when considering the detailed process of the molecular recognition rather than focusing only on the final conformational difference between the free and bound states. As discussed above, most proteins have a rugged energy surface around the bottom of the landscape, leading to distinct population distributions of different conformational substates. The initial selective collision/contact between the partners may decide whether the binding proceeds forward or the occasional complex disassociates. Only in the case of the approximately "correct" conformational state (which provides the initial complex the interactions with enough strength and longevity so that induced fit takes place within a reasonable time) can "conformational selection" occur, although the already chosen conformational sate is significantly distinct from the final bound state. In other words, in the case of molecular recognition between flexible partners, the selectively initial interactions must occur before the event of the conformational adjustment, and therefore the induced fit can be perceived as an extremity of the conformational selection model.

Very recently, Nussinov and her colleagues (Csermely et al., 2010) proposed an extended conformational selection model to integrate the conformational selection and induced fit mechanisms, suggesting that i) the protein binding embraces a repertoire of selection and adjustment processes; and ii) the protein segments whose dynamics are distinct from the rest of the protein can govern conformational transitions and allosteric propagation that accompany the binding processes. A single step of conformational selection generally lowers the entropy barrier and helps to achieve a high degree of specificity (e.g., like the situation under the lock-and-key model). The subsequent mutual conformational rearrangement contributes significantly to the high binding affinity, e.g., the binding interfaces often have a certain degree of flexibility that is advantageous to conformational change and adjustment (Liu et al., 2011; Tao et al., 2010), like the situation in the induced fit mechanism.

The superficial difference between these two mechanisms is whether the binding is "conformational change first" (for the conformational selection) or "binding first" (for the induced fit). However, for a given protein to interact with its ligand, how the binding can be characterized as conformational selection or induced fit? What factors determine which mechanism dominates the binding process? Hammes et al. has proposed a flux criterion by which the sequence of events can be determined quantitatively (Hammes et al., 2009). Through applying the flux calculation to protein-ligand binding, they found that the binding mechanism switches from being dominate by the conformational selection pathway at low ligand concentration to induced fit at high ligand concentration. Interestingly, through establishing a solvable model that describes the conformational transition of the receptor between the inactive (this form prevails in solution in the absence of the ligand) and active (this form is favored while a ligand is loosely bound) forms, Zhou (Zhou, 2010) found that the timescale of the active-inactive conformational transitions has effect on the selection of these two binding mechanisms, i.e., when the active-inactive transition rates increases, the binding mechanism gradually shifts from the conformational selection to induced fit, indicating that the timescale of conformational transitions plays a role in controlling the binding mechanism.

5.3 Examples of protein binding: ubiquitin, aggregation and molecule chaperones

By measuring residual dipolar couplings (RDCs) (Lakomek et al., 2008) resulting from partial alignment in a large number of media, the conformational ensemble of ubiquitin was calculated up to the microsecond timescale at atomic resolution (Lange et al., 2008). The most striking feature of the ensemble is the presence of "bound" conformations in the free form of ubiquitin (Figure 6), which result from the large scale concerted pincer-like motions of the binding interfaces. All the backbone conformations in the available 46 X-ray crystallographic structures of ubiquitin in complex with various binding partners are observed in the solution ensemble, despite the absence of any crystallographic information in ensemble refinement. This provides direct evidence that the conformational ensemble of ubiquitin in its bound states does indeed exist in the solution condition without including protein interaction partners, and that conformational selection is important for protein-protein binding of ubiquitin. However, subsequent conformational changes may be induced fit processes, especially in the side chains, after the initial binding via conformational selection. Accordingly, we could conclude that the side chain flexibility contributes to the high binding affinity through induced fit, while the conformational selection and highly rigid hot spot residues (Ma et al., 2001) contribute to the high specificity through lowering entropy barrier.

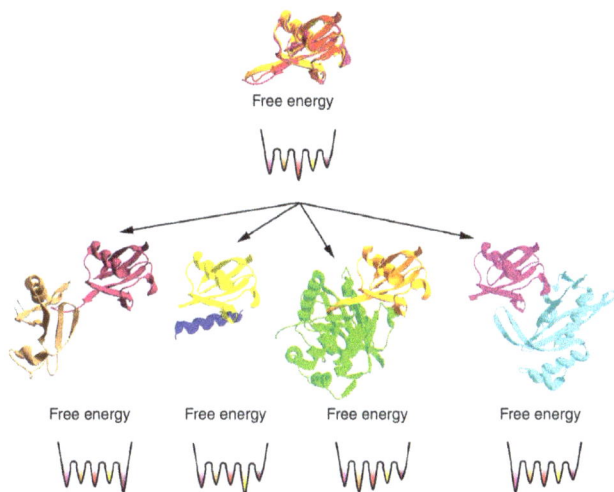

Fig. 6. An example of the conformational selection mechanism responsible for protein-protein binding of ubiquitin. The structures on the top show the NMR-derived conformational ensemble of ubiquitin. Shown in the middle are five ubiquitin crystallographic structures in complex with different ligand proteins (PDB 1F9J, 1S1Q, 1XD3, 2D36 and 2G45). The hypothetical free energy landscapes are shown in the bottom. The color of the free energy well is the same as that of the ubiquitin crystal structure corresponding to the conformational state located within this well. The deeper the well is, the larger population of the corresponding conformation state resides in it. The energy barriers separating the conformations are not known. This figure is modified from (Boehr et al., 2009)

There are several "misfolding diseases" that arise from possible mistakes in protein folding. These diseases, i.e., amyloid (Alzheimer's disease) and prion diseases (BSE or "mad cow" disease, scrapie, Creutzfeld-jacob disease (Serban et al., 1990)), are characterized by abnormal accumulation of aggregated proteins in the brain and other tissues. For example, the formation of amyloid plaque are thought to be related to the conformational transition of a N-terminal structural region of Aβ protein, that is, an easy transition from the native α-helical to non-native, prone-to-plaque-formation β-stranded structures (Kosik, 1992). The prion protein also undergoes a conformational transition from a normal cellular isoform containing a preponderance of helical secondary structure to a scrapie isoform containing a greater proportion of β-sheet (Inouye & Kirschner, 1997; Pan et al., 1993). At first glance, the formation of the pathogenic plaque is due to the misfolding of relevant proteins. However, the so-called "misfolding" is in fact the finally observed already folded structure in the plaque but rather than the cause of the protein aggregation. Conformational analysis demonstrates that the Aβ peptide segment exists as a mixture of rapidly equilibrating extended conformers, including the "native" conformational state and the "natively unfolded" states (Weinreb et al., 1996). Some of the natively unfolded states may initiate protein-protein interaction, forming a fibril seed facilitating plaque growth. The flexible nature of Aβ or prion proteins determines that the bottoms of their funnel-like landscapes, despite ruggedness, are relatively shallow, allowing easy transition between conformational substates. It has been proposed that flexible protein interface regions are more prone to aggregation than other surface regions (Pechmann et al., 2009). Therefore, some of the conformers that have proper flexible surfaces, although having a low population time, can interact and aggregate together, biasing the equilibrium towards the plaque conformation and deepening the energy well in which the plaque state is located. Therefore, aggregation is a manifestation of conformational selection of higher energy, less populated monomer states leading to highly polymorphic aggregate species, and is now increasingly considered as the "side-effect" of protein binding (Csermely et al., 2010). As such, the "misfolding decreases" should be called the "diseases of false recognition and binding".

It seems that protein aggregation is inevitable in the living cell, and therefore there must be some mechanisms capable of repairing or preventing the aggregation. Actually, molecular chaperones can perform this function through temporarily covering aggregation-sensitive surfaces of the relevant proteins. Molecular chaperone is the common name of a diverse family of proteins which help other proteins to fold, refold after transport through biological membrane or maintain the folded state under conditions of stress (Demchenko, 2001). The common feature of molecular chaperones is that they interact only with unfolded or partially folded but not with native proteins, while the formed complex can dissociate into chaperone and a folded protein molecule. The chaperones are not enzymes and do not accelerate the folding reaction, but facilitate the assembly of protein complexes through the temporal occlusion of particular hydrophobic sites in polypeptide sequence, allowing the self-assembly of structure to proceed with the formation of folding determinants at sites that are free from chaperone protection. Molecular chaperones realize their functions through altering the flexibility of the target proteins or through protecting the hydrophobic and flexible interfaces of the potential aggregation-sensitive proteins, causing the distributions and redistributions of different conformational substates along the rugged surface of the dynamic energy landscape.

5.4 Thermodynamics of protein-ligand interaction

The complete understanding of molecular recognition of the proteins and ligands requires characterization of the binding energetics. Like the situation of protein folding, a quantitative description of the forces that drive the protein binding also requires characterization and determination of the complete thermodynamic profile, including the binding free energy (ΔG), the enthalpy (ΔH) and entropy (ΔS) of binding and the heat capacity change (ΔC_p) (Chaires, 2008; Perozzo et al., 2004). These thermodynamic data contain information crucial not only for elucidating the binding mechanism but also for rational drug design through relating the thermodynamic data to the structural data, which alone can not fully describe the driving forces for binding or predict accurately the binding affinity (Weber & Salemme, 2003).

The only direct way to measure the heat change during complex formation at constant temperature is implemented by isothermal titration calorimetry (ITC) (Freire et al., 1990; Wiseman et al., 1989). In this method one binding partner is titrated into a solution containing the interacting partner, thereby generating or absorbing heat. This heat, reflected by temperature difference between a sample cell and a reference cell, is directly measurable and can be quantified by the calorimeter. Under appropriate conditions data analysis from a single experiment of the ITC allows the simultaneous determination of the equilibrium binding constant (K_b), the standard Gibbs free energy change (ΔG), the changes in enthalpy (ΔH) and entropy (ΔS), as well as the stoichiometry (n) of the association event. Moreover, experiments performed at different temperatures yield the heat capacity change (ΔC_p) of the binding reaction (Chen & Wadso, 1982; Cooper & Johnson, 1994; Perozzo et al., 2004).

The basic thermodynamic relationships are summarized below:

$$\Delta G = -RT \ln K_b \tag{1}$$

$$\Delta G = \Delta H - T\Delta S \tag{2}$$

$$\Delta H_{VH} = -R\left(\frac{\delta \ln K_b}{\delta\,(1/T)} \right) \tag{3}$$

$$\Delta C_p = \left(\frac{\delta \Delta H}{\delta T} \right)_p \tag{4}$$

$$\Delta H_T = \Delta H_r + \Delta C_p (T-T_r) \tag{5}$$

$$\Delta S_T = \Delta S_r + \Delta C_p \ln\left(\frac{T}{T_r} \right) \tag{6}$$

$$\Delta G = \Delta H_r - T\Delta S_r + \Delta C_p \left[(T-T_r) - T\ln\left(\frac{T}{T_r} \right) \right] \tag{7}$$

where the ΔG, ΔH, ΔS and ΔC_p are the changes in free energy, enthalpy, entropy and heat capacity, respectively; K_b and R are the binding constant and universal gas constant, respectively; T and T_r refer to the temperature and an arbitrary reference temperature, respectively; and the thermodynamic parameters subscripted with "T" or "r" refer to those temperatures. The equilibrium binding constant K_b, which is defined as the ratio of actual product concentration to reactant concentrations, can be used to calculate the Gibbs free energy change ΔG by equation 1. The change in binding enthalpy of protein-ligand interactions can be determined accurately by the ITC method or indirectly from the temperature dependence of equilibrium binding constants and application of the van't Hoff relationship (ΔH_{VH}; equation 3). Classically, free energy change can be parsed into its enthalpic (ΔH) and entropic components ($-T\Delta S$) as described by equation 2. Equation 4 indicates that the change in enthalpic value is temperature dependent, relating to a nonzero heat capacity change ΔC_p. Equation 7 shows that enthalpy and entropy changes depend on temperature through the nonzero heat capacity change ΔC_p. The relations of temperature dependent ΔC_p to changes in the enthalpy and entropy are described by equation 5 and 6, respectively.

5.4.1 Concepts of the thermodynamics parameters

Gibbs free energy ΔG, a concept that was originally developed in the 1873 by the American mathematician Josiah Willard Gibbs (Gibbs, 1873), can be seen as the approximation of the chemical potential, which is minimized when a system reaches equilibrium at constant pressure and temperature. The Gibbs free energy of binding reaction is the most important thermodynamic description of binding because it determines the stability of any given biological complex. Free energy changes are a function of states, i.e., their values are defined merely by the initial and final thermodynamic states, regardless of the pathway connecting them. Free energy is the key parameter since its value under a particular set of reactant concentrations dictates the direction of the bimolecular equilibria. The negative sign of ΔG means that the binding reaction or conformational transition will proceed spontaneously to an extent governed by the magnitude of ΔG; the positive sign means that the energy is needed to drive the reaction to form a product, with the magnitude of ΔG specifying the amount of the required energy. The free energy is a balance between enthalpy and entropy, e.g., the observed ΔG can be the same with completely opposing contribution by enthalpy and entropy (see equation 2). An interaction with positive ΔH and ΔS (binding dominated by hydrophobic effect) can produce the same ΔG as an interaction with negative ΔH and ΔS (when specific interactions dominate). Such a enthalpy-entropy compensation makes the binding free energy relatively insensitive to change in molecular details of the interaction process, and thus the consideration of the ΔH and ΔS are crucial for a detailed understanding of the free energy of binding (Eftink et al., 1983; Lumry & Rajender, 1970; Perozzo et al., 2004; Williams et al., 1993).

The enthalpy is a measure of the total energy of a thermodynamic system. It includes the internal energy, which is the energy required to create a system, and the amount of energy required to make room for it by displacing its environment and establishing its volume and pressure. In many chemical, biological, and physical measurements, the enthalpy is the preferred way of expression of system energy changes because it simplifies descriptions of energy transfer. The total enthalpy, H, of a system cannot be measured directly and, as such, the change in enthalpy, ΔH, is a more useful quantity than its absolute value, which is equal to the sum of non-mechanical work done on it and the heat supplied to it. The ΔH is positive or negative in endothermic or exothermic process, respectively. In the protein-solvent system, the enthalpy change reflects the amount of heat energy required to achieve a particular state. For binding reactions, negative enthalpy values are common (but not omnipresent), reflecting a tendency for the system to fall to lower energy levels by bond formation. The binding enthalpy in its strict sense is considered as the noncovalent bond formation at binding interface. However, the heat effect of a binding reaction is a global property of the entire system, including contribution from solvent and protons (Cooper & Johnson, 1994). Therefore, the change in enthalpy of binding must be the result of the formation and breaking of many individual bonds, including the loss of protein-solvent and ligand-solvent hydrogen bonds, electrostatic and van der Waals interactions, the formation of noncovalent bonds between the protein and ligand, and the solvent reorganization near the protein surfaces. These individual components may make either favorable or unfavorable contributions, and the result is a combination of these contributions, with specific interactions dominating the binding enthalpy.

The entropy is a measure of the tendency of a process, such as a chemical reaction, to be entropically favored, or to proceed in a particular direction. It determines that thermal energy always flows spontaneously from regions of higher temperature to regions of lower temperature, in the form of heat. These processes reduce the state of order of the initial systems, and therefore the entropy can be seen as an expression of the disorder or randomness. In the protein-solvent system, entropy measures how easily that energy might be distributed among various molecular energy levels and represents all the other positive and negative driving forces (in addition to enthalpy change) contributing to the free energy. The total entropy change (ΔS_{tot}) of protein-ligand binding can be expressed as the sum of the three entropic terms: the solvent entropy ΔS_{solv}, conformational entropy ΔS_{conf} and the rotation and translation entropy $\Delta S_{r/t}$:

$$\Delta S_{tot} = \Delta S_{solv} + \Delta S_{conf} + \Delta S_{r/t} \tag{8}$$

where the ΔS_{solv} describes the change in entropy resulting from solvent release upon binding; the ΔS_{conf} is a configurational term reflecting the reduction in rotational degrees of freedom around torsion angles of protein and ligand; and the $\Delta S_{r/t}$ represents the loss in translational and rotational degrees of freedom of the acceptor and ligand upon complex formation. The most important contribution to the ΔS_{tot} arises from the solvation term ΔS_{solv}, primarily due to the release of well-bound water molecules on the surfaces of the protein and ligand, and the accompanying burial of the hydrophobic surface area. Since the entropy of hydration of polar and apolar groups is large, the burial of solvent accessible surface area and the solvent release upon binding often make a large, favorable contribution to the entropy increase. The ΔS_{conf} and $\Delta S_{r/t}$ make unfavorable contributions to total entropy, i.e., causing entropy reduction. Although the positive entropy changes resulting from a natural tendency for disruption of order and exclusion of water molecules are common for binding reactions (Chaires, 2008), all binding reactions would have to overcome the inescapable entropic penalties (Amzel, 1997, 2000; Brady & Sharp, 1997) (i.e., the negative $\Delta S_{r/t}$ and ΔS_{conf} upon complex formation) through large solvent entropy gain (positive ΔS_{solv}) or favorable protein-ligand interactions (negative ΔH) if binding is to occur.

The heat capacity (ΔC_p) is the measurable physical quantity that characterizes the amount of heat required to change a substance's temperature by a given amount. The removal of protein surface area from contact with solvent often results in a large negative ΔC_p. Therefore, the heat capacity provides a link between thermodynamic data and structural information of macromolecules through the correlation of ΔC_p and burial of surface area (Perozzo et al., 2004).

5.4.2 Enthalpy-entropy compensation

The phenomenon of enthalpy-entropy compensation has been widely observed in biological systems by using the thermodynamic methods (Eftink et al., 1983; Lumry & Rajender, 1970; Perozzo et al., 2004). Equation 2 shows that the change in free energy is characterized by a linear relationship between the changes in enthalpy and entropy. For example, favorable changes in binding enthalpy are often compensated for by unfavorable changes in entropy and vice versa. In a binding process, the increased bonding, which results in more negative ΔH, will be at the expense of increased order, leading to more negative ΔS. However, if molecular binding is coupled with local or global protein folding (e.g. in the intrinsically unstructured protein segments or proteins (Tsai et al., 2001; Wright & Dyson, 1999)), the hydrophobic effect could become much more significant than the entropy loss caused by the loss of conformational entropy and rotational and translational entropy of the protein and the ligand. Furthermore, when the binding is stronger and more specific with higher positive interaction enthalpy contribution to the binding free energy, the entropy contribution can become large and positive (Demchenko, 2001). The consequences of the enthalpy-entropy compensation are i) that it does not bring out dramatic change in the binding free energy and as such leads to only small changes in binding affinity over a range of temperatures; ii) that it makes it difficulty to distinguish between the entropy-driven or enthalpy-driven binding processes and, therefore, cautions are required to explain the experimental data, design drug, or plan engineering experiments. The factors affecting the enthalpy–entropy compensation include the properties of the solvent (water), the architecture of the ligand-binding site/pocket/cavity, the molecular structure of the ligand, and the changes in intermolecular forces in the binding process (Dunitz, 1995; Gilli et al., 1994). The two thermodynamic parameters, ΔH and ΔS, are correlated with each other through a bridge of the ΔC_p. In order to increase binding affinity of a ligand to a protein of interest, for example, in the rational drug design, the ideal optimization strategy requires refining the enthalpic or entropic contributions to result in a minimal entropic or enthalpic penalty but induce the largest lowering in ΔG, thereby defeating the deleterious effects of enthalpy–entropy compensation at the thermodynamic level.

5.4.3 Enthalpy-driven, entropy-driven or their combination?

Although the entropy-enthalpy compensation weakens the effect on lowering the total free energy of binding, it is important to distinguish whether a binding process is a entropy- or enthalpy-driven one in fields of the medicinal chemistry and drug design because this will facilitate the understanding of the binding mechanisms and help to improve binding affinity or specificity by modifying the acceptor or ligand.

In the case of the simple lock-and-key binding model, we speculate that the entropy change would make a substantial contribution to the binding free energy due to the structural rigidity and the perfect surface complementary between the acceptor and ligand. The large positive ΔS_{solv} is gained when water molecules, which in the unliganded acceptor form a well-defined network in ligand-binding pockets or cavities, are displaced upon ligand binding. Intuitively, such an ideal conformationally selective binding between rigid partners results in only very little loss in ΔS_{conf} and $\Delta S_{r/t}$ since there is no large backbone conformation change and only some interacting side chains are restricted, thereby resulting in a large positive ΔS_{tot}. The subsequent step of binding is the formation of some noncovalent bonds between the acceptor and ligand. Because of the lack of large conformational adjustment, it seems that only a small amount of enthalpy is required to compensate for the minor entropy cost. Nevertheless, the types and strength of the bonds formed between the protein and ligand contribute to the binding affinity. Our speculation is in line with the recent study by Odriozola and coworkers (Odriozola et al., 2008), who showed that the key-lock assembly is solely driven by the entropic contribution even in the absence of attractive forces, pointing out the importance of solvent contribution in the underlying mechanisms of substrate-protein assembly processes.

In the case of the induced fit model, the initial contact between the interacting partners exclude limited amount of water molecules compared to that in lock-and-key model due to the lack of surface complementary between the protein and ligand. Therefore, the ΔS_{solv} is smaller than that in the lock-and-key model. In order to achieve a high binding affinity, a full contact between the interacting partners should be established, and this is accompanied by the conformational adjustment and ordering of the interacting regions and as thus, the large loss in ΔS_{conf}. It is possible that for the induced fit binding, the solvent entropy gain can not overcome the loss in the conformational and rotational and translational entropy terms, thus resulting in a negative ΔS_{tot} (equation 8). Such an entropy cost must be compensated for by the large negative enthalpy arising from the formation of extensive hydrogen bonding, electrostatic and van der Waals interactions. Taken together, we consider that the induced fit binding is an enthalpy-driven process because the dominant driving force is the large negative enthalpy, which compensates for the entropy penalty and renders the binding thermodynamically favorable.

In the case of the conformational selection model, it appears that the hydrophobic interaction makes a large contribution to lowering binding free energy, just like the situation in the lock-and-key model, which, as described above (section 5.2), is the extreme of the conformational selection mechanism. Theoretically, the step of selective binding of ligand to a structurally favorable state of the receptor is an optimal strategy since it maximizes the ΔS_{solv} while minimizing the cost of the inevitable entropy penalty arising from the ΔS_{conf} and $\Delta S_{r/t}$ to achieve a as large as possible ΔS_{tot}. Therefore, the conformational selection plays a role in lowering the entropic barrier (Lange et al., 2008; Perica & Chothia, 2010) and contributes to the binding specificity. However, because of the flexibility nature of proteins, especially in the ligand-binding regions, the binding will have to overcome the inescapable entropic penalties. The subsequent step of conformational adaption, which is driven by the enthalpic effect of bond formation, contributes not only to further lowering the free energy, but also to binding affinity. As discussed in section 5.2, the conformational selection mechanism is actually a synthesis of the lock-and-key and induced fit models, and therefore the different stages of molecule recognition can be driven either by the entropy or enthalpy, both making variable contributions to lowering the free energy of binding. The essence of the conformational selection is a combinatorial effect of enthalpy and entropy, which not only guarantees simultaneously the binding affinity and specificity of a single or multiple ligands but also accommodates mutations through shift in the dynamic energy landscape and therefore, is evolutionarily advantageous (Ma et al., 2001).

5.4.4 Case studies on entropy-, enthalpy-, and combination-driven bindings

The thermodynamic principle of enthalpy-entropy compensation (Chaires, 2008) has promoted the Freire group to optimize HIV protease inhibitor binding through taking into account the thermodynamics of the binding interactions (Ohtaka & Freire, 2005; Ohtaka et al., 2004; Velazquez-Campoy et al., 2000a; Velazquez-Campoy et al., 2000b). The results show that, although the interaction of the enthalpically or entropically optimized inhibitors with the target protease may have similar free energies, these inhibitors might have different specificities or different pharmacological properties.

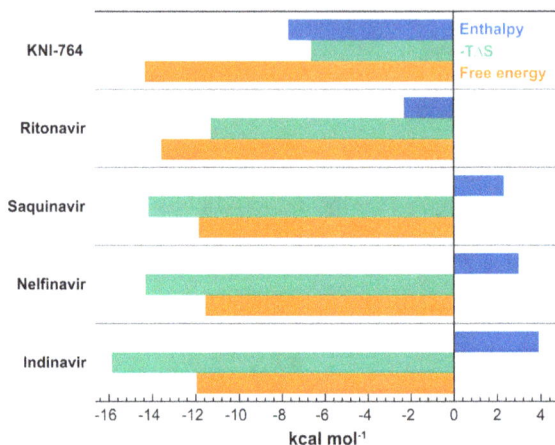

Fig. 7. Thermodynamic profiles of the bindings of the five inhibitors to HIV protease. Free energy, enthalpy and $-T\Delta S$ are shown as orange, blue and green horizontal bars, respectively. This figure is cited from (Chaires, 2008).

Figure 7 shows the thermodynamic parameters of the binding of several inhibitors to HIV protease. For the first-generation inhibitors, i.e., indinavir, nelfinavir, saquinavir, and ritonavir, the binding is entropically driven due to the predominantly positive entropy gain. Since these inhibitors are highly structurally constrained and the protease itself is relatively rigid, the ΔS_{solv} arising from the hydrophobic effect makes a substantial contribution to the binding entropy (ΔS_{tot}). Interestingly, for three (indinavir, nelfinavir and saquinavir in Figure 7) of the four inhibitors, their binding is accompanied by an unfavorable positive enthalpy change. Although such an enthalpy cost can be compensated for by the large entropy gain, the increase in enthalpy is detrimental to binding affinity, as shown by the relatively higher free energy compared to that for ritonavir, whose binding is accompanied by a small favorable negative enthalpy change. In the case of a newer-generation inhibitor, KNI-764, its binding is tighter than bindings of the first-generation inhibitors to the HIV protease by at least 1 kcal mol^{-1}. However, as shown in Figure 7, such a lower free energy is the outcome of the almost equal favorable enthalpic and entropic contributions, suggesting that its binding is driven by both the hydrophobic (entropy effect) and favorable intermolecular forces (enthalpy contribution). Such a binding process may be a conformational selection followed by a induced fit, in line with the extended conformational selection mechanism proposed by (Csermely et al., 2010). The bindings of a series of inhibitors to the HIV protease, which are characterized by being from the pure entropy-driven to entropy-driven plus weak enthalpic contribution to the combinatorial entropy- and enthalpy-driven, suggest that the extremely high binding affinity requires a favorable binding enthalpy, and therefore the enthalpic optimization is one of the most important principles in drug design (Chaires, 2008; Ohtaka et al., 2004). It should be noted that the optimal binding enthalpy does not correlate simply with the number of interactions, but instead reflects mostly the strength and quality of the noncovalent interactions (such as hydrogen bonding, electrostatic and van der Waals interactions) of the inhibitor with the protease.

Temp.	TK+dT[a]			TK:dT+ATP[b]			TK+dT/ATP[c]		
(°C)	ΔG[d, e]	ΔH[d]	$T\Delta S$[d]	ΔG[d, e]	ΔH[d]	$T\Delta S$[d]	ΔG[d, e]	ΔH[d]	$T\Delta S$[d]
10	−7.1	−13.6	−6.5	−8.9	−11.7	−2.8	−18.9	−25.4	−6.5
15	−7.0	−15.8	−8.8	−8.9	−12.1	−3.2	−18.9	−27.9	−9.0
20	−7.1	−17.4	−10.2	−9.1	−13.0	−3.9	−17.2	−30.5	−13.3
25	−7.2	−19.1	−11.9	−9.0	−13.8	−4.8	−16.8	−33.1	−16.3

[a] TK was titrated with the dT.
[b] TK:dT complex was titrated with the ATP.
[c] TK was titrated with a 1:1 mixture of dT and ATP.
[d] The unit was in kcal mol⁻¹.
[e] ΔG was calculated from equation 1, where the K_b is the binding constant determined by ITC.

Table 1. Thermodynamic parameters of binding of thymidine (dT) and ATP to HSV1 TK at pH 7.5. Data in this table are cited from (Perozzo et al., 2004).

Thermodynamic parameters of the binding of thymidine (dT, which is the substrate) and adenosine triphosphate (ATP, which is a cofactor) to the thymidine kinases (TK) from Herpes simplex virus type-1 (HSV1) were determined using the ITC method (Perozzo et al., 2004; Perozzo et al., 2000; Pilger et al., 1999). In the ternary complex TK:dT:ATP, dT and ATP are located in separate, well defined binding pockets of the enzyme. Table 1 lists the thermodynamic parameters for binding of dT and ATP to TK. The "TK+dT" indicates that the TK is titrated with the dT; the "TK:dT+ATP" indicates that using the ATP to titrate the TK:dT binary complex; and the "TK+dT/ATP" indicates that TK is titrated with a 1:1 mixture of dT and ATP. All these three groups of binding reactions at a temperature range of 10-25 °C are an exothermic reaction driven by a large negative change in enthalpy, which is favorable enough to compensate for the unfavorable decrease in the entropy of the system. As expected, the group "TK+dT/ATP" yields a heat change approximately equals the sum of the heat changes for groups "TK+dT" and "TK:dT+ATP". An interesting phenomenon is that the ΔH and $T\Delta S$ of all these three reactions depend strongly on the temperature, e.g., their values decrease when temperature increases, while ΔG is almost insensitive to temperature due to the entropy-enthalpy compensation. The high flexibility caused by high temperature facilitates mutual conformational adaption between interacting partners, thereby resulting in increased bonds and as such, more negative ΔH, which is counterbalanced by more negative ΔS caused by increased order of binding partner after binding. The ΔS_{tot} of the three groups of reactions at 25 °C are further decomposed into ΔS_{solv}, ΔS_{conf} and $\Delta S_{r/t}$ and are listed in Table 2. It is evident that the favorable positive ΔS_{solv} is overcompensated by the large unfavorable negative ΔS_{conf}. The contribution by $\Delta S_{r/t}$, although unfavorable, is one or two orders of magnitude smaller than the other entropy terms. Interestingly, the sum of the decomposed entropy terms of "TK+dT" and "TK:dT+ATP" is approximately equal to the corresponding change in entropy terms of "TK+dT/ATP". The large positive solvent entropy ΔS_{solv} originates from the water release upon binding. The larger the interacting interfaces or binding cavities are, the more ordered water molecules will be displaced, and the larger the positive ΔS_{solv} will be gained. For example, binding of the larger dT to TK yields a larger positive ΔS_{solv} (92 cal K⁻¹ mol⁻¹) than binding of relatively smaller ATP to TK (36 cal K⁻¹ mol⁻¹). Similarly, the magnitude of the ΔS_{conf} is related to the area of the rigidified interfaces and the number of the fixed side chains/main chains upon binding. The large negative ΔS_{tot} needs a larger negative ΔH to overcompensate for to lower the binding free energy, suggesting an induced fit binding driven mainly by intermolecular forces. This is in line with the observation that substrate binding to HSV1 TK leads to a conformational closing of the binding sites, bringing thymidine and ATP into an orientation appropriate for catalysis (Perozzo et al., 2004).

Reaction	ΔS_{tot} (cal K^{-1} mol^{-1})	ΔS_{solv} (cal K^{-1} mol^{-1})	ΔS_{conf} (cal K^{-1} mol^{-1})	$\Delta S_{r/t}$ (cal K^{-1} mol^{-1})
TK+dT	−39.9	92.0	−124.0	−8.0
TK:dT+ATP	−16.1	36.0	−44.0	−8.0
TK+dT/ATP	−54.7	133.0	−172.0	−16.0

Table 2. Decomposition of total entropy changes ΔS_{tot} into ΔS_{solv}, ΔS_{conf} and $\Delta S_{r/t}$ for substrate binding to HSV1 TK at 25 °C. Data in this table are cited from (Perozzo et al., 2004).

The difference in the driving force between induced fit/binding-induced folding and conformational selection/constitutive binding can be illustrated by studies on the binding of the kinase-inducible activation (KID) domain of the cAMP response element-binding protein (CREB) to the KIX domain (residues 586-672) of CREB binding protein (CBP), and binding of the proto-oncogene c-Myb to the KIX domain of CBP (Wright & Dyson, 1999). The KID domain binds with high affinity to the KIX domain only when the Ser133 is phosphorylated to form the pKID. The pKID domain is observed to be disordered in its free state but folds into a pair of orthogonal helices upon binding to the KIX domain. Such a binding is enthalpy-driven and entropically disfavored, with the entropic and enthalpic contributions being −6.0 cal K^{-1} mol^{-1} and −10.6 kcal mol^{-1}, respectively. The entropic penalty of the folding transition (from coil to helix) is compensated for by the large negative enthalpy of binding, attributed partly to the complementary intermolecular hydrogen bonds formed between the phosphor-Ser133 in KID and the Tyr and Lys side chains in the target protein KIX (Radhakrishnan et al., 1997). In the unphosphorylated state, binding of KID to KIX is very weak (Wright & Dyson, 1999) since the smaller enthalpy of binding (lacking the large number of hydrogen bonds formed by phosphor-Ser133) can not compensate for the entropic cost of the folding transition. Here the binding-induced folding is an extreme case of induced fit. In order to guarantee the binding affinity, there must be strong enough inter-molecular interactions to stabilize the folded structure in the complex. In contrast to pKID, the activation domain of the proto-oncogene c-Myb binds constitutively, i.e. without requiring activation by phosphorylation, to the same site on the KIX domain as pKID does. Unlike pKID, the c-Myb activation domain can spontaneously adopt the helical conformation, even as an isolated peptide (Parker et al., 1999). Binding to KIX is now entropically favored ($\Delta S = 7.5$ cal mol^{-1} K^{-1}), but the enthalpy change ($\Delta H = -4.1$ kcal mol^{-1}) is smaller than for pKID, presumably because the hydrogen bonding interactions made by the phosphoryl moiety of pKID are absent for c-Myb (Parker et al., 1999). Therefore, such a binding can be explained by the conformational selection mechanism driven mainly by the hydrophobic force. Furthermore, the negative binding enthalpy, although smaller than that of pKID binding, is able to consolidate binding affinity. Taken together, we conclude that the binding of different activation domains to the same site of the target protein can either be enthalpy- or entropy-driven, depending on the initial structure and the flexibility of the activation domains. Moreover, in the case of binding-induced folding, certain structural factors capable of stabilizing the folded structure in the complex are required, i.e., the enthalpic contribution must be favorable enough to compensate for the large entropic loss.

6. Energy landscape contains both the folding and binding funnels

Protein folding and binding are essentially similar processes because of their common essence: the recognition and organization/reorganization of amino acids, either within a single protein or between different proteins. The sole difference between folding and binding is the presence and absence of the chain connectivity between their components, leading to two different terms, i.e., "the intramolecular" and "the intermolecular" recognition (Ma et al., 1999; Tsai et al., 1998). Most studies generally focus on either one process or the other: studies of protein folding focus merely on a single polypeptide; studies of protein binding involve at least two chains. In this section we will show how protein folding and binding are synthesized together by the energy landscape theory to form a common funnel model.

Under the background of statistical mechanical theory, both the protein folding and binding are driven by a decrease in total Gibbs free energy, which is dictated by a mechanism of the delicate balance of opposing effect, i.e., the entropy-enthalpy compensation. The requirement for lowering free energy while reducing conformational space determines that the energy landscape of a protein folding must be funnel-shaped. Analogously, the spontaneous protein-ligand association also lowers the free energy of the system composed of protein, ligand and solvent, while reducing the conformational entropy and rotational and translational entropy of the protein and ligand. Thus the energy landscape of the protein-ligand binding should also be funnel-like, which we term protein binding funnel.

Two hypotheses can be used to illustrate the similarity of the protein folding and binding: i) the hierarchical building block model (Ma et al., 1999; Tsai et al., 1999a) and ii) the folding funnel model. In the building block model, protein folding is viewed as the consequence of a combinatorial assembly of a set of transient "building blocks". The formation of any building block in a given sequence can be described and guided by a microfunnel-like energy landscape. The mutual recognition between building blocks resembles a fusion of two microfunnel-like landscapes. A compact, stable hydrophobic folding unit is located at the bottom of a subfunnel-like landscape. Such a hydrophobic folding unit, in turn, serves as the basic unit in building a functional single domain or multi-domain protein, an oligomer, or a functional complex. The entire process may be seen as a sequential fusion and modification of these individual microfunnels. In the case of protein binding, the folded monomers for protein-protein association correspond to the building blocks in protein folding. In both cases, structural entities associate with each other and this process is driven by fusing two funnels into a higher dimensional funnel, regardless of the chain connectivity. Therefore, the essence of the hierarchical building block model is that a series of microfunnel fusion events lower the Gibbs free energy of the system and finally achieve a global free energy minimal state or protein-ligand complex.

In the folding funnel model, the first step is the fast hydrophobic collapse driven by the entropy. Concretely, the water molecules squeeze the peptide chain to collapse into molten globule intermediates and exclude themselves from the interior of the collapsed protein, resulting in the maximum of the solvent entropy. At the same time, the heat energy is required to strip the water molecules from the polypeptide surface, leading to an endothermic reaction and therefore, an enthalpy increase. However, in order to lower the free energy of the system, the increase in entropy must be large enough to overcompensate for the increase in enthalpy (see equation 2), putting this step under entropy control (Griko, 2000). In the molten globule intermediates, some of the native secondary structures may have been formed, while many native contacts, or close residue-residue interactions present in the native state, have yet to form. Therefore, the subsequent step of protein folding is to establish these native inter-atomic contacts through conformational adjustment and adaption, which may be a slow process because of the many times trial attempts of bond breaking and formation. Finally, the formed tertiary native interactions reduce the conformational flexibility, leading to the loss of the conformational entropy of the protein. Although the native state has lower entropy than the intermediate molten globule, this step also leads to the formation of a large number of favorable specific interactions and therefore, the large negative enthalpy, putting it under enthalpy control. Interestingly, the formation of the protein-protein or protein-ligand complex is similar to the protein folding process. The initial contact of the protein and its ligand will inevitably displace water molecules around the interfaces of both acceptor and ligand. Therefore, this step is driven by the increase of the solvent entropy in a way similar to the hydrophobic collapse of protein folding. An ideal "hydrophobic collapse" may occur for the lock-and-key and to a lesser extent, the conformational selection processes because the interacting interfaces are generally hydrophobic and match well between the protein and ligand. This process can be considered to occur at the bottom of the relatively smooth folding funnel (in the case of lock-and-key model for rigid proteins) or the bottom of the tier-0 wells/microfunnels of the rugged energy landscape (in the case of conformational selection model; see Figure 4). Although the initial trial collisions and contacts may be time-consuming, the correct orientation can lead to a rapid change of the shape of funnel or microfunnel, i.e., the downward extension of the bottom of the funnel/microfunnel driven by the entropy gain of the system. Similar to what occurs to the protein folding process, the next step is the conformational adjustment and mutual adaption between interacting partners driven by the negative enthalpy, resulting in the final formation of favorable specific bonds. This step further downwards extends and narrows the binding funnel, shifting the equilibrium towards the bound sate. The induced fit may occur along the very rugged floor of the folding funnel that includes many free energy minimum wells (i.e., the tier-1 or tier-2 wells as described in section 4.4.2) separated by small energy barriers, thereby allowing easy conformational transition between different substates. This is in agreement with the hypothesis that the fast conformational transition rate is beneficial to induced fit binding (Zhou, 2010). The fast and easy conformational transition may increase the probability of the conformational substates appropriate for ligand binding. As argued by (Tsai et al., 2001), these conformational substates, although having a low population time, still have higher concentrations than the other substates, thus facilitating the selective binding of the ligand. This also suggests that the conformational selection process may play a relatively important role in the induced fit binding. Although the initial solvent entropy gain is small and contributes little to lowering the free energy, such an entropy change triggers the deepening of the small binding funnel, allowing the dominant driving force, the negative enthalpy, to further lower the binding free energy. In this process, the entropy penalty arising from the ordering of interacting partners is counterbalanced by the large negative enthalpy arising from bond formations, analogous to the enthalpy-driven conformational adjustment step occurring in the folding funnel. Taken together, although different proteins may have different landscapes around the bottom of their folding funnels (e.g., smooth, rugged or very rugged floors), the binding event occurring in one of the energy wells can further extend and deepen the single well into a binding funnel. Binding is also driven by the decrease in total Gibbs free energy through the entropy-enthalpy compensation, with different steps being dominated by either the entropy or enthalpy. If the ligand is a protein, the binding funnel can be viewed as the consequence of fusion of two individual microfunnels or wells; if the ligand is a small compound, the binding funnel can be viewed as extension and deepening of a specific energy well located at the bottom of the protein folding funnel.

7. Conclusion

We conclude that the protein folding and binding are similar processes that are commonly driven by the decrease in total Gibbs free energy of the system, which is dictated by a mechanism of the delicate balance of opposing effects involving entropy and enthalpy contributions, i.e., the effect of entropy-enthalpy compensation. Such a mechanism is described vividly by the free energy landscape theory, which characterizes the protein folding as going down the minimally frustrated, funnel-like energy landscape via multiple pathways towards native states located at the bottom of the funnel, and the protein binding as fusion of the microfunnels/energy wells or as downwards extending the free energy well into a binding funnel at the bottom of the folding funnel. Furthermore, the extent of the ruggedness of the energy landscape, which governs the dynamic behavior of the protein, is also dictated by the entropy-enthalpy compensation. Therefore, the protein folding, binding and its dynamic behavior are all essentially thermodynamically controlled processes that are governed by entropy-enthalpy compensation. The tendency to maximize the entropy of the protein-solvent system is the original driving force for protein folding, binding and dynamics, while the enthalpy change, an opposing force that tends to drive the system to become ordered, can compensate for the entropy change to ultimately allow the system to reach equilibrium at the free energy minimization, either global or local. Therefore, the description of protein folding, binding and dynamics by the leveling free-energy landscape is consistent with the second law of thermodynamics (Sharma et al., 2009). This law, generally understood as the entropy maximum principle, dictates and underlies not only the protein folding and binding, but also other life phenomena (Erwin, 1944) such as the origin of life (Doolittle, 1984; L. James, 1979), biological evolution (John, 2003), cancer occurrence (Ji & Liu, 2011b), among others. Interestingly, entropy maximum principle or the thermodynamics second law can also be used to explain and describe many of the economic and social phenomena. Summarily, a deep understanding of the thermodynamic entropy-enthalpy compensation based on the dynamic energy landscape view can not only help to understand the nature of the forces that drive protein folding and complex formation, but also facilitate the function, protein engineering and drug design studies in the post-genomic era.

8. Acknowledgement

We thank Ms. Sara A. Barton for her careful reading and useful comments. Research described was supported by grants from National Natural Science Foundation of China (No.

31160181 and 30860011) and Yunnan province (2007PY-22 and 2011CI123), and by the foundation for Key Teacher of Yunnan University.

9. References

Agashe, V.R.; Shastry, M.C.R. & Udgaonkar, J.B. (1995). Initial hydrophobic collapse in the folding of barstar. *Nature*, 377, 754-757.

Amadei, A.; Linssen, A.B.M. & Berendsen, H.J.C. (1993). Essential dynamics of proteins. *Proteins: Struct Funct Genet*, 17, 412-425.

Aman, M.J.; Karauzum, H.; Bowden, M.G. & Nguyen, T.L. (2010). Structural model of the pre-pore ring-like structure of panton-valentine leukocidin: providing dimensionality to biophysical and mutational data. *J Biomol Struct Dyn*, 28, 1-12.

Amzel, L.M. (1997). Loss of translational entropy in binding, folding, and catalysis. *Proteins: Struct Funct Genet* 28, 144-149.

Amzel, L.M. (2000). Calculation of entropy changes in biological processes: folding, binding, and oligomerization. *Methods Enzymol*, 323, 167-177.

Anfinsen, C.B.; Haber, E.; Sela, M. & White, F.H.J. (1961). The kinetics of formation of native ribonuclease during oxidation of the reduced polypeptide chain. *Proc Natl Acad Sci USA*, 47, 1309-1314.

Anfinsen, C.B. & Scheraga, H.A. (1975). Experimental and theoretical aspects of protein folding. *Adv Protein Chem*, 29, 205-300.

Ansari, A.; Berendzen, J.; Bowne, S.F.; Frauenfelder, H.; Iben, I.E.T.; Sauke, T.B.; Shyamsunder, E. & Young, R.D. (1985). Protein states and protein quakes. *Proc Natl Acad Sci USA*, 82, 5000-5004.

Austin, R.H.; Beeson, K.W.; Eisenstein, L.; Frauenfelder, H. & Gunsalus, I.C. (1975). Dynamics of ligand binding to myoglobin. *Biochemistry*, 14, 5355-5373.

Baker, D. & Sali, A. (2001). Protein structure prediction and structural genomics. *Science*, 294, 93-96.

Berendsen, H.J.C. (2011). The relevance of distance statistics for protein folding. *J Biomol Struct Dyn*, 28, 599-602.

Berger, C.; Weber-Bornhauser, S.; Eggenberger, J.; Hanes, J.; Pluckthun, A. & Bosshard, H.R. (1999). Antigen recognition by conformational selection. *FEBS Lett*, 450, 149-153.

Boehr, D.D.; Nussinov, R. & Wright, P.E. (2009). The role of dynamic conformational ensembles in biomolecular recognition. *Nat Chem Biol*, 5, 789-796.

Bosshard, H.R. (2001). Molecular recognition by induced fit: How fit is the concept? *News Physiol Sci*, 16, 171-173.

Bowie, J.U. (2005). Solving the membrane protein folding problem. *Nature*, 438, 581-589.

Brady, G.P. & Sharp, K.A. (1997). Entropy in protein folding and in protein-protein interactions. *Curr Opin Struct Biol*, 7, 215-221.

Bryngelson, J.D.; Onuchic, J.N.; Socci, N.D. & Wolynes, P.G. (1995). Funnels, pathways, and the energy landscape of protein Folding: A synthesis. *Proteins: Struct Funct Genet*, 21, 167-195.

Chaires, J.B. (2008). Calorimetry and thermodynamics in drug design. *Annu Rev Biophys*, 37, 135-151.

Chan, H.S. (2011). Short-range contact preferences and long-range indifference: Is protein folding stoichiometry driven? *J Biomol Struct Dyn*, 28, 603-606.

Chen, A. & Wadso, I. (1982). Simultaneous determination of delta G, delta H and delta S by an automatic microcalorimetric titration technique: application to protein ligand binding. *J Biochem Biophys Meth*, 6, 307-316.

Cooper, A. & Johnson, C.M. (1994). Introduction to microcalorimetry and biomolecular energetics. *Methods Mol Biol*, 22, 109-124.

Csermely, P.; Palotai, R. & Nussinov, R. (2010). Induced fit, conformational selectionand independent dynamic segments: an extended view of binding events. *Trends Biochem Sci* 35, 539-546.

Demchenko, A.P. (2001). Recognition between flexible protein molecules: induced and assisted folding. *J Mol Recognit*, 14, 42-61.

Dill, K.A. (1985). Theory for the folding and stability of globular proteins. *Biochemistry*, 24, 1501-1509.

Dill, K.A.; Fiebig, K.M. & Chan, H.S. (1993). Cooperativity in protein-folding kinetics. *Proc Natl Acad Sci USA*, 90, 1942-1946.

Dill, K.A.; Bromberg, S.; Yue, K.; Fiebig, K.M.; Yee, D.P.; Thomas, P.D. & Chan, H.S. (1995). Principles of protein folding - A perspective from simple exact models. *Protein Sci*, 4, 561-602.

Dill, K.A. & Chan, H.S. (1997). From Levinthal to pathways to funnels. *Nat Struct Biol* 4, 10-19.

Dill, K.A. (1999). Polymer principles and protein folding. *Protein Sci*, 8, 1166-1180.

Dill, K.A.; Ozkan, S.B.; Weikl, T.R.; Chodera, J.D. & Voelz, V.A. (2007). The protein folding problem: when will it be solved? *Curr Opin Struct Biol*, 17, 342-346.

Dobson, C.M. (2000). The nature and significance of protein folding. In: *Mechanisms of Protein Folding (2nd ed.)*, R.H. Pain, (Ed.), Oxford University Press, Oxford, UK.

Doolittle, R. (1984). The probability and origin of life. In: *Scientists Confront Creationism*, L.R. Godfrey, (Ed.), pp. 85.

Duan, Y. & Kollman, P.A. (1998). Pathways to a protein folding intermediate observed in a 1-microsecond simulation in aqueous solution. *Science*, 282, 740-744.

Dunitz, J.D. (1995). Win, some, lose some: enthalpy-entropy compensation in weak intermolecular interactions. *Chem Biol*, 2, 709-712.

Eftink, M.R.; Anusiem, A.C. & Biltonen, R.L. (1983). Enthalpy-entropy compensation and heat capacity changes for protein-ligand interactions: general thermodynamic models and data for the binding of nucleotides to ribonuclease A. *Biochemistry*, 22, 3884-3896.

Erwin, S. (1944). *What is fife - the physical aspect of the living cell*, Cambridge University Press, Cambridge.

Fischer, E. (1894). Einfluss der configuration auf die wirkung der enzyme. *Ber Dtsch Chem Ges*, 27, , 2984-2993.

Foote, J. & Milstein, C. (1994). Conformational isomerism and the diversity of antibodies. *Proc Natl Acad Sci USA*, 91, 10370-10374.

Frauenfelder, H.; Petsko, G.A. & Tsernoglou, D. (1979). Temperature-dependent X-ray diffraction as a probe of protein structural dynamics. *Nature*, 280, 558-563.

Frauenfelder, H.; Sligar, S.G. & Wolynes, P.G. (1991). The energy landscapes and motions of proteins. *Science,* 254, 1598-1603.

Freire, E.; Mayorga, O.L. & Straume, M. (1990). Isothermal titration calorimetry. *Anal Chem,* 62, 950A-959A.

Freire, E. (1999). The propagation of binding interactions to remote sites in proteins: Analysis of the binding of the monoclonal antibody D1.3 to lysozyme. *Proc Natl Acad Sci USA,* 96, 10118-10122.

Galzitskaya, O.V.; Lobanov, M.Y. & Finkelstein, A.V. (2011). Cunning simplicity of a stoichiometry driven protein folding thesis. *J Biomol Struct Dyn,* 28, 595-598.

Gibbs, J.W. (1873). A method of geometrical representation of the thermodynamic properties of substances by means of surfaces. *Trans Conn Acad Arts Sci,* 2, 382-404.

Gilli, P.; Ferretti, V.; Gilli, G. & Borea, P.A. (1994). Enthalpy-entropy compensation in drug receptor binding. *J Phys Chem B,* 98, 1515-1518.

Griko, Y.V. (2000). Energetic basis of structural stability in the molten globule state: a-lactalbumin. *J Mol Biol,* 297, 1259-1268.

Gruebele, M. (2005). Downhill protein folding: evolution meets physics. *C R Biol,* 328, 701-712.

Gulukota, K. & Wolynes, P. (1994). Statistical mechanics of kinetic proofreading in protein folding *in vivo. Proc Natl Acad Sci USA,* 91, 9292-9296.

Hammes, G.G.; Chang, Y.C. & Oas, T.G. (2009). Conformational selection or induced fit: A flux description of reaction mechanism. *Proc Natl Acad Sci USA,* 106, 13737-13741.

Harrison, S.C. & Durbin, R. (1985). Is there a single pathway for the folding of a polypeptide chain? *Proc Natl Acad Sci USA,* 82, 4028-4030.

Henzler-Wildman, K.A. & Kern, D. (2007). Dynamic personalities of proteins. *Nature,* 450, 964-972.

Huang, G.S. & Oas, T.G. (1995). Structure and stability of monomeric l repressor: NMR evidence for two-state folding. *Biochemistry* 34, 3884-3892.

Inouye, H. & Kirschner, D.A. (1997). X-ray diffraction analysis of scrapie prion: intermediate and folded structures in a peptide containing two putative a-helices. *J Mol Biol,* 268, 375-389.

Itzhaki, L.S.; Otzen, D.E. & Fersht, A.R. (1995). The structure of the transition state for folding of chymotrypsin inhibitor 2 analysed by protein engineering methods: Evidence for a nucleation-condensation mechanism for protein folding. *J Mol Biol,* 254, 260-288.

Jackson, S.E. & Fersht, A.R. (1991). Folding of chymotrypsin inhibitor 2.1. Evidence for a two-state transition. *Biochemistry,* 30, 10428-10435.

Jacob, J.; Schindler, T.; Balbach, J. & Schmid, F.X. (1997). Diffusion control in an elementary protein folding reaction. *Proc Natl Acad Sci USA,* 94, 5622-5627.

James, L. (1979). *GAIA - A new look at life on earth,* Oxford University Press, Oxford.

James, L.C.; Roversi, P. & Tawfik, D.S. (2003). Antibody multispecificity mediated by conformational diversity. *Science,* 299, 1362-1367.

Janin, J. (1995). Protein-protein recognition. *Prog Biophys Mol Biol,* 64, 145-166.

Ji, X.L. & Liu, S.Q. (2011a). Is stoichiometry-driven protein folding getting out of thermodynamic control? *J Biomol Struct Dyn,* 28, 621-623.

Ji, X.L. & Liu, S.Q. (2011b). Thinking into mechanism of protein folding and molecular binding. *J Biomol Struct Dyn,* 28, 995-996.

John, A. (2003). *Information theory and evolution,* World Scientific, New Jersey.

Kapon, R.; Nevo, R. & Reich, Z. (2008). Protein energy landscape roughness. *Biochem Soc Trans,* 36, 1404-1408.

Karplus, M. & Weaver, D.L. (1976). Protein-folding dynamics. *Nature,* 260, 404-406.

Karplus, M. & Weaver, D.L. (1994). Protein folding dynamics: the diffusion - collision model and experimental data. *Protein Sci,* 3, 650-668.

Kauzmann, W. (1954). Denaturation of proteins and enzymes. In: *The mechanism of enzyme reaction,* W.D. McElroy, & B. Glass, (Eds.), pp. 70-120, Johns Hopkins Press, Baltimore.

Kauzmann, W. (1959). Some factors in the interpretation of protein denaturation. *Adv Protein Chem,* 14, 1-63.

Kim, P.S. & Baldwin, R.L. (1982). Specific intermediates in the folding reactions of small proteins and the mechanism of folding. *Annu Rev Biochem,* 51, 459-489.

Koshland, D.E.J. (1958). Application of a theory of enzyme specificity to protein synthesis. *Proc Natl Acad Sci USA,* 44, 98-104.

Kosik, K.S. (1992). Alzheimer's disease: a cell biological perspective. *Science,* 256, 780-783.

Kubelka, J.; Hofrichter, J. & Eaton, W.A. (2004). The protein folding "speed limit". *Curr Opin Struct Biol,* 14, 76-88.

Lakomek, N.A.; Lange, O.F.; Walter, K.F.; Farès, C.; Egger, D.; Lunkenheimer, P.; Meiler, J.; Grubmüller, H.; Becker, S.; de Groot, B.L. & Griesinger, C. (2008). Residual dipolar couplings as a tool to study molecular recognition of ubiquitin. *Biochem Soc Trans,* 36, 1433-1437.

Lange, O.F.; Lakomek, N.A.; Fares, C.; Schroder, G.F.; Walter, K.F.A.; Becker, S.; Meiler, J.; Grubmuller, H.; Griesinger, C. & de Groot, B.L. (2008). Recognition dynamics up to microseconds revealed from an RDC-derived ubiquitin ensemble in solution. *Science,* 320:, 1471-1475.

Leopold, P.E.; Montal, M. & Onuchic, J.N. (1992). Protein folding funnels: A kinetic approach to the sequence-structure relationship. *Proc Natl Acad Sci USA,* 89, 8721-8725.

Levinthal, C. (1968). Are there pathways for protein folding? *J Chim Phys,* 65, 44-45.

Levinthal, C. (1969). How to fold graciously. In: *Mossbauer spectroscopy inbiological systems,* P. Debrunner, J. Tsibris, & E. Munck, (Eds.), pp. 22-24, University of Illinois Press, Urbana.

Liu, S.Q.; Fu, Y.X. & Liu, C.Q. (2007a). Molecular motions and conformational transition between different conformational states of HIV-1 gp120 envelope glycoprotein. *Chin Sci Bull,* 52, 3074-3088.

Liu, S.Q.; Liu, C.Q. & Fu, Y.X. (2007b). Molecular motions in HIV-1 gp120 mutants reveal their preferences for different conformations. *J Mol Graphics Model,* 26, 306-318.

Liu, S.Q.; Liu, C.Q. & Fu, Y.X. (2007c). Molecular motions in HIV-1 gp120 mutants reveal their preferences for different conformations. *J Mol Graphics Modell,* 26, 306-318.

Liu, S.Q.; Liu, S.X. & Fu, Y.X. (2007d). Dynamic domains and geometrical properties of HIV-1 gp120 during conformational changes induced by CD4-binding. *J Mol Model*, 13, 411-424.

Liu, S.Q.; Liu, S.X. & Fu, Y.X. (2008). Molecular motions of human HIV-1 gp120 envelope glycoproteins. *J Mol Model*, 14, 857-870.

Liu, S.Q.; Meng, Z.H.; Fu, Y.X. & Zhang, K.Q. (2010). Insights derived from molecular dynamics simulation into the molecular motions of serine protease proteinase K. *J Mol Model*, 16, 17-28.

Liu, S.Q.; Meng, Z.H.; Fu, Y.X. & Zhang, K.Q. (2011). The effect of calciums on the molecular motions of proteinase K. *J Mol Model*, 17, 289-300.

Lum, K.; Chandler, D. & Weeks, J.D. (1999). Hydrophobicity at small and large length scales. *J Phys Chem B*, 103, 4570-4577.

Lumry, R. & Rajender, S. (1970). Enthalpy-entropy compensation phenomena in water solutions of proteins and small molecules: a ubiquitous property of water. *Biopolymers*, 9, 1125-1227.

Ma, B.; Kumar, S.; Tsai, C.J. & Nussinov, R. (1999). Folding funnels and binding mechanisms. *Protein Eng*, 12, 713-720.

Ma, B.; Wolfson, H.J. & Nussinov, R. (2001). Protein functional epitopes: hot spots, dynamics and combinatorial libraries. *Curr Opin Struct Biol*, 11, 364-369.

Magg, C.; Kubelka, J.; Holtermann, G.; Haas, E. & Schmid, F.X. (2006). Specificity of the initial collapse in the folding of the cold shock protein. *J Mol Biol*, 360, 1067-1080.

Maity, H.; Maity, M.; Krishna, M.M.; Mayne, L. & Englander, S.W. (2005). Protein folding: the stepwise assembly of foldon units. *Proc Natl Acad Sci USA*, 102, 4741-4746.

Matouschek, A.; Kellis, J.T.J.; Serrano, L. & Fersht, A.R. (1989). Mapping the transition state and pathway of protein folding by protein engineering. *Nature*, 340, 122-126.

Matthews, B.W. (2011). Stoichiometry versus hydrophobicity in protein folding. *J Biomol Struct Dyn*, 28, 589-591.

Mittal, A.; Jayaram, B.; Shenoy, S.R. & Bawa, T.S. (2010). A Stoichiometry driven universal spatial organization of backbones of folded proteins: Are there Chargaff's rules for protein folding? *J Biomol Struc Dyn*, 28, 133-142.

Mittal, A. & Jayaram, B. (2011a). Backbones of folded proteins reveal novel invariant amino-acid neighborhoods. *J Biomol Struc Dyn*, 28, 443-454.

Mittal, A. & Jayaram, B. (2011b). The newest view on protein folding: stoichiometric and spatial unity in structural and functional diversity. *J Biomol Struc Dyn*, 28, 669-674.

Monod, J.; Wyman, J. & Changeux, J.P. (1965). On the nature of allosteric transitions: a plausible model. *J Mol Biol*, 12, 88-118.

Moult, J.; Pedersen, J.T.; Judson, R. & Fidelis, K. (1995). A large-scale experiment to assess proteinstructure prediction methods. *Proteins: Struct Funct Genet*, 23, ii-iv.

Noe, F.; Schutte, C.; Vanden-Eijnden, E.; Reich, L. & Weikl, T.R. (2009). Constructing the equilibrium ensemble of folding pathways from short off-equilibrium simulations. *Proc Natl Acad Sci USA*, 106, 19011-19016.

Nolting, B. & Agard, D.A. (2008). How general is the nucleation-condensation mechanism? *Proteins*, 73, 754-764.

Odriozola, G.; Jiménez-Ángeles, F. & Lozada-Cassou, M. (2008). Entropy driven key-lock assembly. *J Chem Phys*, 129, 111101-111104.

Ohtaka, H.; Muzammil, S.; Schon, A.; Velazquez-Campoy, A.; Vega, S. & Freire, E. (2004). Thermodynamic rules for the design of high affinity HIV-1 protease inhibitors with adaptability to mutations and high selectivity towards unwanted targets. *Int J Biochem Cell Biol*, 36, 1787-1799.

Ohtaka, H. & Freire, E. (2005). Adaptive inhibitors of the HIV-1 protease. *Prog Biophys Mol Biol*, 88, 193-208.

Onuchic, J.N.; Wolynes, P.G.; Luthey-Schulten, Z. & Socci, N.D. (1995). Towards an outline of the topography of a realistic protein folding funnel. *Proc Natl Acad Sci USA*, 92, 3626-3630.

Onuchic, J.N.; Luthey-Schulten, Z. & Wolynes, P.G. (1997). Theory of protein folding: The energy landscape perspective. *Ann Rev Phys Chem*, 48, 545-600.

Onuchic, J.N. & Wolynes, P.G. (2004). Theory of protein folding. *Curr Opin Struct Biol*, 14, 70-75.

Otlewski, J. & Apostoluk, W. (1997). Structural and energetic aspects of protein-protein recognition. *Acta Biochim Pol*, 44, 367-387.

Ozkan, S.B.; Wu, G.A.; Chodera, J.D. & Dill, K.A. (2007). Protein folding by zipping and assembly. *Proc Natl Acad Sci USA*, 104, 11987-11992.

Pan, K.M.; Baldwin, M.; Nguyen, J.; Gasset, M.; Serban, A.; Groth, D.; Mehlhorn, I.; Huang, Z.; Fletterick, R.J.; Cohen, F.E. & Prusiner, S.B. (1993). Conversion of a-helices into b-sheets features in the formation of the scrapie prion proteins. *Proc Natl Acad Sci USA*, 90(10926-10966).

Parker, D.; Rivera, M.; Zor, T.; Henrion-Caude, A.; Radhakrishnan, I.; Kumar, A.; Shapiro, L.H.; Wright, P.E.; Montminy, M. & Brindle, P.K. (1999). Role of secondary structure in discrimination between constitutive and inducible activators. *Mol Cell Biol*, 19, 5601-5607.

Pechmann, S.; Levy, E.D.; Tartaglia, G.G. & Vendruscolo, M. (2009). Physicochemical principles that regulate the competition between functional and dysfunctional association of proteins. *Proc Natl Acad Sci USA*, 106, 10159-10164.

Perica, T. & Chothia, C. (2010). Ubiquitin - molecular mechanisms for recognition of different structures. *Curr Opin Struct Biol*, 20, 367-376.

Perozzo, R.; Jelesarov, I.; Bosshard, H.R.; Folkers, G. & Scapozza, L. (2000). Compulsory order of substrate binding to Herpes simplex virus type 1 thymidine kinase. A calorimetric study. *J Biol Chem*, 275, 16139-16145.

Perozzo, R.; Folkers, G. & Scapozza, L. (2004). Thermodynamics of protein-ligand interactions: history, presence, and future aspects. *J Recept Signal Transduct Res*, 24, 1-52.

Pieper, U.; Eswar, N.; Davis, F.P.; Braberg, H.; Madhusudhan, M.S.; Rossi, A.; Marti-Renom, M.; Karchin, R.; Webb, B.M.; Eramian, D.; Shen, M.Y.; Kelly, L.; Melo, F. & Sali, A. (2006). MODBASE: a database of annotated comparative protein structure models and associated resources. *Nucleic Acids Res*, 34, D291-295.

Pilger, B.D.; Perozzo, R.; Alber, F.; Wurth, C.; Folkers, G. & Scapozza, L. (1999). Substrate diversity of Herpes simplex virus thymidine kinase-impact of the kinematics of the enzyme. *J Biol Chem*, 274, 31967-31973.

Pitera, J.W. & Swope, W. (2003). Understanding folding and design: replica-exchange simulations of "Trp-cage" miniproteins. *Proc Natl Acad Sci USA*, 100, 7587-7592.

Plaxco, K.W.; Simons, K.T. & Baker, D. (1998). Contact order, transition state placement and the refolding rates of single domain proteins. *J Mol Biol*, 277, 985-994.

Plotkin, S.; Wang, J. & Wolynes, P.G. (1996). Correlated energy landscape model for finite, random heteropolymers. *Phys Rev E Stat Nonlin Soft Matter Phys*, 53, 6271-6296.

Rackovsky, S. & Schraga, H.A. (2011). On the information content of protein sequences. *J Biomol Struct Dyn*, 28, 593-594.

Radhakrishnan, I.; PeÂrez-Alvarado, G.C.; Parker, D.; Dyson, H.J.; Montminy, M.R. & Wright, P.E. (1997). Solution structure of the KIX domain of CBP bound to the transactivation domain of CREB: A model for activator:coactivator interactions. *Cell*, 91, 741-752.

Sali, A. & Blundell, T.L. (1993). Comparative protein modelling by satisfaction of spatial restraints. *J Mol Biol*, 234, 779-815.

Sali, A.; Shakhnovich, E. & Karplus, M. (1994). Kinetics of protein folding. A lattice model study of the requirements for folding to the native state. *J Mol Biol*, 235, 1614-1636.

Sarma, R.H. (2011). A conversation on protein folding. *J Biomol Struct Dyn*, 28, 587-588.

Schindler, T.; Herrler, M.; Marahiel, M.A. & Schmid, F.X. (1995). Extremely rapid protein folding in the absence of intermediates. *Nat Struct Biol*, 2, 663-673.

Schuler, B.; Lipman, E.A. & Eaton, W.A. (2002). Probing the free-energy surface for protein folding with single-molecule fluorescence spectroscopy. *Nature*, 419, 743-747.

Serban, D.; Taraboulos, A.; DeArmond, S.J. & B., P.S. (1990). Rapid detection of Creutzfeldt-Jakob disease and scrapie prion proteins. *Neurology*, 40, 110-117.

Sharma, V.; Kaila, V.R.I. & Annila, A. (2009). Protein folding as an evolutionary process. *Physica A*, 388, 851-862.

Shea, J.E.; Onuchic, J.N. & Brooks, C.L.I. (1999). Exploring the origins of topological frustration: Design of a minimally frustrated model of fragment B of protein A. *Proc Natl Acad Sci USA*, 96, 12512-12517.

Sklenovský, P. & Otyepka, M. (2010). *In silico* structural and functional analysis of fragments of the ankyrin repeat protein p18[INK4c]. *J Biomol Struct Dyn*, 27, 521-539.

Sosnick, T.R.; Mayne, L. & Englander, S.W. (1996). Molecular collapse: The rate-limiting step in two-state cytochrome *c* folding. *Proteins Struct Funct Genet*, 24, 413-426.

Sosnick, T.R.; Dothager, R.S. & Krantz, B.A. (2004). Differences in the folding transition state of ubiquitin indicated by f and c analyses. *Proc Natl Acad Sci USA*, 101, 17377-17382.

Soundararajan, V.; Raman, R.; Raguram, S.; Sasisekharan, V. & Sasisekharan, R. (2010). Atomic interaction networks in the core of protein domains and their native folds. *Plos One*, 5, e9391.

Stillinger, F.H. (1973). Structure in aqueous solutions of nonpolar solutes from the standpoint of scaled-particle theory. *J Solution Chem*, 2, 141-158.

Tao, Y.; Rao, Z.H. & Liu, S.Q. (2010). Insight derived from molecular dynamics simulation into substrate-induced changes in protein motions of proteinase K. *J Biomol Struct Dyn*, 28, 143-157.

Tobi, D. & Bahar, I. (2005). Structural changes involved in protein binding correlate with intrinsic motions of proteins in the unbound state. *Proc Natl Acad Sci USA*, 102, 18908-18913.

Todd, M.J. & Freire, E. (1999). The effects of inhibitor binding on the structural stability and cooperativity of the HIV-1 protease. *Proteins: Struct Funct Genet*, 36, 147-156.

Tsai, C.J.; Xu, D. & Nussinov, R. (1998). Protein folding via binding and vice versa. *Fold Des*, 3, R71-R80.

Tsai, C.J.; Kumar, S.; Ma, B. & Nussinov, R. (1999a). Folding funnels, binding funnels, and protein function. *Protein Sci*, 8, 1181-1190.

Tsai, C.J.; Ma, B. & Nussinov, R. (1999b). Folding and binding cascades: Shifts in energy landscapes. *Proc Natl Acad Sci USA*, 96, 9970-9972.

Tsai, C.J.; Ma, B.; Sham, Y.Y.; Kumar, S. & Nussinov, R. (2001). Structured disorder and conformational selection. *Proteins: Struct Funct Genet*, 44, 418-427.

Velazquez-Campoy, A.; Luque, I.; Todd, M.J.; Milutinovich, M.; Kiso, Y. & Freire, E. (2000a). Thermodynamic dissection of the binding energetics of KNI-272, a potent HIV-1 protease inhibitor. *Protein Sci*, 9, 1801-1809.

Velazquez-Campoy, A.; Todd, M.J. & Freire, E. (2000b). HIV-1 protease inhibitors: enthalpic versus entropic optimization of the binding affinity. *Biochemistry*, 39, 2201-2207.

Voelz, V.A.; Bowman, G.R.; Beauchamp, K. & Pande, V.S. (2010). Molecular simulation of ab Initio protein folding for a millisecond folder NTL9(1-39). *J Am Chem Soc*, 132, 1526-1528.

Wales, D.J. (2003). *Energy Landscapes*, Cambridge University Press, Cambridge.

Weber, P.C. & Salemme, F.R. (2003). Applications of calorimetric methods to drug discovery and the study of protein interactions. *Curr Opin Struct Biol*, 13, 115-121.

Weinreb, P.H.; Zhen, W.; Poon, A.W.; Conway, K.A. & Lansbury, P.T.J. (1996). NACP, a protein implicated in Alzheimer's disease and learning, is natively unfolded. *Biochemistry*, 35, 13709-13715.

Williams, D.H.; Searle, M.S.; Mackay, J.P.; Gerhard, U. & Maplestone, R.A. (1993). Toward an estimation of binding constants in aqueous solution: studies of associations of vancomycin group antibiotics. *Proc Natl Acad Sci USA*, 90, 1172-1178.

Wiseman, T.; Williston, S.; Brandts, J.F. & Lin, L.N. (1989). Rapid measurement of binding constants and heats of binding using a new titration calorimeter. *Anal Biochem*, 179, 131-137.

Wolynes, P.G.; Onuchic, J.N. & Thirumalai, D. (1995). Navigating the folding routes. *Science*, 267, 1619-1620.

Wright, P.E. & Dyson, H.J. (1999). Intrinsically unstructured proteins: re-assessing theprotein structure-function paradigm. *J Mol Biol*, 293, 321-331.

Wu, Z.; Elgart, V.; Qian, H. & Xing, J. (2009). Amplification and detection of single-molecule conformational fluctuation through a protein interaction network with bimodal distributions. *J Phys Chem B*, 113, 12375-12381.

Yang, W.Y. & Gruebele, M. (2003). Folding at the speed limit. *Nature*, 423, 193-197.

Yon, J.M. (2002). Protein folding in the post-genomic era. *J Cell Mol Med,* 6, 307-327.

Zhou, H.X. (2010). From induced fit to conformational selection: A continuum of binding mechanism controlled by the timescale of conformational transitions. *Biophys J,* 98, L15-L17.

Zhou, R.; Huang, X.; Margulis, C.J. & Berne, B.J. (2004). Hydrophobic collapse in multidomain protein folding. *Science,* 305, 1605-1609.

10

Protein Engineering with Non-Natural Amino Acids

Aijun Wang, Natalie Winblade Nairn,
Marcello Marelli and Kenneth Grabstein
Allozyne
USA

1. Introduction

Among all biomolecules in living cells, proteins represent a group with highly divergent biological functions. These include maintaining structural integrity, catalyzing most chemical reactions *in vivo*, regulating cellular responses via signal transduction, and targeting foreign molecules via the immune system. Because of these diverse functions and associated applications, protein engineering has been a field for intense academic research and biopharmaceutical drug development. However, proteins made by nature are generally composed of only 20 canonical amino acids. The recently developed technology to incorporate non-natural amino acids (NNAAs) into proteins at specifically chosen sites represents an extremely powerful tool for protein engineering. Unlike traditional protein engineering with the 20 canonical amino acids as building blocks, protein engineering with NNAAs has nearly unlimited novel side-chain structures to choose from. Coupled with rapid advances in protein structure computation and bioorthogonal chemical reactions, proteins with tailored properties or novel functions beyond natural or directed evolution are becoming a reality.

In this book chapter, we will first review technologies being developed to incorporate NNAAs into proteins. While NNAAs can be incorporated via chemical synthesis, or *in vitro* cell-free protein translation systems, we will focus our review on cell-based protein expression systems because of their capacity for large scale manufacturing. Examples of different ways to assign genetic codons for NNAAs will be reviewed and compared with each other in terms of incorporation efficiency at desired site(s), sequence fidelity of the engineered protein, protein yield, and scalability. Then we will provide examples of chemical conjugation reactions that are orthogonal to the naturally occurring functional groups present in proteins. Topics include reaction kinetics, effect of the reaction conditions on protein, and stability of the final conjugate. At the end, we will highlight a few potential applications of proteins engineered with NNAAs. A special focus will be on therapeutic proteins for clinical use. It is our hope that with this review, more scientists in the protein engineering field will harness this powerful technology and apply it to meet the challenges in their work.

2. Methodologies for incorporating non-natural amino acids into proteins

Protein translation is the process where the genetic information, stored as triplet codon sequence in the mRNA, is used to dictate the polypeptide synthesis. The components required in protein translation were first postulated in the original Adaptor hypothesis by Crick in 1958, and were subsequently identified experimentally (Ibba *et al.*, 2000). During the course of protein synthesis, a ribosome moves along an mRNA molecule and pairs each triplet codon with the adaptor aminoacyl-tRNA containing the complementary anticodon, and catalyzes the formation of a peptide bond with the growing polypeptide chain (Figure 1).

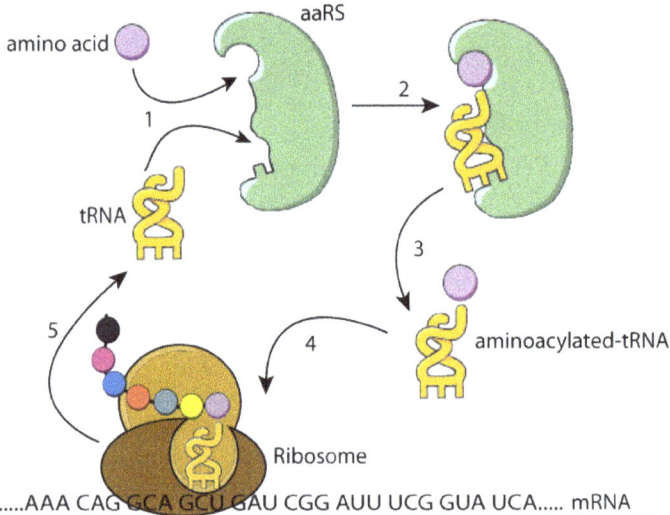

Fig. 1. Natural protein translation system. Aminoacyl-tRNA synthetases (aaRSs) play a key role in maintaining the fidelity of protein translation by demonstrating strict specificity for their cognate amino acid and their tRNA. (1) Each aaRS recognizes unique structural elements of its cognate tRNA(s) and amino acid. (2) The aaRS joins the amino acid to the tRNA, forming the aminoacyl-tRNA (3) that is released into the cytosol. (4) As the ribosome moves along an mRNA molecule, codons are paired with aminoacylated tRNAs containing the complementary anticodons. The ribosome catalyses the sequential peptide-bond formation of amino acids delivered by tRNAs, generating the nascent protein. At the end of translation, release factors occupy stop codons and trigger the termination of protein synthesis. (5) Deaminoacylated tRNA is recycled for further use.

Translation is a highly regulated process, and protein sequence fidelity is largely controlled at two steps: 1) the correct coupling of the amino acid with the tRNA to generate the aminoacyl-tRNA; 2) the correct pairing of the anticodon in the aminoacyl-tRNA with the codon in mRNA. The former is governed by the function of the aminoacyl-tRNA synthetases (aaRSs). The substrate selectivity for tRNA is accomplished through structural recognition elements that are unique to each tRNA and often include the anticodon site. The substrate specificity for the amino acid is achieved by a highly selective amino acid binding pocket. Some aaRSs cannot distinguish between structurally similar amino acids, but they have evolved an editing function to eliminate mis-acylated tRNAs. In most cells, there are 20 aaRS enzymes, one for each amino acid. The 20 canonical amino acids are encoded by 61 codons, with most amino acids encoded by more than one codon. Protein translation is terminated at one of the three stop codons by the protein release factors (RF). It should be mentioned that there are exceptions to nearly every rule in the general description of the translation process. These exceptions actually provide some of the tools that protein engineers have used to introduce NNAAs into proteins, and will be described in subsequent sections.

Several methods have been developed for genetically encoding NNAAs into recombinant proteins. These methodologies range from simple exploitation of the promiscuity of the endogenous protein translation system in the host cells, to the introduction of an exogenous aaRS/tRNA pair specifically designed for a NNAA. All these methods manipulate the natural protein translation process and rely on the reassignment of a genetic codon (such as a sense codon or a stop codon) to the NNAA of interest. As a result, incorporation of the NNAA is intrinsically in competition with certain natural processes: for example, competition from the natural amino acid for a sense codon, or from the release factor for a stop codon. This competition impacts on the efficiency of NNAA incorporation, sequence fidelity as well as the expression yield of target proteins. The ability to reduce or totally eliminate such competition will significantly expedite protein engineering with NNAAs. Recent progress in synthetic biology, which enables the generation of artificial cells with synthetic genomes, shows great promise for such a goal. In this section, we will first describe methods to evolve aaRSs and tRNAs for NNAAs, and then review NNAA incorporation methods based on codon reassignment.

2.1 Methods for evolution of aaRSs and tRNAs for NNAAs

Site-specific introduction of NNAAs requires the expression of an aaRS/tRNA pair specific to a NNAA in the expression host. While this pair must integrate seamlessly with the host translational apparatus, controlled insertion of NNAAs at desired sites requires that it be orthogonal to the host cell. That is, the aaRS/tRNA pair should have the following properties: the introduced aaRS should aminoacylate only NNAA to its cognate tRNA, but not any of the host tRNAs. This avoids mis-incorporation of the NNAA at other codons; in turn, the introduced tRNA should be activated only by its cognate aaRS, but not any of the host aaRSs. This ensures that only the NNAA is incorporated at its assigned codon. The orthogonality of the aaRS/tRNA pair within the expression host cell is essential to maintain sequence fidelity of the target proteins. By exploiting the evolutionary divergence between species, orthogonal aaRS/tRNA pairs have been obtained by transferring an aaRS/tRNA pair from one species into another. An alternate approach requires the re-engineering of recognition elements of the aaRS and its tRNA *de novo* to achieve orthogonality (Neumann *et al.*, 2010a).

While most cells contain 20 aaRSs, some species have evolved additional aaRSs to allow for the utilization of other naturally occurring, but non-canonical amino acids. Pyrrolysine (Pyl) is used by several species of the methanogenic archea, *Methanosarcina*. In its native context, the PylRS/tRNA evolved alongside the 20 canonical aaRSs and developed natural specificity for its cognate amino acid, making it an ideal pair for transfer into other expression hosts. Indeed, the PylRS/tRNA has been shown to be fully orthogonal in *E. coli*, yeast and mammalian cells (reviewed by Voloshchuk & Montclare, 2010). Furthermore, Pyl is naturally encoded by the amber stop codon, the most commonly reassigned codon to specify a NNAA, and thus requires no engineering to adapt it to this purpose. However, the Pyl-tRNA is not hardwired for amber codon suppression, and can be adapted to decode a number of other codons (Ambrogelly *et al.*, 2007). In addition, the PylRS has been shown to readily accept a variety of side chain structures (Polycarpo *et al.*, 2006;Yanagisawa *et al.*, 2008;Neumann *et al.*, 2008;Chen *et al.*, 2009;Nguyen *et al.*, 2009;Li *et al.*, 2009;Hancock *et al.*, 2010;Ou *et al.*, 2011;Plass *et al.*, 2011;Takimoto *et al.*, 2011) as well as a set of non-alpha amino derivatives (Kobayashi *et al.*, 2009). This feature makes the PylRS/tRNA pair an ideal candidate for site specific integration of NNAAs.

2.1.1 Positive/negative selection

Alternating positive and negative screens were pioneered by Schultz's group to evolve orthogonal aaRS/tRNA pairs from large mutational libraries for NNAA incorporation at a stop codon (reviewed by Liu & Schultz, 2010). Figure 2 illustrates the application for evolving aaRSs: First, an aaRS mutant library, along with its cognate tRNA and a conditionally required gene, whose open reading frame is interrupted by a stop codon, is introduced to the host cells for positive selection. Cells are grown the presence of NNAA and only cells with a functional aaRS/tRNA pair (for either NNAA or canonical amino acids) will enable the expression of the essential gene for survival. The selected aaRS mutants (along with the suppressor tRNA) are then shuffled into the host containing a toxic gene (containing an in frame stop codon) for negative selection. When cells are grown in the absence of NNAA, cells containing the aaRS variants that activate canonical amino acids will express the full-length lethal protein and be eliminated; cells containing aaRSs that activate NNAA but not canonical amino acids will survive. The resulting aaRS mutants will be shuffled back for multiple iterations of positive and negative selections until the library converges and/or the desired functionality of the aaRS is obtained. While the alternating positive and negative screen has proven to be extremely successful, the strategy requires plasmid shuttling that is time consuming and affects library diversity. In addition, the common practice of using antibiotic resistance genes for positive selection, while effective, can lead to unexpected variability of identified mutants (Pastrnak & Schultz, 2001;Young *et al.*, 2010). Furthermore, this scheme does not eliminate the mis-acylation of host tRNA(s) with the NNAA by some aaRS variants, leading to potential non-orthogonality of the aaRS/tRNA pair and the introduction of NNAA at undesired sites.

The positive/negative selection (or its various modified formats) is the most widely practiced method for evolving orthogonal aaRS/tRNA pairs. While it was initially developed for an *E. coli* expression system, its principle has been applied successfully in yeast (Chin *et al.*, 2003). Development of orthogonal aaRS/tRNA pairs for mammalian expression faces some additional challenges. Due to the low efficiency of mammalian cell transfection and their relatively slow growth, it is impractical to screen large libraries and evolve aaRSs directly in mammalian cells. Instead, aaRS/tRNA pairs derived from prokaryotes and developed for yeast hosts have been adapted successfully for mammalian expression. For example, both the *E. coli* TyrRS/tRNA pair and *E. coli* TyrRS paired with *Bacillus stearothermophilus* (*B.St*) tRNA were successfully transferred to mammalian cells (Liu *et al.*, 2007;Wang *et al.*, 2007b;Takimoto *et al.*, 2009)

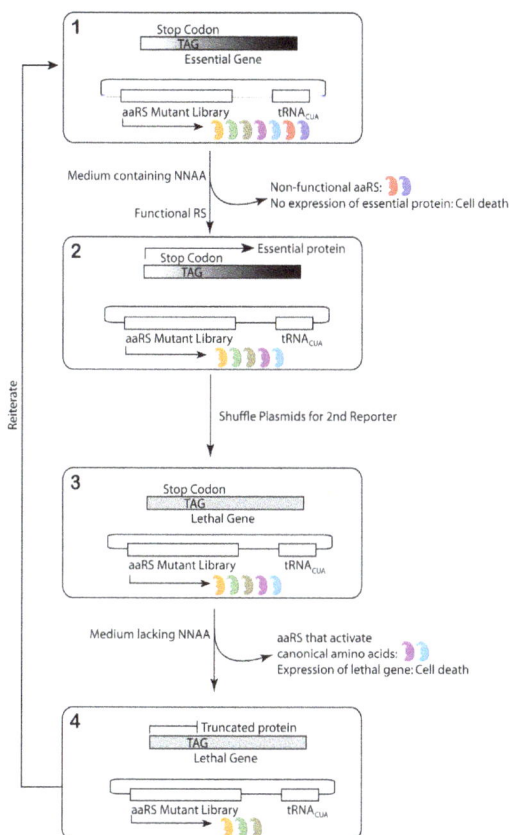

Fig. 2. An alternating positive and negative screen for evolving orthogonal aaRSs and tRNAs. (1) A library of aaRS mutants and its cognate tRNA$_{CUA}$ are introduced into an expression host containing a conditionally essential gene (e.g. antibiotic resistance marker) with an amber stop codon interrupting its open reading frame and grown in medium containing NNAA. (2) Cells containing functional aaRSs express the full-length essential gene and are able to survive under selective conditions (e.g., presence of antibiotic). (3) Surviving cells are harvested and the aaRS/tRNA$_{CUA}$ variants are introduced into cells containing a gene (with an amber codon) that when expressed is lethal (e.g. Barnase), and grown in the absence of NNAA. (4) Cells containing orthogonal NNAA aaRS/tRNA$_{CUA}$ pairs express truncated non-functional Barnase and get selected for the next round of screening.

2.1.2 Green fluorescent protein (GFP)-based selection

A more streamline scheme utilizing GFP for both positive and negative selection was developed to evolve aaRS/tRNA pair in *E. coli* (Santoro *et al.*, 2002). The method depends on antibiotic resistance and the expression of GFP, and enables high throughput screening of library variants by FACS. A single plasmid carrying two reporter genes, the chloramphenicol resistance gene containing an amber codon, and a GFP gene that depends on amber suppression for its expression, was used to select *Methanococcus jannaschii* (*Mj.*) TyrRS mutants in the presence of its cognate tRNA$_{CUA}$. Positive selection was performed in medium containing chloramphenicol and a NNAA, and the isolation of live green bacteria led to bacteria cells carrying functional *Mj.* TyrRS. Negative selection was performed in medium lacking both the antibiotic and the NNAA. By isolating non-fluorescent cells, *Mj.* TyrRS mutants that incorporated natural amino acids were eliminated. The resulting clones were then retested for growth on chloramphenicol in the presence of the NNAA. This screen led to the identification of mutations that preferred *p*-isopropyl-Phe, *p*-amino-Phe, and *O*-allyl-Tyr (Santoro *et al.*, 2002).

2.1.3 Surface-display-based selection

Phage (or yeast, or any cell surface) display utilizes the special properties of NNAA-containing peptides displayed on the cell surface for screening and selection of functional aaRSs (or tRNAs) from a vast number of variants in a library. NNAAs with special binding properties to a particular type of immobilized resin or ligand may be used for selection. For example, proteins containing a boronate amino acid, which forms strong and reversible complexes with polyols and *cis*-diols, could be selected on *N*-methylglucamine resin (Brustad *et al.*, 2008). Alternatively, a monoclonal antibody that binds specifically to a NNAA-containing peptide may be used for selection (Pastrnak & Schultz, 2001); conjugation of a florescent tag to the NNAA (containing a special chemical reactive group) displayed on the cell surface, followed by cell sorting via flow cytometry, may also lead to selection of the aaRS or tRNA. In fact, MetRS mutants with improved activities toward azidonorleucine and better discrimination against Met were generated via surface labelling and cell sorting (Tanrikulu *et al.*, 2009). The surface-display based selection can be applied to evolve aaRS/tRNA for either a sense codon or a stop codon. The limitation of the method is that the method may not be readily applicable to any NNAA, as it relies on the special properties of the NNAA or on antibody reagents.

2.1.4 Rational design for selection

Rational design predicts structural changes needed for NNAA activation based on a solved (or simulated) aaRS structure for its cognate amino acid. By computationally simulating the binding (or editing) sites and binding energy for a NNAA, putative mutations that favour the introduction of a NNAA over the natural amino acid can be identified and tested experimentally. This approach avoids the laborious generation and screening of a large aaRS mutation library, and is expected to be utilized more frequently as structural prediction algorithms improve. An example of this approach was the successful development of an orthogonal yeast PheRS/tRNA pair for the incorporation of a NNAA at an amber stop codon in *E. coli*. By sequence and structural analysis of PheRS from various species, a key residue in yeast PheRS (T415) was identified that could relax the specificity of the amino acid binding pocket and allow for the incorporation of NNAAs (Kwon *et al.*, 2006). In combination with a yeast amber suppressor Phe-tRNA$_{CUA-UG}$ (which was designed to limit the cross-reactivity of the tRNA with the *E. coli* LysRS), the mutant PheRS (T415G or A) enabled the incorporation of several Trp or Phe analogs in a site specific manner, with greater than 98% NNAA occupancy at the amber site (Kwon *et al.*, 2006; Kwon & Tirrell, 2007).

Similarly, rational design on tRNA can lead to a smaller number of candidates for the generation of orthogonal tRNAs. For example, the amber suppressor Trp-tRNA$_{CUA}$ from yeast was found to be mis-acylated by endogenous LysRS in *E. coli* (Hughes & Ellington, 2010). Analysis of the tRNA structure revealed that the yeast amber suppressor Trp-tRNA$_{CUA}$ shares 73% sequence homology with *E. coli* Lys-tRNA$_{UUU,}$ and contains several key identity elements for the recognition of *E. coli* Lys-tRNA$_{UUU}$ by its cognate aaRS. Furthermore, structural analysis revealed that the modification of the G:C content of the anticodon stem may reduce the structural flexibility of this stem and eliminate mis-acylation. Combined with experimental testing, an orthogonal and functional yeast Trp-tRNA$_{CUA}$ was obtained for the incorporation of NNAAs (Hughes & Ellington, 2010).

2.2 Methods for incorporation of NNAAs into recombinant proteins

2.2.1 Reassigned sense codon

The simplest method to introduce NNAAs into proteins exploits the promiscuity of aaRSs of the host cells. That is, when a particular aaRS can charge a NNAA non-specifically to its cognate tRNA, the activated NNAA-tRNA may be utilized by the ribosomal translation machinery, and the NNAA added to the nascent polypeptide chain according to the sense codon specified for the canonical amino acid. The sense codon is reassigned to the NNAA, resulting in a recombinant protein composed of 19 canonical amino acids and one NNAA. Thus, this is a residue-specific global substitution method.

The successful substitution of a canonical amino acid with a NNAA relies on the use of auxotrophic expression hosts deficient in the biosynthesis of that amino acid. Employment of such hosts limits competition from the canonical amino acid for the reassigned sense codon, and improves the incorporation efficiency and yield of target proteins. For example, the MetRS has a relatively relaxed substrate binding pocket, and a number of Met analogs have been successfully incorporated into target proteins (reviewed by Hendrickson *et al.*, 2004;Voloshchuk & Montclare, 2010). A special consideration worth mentioning when using Met analogs for protein engineering is the *in vivo* processing of NNAAs at the N-terminus of proteins. It is well know that most protein translation starts with a Met codon, and that Met at the N-terminus of proteins is subjected to enzyme-mediated processing, resulting either in the retention or cleavage of N-terminal Met. The rule for such N-terminal processing depends on the identity of the 2nd amino acid in the protein sequence. A set of similar but not identical rules have been established for the N-terminal processing of NNAAs in recombinant interferon (Wang *et al.*, 2008) as well as other proteins (unpublished data). As a result, the N-terminal NNAA can be either retained or cleaved by choosing an appropriate 2nd residue (Table 1).

2nd residual	Aha cleavage%	Hpg cleavage%
Ala	96	91
Ser	80	80
Gly	52	33
Gln	8	0
Glu	0	0
His	0	0

Table 1. N-terminal processing of Met analog Aha or Hpg (structures provided in Fig 4) in *E. coli*.

Besides the Met codon, many other sense codons (such as codons for Pro, Tyr, Phe, Leu, Val etc.) have also been reassigned, and more than 60 NNAAs have been incorporated into proteins via global substitution (reviewed by Hendrickson *et al.*, 2004;Voloshchuk & Montclare, 2010). In many cases, the intracellular level of wild-type aaRS may need to be augmented in order to compensate for the low activity of native aaRS towards the NNAA. In other cases, engineered aaRSs with higher catalytic activity towards the NNAA could be supplied to the expression host cells. An impressive example is with a MetRS mutant, which was identified via cell-surface display. The mutant MetRS activated azidonorleucine two-fold more efficiently than Met. As a result, the target protein was expressed with azidonorleucine at higher levels than with Met (Tanrikulu *et al.*, 2009). By utilizing auxotrophic host cells for two or more canonical amino acids, it may be possible to produce recombinant proteins with more than one type of NNAA by this strategy. Furthermore, the global substitution method may be applied in other expression hosts, such as mammalian cells.

It should be mentioned that while native or mutant aaRSs can utilize NNAAs to replace a canonical amino acid *in vivo*, most NNAAs don't support the normal growth and expansion of host cells, presumably due to the impairment of some protein functions when a NNAA is incorporated. As a result, cell density (and thus protein expression yield) is affected during the induction phase when a NNAA is added into the media. Adaption of host strains to the NNAA may provide a solution. By gradually increasing the amount of NNAA in the growth media until complete replacement was accomplished, *E. coli* hosts capable of surviving in media containing 4-fluoro-Trp were evolved (Bacher & Ellington, 2001). These evolved *E. coli* hosts contain 3 mutant enzymes involved in the uptake and utilization of Trp, and are capable of fully substituting 4-fluoro-Trp in their proteomes. Utilization of such adapted expression hosts represents another approach to improve the yield and NNAA incorporation efficiency.

The main limitation of this technology is that the NNAA will replace the canonical amino acid throughout the protein sequence, which may restrict its application if such global substitution is undesired. One solution to this problem is to mutate undesired sites with other canonical amino acids so that the desired site(s) is reserved for NNAA. To this end, functional target proteins lacking a particular amino acid have been generated, such as Met-free human interferon beta or scFv (unpublished data) and Phe-free GFP (Goltermann *et al.*, 2010). With these proteins, the NNAA can be site-specifically incorporated at any desired site. Alternatively, site-specific incorporation can be accomplished via the introduction of an aaRS/tRNA pair specifically designed for the NNAA, as described in the following sections.

The global substitution technology offers many advantages, which include: 1) the efficiency of NNAA incorporation, which can be near 100% at lab scale using a media exchange procedure. At production scale with a media depletion procedure, the incorporation efficiency can still reach 85-90% (unpublished data); 2) the target protein sequence is tightly controlled. In general, the NNAA incorporates only at the reassigned sense codon; the protein sequence at all other codons is strictly controlled and not affected by the reassigned sense codon; 3) target protein expression level is generally very high, even for proteins containing multiple copies of the NNAA. Protein yields at 10-100mg/L have been reported (Link & Tirrell, 2005) in shaker flasks, and yields of up to several gram/L can be achieved via high-density fermentation in bioreactors (unpublished data); 4) the expression system is simple and well-controlled, and the fermentation process is highly scalable. Thus, this technology is amendable for large-scale manufacture.

2.2.2 Wobble codon

Wobble codons refer to codons that are decoded by tRNA via non-classical Watson-Crick base-pairing. As there are 61 sense codons to encode 20 canonical amino acids, but fewer than 50 tRNAs in many organisms (Sharp et al., 2005) (for a tRNA database on different organisms: http://gtrnadb.ucsc.edu/), some tRNA must pair with more than one codon. The non-classical (or wobble) pairing is enabled through modification at the tRNA's 1st anticodon base (which pairs with the 3rd base to the codon triplet), as proposed in the "Wobble Hypothesis"(Crick, 1966). For example, many organisms have only one Phe-tRNA$_{GAA}$ to decode two codons for Phe: UUU and UUC. As a result, the GAA anticodon binds to the UUC codon via Watson-Crick base-pairing, and to the UUU codon via "wobble" base-pairing. While many more modifications on tRNA have since been identified over the last 40 years, and the "Wobble Hypothesis" has been modified and expanded, the essential premise remains the same: through various base modifications, the anticodon structure is shaped in a way that enables it to bind to cognate and wobble bases at the ribosome (Agris et al., 2007).

Because of the wobble pairing between codon and anticodon, one tRNA may pair with several codons, and a given codon may pair with more than one tRNAs. However, pairing at a wobble codon may be less effective to support protein translation (Curran & Yarus, 1989;Sharp et al., 2005). Taking advantage of this property, a wobble codon may be assigned to a NNAA to generate recombinant protein that contains 20 canonical amino acids and one NNAA, as described below.

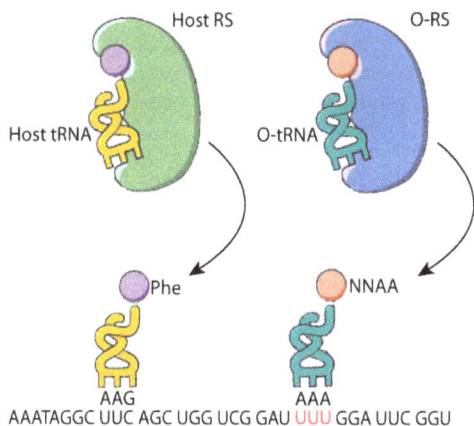

Fig. 3. Site-specific integration of NNAAs at wobble codons. Phe is normally encoded by two codons UUC and UUU, with both codons recognized by the tRNA$_{GAA}$. By expressing an orthogonal aaRS/tRNA$_{AAA}$ (O-RS and O-tRNA) pair, with specificity for a NNAA and containing the "AAA" anticodon, efficient introduction of the NNAA at UUU codons can be achieved.

To assign a wobble codon to a NNAA, a heterologous PheRS/tRNA$_{AAA}$ pair from the yeast S. cerevisiae was introduced into an E. coli Phe auxotrophic expression host containing the target gene under a strong inducible promoter (Kwon et al., 2003). In this system, the yeast

PheRS was engineered to activate 2-naphthylalanine to the yeast Phe-tRNA$_{AAA}$, and the target gene was designed with a UUC codon for all desired Phe sites, and a UUU wobble codon was reserved at specific sites for 2-naphthylalanine (Figure 3).

With this method, Phe can be essentially quantitatively assigned to the UUC codon, and NNAA to the UUU wobble codon at approximately 80% efficiency in *E. coli* expression host (Link & Tirrell, 2005). Furthermore, multiple copies of a NNAA can be introduced site-specifically into a target protein. The expression level of target protein, however, is typically an order of magnitude lower than the global sense codon reassignment method, on the order of 2-5 mg DHFR per liter with shaker flasks. The applicability of this method in other expression hosts such as yeast and mammalian cells remains to be investigated.

2.2.3 Bias codon

It is well known that degenerate codons are not used with equal frequency in many organisms. The preferred codons are the ones that match the tRNAs that are most abundant in the cells, and this codon preference is more evident in rapidly multiplying organisms or in highly expressed genes. Presumably, codon usage and tRNA gene content coevolved to match each other, and in doing so, organisms effectively minimize the energy and time costs for protein translation and cell growth (Higgs & Ran, 2008).

The preferred codons differ between organisms, and even between different tissues or cell types of the same organism. The cellular content of tRNA species is a determining factor on the rates and amounts of protein synthesized (Gu *et al.*, 2004). As a consequence, recombinant protein production in heterologous host cells is often codon-optimized to match the preferred host cell codon bias (The codon usage database for different organisms and codon analysis of a given gene can be found at: http://www.kazusa.or.jp/codon/).

The biased codon usage provides another method to introduce NNAAs into recombinant proteins. For example, out of the six degenerate codons for Arg, AGG and AGA are rarely used in *E. coli*. Introduction of an orthogonal ArgRS/tRNA$_{CCU}$ pair into an *E. coli* expression host may enable the NNAA-activated orthogonal tRNA$_{CCU}$ to compete effectively with the low endogenous *E. coli* Arg-tRNA$_{CCU}$. This method has been proven feasible with an *in vitro* cell-free based system, where chemically synthesized NNAA-tRNA$_{CCU}$ was incorporated at AGG codons (Hohsaka *et al.*, 1994). The method could be adapted to *E. coli* cell-based expression system if orthogonal ArgRS/tRNA pair can be engineered.

Similarly, a bias codon may be assigned to a NNAA in mammalian cells that exhibit codon bias. For example, through study of human papillomavirus gene expression in different mammalian cells, Frazer and his colleagues have found that papillomavirus protein expression is determined by the codon usage and tRNA availability. Substantial differences in the tRNA pools were discovered between differentiated and undifferentiated keratinocytes (Zhao et al., 2005), and the observed bias in their tRNA may be the reason that papillomavirus replicates exclusively in epithelial cells. In CHO and Cos1 cells, wild-type papillomavirus L1 gene is effectively transcribed, but not translated. Introduction of an extra copy of Ser-tRNA$_{CGA}$ to the CHO or Cos1 (but not the empty vector) enabled the expression of L1 protein (Gu et al., 2004). Thus, it seems that TCG is a bias codon in these mammalian hosts and thus might be assigned to a NNAA.

As the codon bias phenomenon is wide-spread in different organisms, utilization of such codons for site-specific incorporation of NNAAs could be applied in many expression host cells. The limitation would be the engineering of the orthogonal aaRS/tRNA pair that matches the bias codon. The competition from a canonical amino acid for the bias codon could be minimized by using auxotrophic host cells, and/or by limiting the canonical amino acid concentration in the induction medium. Deletion (or conditional inactivation) of the tRNA genes corresponding to the bias codon in the host genome may abolish the competition. Recent advances in synthetic biology to generate artificial organisms may improve this method dramatically, as it holds promise to express target proteins at high level and with different NNAA molecules (see section 2.2.6).

2.2.4 Stop codon

Generally, protein translation terminates at one of the three stop codons (encoded by UAG (amber), UAA (ochre) and UGA (opal)) by the action of protein release factors (RF). However, occasional read-through of a stop codon with an amino acid has been observed to happen naturally in a variety of species. The suppression is caused by either mutations in the tRNA anticodon or mismatches of the codon-anticodon (Beier & Grimm, 2001). Utilization of stop codon suppression represents another way to engineer proteins containing NNAAs, and generally involves the introduction of an orthogonal aaRS/tRNA pair.

A growing number of orthogonal aaRS/tRNA pairs have been developed to introduce NNAAs site-specifically at amber codons, as it is the least frequently used stop codon in both eukaryotic (23% in humans) and prokaryotic genomes (7% in E. coli) (Liu et al., 2007). However, ochre and opal suppressors have been evolved for the introduction of NNAAs as well (Köhrer et al., 2004). So far, over 70 NNAAs have been site-specifically incorporated into target proteins (Liu & Schultz, 2010). Typically, over 95% NNAA incorporation efficiency (defined as occupancy rate of NNAA in the full-length product) at the desired site can be obtained, making it one of the most frequently used methods for NNAA incorporation.

The main limitation of this method is protein yield. In general, yields in the range of 1-10 mg/L have been obtained, but 0.5-0.8 g/L has been reported using optimized fed-batch E. coli systems (Cho et al., 2011). One reason for the limited yields is the low overall suppression efficiency (defined as the percentage of stop codon read-though), generally in the range of 10-40%. The low suppression efficiency also limits the number of integration sites for NNAAs in a given gene. As the number of inserted stop codons increases, the overall suppression efficiency decreases multiplicatively. Several approaches have been tested to increase the overall suppression efficiency.

One way to increase suppression efficiency is to eliminate the competition from RF. In prokaryote organisms, RF1 terminates gene translation at UAA and UAG stop codons, RF2 at UAA and UGA stop codons, and RF3 facilitates the function of RF1 and RF2 (Nakamura et al., 2000). The principle of deleting RF1 in E. coli for amber stop codon reassignment was demonstrated by Sakamoto and his colleagues (Mukai et al., 2010). As expected, deletion of the prfA gene, which encodes RF1, is lethal to the organism. The prfA- bacteria were rescued by 1) mutation of seven essential genes containing amber UAG stop codon to UAA stop codon and supplementation of these mutant genes with a bacterial artificial chromosome

(BAC) plasmid; 2) introduction of an amber suppressor tRNA such as Gln-tRNA or Tyr-tRNA. The requirement for an amber suppressor tRNA with a canonical amino acid can be extended to amber suppressor tRNA designated for a NNAA. In fact, when an archaebacterial TyrRS/tRNA$_{CUA}$ pair (engineered for 3-iodo-L-Tyr) was introduced to replace the E. coli amber suppressors, 3-iodo-L-Tyr was incorporated to the target protein containing six copies of the amber stop codon (Mukai et al., 2010). The target protein expression level was not reported, but the full-length protein expression was demonstrated by western blot analysis. Such a result is very impressive given that six amber codons were read in the target protein, and it was only made possible by eliminating competition from the RF-1. However, the introduction of multiple mutated components in an E. coli host strain may be problematic for commercial manufacturing. The advent of designer organisms, containing artificial genomes, may provide the right solution to address these concerns and offer unprecedented potential for NNAA applications.

Deleting RF is not feasible in yeast and mammalian cells since eukaryotes have a single RF that functions for all three stop codons. One can minimize competition from RF by increasing the overall abundance of functional tRNA levels. However, the expression of most heterologus (usually prokaryotic) tRNA in eukaryotes is problematic due to their divergent transcriptional regulation. Eukaryotic tRNAs are transcribed by RNA Polymerase III (Pol-III) that is regulated by intragenic promoter regions known as A- and B- box elements. Most prokaryotic tRNAs do not contain these consensus sequences and are poorly transcribed in eukaryotic cells. Careful selection of a prokaryotic tRNA that contains consensus A- and B- box elements, or are permissive to mutation in order to include them, has been used successfully (Sakamoto et al., 2002;Zhang et al., 2004). However, the strict recognition elements of aaRSs for tRNAs often prevents modifications of the tRNA sequence to introduce these elements. Several extragenic promoters for Pol-III have been shown to allow for high levels of heterologous tRNA expression in both yeast and mammalian cells. In mammalian cells, the promoters for the small nuclear RNA U6 and H1 have been used successfully (Mukai et al., 2008). But, RNA expressed by the U6 promoter has been shown to mislocalize to nucleoli (reviewed by Haeusler & Engelke, 2006) and may affect the processing and targeting of the transcript, both important processes required for the generation of functional tRNA. In yeast, RNA Pol-III -based promoters for the tRNA (e.g. SUP4), and other RNAs (RPR1 and SNP52), were shown to efficiently express tRNA genes. Levels of expression could be increased by coupling these promoters with RNA Pol-II-based promoters (e.g. PGK1). Interestingly, correct processing of the tRNA, and not absolute expression levels, is essential to generate functional tRNA and increase NNAA incorporation (reviewed by Wang et al., 2009).

Another limiting factor to incorporate NNAAs at amber codon is a mechanism that degrades mRNA containing premature stop codons, called nonsense mediated mRNA decay (NMD). Since the most prevalent site-specific integration method includes amber codon read-through, the degradation of the target mRNA can have a significant impact on the yield. The importance of this process was highlighted when Wang and co-workers developed a yeast strain lacking an essential component of the NMD mechanism, UPF1. Strains lacking this gene exhibited a two-fold increase in target gene expression when compared with wild type cells (Wang & Wang, 2008). A similar approach has not been accomplished in higher eukaryotic cells.

2.2.5 Four-base-pair codon or non-natural base pair

The discovery of frame shift suppressor tRNA has opened another door to expand codons for NNAAs. In principle, a four-base codon can be used at a desired site in a target gene and read by a complementary suppressor tRNA (containing a four-base anticodon), leading to a reading frame shift and synthesis of a full length protein. Four-base codons have been used to incorporate L-homoglutamine in *E. coli*, using the orthogonal mutant *Pyrococcus horikoshii* LysRS/tRNA$_{AGGA}$ pair (Anderson *et al.*, 2004). The observed incorporation efficiency is similar to those seen for amber codon suppression (>95%) but the overall yields, in the order of 1-2 mg/L, are significantly lower. The yield of target protein with NNAA is affected by the competition of the host tRNA (containing a triplet codon) for the first 3 bases in the 4-base codon. As such, competition gives rise to a different reading frame and a protein of different sequence. To minimize this competition by the host tRNA, the 4-base codon often targets a rare codon (Anderson *et al.*, 2004). When the *P. horikoshii* LysRS/tRNA$_{AGGA}$ pair was co-expressed with an *Mj*.TyrRS/tRNA$_{CUA}$ pair for *O*-methyl-L-Tyr, target protein containing two NNAAs was generated with no appreciable loss in yield or efficiency (Anderson *et al.*, 2004). Additional variants of this approach using five-base codons have been successfully demonstrated *in vitro* (reviewed by Hohsaka & Sisido, 2002). Together, the use of four- and five-base codons will greatly increase the number determinants for NNAAs and provides one method that skirts the limitations of stop codon suppression and allows for the introduction of multiple NNAAs into a single protein. However, this approach suffers from the relative inefficiency of ribosomes to translate an extended codon. But, one can envision the development of modified, orthogonal ribosomes with increased efficiency at these codons, as described in section 2.2.7.

An alternate strategy to encode NNAAs involves the generation of codons made of new base pairs. Pioneering work by Benner and colleagues showed that new codon-anticodon pairs could be functionally introduced to support translation. By introducing a new letter to the genetic lexicon, the number of coding triplets can effectively be increased from 64 to 216. Using two complementary base-pairs, termed isoC and isoG to generate a new codon, Benner and colleagues introduced *p*-iodo-Tyr into a short peptide sequence in a cell free expression system utilizing chemically synthesized mRNA and tRNA (Bain *et al.*, 1992). But to expand the capacity for protein expression this technology requires an expansion into *in vivo* systems. This is a significant hurdle as the new base-pairs need to be orthogonal to the existing nucleotides and be accepted by DNA and RNA polymerases for faithful replication and transcription. Some of these challenges are beginning to be addressed: Benner and colleagues showed that the ribosome could translate a 65th codon (Bain *et al.*, 1992). In addition, recent progress in the design of new nucleotide pairs has been made that are specific and stable, like 2-amino-6-(2-thienyl)purine and pyridine-2-one, and can be faithfully integrated by polymerases into tRNA and mRNA, and be translated *in vitro (Hirao et al.*, 2002;*Hirao et al.*, 2004) . By combining these features with an orthogonal aaRS for the aminoacylation of tRNA with a synthetic base at its anticodon all the elements for *in vivo* transcription-translation would be in place.

2.2.6 Reserved codon in artificial cells

Incorporation of NNAAs into recombinant protein *in vivo* requires the reassignment of a codon (such as a sense codon or a stop codon) to the NNAA, and this inevitably introduces competition from the existing endogenous components: the canonical amino acid for the sense codon and protein release factor for the stop codon. Such competition will not only reduce the NNAA incorporation efficiency, but may also reduce the target protein expression yield. Recent breakthroughs in *de novo* synthesis and implantation of an entire prokaryotic genome to create a living artificial organism (Gibson *et al.*, 2010) provide the ultimate solution for the codon competition problem.

Generation of an artificial synthetic organism via synthetic biology has many applications, as the entire genome can be tailor-designed. One such application is for NNAA technology by reserving specific codons for NNAAs and by removing the endogenous competition components. For example, the host genome may be designed with fewer than the 61 sense codons. The tRNA genes might also be minimized, potentially avoiding tRNAs such as those with wobble base-pair properties as much as possible. Such design may enable the complete reassignment of particular sense codons to a NNAA. To make better use of stop codons, host genes might be designed with as few as one type of stop codon so that the unused ones may be reserved exclusively to encode the NNAA. This will avoid the toxic effect often observed in the existing expression system with stop codon suppression, presumably caused by the functional inactivation of essential genes due to NNAA-mediated translation extension at stop codons. Furthermore, a release factor (RF) may be eliminated from the host genome to avoid premature translation termination of target proteins at the stop codon intended for the NNAA. In prokaryotes, the UAA stop codon is recognized by both RF1 and RF2, while the other two stop codons are each recognized by a single RF (UAG by RF1 and UGA by RF2). RF1 may be deleted from the artificial genome if all host genes are terminated with UGA/UAA, reserving UAG for the NNAA; alternatively, RF2 may be deleted if all artificial host genes are terminated with UAG/UAA, so that UGA is reserved for the NNAA. Deletion of RF for eukaryote organisms may need some manipulation of the wild-type eRF1, as eukaryote cells have only one eRF1 for all three stop codons.

It should be mentioned that for the majority of current expression systems, the NNAA is supplied in the culture media, from where it is transported into host cells and taken up by the aaRS. However, some NNAAs may be synthesized in the host cells by manipulating metabolic pathway(s). For example, L-homoalanine can be made biosynthetically in large quantity when a mutant glutamate dehydrogenase and an exogenous threonine dehydratase were introduced to threonine-hyperproducing *E. coli* (Zhang *et al.*, 2010). By introducing a biosynthetic pathway for the *p*-amino-Phe (pAF) from *S. venezuelae* into *E. coli* host (which also contains an orthogonal pair of pAF-RS/Tyr-tRNA$_{CUA}$), target protein with *p*-AF incorporated at the amber codon was produced (Mehl *et al.*, 2003). With deep understanding of various metabolic pathways, synthetic cells may be designed to integrate NNAA biosynthesis with NNAA incorporation, thus substantially reduce the cost of the target proteins.

2.2.7 Orthogonal ribosome

Incorporation of NNAAs via orthogonal ribosomes was lead by Chin and his colleagues, to overcome cell toxicity and low suppression efficiency associated with the amber suppression method, and to expand the numbers of NNAAs that can be introduced to target proteins. Through screening of a ribosome mutational library, they have identified ribosome variants that recognize unique ribosome binding sequences (o-RBS). The orthogonal ribosome (o-Ribo) functions in parallel with, but independently from endogenous ribosome in host cells: o-Ribo does not translate cellular mRNAs, and the target gene cloned with o-RBS is not translated by the endogenous ribosome (Rackham & Chin, 2005). Subsequent evolution identified a mutant Ribo-x (containing U531G and U534A mutation in 16s rRNA) that improved the amber tRNA suppression of the target gene (Wang et al., 2007a). By combining the Ribo-x, o-RBS-target gene that contains an amber codon, and an orthogonal aaRS/tRNA pair for NNAA into E. coli, these investigators were able to significantly increase the amber suppression efficiency (from 20% to >60% for a single amber codon, or from <1% to >20% for double amber codons) (Wang et al., 2007a). In addition, a ribosome variant (Ribo-Q1) that efficiently decodes both quadruplet codons and amber codons was evolved (Neumann et al., 2010b). By introducing two mutually orthogonal aaRS/tRNA pairs in E. coli host containing Ribo-Q1, they could incorporate two different NNAAs molecules into one target protein (Neumann et al., 2010b).

The protein yield obtained with orthogonal ribosome system is very impressive when compared with results obtained with wild-type ribosome. For example, full-length protein with two NNAA molecules (assigned to amber codon and a quadruplet codon, respectively) was produced at 0.5 mg/L (Neumann et al., 2010b). However, the expression system contains multiple components, and its application in large scale manufacturing remains to be explored.

3. Orthogonal chemical reactions at the side chain of non-natural amino acids

One of the major advantages of incorporating NNAAs into proteins is the ability to introduce orthogonal, site-specific chemistries for modification of the recombinant protein. These modifications are often desired for applications such as PEGylation, drug conjugation, or labelling (see section 4).

Techniques for chemical modification of proteins have steadily increased in their sophistication. Early methods used electrophilic addition to the ε-amine groups of lysines. However, the large number of reactive lysines in a protein gives rise to a heterogeneous reaction product. For instance, PEG-Intron™ contains 15 positional isomers of PEGylation sites, in addition to multi-PEGylated species (Wang et al., 2002). The various isomers possess 6-37% bioactivity compared to the unPEGylated protein. Conjugation technology then shifted to reductive amination at the N-terminus or thiol-specific additions at engineered cysteine residues (Harris & Chess, 2003). However, even these approaches do not result in complete limitation to a single reaction site, and they also introduce significant limitations in conjugation site and/or requirements for more protein engineering.

With the introduction of NNAAs at defined sites in a protein, a whole new world is opened up to the conjugation chemist (Sletten & Bertozzi, 2009). A new array of functional groups is available and reactions that are orthogonal to (do not react with) the functional groups in canonical amino acids present the opportunity for truly site specific conjugations. The requirements are that the necessary functional group can be incorporated into the recombinant protein, the complementary functional group can be installed on the other molecule to be conjugated, the reaction can occur in conditions acceptable for the protein (typically aqueous, 5-37°C, mild pH, preferably in air), the reaction does not cause unacceptable side reactions (e.g., metal-catalyzed degradation of amino acids, disulfide shuffling of a folded protein), and that the chemistry is suitably bioorthogonal. Useful functional groups available in commonly incorporated NNAAs include azides, alkynes, alkenes, ketones, and aryl halides (Figure 4). The following sections highlight the most frequently harnessed NNAA conjugation chemistries in this rapidly evolving field.

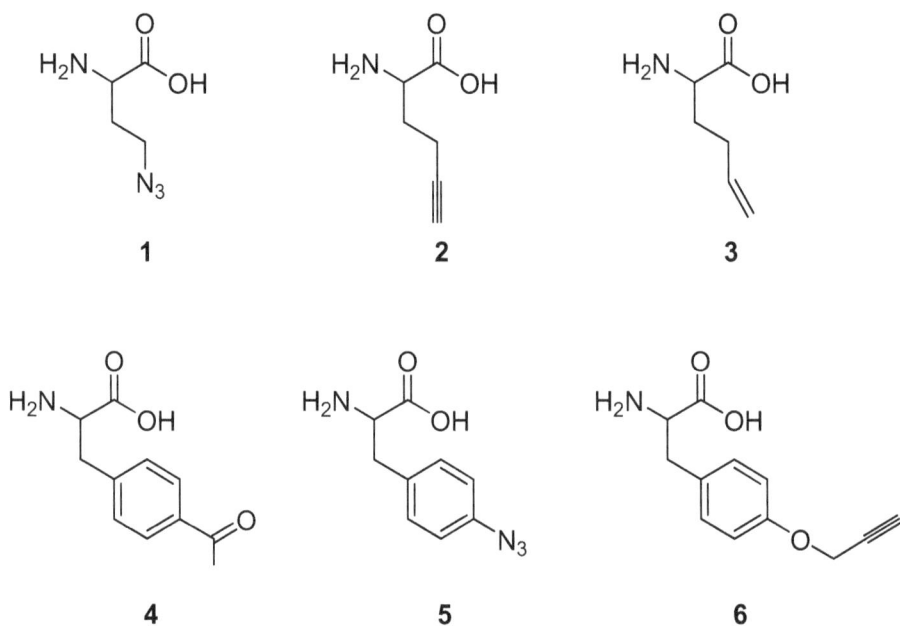

Fig. 4. NNAAs frequently used for bioconjugation. Met analogs: 1 azidohomoalanine (Aha, azide group), 2 homopropargylglycine (Hpg, alkyne group), 3 homoallylglycine (Hag, alkene group). Phe analogs: 4 p-acetyl-Phe (ketone group), 5 p-azido-Phe (azide group), 6 p-propargyloxy-Phe (alkyne group).

3.1 Copper-catalyzed azide-alkyne cycloaddition ("click" chemistry)

Copper-catalyzed azide-alkyne cycloaddition (often referred to as "click" chemistry or CuAAC) was first reported simultaneously by both Sharpless and Meldal and co-workers (Rostovtsev et al., 2002;Tornoe et al., 2002). Click chemistry is a term coined by the Sharpless group referring to a philosophy of efficient chemical synthesis, but has since been co-opted to also refer specifically to CuAAC. CuAAC joins a terminal alkyne and an azide, via a Cu(I) catalyst, to form an aromatic triazole ring (Figure 5). Huisgen had previously demonstrated catalyst-free cycloaddition of azides and alkynes at high temperature (e.g., 100°C), but with the addition of the copper catalyst, the reaction occurs readily at room temperature and forms only the 1,4 regioisomer. It is essential that the alkyne is terminal, so the copper acetylide can be formed as a reaction intermediate.

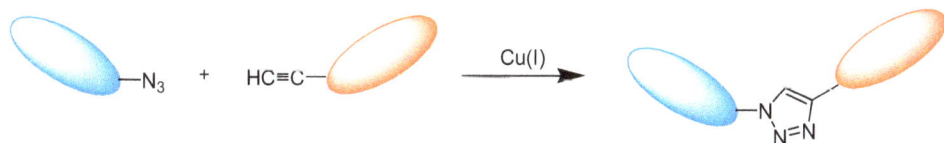

Fig. 5. Copper-catalyzed azide-alkyne [3+2] cycloaddition ("click" chemistry or CuAAC). An azide and a terminal alkyne combine to form an aromatic triazole linkage.

There are several advantages to CuAAC for bioconjugation. Azides and alkynes do not react with canonical amino acids, are kinetically stable, and yet are highly primed to react when presented with the appropriate catalyst. Either functional group is readily installed by the simple codon reassignment method with commercially available NNAAs. At physiological temperatures, the reaction does not occur in the absence of the catalyst. Importantly, the resultant linkage between the protein and the conjugated molecule is a triazole group, which is highly stable due to its aromaticity. The reaction can be performed in many aqueous buffers at a wide range of pHs, in many organic solvents, in the presence of SDS (Grabstein et al., 2010), and, in most cases, in the presence of oxygen. Importantly, the reaction displays especially rapid kinetics (Presolski et al., 2010).

As stated, CuAAC is catalyzed by Cu(I). Cu(I) is not stable in water, but rapidly disproportionates into Cu(II) and Cu(0). Cu(I) salts such as CuBr and $[Cu(NCCH_3)_4][PF_6]$ have been used successfully, though the literature suggests ultra high purity may be required (Link et al., 2004). More frequently, the cupric salt $CuSO_4$ is used in conjunction with a reducing agent such as TCEP or elemental copper. Thiol-containing reducing agents such as DTT can also be used (Grabstein et al., 2010). Finn and colleagues have recently developed the use of aminoguanidine as an additive to prevent deleterious side reactions when using sodium ascorbate as a reducing agent (Hong et al., 2009).

A ligand is typically added to greatly increase reaction yields. The ligand is thought to complex and stabilize the Cu(I) and intermediates and also to protect the protein from deleterious effects of copper. The most commonly used ligand is TBTA, shown in Figure 6. Significant protein breakdown can be seen after reaction with no ligand, whereas breakdown can be prevented by TBTA (unpublished data). TBTA is very hydrophobic, but solubility is not required for the reaction. Many other ligands have been investigated in the literature, including water soluble analogs such as THPTA and benzimidazoles. Finn and colleagues have had good success with a water soluble bathophenanthroline ligand (Gupta et al., 2005), but this requires careful removal of oxygen from the system. A thorough structure function investigation of ligand acceleration generated recommendations for different systems (Presolski et al., 2010). Dilute reactions in aqueous media have very different requirements than concentrated reactions in organic solvents due to the varying degree of copper binding competition in the reaction. The literature and the authors' own experience suggests that the best ligand must be determined empirically for any given reaction, which will depend on the protein, solvent, and other conditions.

While CuAAC can be readily employed in many systems, there are some side reactions that can be problematic. First, copper is well known to create oxidative species in aqueous solution, including hydrogen peroxide and free radicals. These effects can be mediated by choice of ligand, reducing agent, and other reaction conditions. Copper's toxicity can also be problematic for reactions on live cells or in living organisms. Second, the presence of reducing agents (or even the copper itself) can disrupt disulfide bonds present in a folded protein. This can lead to misfolding or disulfide-linked aggregates, though it has been shown that some proteins such as antibodies can be exposed to reducing agents and then oxidized to re-form the correct disulfide bonds (Sun *et al.*, 2005). Third, Back and co-workers have shown that reducing agents can result in azide reduction to an amine group or can cleave the peptide bond C-terminal to Aha, wherein the Aha is modified to a homoserine lactone which can then react with amine groups (Back *et al.*, 2005). High pH can also result in direct replacement of the azide group with a hydroxyl group. We have developed reaction conditions that avert all of the above issues, apart from toxicity in living systems, for various protein conjugation systems (unpublished data). Recent publications have also demonstrated conditions that allow short-term CuAAC in the presence of live cells or organisms (Hong *et al.*, 2010; del Amo *et al.*, 2010).

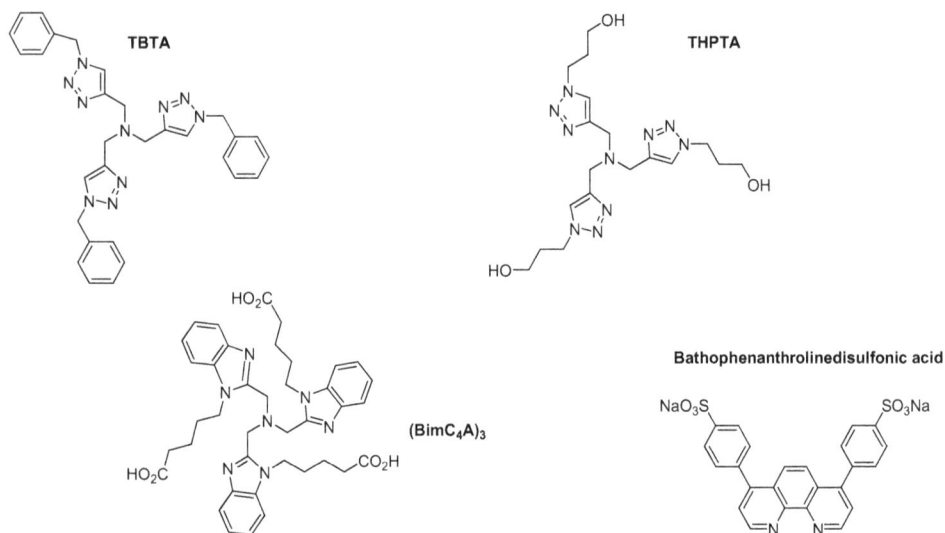

Fig. 6. Popular ligands for CuAAC

3.2 Strain-promoted azide-alkyne cycloaddition

Strain-promoted azide-alkyne cycloaddition (SPAAC) is popular as a metal-free alternative to CuAAC (Jewett & Bertozzi, 2010). While CuAAC uses a copper catalyst to access the reactivity of the alkyne, the [3+2] cycloaddition activation barrier can also be lowered by placing strain on the alkyne group. Bertozzi and colleagues first pioneered this approach by using a highly strained (18 kcal/mol of ring strain) cyclooctyne ring (Agard *et al.*, 2004). As shown in Figure 7, the alkyne is internal rather than terminal, the resultant linkage is comprised of two fused rings, and no metals, reducing agents, or ligands are necessary. SPAAC is of particular interest for labelling applications in living systems in order to avoid a toxic copper catalyst.

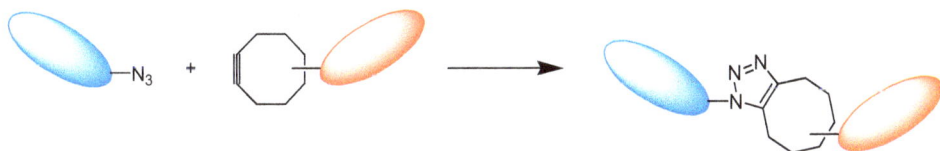

Fig. 7. Strain-promoted azide-alkyne [3+2] cycloaddition (SPAAC). An azide and a strained cyclooctyne ring combine to form a fused ring linkage.

SPAAC has undergone significant improvements beyond the slow reactivity and difficult synthesis of the original cyclooctyne ring. Installation of electron withdrawing groups on the cyclooctyne or additional ring strain provided by fusing ring(s) to the cyclooctyne has improved reaction kinetics, and a practical focus has improved synthetic accessibility of the starting materials. A sampling of structures is given in Figure 8. An initial advance occurred with DIFO, where the introduction of two propargylic fluorine atoms resulted in more than an order of magnitude increase in the second-order rate constant. Second generation DIFO analogs displayed similar reactivity but synthetic yields were improved from 1% in 12 steps to 28-36% in 6 steps. Increasing ring strain by fusing two aromatic rings with the cyclooctyne has also increased reactivity to similar levels, such as with DIBO. Addition of a nitrogen atom (DIBAC) or amide group (BARAC) within the central cyclooctyne ring increases the reactivity even further (Debets et al., 2010;Jewett et al., 2010). Alternatively, ring strain introduced by fusion of a 3-membered ring with the cyclooctyne generates a bicyclononyne (BCN) with similarly high reactivity (Dommerholt et al., 2010). This molecule was made in 4 steps with an overall synthetic yield of 46% and, due to its symmetry, its cycloaddition reaction product is a single regioisomer. These improved reagents have now brought SPAAC reactivity to a similar level as CuAAC. A much more water soluble reagent was generated by the addition of 3 heteroatoms to the single cyclooctyne ring (DIMAC), but the reactivity was significantly lower than other recent analogs.

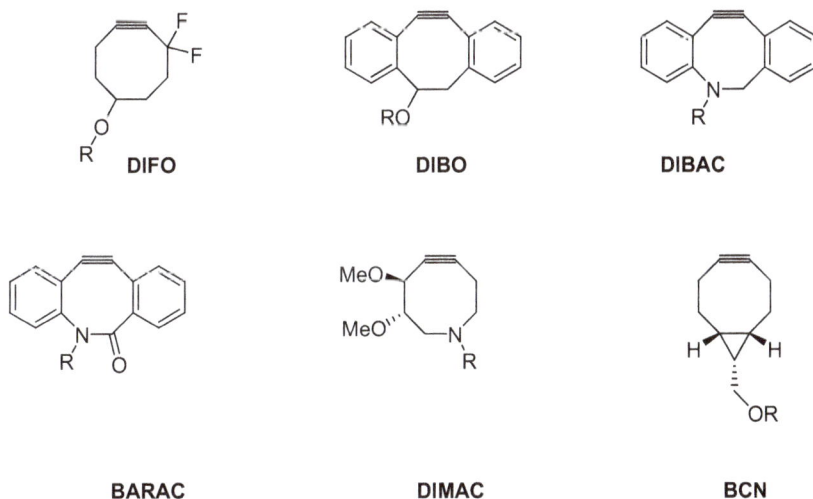

Fig. 8. Improved strained alkynes for use in SPAAC.

SPAAC can expand the application of azide-alkyne cycloaddition to metal-sensitive systems, but it does present some challenges. Unfortunately, many of the modifications that reduce the activation barrier enough for fast reaction rates also reduce the barrier for side reactions with canonical functional groups such as cysteinyl thiols. This loss of orthogonality can be addressed by tailoring of the activated alkyne and reaction conditions, but ultimately the stringency of the orthogonality requirement is dependent upon the application. In addition, the cyclooctyne ring is a large hydrophobic structure with limited aqueous solubility. It has been shown to nonspecifically bind to proteins even in the presence of SDS (Sletten and Bertozzi, 2008). The fusion of additional rings only serves to magnify this problem (Agard *et al.*, 2006;Codelli *et al.*, 2008). Commercial availability of cyclooctyne reagents is also limited. For applications such as therapeutic protein conjugates, other considerations include the stability of the activated alkyne, its synthetic availability and purity, and the production of regioisomers in the reaction product. Importantly, the final fused ring linker that is generated is bulky, rigid, hydrophobic, and still contains some strain. Whether this presents any challenges for immunogenicity or stability of a therapeutic conjugate remains to be investigated.

3.3 Carbonyl chemistries

Carbonyl chemistries for creating oxime or hydrazone linkages are well established methods for protein conjugation. With this approach, a ketone or aldehyde handle is reacted with an aminooxy or hydrazide group to form an oxime or hydrazone linkage (Figure 9). The reaction is extremely selective with the exception that N-terminal cysteines will undergo thiazolidine formation. The reaction is typically performed around pH 4-5 to facilitate protonation of the carbonyl but maintain the nucleophilicity of the aminooxy group. No metal catalysts are required and the final linker is a small inconspicuous group. The recent discovery of aniline as a nucleophilic catalyst of this reaction was key to extending the utility of this reaction, even making conjugation at neutral pH viable (Dirksen & Dawson, 2008).

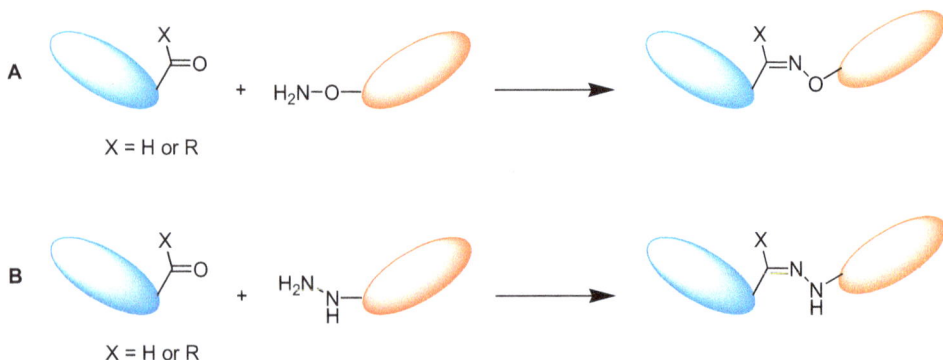

Fig. 9. Carbonyl chemistries, X = H for aldehydes, X = R for ketones. (A) A carbonyl and an aminooxy group combine to form an oxime linkage. (B) A carbonyl and a hydrazide group combine to form a hydrazone linkage.

The final oxime and hydrazone linkages can degrade, undergoing hydrolysis to regenerate the starting materials (Kalia & Raines, 2008). Hydrazone half lives can be on the order of minutes to hours, so oximes are the linkage of choice for stable bioconjugates. The half life of a simple oxime was reported as ~25 day at pD 7.0 but dropped to 15.7h at pD 5.0 due to the acid catalysis of this degradation. The chemistry of neighboring groups (electron withdrawing or donating, aromaticity) can be tailored for desired stability. Importantly, researchers at Ambrx have recently reported stability for more than 1 year at pH 7.3 of a PEGylated growth hormone containing an oxime linkage through p-acetyl-Phe (Cho et al., 2011). On the other hand, the instability of the hydrazone can be used as a tool, such as for temporary labelling or purification purposes. The ability of the final linkage to hydrolyze can also be a potential advantage if a therapeutic conjugate is desired to eventually degrade in vivo. Another problem when working with proteins containing carbonyl groups is reversible imine formation between the carbonyl group and lysinyl amino groups.

While the optimal reaction pH of 4-5 is not extreme, it is not desirable for some proteins, especially if this is near the pI. The reaction is often performed at a suboptimal pH in order to keep the protein in solution. When hGH (pI ~5) was PEGylated by oxime chemistry, 15% DMF was added and the reaction was performed at pH 6 to prevent precipitation, necessitating an amazing 10 days at 30°C for the reaction (Peschke et al., 2007). When the same investigators utilized CuAAC, the reaction took place over 22h with a much higher final yield of protein. The final conjugate also had a higher in vitro potency after CuAAC, perhaps due to the suboptimal conditions the protein was exposed to for the oxime conjugation.

3.4 Staudinger ligation

Bertozzi and colleagues have pioneered the Staudinger ligation as another orthogonal conjugation chemistry reactive to azide-containing NNAAs (Saxon & Bertozzi, 2000). In this reaction, a triarylphosphine moiety forms an aza-ylide intermediate (iminophosphorane) with the NNAA (Figure 10). A methyl ester group ortho to the phosphorous atom provides an electrophilic trap for the nitrogen, causing rearrangement to the final amide linkage.

Raines and colleagues have since developed a traceless Staudinger ligation (Figure 10), wherein a phosphinothioester reacts with the azide to form an aza-ylide, which undergoes a rearrangement to form an amide bond with release of the phosphine oxide (Tam & Raines, 2009). Thus, no "trace" of the phosphine group remains in the conjugated product. Various water soluble phosphinothiols are now available, incorporating amine groups that also favor the ligation over hydrolysis.

The Staudinger ligation has benefits including the lack of a metal catalyst. No ligation occurs with canonical amino acids and the reaction has been shown to work even on crude cell lysates. Particularly appealing in the case of the traceless ligation is that a simple amide bond is the reaction product.

Challenges with the Staudinger ligation include competing side reactions and the nature of the phosphine reagent. The design of the phosphine group is important to drive ligation rather than the competing classical Staudinger reaction. In the classical reaction, hydrolysis of the aza-ylide results in reduction of the azide to an amine and release of the phosphine oxide. The ligation works best around pH 8, presumably due to higher hydrolysis competition at acidic pH. Phosphines are also known to reduce disulfide bonds. Fortunately, triarylphosphines are mild reducing agents, and were shown not to induce significant disulfide reduction on a cell surface (Saxon & Bertozzi, 2000). Aryl phosphines are, however, very bulky (affecting reaction rates on protein surfaces), hydrophobic (affecting solubility of the reagent), and still prone to oxidation by air (requiring special handling). In the case of the nontraceless ligation, the aryl phosphine leaves a bulky hydrophobic signature in the linkage.

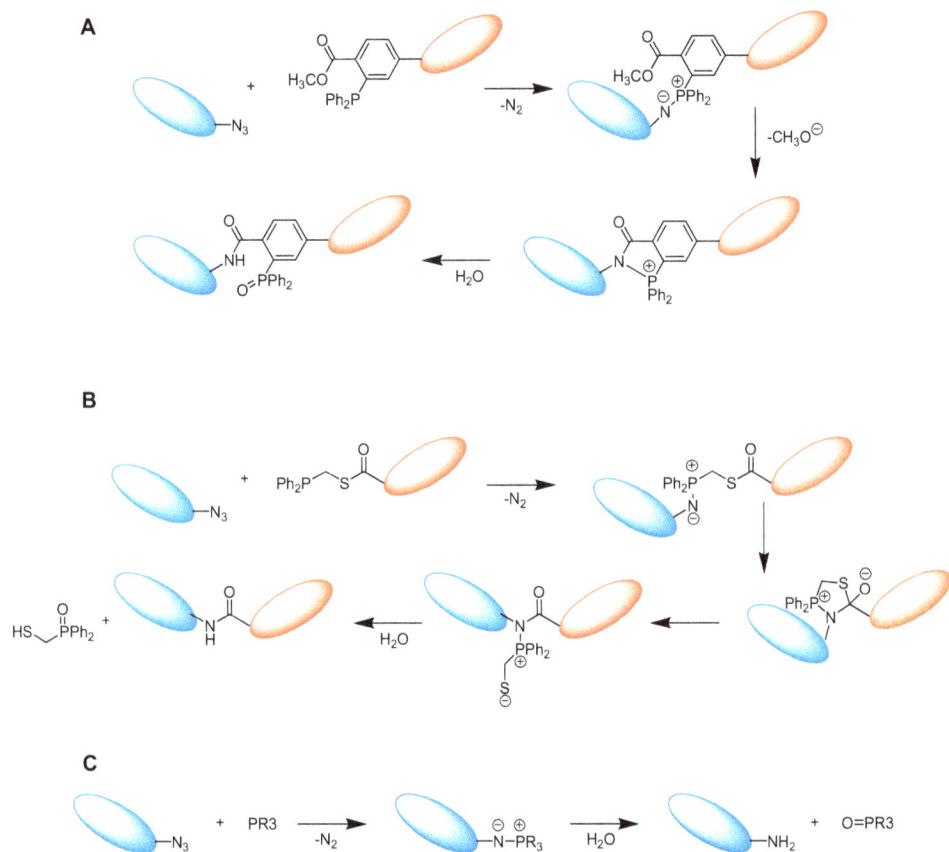

Fig. 10. (A) Staudinger ligation – an azide and a triarylphosphine moiety bearing an electrophilic trap combine to form an amide linkage bearing the triarylphosphine oxide. (B) Traceless Staudinger ligation – an azide and a phosphinothioester combine to form an amide linkage. (C) Classical Staudinger reaction – an azide reacts with a phosphine to form an amine and a phosphine oxide.

Reaction kinetics are perhaps the most significant shortcoming of the Staudinger ligation. The rate is orders of magnitude slower than other available chemistries. Attempts to accelerate the reaction by increasing the electron density of the phosphine substituent also increased the air instability of the reagent. The slow reaction rates, combined with air instability and competing hydrolysis, may be why the Staudinger ligation is used less frequently than other conjugation chemistries.

3.5 Other reactions

Due to the abundant applications, new chemistries are constantly in development for site specification conjugation to proteins containing NNAAs. Summarized below are a number of interesting approaches, some of which await further refinement before implementation as a readily available bioorthogonal strategy.

3.5.1 Metal-free

A tandem 1,3-dipolar cycloaddition-retro-Diels-Alder (tandem crDA) has recently emerged as an intriguing metal-free alternative to CuAAC (van Berkel et al., 2008). In this reaction, an azide group is reacted with a strained and electron deficient oxanorbornadiene derivative. The azide undergoes [3+2] cycloaddition with one of the double bonds, which is followed by elimination of a furan in a retro-Diels-Alder reaction, leading to formation of a stable 1,2,3-triazole. While an elegant approach, the reaction can take days to complete and side reactions with basic canonical amino acids can occur. Most importantly, the addition can also take place over the unsubstituted double bond in the strained ring; this results in elimination of the desired substituent and the conjugate is not formed. Hopefully further refinements will enable this chemistry for more general use.

Cycloadditions between alkenes and tetrazines or tetrazoles have also been an active area of research for bioorthogonal conjugation. In the inverse electron demand Diels-Alder cycloaddition, a strained alkene dienophile undergoes cycloaddition with a tetrazine, leading to a fused ring conjugate and elimination of N_2 (Blackman et al., 2008;Devaraj et al., 2008). The reaction is fast, selective, and metal-free. However, the reaction requires an alkene (such as *trans*-cyclooctene or norbornene) with much more strain than NNAAs that have been incorporated *in vivo* (such as homoallylglycine). One solution has been to use a tetrazole that can be photoactivated to produce a nitrile imine intermediate that then reacts with unactivated alkenes (Song et al., 2008). This has been used successfully with a Z-domain protein containing O-allyl-Tyr expressed from E. coli.

The cycloadditions of strained alkynes or alkenes with nitrile oxide have also been investigated recently (Gutsmiedl et al., 2009). The reaction also occurred with an unstrained alkyne and *in situ* formation of the nitrile oxide by hypervalent iodine, but was sluggish (Jawalekar et al., 2011). This approach suffers from a lack of incorporatable NNAAs containing the ideal functional groups and also the nitrile oxide can undergo side reactions with other functional groups or homodimerize with itself.

As an alternative to SPAAC, appendage of electron-withdrawing groups neighboring an alkyne can also enable azide alkyne cycloaddition without a metal catalyst (Li et al., 2004). Similarly to SPAAC, refinement of the local environment of the alkyne is necessary to promote a desirable reaction rate while maintaining orthogonality, but this approach avoids the use of bulky rings in the reactants.

Various photoactivated chemistries have been used, either with photoreactive NNAAs or by generation of reactive functional groups through irradiation. However, photoirradiation is well known to generate other reactive species such as free radicals that could cause side reactions with proteins and can also be difficult to scale up for a manufacturing process, and thus is not discussed in depth in this practical review.

Thiol-ene coupling to form a thioether linkage between thiols and alkenes has also gained recent popularity (Dondoni, 2008). Photoirradiation or free radical initiators are used to generate a reactive thiyl radical. This approach has been recently used to generate various bovine serum albumin conjugates. A similar thiol-yne coupling with alkynes can also be utilized (Minozzi *et al.*, 2011). For the application of thiol-ene/yne chemistry to NNAA-containing proteins, it would be possible to incorporate an alkene- or alkyne- containing NNAA and conjugate it to a thiol-containing substituent. However, given the prevalence and importance of cysteine residues in most proteins, and the side reactions to be expected by deliberate free radical generation, the generalizability of this approach for bioorthogonal reactions with NNAA-containing proteins is limited.

3.5.2 Metal-catalyzed

The family of palladium-catalyzed cross coupling reactions has great potential for bioorthogonal conjugations. These include couplings between an aryl or vinyl halide and an alkyne (Sonogashira), alkene (Heck), or boronic acid (Suzuki) group (Genet & Savignac, 1999). The Sonogashira coupling typically also includes a Cu(I) cocatalyst. The simple carbon-carbon linkage that is formed is very appealing with respect to stability and bulk. Classically these reactions are performed in organic solvent, under inert atmosphere, and with significant heat. However, recent advances in this rapidly evolving area have enabled these reactions for mild aqueous conditions.

These palladium-catalyzed reactions have now been employed for protein modification with varying success. Ras protein containing 4-iodo-Phe has been conjugated, via Heck or Sonogashira couplings, with vinylated or propargylated biotin (Kodama *et al.*, 2007). The reactions yielded only 2 - 25% product and the addition of DMSO, $MgCl_2$, and tyramine was required to reduce protein degradation and to suppress cysteine interference. More success has been seen with Suzuki couplings, performed with up to quantitative yields with 4-iodo-Phe-containing proteins (Ojida *et al.*, 2005;Chalker *et al.*, 2009). Issues with protein precipitation, dehalogenation, and cysteine interference could still be seen. Less success was seen when coupling a boronic acid-containing protein by Suzuki coupling (Brustad *et al.*, 2008). An important limitation to all Pd-catalyzed reactions is that the catalyst is easily poisoned by thiols and the presence of free cysteines (and potentially even cystines or methionines) can often block the reaction completely.

Ruthenium has emerged as a different metal that can catalyze the cycloaddition of azides and alkynes (Zhang *et al.*, 2005). Unlike CuAAC, the reaction is catalyzed by the metal in the +2 oxidation state and internal or terminal alkynes can be used. Also, depending on the catalyst complex used, the reaction product can be the 1,4- or the 1,5- disubstituted regioisomer of the triazole. One potential side reaction is oligomerization of the alkyne. RuAAC has been used to ligate pancreatic ribonuclease with a *cis*-peptide bond surrogate (Tam *et al.*, 2007).

Olefin cross-metathesis, the exchange of fragments between double-bond-containing molecules, has recently been applied to protein conjugations (Lin *et al.*, 2009). This has been enabled by the development of ruthenium-based catalysts that are air stable and active in water. Davis and co-workers recently demonstrated the first cross-metathesis coupling reactions with a protein (Lin *et al.*, 2010). A protein containing *S*-allylcysteine reacted quantitatively with alkene-containing glycosides after short times at 25-37°C. The reactions, however, took place in 30% *tert*-butyl alcohol and required addition of $MgCl_2$ as a mild Lewis acid to prevent non-productive chelation. The proximity of the sulfur seems to be critical for rate enhancement. Unfortunately, homoallylglycine, which can be readily incorporated into proteins, is much less reactive. Olefin metathesis is attractive due to the simple alkene bond that is formed between the substituents. Potential issues with olefin cross metathesis include self-metathesis homodimerization side reactions and interference from sulfur atoms (cysteine, cystine, methionine).

4. Potential applications of proteins engineered with non-natural amino acids

Protein engineering with NNAAs harnesses the vast divergent array of new protein building blocks and brings forth many new applications, such as monitoring protein functions *in vivo* via probe-labelling, trapping transient protein-protein interactions via cross-linking, facilitating structural determination and proteomic studies for newly synthesized proteins. In addition, it may also find practical applications in protein purification, materials science, diagnostics, drug screening, and green energy. In this section, we will highlight its applications on protein therapeutics and protein functional evolution.

4.1 Protein therapeutics

Protein therapeutics represent a $72 billion market with more than 130 products approved (Leader *et al.*, 2008). The ability of complex proteins to exquisitely target disease pathways continues to drive rapid development despite their high cost of goods. While the proteins themselves can offer a strong therapeutic effect, additional modifications such as polymer or drug conjugation can improve or add new pharmacodynamic effects. A number of conjugated biotherapeutics are already approved for clinical use, including PEGylated proteins, antibody-drug conjugates, and saccharide-protein conjugate vaccines.

First generation conjugate therapeutics suffer from a lack of control over the conjugation site and current research is focused on enabling precise direction to a chosen site of modification. The expression of proteins with NNAAs at predetermined locations lends itself readily to these applications. While many relevant NNAA bioconjugations have been reported in the literature, a number of issues must be addressed when considering translation to a clinical product. The conjugation reaction must use conditions that are amenable to scale-up, use reagents that can be obtained with sufficient purity and scale, have cost effective reactant ratios with high yields, and avoid side reactions that can be problematic at even very low levels. In addition, the nature of the linkage that is formed between the protein and the conjugated species is important with respect to shelf life and *in vivo* stability as well as to immunogenicity of the product.

In the following sections we provide an overview of NNAA conjugate therapeutic approaches that are readily contemplated or have already been reduced to practice. We also provide an example of therapeutic effect derived from incorporation of an unmodified NNAA.

4.1.1 Modifications for improved pharmacokinetics

Increasing the residence time of protein therapeutics in the body has been a focus of biotherapeutic development for some time. Because these pharmaceuticals are typically administered by injection, decreasing the dosing frequency from daily to weekly or monthly is highly desirable. This increases patient comfort and compliance, and the maintenance of a more steady plasma concentration of the drug, rather than frequent peaks and troughs, can increase the therapeutic effect (Harris & Chess, 2003).

Conjugation of polyethylene glycol (PEG) is the current gold standard for protein half-life extension (Fishburn, 2008). There are now at least eight approved PEGylated protein therapeutics in the US. PEGylation of proteins is known to decrease renal clearance, decrease enzymatic degradation, reduce immunogenicity of the protein, and increase serum half-life due to the steric hindrance of a PEG "cloud" around the protein. Upon PEGylation, the half-life of a protein may increase by a few-fold to even a thousand-fold. Consequently, even though PEGylation typically results in decreased *in vitro* activity, *in vivo* efficacy is usually greatly enhanced. Alternative polymers for conjugation are under investigation, including hydroxyethyl starch.

The site of conjugation to a protein has important implications. PEGylation near a binding or catalytic site of the protein can strongly reduce binding affinity and *in vitro* potency. As discussed in sections 3 and 4.1.3, protein PEGylation can occur through reactive groups such as lysinyl amines or cysteinyl thiols, but this provides poor control over the site of and number of modification(s). Even reductive amination, which is designed to be selective for the N-terminal amine due to its lower pKa, still results in PEGylation at additional sites on the protein (Gronke et al., 2010). While multi-PEGylated species can often be purified from mono-PEGylated species, separation of positional PEG isomers (monoPEGylation at different sites on the protein) is unfeasible in commercial production. The resultant heterogeneous product contains isomeric molecules with a range of potencies and provides a challenge for batch-to-batch reproducibility. Thus, the field is rapidly moving toward true site specific modification to allow production of a single designed homogenous product.

PEGylation has now been performed through NNAA conjugation, and a few companies are advancing these products in the clinic. Allozyne has reported a PEGylated interferon beta-1b (IFNβ) produced by incorporating a single Aha as a reassigned sense codon for Met in *E. coli* and then attaching a PEG-alkyne via CuAAC (Nairn et al., 2007). We have since produced a PEG-IFNβ candidate under cGMP at >800L fermentation scale (unpublished data). Up to 98% conjugation of Aha-containing IFNβ was achieved using a low molar excess of reactive 40 kDa branched PEG. No PEGylation was seen at any site besides the Aha and residual copper in the final product was <0.1 ppm with no special efforts for removal. After 18 months storage at 5°C, there was no detected unPEGylated IFNβ in the Drug Product, indicating a very stable linkage. Phase Ia trials showed the resultant half-life extension may allow monthly dosing, whereas the unPEGylated analog Betaseron® is dosed every other day. Ambrx recently reported clinical data of a PEGylated human growth hormone that supports weekly dosing. This molecule was made by amber suppression in *E. coli* with incorporation of *p*-acetyl-Phe at position 35 followed by reaction with linear 30 kDa aminooxy-PEG to create an oxime linker (Cho *et al.*, 2011). As has been seen for other PEGylated proteins, variations in the PEGylation site had a significant effect on binding affinity and *in vitro* bioactivity; a systematic comparison *in vitro* and *in vivo* was performed in order to select the site taken into the clinic. Fermentation was performed at >1000L scale with 0.5-0.8g/L titers and the linkage was shown to be stable for more than a year at 5°C.

4.1.2 Bispecific protein conjugates

Bispecific and multivalent proteins are receiving growing interest as therapeutic candidates (Marvin & Zhu, 2005;Filpula, 2007). Antibody-like molecules with specificity for two different epitopes show promise for targeted cancer therapy, such as by bringing cancer cells and effector cells into close contact. Combination of two antibody binding domains also has potential for immunotherapy, such as by simultaneously binding two cytokines or to regulate multiple pathways by simultaneously binding cell surface receptors. Other applications include toxin conjugates, where a proteinaceous toxin is linked to a targeting protein such as an antibody binding domain or a targeting ligand. As another approach, multivalency of the same protein can improve avidity, targeting, and function.

Bispecific or multivalent proteins are typically constructed by expression of genetic fusions or by conjugation through an engineered cysteine residue. Limitations of genetic fusions can include mis-folding, mis-pairing of antibody chains, low fermentation yields, and a lack of choices for relative orientation between the proteins, as the linkage can only occur between the N-terminus of one protein and the C-terminus of the other (Griffiths *et al.*, 2004;Marvin & Zhu, 2005). Engineered cysteine residues can result in mis-disulfide bond formation (Kim *et al.*, 2008) and a complex reaction product (Natarajan *et al.*, 2005).

The use of NNAAs for protein-protein coupling enables production of a single conjugate with any desired orientation. It requires that the two target proteins be produced in separate fermentations with 2 complementary NNAAs. This can allow independent folding of the two proteins prior to conjugation if desired to avoid mis-folding events.

There are a few examples of protein-protein conjugation through incorporated NNAAs. DHFR and GFP (or 2 GFPs) have been directly conjugated through CuAAC between incorporated *p*-propargyloxy-Phe and *p*-azido-Phe residues (Bundy & Swartz, 2010). Up to 62% conjugation yield was achieved with a 1:1 molar reactant ratio. Not surprisingly, use of a ligand to protect the protein from the copper and removal of the copper prior to assay were important to demonstrate that bioactivity was retained. An scFv and a luciferase reporter protein have also been conjugated by CuAAC between incorporated Hpg and *p*-azido-Phe residues (Patel *et al.*, 2009). Another interesting approach was to react *p*-acetyl-Phe genetically incorporated in an anti-HER2 Fab domain with hydroxylamine biotin to form an oxime linkage (Hutchins *et al.*, 2011). This conjugate was then assembled into a tetravalent species through incubation with deglycosylated avidin. By incorporating the NNAA at various sites, the authors showed that the orientation of the protein assemblage is important, with 5-50 fold increased in phosphorylation inhibition versus the monovalent parent Fab depending on the conjugation site.

4.1.3 Antibody-drug conjugates and radioimmunoconjugates

The chemical conjugation of highly potent cytotoxic drugs to monoclonal antibodies has proven to be effective means of targeted delivery of chemotherapeutics that would otherwise be too toxic for systemic administration (Alley *et al.*, 2010). The key attributes of antibody-drug conjugates (ADCs) include the specificity of the antibody for the tumor, the potency of the toxin, and the linker, which must release the toxin conditionally upon internalization into the tumor cell. ADCs are typically linked via acylation of lysines, alkylation of reduced interchain disulfides, or alkylation of introduced unpaired Cys residues. Each of these methods leads to some degree of heterogeneity of the product. Modification at native Cys residues produces a mixture of molecules including 0, 2, 4, 6, and 8 drugs, where the final conjugate also contains less stabilizing disulfide bonds than the parent (Sun *et al.*, 2005). The therapeutic index will be highest for a molecule with the ideal number of drugs attached, for example, the maximum tolerated dose and clearance rate improved by 4-5 fold when comparing ADCs containing exactly 2 auristatin analogs versus 8 (Hamblett *et al.*, 2004). One approach to limit product heterogeneity is to replace some of the natural cysteines with serines (McDonagh *et al.*, 2006). Engineering in new Cys residues has also allowed more control over the degree of conjugation, but the reaction product still carries a distribution of species (Junutula *et al.*, 2008).

The ability to introduce a NNAA at any location in an antibody (or Fab, scFv, etc.) provides several advantages for ADC construction. The NNAA creates specific sites where the toxin is conjugated, leaving all other residues and the disulfide bonds unmodified. These sites can be positioned at favorable locations that are permissive for full function of the antibody. The number of toxins per antibody can also be efficiently adjusted without sacrificing the yield or homogeneity of the product. There are, to our knowledge, no literature reports of ADC construction with genetically incorporated NNAAs, but Allozyne and Ambrx are both pursuing undisclosed ADC targets for oncology.

Radioimmunoconjugates for radioimmunotherapy (RIT) are similar to ADCs, but a chelating agent for a radionuclide is conjugated instead of a toxin (Martin *et al.*, 2010). The design principles are similar to ADCs, where the use of NNAAs to control the location and number of attached chelating agents can be expected to substantially improve the product. One significant advantage of RIT is that internalization is not required. Thus, there are a greater number of available targets. In addition, smaller targeting proteins, such as scFvs may be used, which should provide for better tumor penetration.

4.1.4 Cancer vaccines

Cancer vaccines hold great promise to treat metastatic tumors effectively and to elicit long-term protection against recurrence (Farkas & Finn, 2010). However, because most tumors are derived from normal cells, the identification of true tumor specific antigens (absent in all normal tissues) is difficult. Instead, many tumor associated antigens, especially those self-proteins over-expressed in tumor, have been identified and used for immunotherapy. While it has been shown that cancer vaccines can be effective, it is still a challenge to elicit effective immunity against the over-expressed tumor associated antigens because of self-tolerance.

Introduction of NNAAs into tumor antigens offers potential strategies to break such self-tolerance, since the NNAA may be recognized as a foreign entity by the immune system. Such a theory is supported by animal data (Grunewald *et al.*, 2009): when the wild-type murine tumor necrosis factor-α was used as an immunogen, mice failed to generate IgG antibodies; however, when the mutant protein containing only one NNAA (*p*-nitro-Phe) was used, sustained polyclonal IgG antibodies against both the wild-type as well as NNAA-containing protein antigens were generated. The resulting immune response protected mice from severe endotoxemia induced by lipopolysaccharide (LPS) challenge. Therefore, vaccination with NNAA-modified self-protein can break self-tolerance due to epitope spreading, resulting in polyclonal antibodies against the native protein.

Incorporation of NNAAs into antigens may also improve vaccine efficacy by increasing the antigen stability *in vivo*. This may be of special importance to peptide-based vaccines due to the short half-life of peptides. Peptides designed with NNAAs may adopt bioactive conformation with enhanced stability, and thus elicit immune response to wild-type tumor antigens (Vichier-Guerre *et al.*, 2004;Corzana *et al.*, 2011).

4.2 Functional evolution of proteins with NNAAs

Engineering proteins for desired biophysical properties, enhanced or novel functions is of central importance, not only for basic scientific research, but also for biopharmaceutical drug development. The availability of the vast number of NNAAs as building blocks greatly expands our ability to customize proteins for different purposes. Several strategies for functional evolution of proteins with NAA are described below.

Rational-design (based on the known structural information of target proteins) is an efficient way to evolve proteins with NNAAs. For example, based on the hyper-hydrophobic and near isosteric properties of fluorinated amino acids to their natural counterparts, trifluoroleucine (Tfl) introduced to leucine zipper peptides (replacing all Leu residues) has been shown to increase the stability while maintaining their DNA-binding function (Tang *et al.*, 2001). Molecular dynamics simulation of the thermodynamic properties of the fluorinated leucine zipper peptide indicated that Tfl-substituted structure is indeed more stable than the wild-type, correlating well with experimental data (Tang *et al.*, 2001). Another impressive example involves replacing a single amino acid (Tyr 309) in the bacterial enzyme phosphotriesterase with NNAAs (L-(7-hydroxycoumarin-4-yl)ethylglycine and L-(7-methycoumarin-4-yl)ethylglycine amino acid) (Ugwumba *et al.*, 2011). This enzyme catalyzes the hydrolysis of the pesticide paraoxon with very fast turnover rate, which was thought to be near the evolutionary limit. However, through a single, rationally designed mutation with NNAA, which increased the electrostatic repulsion of the negatively charged hydrolysis product, and thus improved the rate-limiting product release step, about a 10-fold improvement of this already highly efficient enzyme was achieved. Such result is in great contrast to the difficulty in improving this enzyme's activity through screening large mutation library with canonical amino acids.

In the absence of structural information required for rational design, library screening methods can be utilized to evolve NNAA-containing proteins for the desired properties or functions. Briefly, a target gene containing various mutations is expressed in the presence of the NNAA, and the resultant NNAA-containing protein variants are screened and selected for the desired properties. While the designs of library (ranging from total random mutation library, to the limited mutation library at a few codons), the methods to introduce NNAAs into proteins (as reviewed in previous sections), as well as the library screening strategies can vary from case to case, the basic principle for functional evolution of proteins with NNAAs can be applied broadly for protein engineering. For example, sense codon global substitution with a NNAA can often result in the loss of a protein's structure and function. By screening random libraries with a sense codon global substitution method to replace Leu with Tfl, a mutant Tfl-containing enzyme with enhanced thermal stability (Montclare & Tirrell, 2006) and a mutant Tfl-containing GFP with increased fluorescence intensity (Yoo *et al.*, 2007) were obtained. By screening a phage-display scFv antibody library, where six residues in the V_H CDR3 were randomized, one scFv containing sulfotyrosine was obtained that binds gp120 more effectively than any similarly evolved scFv molecules containing only canonical amino acids (Liu *et al.*, 2008). By screening a peptide library based on cellular viability, cyclic peptide (containing NNAA) protease inhibitors were derived (Young *et al.*, 2011). Other NNAA-containing proteins, such as boron-containing proteins for carbohydrate binding (Liu *et al.*, 2009), have been evolved as well. Therefore, functional evolution of proteins with NNAAs can offer a selective advantage beyond the traditional directed evolution with only the canonical amino acids.

5. Conclusion

NNAAs greatly expand the amino acid repertoire for protein engineering. The ability to introduce NNAAs into proteins is a very important technical advance in the fields of both discovery research as well as commercial production of medicines and industrial enzymes. There are several established methods to accomplish the incorporation of NNAAs. In most cases, the NNAA serves as a point of protein modification. For biopharmaceutical applications, the ability to introduce the NNAA at a specific position becomes a very important advantage, as protein function is very sensitive to the position of conjugation. These new technologies are currently being used in the development of new and better protein therapeutics that may soon be on the market. The future will likely include a host of new medicines based on this technology, and its many diverse applications.

6. References

Agard,N.J., Baskin,J.M., Prescher,J.A., Lo,A., and Bertozzi,C.R. (2006). A comparative study of bioorthogonal reactions with azides. *ACS Chem. Biol. 1*, 644-648.

Agard,N.J., Prescher,J.A., and Bertozzi,C.R. (2004). A strain-promoted [3 + 2] azide-alkyne cycloaddition for covalent modification of biomolecules in living systems. *J. Am. Chem. Soc. 126*, 15046-15047.

Agris,P.F., Vendeix,F.A., and Graham,W.D. (2007). tRNA's wobble decoding of the genome: 40 years of modification. *J. Mol. Biol. 366*, 1-13.

Alley,S.C., Okeley,N.M., and Senter,P.D. (2010). Antibody-drug conjugates: targeted drug delivery for cancer. *Curr. Opin. Chem. Biol. 14*, 529-537.

Ambrogelly,A., Gundllapalli,S., Herring,S., Polycarpo,C., Frauer,C., and Soll,D. (2007). Pyrrolysine is not hardwired for cotranslational insertion at UAG codons. *Proc. Natl. Acad. Sci. U. S. A 104*, 3141-3146.

Anderson,J.C., Wu,N., Santoro,S.W., Lakshman,V., King,D.S., and Schultz,P.G. (2004). An expanded genetic code with a functional quadruplet codon. *Proc. Natl. Acad. Sci. U. S. A 101*, 7566-7571.

Bacher,J.M. and Ellington,A.D. (2001). Selection and characterization of Escherichia coli variants capable of growth on an otherwise toxic tryptophan analogue. *J. Bacteriol. 183*, 5414-5425.

Back,J.W., David,O., Kramer,G., Masson,G., Kasper,P.T., de Koning,L.J., de Jong,L., van Maarseveen,J.H., and de Koster,C.G. (2005). Mild and chemoselective peptide-bond cleavage of peptides and proteins at azido homoalanine. *Angew. Chem. Int. Ed Engl. 44*, 7946-7950.

Bain,J.D., Switzer,C., Chamberlin,A.R., and Benner,S.A. (1992). Ribosome-mediated incorporation of a non-standard amino acid into a peptide through expansion of the genetic code. *Nature 356*, 537-539.

Beier,H. and Grimm,M. (2001). Misreading of termination codons in eukaryotes by natural nonsense suppressor tRNAs. *Nucleic Acids Res. 29*, 4767-4782.

Blackman,M.L., Royzen,M., and Fox,J.M. (2008). Tetrazine ligation: fast bioconjugation based on inverse-electron-demand Diels-Alder reactivity. *J. Am. Chem. Soc. 130*, 13518-13519.

Brustad,E., Bushey,M.L., Lee,J.W., Groff,D., Liu,W., and Schultz,P.G. (2008). A genetically encoded boronate amino acid. *Angew. Chem. Int. Ed Engl. 47*, 8220-8223.

Bundy,B.C. and Swartz,J.R. (2010). Site-specific incorporation of

Chalker,J.M., Wood,C.S., and Davis,B.G. (2009). A convenient catalyst for aqueous and protein Suzuki-Miyaura cross-coupling. *J. Am. Chem. Soc. 131*, 16346-16347.

Chen,P.R., Groff,D., Guo,J., Ou,W., Cellitti,S., Geierstanger,B.H., and Schultz,P.G. (2009). A facile system for encoding unnatural amino acids in mammalian cells. *Angew. Chem. Int. Ed Engl. 48*, 4052-4055.

Chin,J.W., Cropp,T.A., Anderson,J.C., Mukherji,M., Zhang,Z., and Schultz,P.G. (2003). An expanded eukaryotic genetic code. *Science 301*, 964-967.

Cho,H., Daniel,T., Buechler,Y.J., Litzinger,D.C., Maio,Z., Putnam,A.H., Kraynov,V.S., Sim,B., Bussell,S., Javahishvili,T., Kaphle,S., Viramontes,G., Ong,M., Chu,S., GC,B., Lieu,R., Knudsen,N., Castiglioni,P., Norman,T.C., Axelrod,D.W., Hoffman,A.R., Schultz,P.G., DiMarchi,R.D., and Kimmel,B.E. (2011). Optimized clinical performance of growth hormone with an expanded genetic code. *Proc. Natl. Acad. Sci. U. S. A 108*, 9060-9065.

Codelli,J.A., Baskin,J.M., Agard,N.J., and Bertozzi,C.R. (2008). Second-generation difluorinated cyclooctynes for copper-free click chemistry. *J. Am. Chem. Soc. 130*, 11486-11493.

Corzana,F. *et al.* (2011). Rational design of a Tn antigen mimic. *Chem. Commun. (Camb.) 47*, 5319-5321.

Crick,F.H. (1966). Codon--anticodon pairing: the wobble hypothesis. *J. Mol. Biol. 19*, 548-555.

Curran,J.F. and Yarus,M. (1989). Rates of aminoacyl-tRNA selection at 29 sense codons in vivo. *J. Mol. Biol. 209*, 65-77.

Debets,M.F., van Berkel,S.S., Schoffelen,S., Rutjes,F.P., van Hest,J.C., and van Delft,F.L. (2010). Aza-dibenzocyclooctynes for fast and efficient enzyme PEGylation via copper-free (3+2) cycloaddition. *Chem. Commun. (Camb.) 46*, 97-99.

del Amo,D.S., Wang,W., Jiang,H., Besanceney,C., Yan,A.C., Levy,M., Liu,Y., Marlow,F.L., and Wu,P. (2010). Biocompatible copper(I) catalysts for in vivo imaging of glycans. *J. Am. Chem. Soc. 132*, 16893-16899.

Devaraj,N.K., Weissleder,R., and Hilderbrand,S.A. (2008). Tetrazine-based cycloadditions: application to pretargeted live cell imaging. *Bioconjug. Chem. 19*, 2297-2299.

Dirksen,A. and Dawson,P.E. (2008). Rapid oxime and hydrazone ligations with aromatic aldehydes for biomolecular labeling. *Bioconjug. Chem. 19*, 2543-2548.

Dommerholt,J., Schmidt,S., Temming,R., Hendriks,L.J., Rutjes,F.P., van Hest,J.C., Lefeber,D.J., Friedl,P., and van Delft,F.L. (2010). Readily accessible bicyclononynes for bioorthogonal labeling and three-dimensional imaging of living cells. *Angew. Chem. Int. Ed Engl. 49*, 9422-9425.

Dondoni,A. (2008). The emergence of thiol-ene coupling as a click process for materials and bioorganic chemistry. Angew. Chem. Int. Ed Engl. 47, 8995-8997.

Farkas,A.M. and Finn,O.J. (2010). Vaccines based on abnormal self-antigens as tumor-associated antigens: immune regulation. Semin. Immunol. 22, 125-131.

Filpula,D. (2007). Antibody engineering and modification technologies. Biomol. Eng 24, 201-215.

Fishburn,C.S. (2008). The pharmacology of PEGylation: balancing PD with PK to generate novel therapeutics. J. Pharm. Sci. 97, 4167-4183.

Genet,J.P. and Savignac,M. (1999). Recent developments of palladium(0) catalyzed reaactions in aqueous medium. Journal of Organometallic Chemstry 576, 305-317.

Gibson,D.G. et al. (2010). Creation of a bacterial cell controlled by a chemically synthesized genome. Science 329, 52-56.

Goltermann,L., Larsen,M.S., Banerjee,R., Joerger,A.C., Ibba,M., and Bentin,T. (2010). Protein evolution via amino acid and codon elimination. PLoS. One. 5, e10104.

Grabstein, K., Wang, A., Nairn,N.W., Graddis, T.J. (2010). Methods of modifying polypeptides comprisig non-natural amino acids, US patent 7,829,659

Griffiths,G.L., Chang,C., McBride,W.J., Rossi,E.A., Sheerin,A., Tejada,G.R., Karacay,H., Sharkey,R.M., Horak,I.D., Hansen,H.J., and Goldenberg,D.M. (2004). Reagents and methods for PET using bispecific antibody pretargeting and 68Ga-radiolabeled bivalent hapten-peptide-chelate conjugates. J. Nucl. Med. 45, 30-39.

Gronke, R.S., Jaquez, O. and Kiistala, M. (2010). Challenges of developing a PEGylaed Interferon beta-1a. 239 ACS National Meeting, San Francisco, CA

Grunewald,J. et al. (2009). Mechanistic studies of the immunochemical termination of self-tolerance with unnatural amino acids. Proc. Natl. Acad. Sci. U. S. A 106, 4337-4342.

Gu,W., Li,M., Zhao,W.M., Fang,N.X., Bu,S., Frazer,I.H., and Zhao,K.N. (2004). tRNASer(CGA) differentially regulates expression of wild-type and codon-modified papillomavirus L1 genes. Nucleic Acids Res. 32, 4448-4461.

Gupta,S.S., Kuzelka,J., Singh,P., Lewis,W.G., Manchester,M., and Finn,M.G. (2005). Accelerated bioorthogonal conjugation: a practical method for the ligation of diverse functional molecules to a polyvalent virus scaffold. Bioconjug. Chem. 16, 1572-1579.

Gutsmiedl,K., Wirges,C.T., Ehmke,V., and Carell,T. (2009). Copper-free "click" modification of DNA via nitrile oxide-norbornene 1,3-dipolar cycloaddition. Org. Lett. 11, 2405-2408.

Haeusler,R.A. and Engelke, D.R. (2006) Spatial organization of transcription by RNA polymerase III. Nucleic Acids Res. 34, 4826-4836

Hamblett,K.J., Senter,P.D., Chace,D.F., Sun,M.M., Lenox,J., Cerveny,C.G., Kissler,K.M., Bernhardt,S.X., Kopcha,A.K., Zabinski,R.F., Meyer,D.M., and Francisco,J.A. (2004). Effects of drug loading on the antitumor activity of a monoclonal antibody drug conjugate. Clin. Cancer Res. 10, 7063-7070.

Hancock,S.M., Uprety,R., Deiters,A., and Chin,J.W. (2010). Expanding the genetic code of yeast for incorporation of diverse unnatural amino acids via a pyrrolysyl-tRNA synthetase/tRNA pair. J. Am. Chem. Soc. 132, 14819-14824.

Harris,J.M. and Chess,R.B. (2003). Effect of pegylation on pharmaceuticals. Nat. Rev. Drug Discov. 2, 214-221.

Hendrickson,T.L., de Crecy-Lagard,V., and Schimmel,P. (2004). Incorporation of nonnatural amino acids into proteins. *Annu. Rev. Biochem. 73*, 147-176.

Higgs,P.G. and Ran,W. (2008). Coevolution of codon usage and tRNA genes leads to alternative stable states of biased codon usage. *Mol. Biol. Evol. 25*, 2279-2291.

Hirao,I. *et al.* (2002). An unnatural base pair for incorporating amino acid analogs into proteins. *Nat. Biotechnol. 20*, 177-182.

Hirao,I., Harada,Y., Kimoto,M., Mitsui,T., Fujiwara,T., and Yokoyama,S. (2004). A two-unnatural-base-pair system toward the expansion of the genetic code. *J. Am. Chem. Soc. 126*, 13298-13305.

Hohsaka,T. and Sisido,M. (2002). Incorporation of non-natural amino acids into proteins. *Curr. Opin. Chem. Biol. 6*, 809-815.

Hohsaka,T., Sato,K., Sisido,M., Takai,K., and Yokoyama,S. (1994). Site-specific incorporation of photofunctional nonnatural amino acids into a polypeptide through in vitro protein biosynthesis. *FEBS Lett. 344*, 171-174.

Hong,V., Presolski,S.I., Ma,C., and Finn,M.G. (2009). Analysis and optimization of copper-catalyzed azide-alkyne cycloaddition for bioconjugation. *Angew. Chem. Int. Ed Engl. 48*, 9879-9883.

Hong,V., Steinmetz,N.F., Manchester,M., and Finn,M.G. (2010). Labeling live cells by copper-catalyzed alkyne--azide click chemistry *Bioconjug. Chem. 21*, 1912-1916.

Hughes,R.A. and Ellington,A.D. (2010). Rational design of an orthogonal tryptophanyl nonsense suppressor tRNA. *Nucleic Acids Res. 38*, 6813-6830.

Hutchins,B.M., Kazane,S.A., Staflin,K., Forsyth,J.S., Felding-Habermann,B., Schultz,P.G., and Smider,V.V. (2011). Site-specific coupling and sterically controlled formation of multimeric antibody fab fragments with unnatural amino acids. *J. Mol. Biol. 406*, 595-603.

Ibba,M., Becker,H.D., Stathopoulos,C., Tumbula,D.L., and Soll,D. (2000). The adaptor hypothesis revisited. *Trends Biochem. Sci. 25*, 311-316.

Jawalekar,A.M., Reubsaet,E., Rutjes,F.P., and van Delft,F.L. (2011). Synthesis of isoxazoles by hypervalent iodine-induced cycloaddition of nitrile oxides to alkynes. *Chem. Commun. (Camb.) 47*, 3198-3200.

Jewett,J.C. and Bertozzi,C.R. (2010). Cu-free click cycloaddition reactions in chemical biology. *Chem. Soc. Rev. 39*, 1272-1279.

Jewett,J.C., Sletten,E.M., and Bertozzi,C.R. (2010). Rapid Cu-free click chemistry with readily synthesized biarylazacyclooctynones. *J. Am. Chem. Soc. 132*, 3688-3690.

Junutula,J.R. *et al.* (2008). Site-specific conjugation of a cytotoxic drug to an antibody improves the therapeutic index. *Nat. Biotechnol. 26*, 925-932.

Kalia,J. and Raines, R.T. (2008). Hydrolytic stability of hydrazones and oximes. *Angew.Chem.Int.Ed. 47*, 7523-7526.

Kim,K.M., McDonagh,C.F., Westendorf,L., Brown,L.L., Sussman,D., Feist,T., Lyon,R., Alley,S.C., Okeley,N.M., Zhang,X., Thompson,M.C., Stone,I., Gerber,H., and Carter,P.J. (2008). Anti-CD30 diabody-drug conjugates with potent antitumor activity. *Mol. Cancer Ther. 7*, 2486-2497.

Kobayashi,T., Yanagisawa,T., Sakamoto,K., and Yokoyama,S. (2009). Recognition of non-alpha-amino substrates by pyrrolysyl-tRNA synthetase. *J. Mol. Biol. 385*, 1352-1360.

Kodama,K., Fukuzawa,S., Nakayama,H., Sakamoto,K., Kigawa,T., Yabuki,T., Matsuda,N., Shirouzu,M., Takio,K., Yokoyama,S., and Tachibana,K. (2007). Site-specific functionalization of proteins by organopalladium reactions. *Chembiochem.* 8, 232-238.

Köhrer,C., Sullivan,E.L., and RajBhandary, U.L. (2004) Complete set of orthogonal 21st aminoacyl-tRNA synthetase-amber, ochre and opal suppressor tRNA pairs: concomitant suppression of three different termination codons in an mRNA in mammalian cells. *Nucleic Acids Res.* 32, 6200-6211

Kwon,I. and Tirrell,D.A. (2007). Site-specific incorporation of tryptophan analogues into recombinant proteins in bacterial cells. *J. Am. Chem. Soc.* 129, 10431-10437.

Kwon,I., Kirshenbaum,K., and Tirrell,D.A. (2003). Breaking the degeneracy of the genetic code. *J. Am. Chem. Soc.* 125, 7512-7513.

Kwon,I., Wang,P., and Tirrell,D.A. (2006). Design of a bacterial host for site-specific incorporation of p-bromophenylalanine into recombinant proteins. *J. Am. Chem. Soc.* 128, 11778-11783.

Leader,B., Baca,Q.J., and Golan,D.E. (2008). Protein therapeutics: a summary and pharmacological classification. *Nat. Rev. Drug Discov.* 7, 21-39.

Li,W.T., Mahapatra,A., Longstaff,D.G., Bechtel,J., Zhao,G., Kang,P.T., Chan,M.K., and Krzycki,J.A. (2009). Specificity of pyrrolysyl-tRNA synthetase for pyrrolysine and pyrrolysine analogs. *J. Mol. Biol.* 385, 1156-1164.

Li,Z., Seo,T.S., and Ju,J. (2004). 1,3-Dipolar cycloaddition of azides with electron-deficient alkynes under mild condition in water. *Tetrahedron Letters* 45, 3143-3146.

Lin,Y.A., Chalker,J.M., and Davis,B.G. (2009). Olefin metathesis for site-selective protein modification. *Chembiochem.* 10, 959-969.

Lin,Y.A., Chalker,J.M., and Davis,B.G. (2010). Olefin cross-metathesis on proteins: investigation of allylic chalcogen effects and guiding principles in metathesis partner selection. *J. Am. Chem. Soc.* 132, 16805-16811.

Link,A.J. and Tirrell,D.A. (2005). Reassignment of sense codons in vivo. *Methods* 36, 291-298.

Link,A.J., Vink,M.K., and Tirrell,D.A. (2004). Presentation and detection of azide functionality in bacterial cell surface proteins. *J. Am. Chem. Soc.* 126, 10598-10602.

Liu,C.C. and Schultz,P.G. (2010). Adding new chemistries to the genetic code. *Annu. Rev. Biochem.* 79, 413-444.

Liu,C.C., Mack,A.V., Brustad,E.M., Mills,J.H., Groff,D., Smider,V.V., and Schultz,P.G. (2009). Evolution of proteins with genetically encoded "chemical warheads". *J. Am. Chem. Soc.* 131, 9616-9617.

Liu,C.C., Mack,A.V., Tsao,M.L., Mills,J.H., Lee,H.S., Choe,H., Farzan,M., Schultz,P.G., and Smider,V.V. (2008). Protein evolution with an expanded genetic code. *Proc. Natl. Acad. Sci. U. S. A* 105, 17688-17693.

Liu,W., Brock,A., Chen,S., Chen,S., and Schultz,P.G. (2007). Genetic incorporation of unnatural amino acids into proteins in mammalian cells. *Nat. Methods* 4, 239-244.

Martin,M.E., Parameswarappa,S.G., O'Dorisio,M.S., Pigge,F.C., and Schultz,M.K. (2010). A DOTA-peptide conjugate by copper-free click chemistry. *Bioorg. Med. Chem. Lett.* 20, 4805-4807.

Marvin,J.S. and Zhu,Z. (2005). Recombinant approaches to IgG-like bispecific antibodies. *Acta Pharmacol. Sin.* 26, 649-658.

McDonagh,C.F. Turcott,E., Westendorf,L., Webster,J.B., Alley,S.C., Kim,K., Andreyka,J., Stone,I., Hamblett,K.J., Francisco,J.A., and Carter,P. (2006). Engineered antibody-drug conjugates with defined sites and stoichiometries of drug attachment. *Protein Eng Des Sel 19*, 299-307.

Mehl,R.A., Anderson,J.C., Santoro,S.W., Wang,L., Martin,A.B., King,D.S., Horn,D.M., and Schultz,P.G. (2003). Generation of a bacterium with a 21 amino acid genetic code. *J. Am. Chem. Soc. 125*, 935-939.

Minozzi,M., Monesi,A., Nanni,D., Spagnolo,P., Marchetti,N., and Massi,A. (2011). An insight into the radical thiol/yne coupling: the emergence of arylalkyne-tagged sugars for the direct photoinduced glycosylation of cysteine-containing peptides. *J. Org. Chem. 76*, 450-459.

Montclare,J.K. and Tirrell,D.A. (2006). Evolving proteins of novel composition. *Angew. Chem. Int. Ed Engl. 45*, 4518-4521.

Mukai,T., Hayashi,A., Iraha,F., Sato,A., Ohtake,K., Yokoyama,S., and Sakamoto,K. (2010). Codon reassignment in the Escherichia coli genetic code. *Nucleic Acids Res. 38*, 8188-8195.

Mukai,T., Kobayashi,T., Hino,N., Yanagisawa,T., Sakamoto,K., and Yokoyama,S. (2008). Adding l-lysine derivatives to the genetic code of mammalian cells with engineered pyrrolysyl-tRNA synthetases. *Biochem. Biophys. Res. Commun. 371*, 818-822.

Nairn, N.W., Graddis, T.J., Wang, A., Shanebeck, K., and Grabstein, K. (2007). Site-specific PEGylation of interferon-beta by Cu(I)-catalyzed cycloaddition. *ACS National Meeting 2007, Boston, MA*

Nakamura,Y., Ito,K., and Ehrenberg,M. (2000). Mimicry grasps reality in translation termination. *Cell 101*, 349-352.

Natarajan,A., Xiong,C.Y., Albrecht,H., DeNardo,G.L., and DeNardo,S.J., (2005). Characterization of site-specific ScFv PEGylation for tumor-targeting pharmaceuticals Bioconjug. Chem. *16*, 113-121.

Neumann,H., Peak-Chew,S.Y., and Chin,J.W. (2008). Genetically encoding N(epsilon)-acetyllysine in recombinant proteins. *Nat. Chem. Biol. 4*, 232-234.

Neumann,H., Slusarczyk,A.L., and Chin,J.W. (2010a). De novo generation of mutually orthogonal aminoacyl-tRNA synthetase/tRNA pairs. *J. Am. Chem. Soc. 132*, 2142-2144.

Neumann,H., Wang,K., Davis,L., Garcia-Alai,M., and Chin,J.W. (2010b). Encoding multiple unnatural amino acids via evolution of a quadruplet-decoding ribosome. *Nature 464*, 441-444.

Nguyen,D.P., Lusic,H., Neumann,H., Kapadnis,P.B., Deiters,A., and Chin,J.W. (2009). Genetic encoding and labeling of aliphatic azides and alkynes in recombinant proteins via a pyrrolysyl-tRNA Synthetase/tRNA(CUA) pair and click chemistry. *J. Am. Chem. Soc. 131*, 8720-8721.

Ojida,A., Tsutsumi,H., Kasagi,N., and Hamachi,I. (2005). Suzuki coupling for protein modification. *Tetrahedron Letters 46*, 3301-3305.

Ou,W. *et al.* (2011). Site-specific protein modifications through pyrroline-carboxy-lysine residues. *Proc. Natl. Acad. Sci. U. S. A. 108*, 3301-3305.

Pastrnak,M. and Schultz,P.G. (2001). Phage selection for site-specific incorporation of unnatural amino acids into proteins in vivo. *Bioorg. Med. Chem. 9*, 2373-2379.

Patel,K.G., Ng,P.P., Kuo,C.C., Levy,S., Levy,R., and Swartz,J.R. (2009). Cell-free production of Gaussia princeps luciferase--antibody fragment bioconjugates for ex vivo detection of tumor cells. *Biochem. Biophys. Res. Commun. 390*, 971-976.

Peschke,B., Zundel,M., Bak,S., Clausen,T.R., Blume,N., Pedersen,A., Zaragoza,F., and Madsen,K. (2007). C-Terminally PEGylated hGH-derivatives. *Bioorg. Med. Chem. 15*, 4382-4395.

Plass,T., Milles,S., Koehler,C., Schultz,C., and Lemke,E.A. (2011). Genetically encoded copper-free click chemistry. *Angew. Chem. Int. Ed Engl. 50*, 3878-3881.

Polycarpo,C.R., Herring,S., Berube,A., Wood,J.L., Soll,D., and Ambrogelly,A. (2006). Pyrrolysine analogues as substrates for pyrrolysyl-tRNA synthetase. *FEBS Lett. 580*, 6695-6700.

p-propargyloxyphenylalanine in a cell-free environment for direct protein-protein click conjugation. *Bioconjug. Chem. 21*, 255-263.

Presolski,S.I., Hong,V., Cho,S.H., and Finn,M.G. (2010). Tailored ligand acceleration of the Cu-catalyzed azide-alkyne cycloaddition reaction: practical and mechanistic implications. *J. Am. Chem. Soc. 132*, 14570-14576.

Rackham,O. and Chin,J.W. (2005). A network of orthogonal ribosome x mRNA pairs. *Nat. Chem. Biol. 1*, 159-166.

Rostovtsev,V.V., Green,L.G., Fokin,V.V., and Sharpless,K.B. (2002). A stepwise huisgen cycloaddition process: copper(I)-catalyzed regioselective "ligation" of azides and terminal alkynes. *Angew. Chem. Int. Ed Engl. 41*, 2596-2599.

Sakamoto,K. *et al.* (2002). Site-specific incorporation of an unnatural amino acid into proteins in mammalian cells. *Nucleic Acids Res. 30*, 4692-4699.

Santoro,S.W., Wang,L., Herberich,B., King,D.S., and Schultz,P.G. (2002). An efficient system for the evolution of aminoacyl-tRNA synthetase specificity. *Nat. Biotechnol. 20*, 1044-1048.

Saxon,E. and Bertozzi,C.R. (2000). Cell surface engineering by a modified Staudinger reaction. *Science 287*, 2007-2010.

Sharp,P.M., Bailes,E., Grocock,R.J., Peden,J.F., and Sockett,R.E. (2005). Variation in the strength of selected codon usage bias among bacteria. *Nucleic Acids Res. 33*, 1141-1153.

Sletten,E.M. and Bertozzi,C.R. (2008). A hydrophilic azacyclooctyne for Cu-free click chemistry. *Org. Lett. 10*, 3097-3099.

Sletten,E.M. and Bertozzi,C.R. (2009). Bioorthogonal chemistry: fishing for selectivity in a sea of functionality. *Angew. Chem. Int. Ed Engl. 48*, 6974-6998.

Song,W., Wang,Y., Qu,J., and Lin,Q. (2008). Selective functionalization of a genetically encoded alkene-containing protein via "photoclick chemistry" in bacterial cells. *J. Am. Chem. Soc. 130*, 9654-9655.

Sun,M.M., Beam,K.S., Cerveny,C.G., Hamblett,K.J., Blackmore,R.S., Torgov,M.Y., Handley,F.G., Ihle,N.C., Senter,P.D., and Alley,S.C. (2005). Reduction-alkylation strategies for the modification of specific monoclonal antibody disulfides. *Bioconjug. Chem. 16*, 1282-1290.

Takimoto,J.K., Adams,K.L., Xiang,Z., and Wang,L. (2009). Improving orthogonal tRNA-synthetase recognition for efficient unnatural amino acid incorporation and application in mammalian cells. *Mol. Biosyst. 5*, 931-934.

Takimoto,J.K., Dellas,N., Noel,J.P., and Wang,L. (2011). Stereochemical Basis for Engineered Pyrrolysyl-tRNA Synthetase and the Efficient in Vivo Incorporation of Structurally Divergent Non-native Amino Acids. *ACS Chem. Biol.* Ahead of publication

Tam,A. and Raines,R.T. (2009). Protein engineering with the traceless Staudinger ligation. *Methods Enzymol.* 462, 25-44.

Tam,A., Arnold,U., Soellner,M.B., and Raines,R.T. (2007). Protein prosthesis: 1,5-disubstituted[1,2,3]triazoles as cis-peptide bond surrogates. *J. Am. Chem. Soc.* 129, 12670-12671.

Tang,Y., Ghirlanda,G., Vaidehi,N., Kua,J., Mainz,D.T., Goddard III,W.A., DeGrado,W.F., and Tirrell,D.A. (2001). Stabilization of coiled-coil peptide domains by introduction of trifluoroleucine. *Biochemistry* 40, 2790-2796.

Tanrikulu,I.C., Schmitt,E., Mechulam,Y., Goddard,W.A., III, and Tirrell,D.A. (2009). Discovery of Escherichia coli methionyl-tRNA synthetase mutants for efficient labeling of proteins with azidonorleucine in vivo. *Proc. Natl. Acad. Sci. U. S. A 106*, 15285-15290.

Tornoe,C.W., Christensen,C., and Meldal,M. (2002). Peptidotriazoles on solid phase: [1,2,3]-triazoles by regiospecific copper(I)-catalyzed 1,3-dipolar cycloadditions of terminal alkynes to azides. *J. Org. Chem.* 67, 3057-3064.

Ugwumba,I.N. et al. (2011). Improving a Natural Enzyme Activity through Incorporation of Unnatural Amino Acids. *J. Am. Chem. Soc.* 133, 326-333

van Berkel,S.S., Dirks,A.T., Meeuwissen,S.A., Pingen,D.L., Boerman,O.C., Laverman,P., van Delft,F.L., Cornelissen,J.J., and Rutjes,F.P. (2008). Application of metal-free triazole formation in the synthesis of cyclic RGD-DTPA conjugates. *Chembiochem.* 9, 1805-1815.

Vichier-Guerre,S., Lo-Man,R., Huteau,V., Deriaud,E., Leclerc,C., and Bay,S. (2004). Synthesis and immunological evaluation of an antitumor neoglycopeptide vaccine bearing a novel homoserine Tn antigen. *Bioorg. Med. Chem. Lett.* 14, 3567-3570.

Voloshchuk,N. and Montclare,J.K. (2010). Incorporation of unnatural amino acids for synthetic biology. *Mol. Biosyst.* 6, 65-80.

Wang,A., Winblade,N.N., Johnson,R.S., Tirrell,D.A., and Grabstein,K. (2008). Processing of N-terminal unnatural amino acids in recombinant human interferon-beta in Escherichia coli. *Chembiochem.* 9, 324-330.

Wang,K., Neumann,H., Peak-Chew,S.Y., and Chin,J.W. (2007a). Evolved orthogonal ribosomes enhance the efficiency of synthetic genetic code expansion. *Nat. Biotechnol.* 25, 770-777.

Wang,Q. and Wang,L. (2008). New methods enabling efficient incorporation of unnatural amino acids in yeast. *J. Am. Chem. Soc.* 130, 6066-6067.

Wang,Q., Parrish,A.R., and Wang,L. (2009). Expanding the genetic code for biological studies. *Chem. Biol.* 16, 323-336.

Wang,W., Takimoto,J.K., Louie,G.V., Baiga,T.J., Noel,J.P., Lee,K.F., Slesinger,P.A., and Wang,L. (2007b). Genetically encoding unnatural amino acids for cellular and neuronal studies. *Nat. Neurosci.* 10, 1063-1072.

Wang,Y.S., Youngster,S., Grace,M., Bausch,J., Bordens,R., and Wyss,D.F. (2002). Structural and biological characterization of pegylated recombinant interferon alpha-2b and its therapeutic implications. *Adv. Drug Deliv. Rev.* 54, 547-570.

Yanagisawa,T., Ishii,R., Fukunaga,R., Kobayashi,T., Sakamoto,K., and Yokoyama,S. (2008). Multistep engineering of pyrrolysyl-tRNA synthetase to genetically encode N(epsilon)-(o-azidobenzyloxycarbonyl) lysine for site-specific protein modification. *Chem. Biol.* 15, 1187-1197.

Yoo,T.H., Link,A.J., and Tirrell,D.A. (2007). Evolution of a fluorinated green fluorescent protein. *Proc. Natl. Acad. Sci. U. S. A 104*, 13887-13890.

Young,T.S., Ahmad,I., Yin,J.A., and Schultz,P.G. (2010). An enhanced system for unnatural amino acid mutagenesis in E. coli. *J. Mol. Biol. 395*, 361-374.

Young,T.S., Young,D.D., Ahmad,I., Louis,J.M., Benkovic,S.J., and Schultz,P.G. (2011). Evolution of cyclic peptide protease inhibitors. *Proc. Natl. Acad. Sci. U. S. A.*

Zhang,K., Li,H., Cho,K.M., and Liao,J.C. (2010). Expanding metabolism for total biosynthesis of the nonnatural amino acid L-homoalanine. *Proc. Natl. Acad. Sci. U. S. A 107*, 6234-6239.

Zhang,L., Chen,X., Xue,P., Sun,H.H., Williams,I.D., Sharpless,K.B., Fokin,V.V., and Jia,G. (2005). Ruthenium-catalyzed cycloaddition of alkynes and organic azides. *J. Am. Chem. Soc. 127*, 15998-15999.

Zhang,Z., Alfonta,L., Tian,F., Bursulaya,B., Uryu,S., King,D.S., and Schultz,P.G. (2004). Selective incorporation of 5-hydroxytryptophan into proteins in mammalian cells. *Proc. Natl. Acad. Sci. U. S. A 101*, 8882-8887.

Zhao,K.N., Gu,W., Fang,N.X., Saunders,N.A., and Frazer,I.H. (2005). Gene codon composition determines differentiation-dependent expression of a viral capsid gene in keratinocytes in vitro and in vivo. *Mol. Cell Biol. 25*, 8643-8655.

11

Applications of Bioinformatics and Experimental Methods to Intrinsic Disorder-Based Protein-Protein Interactions

Xiaolin Sun[1], William T. Jones[1] and Vladimir N. Uversky[2,3]
[1]The New Zealand Institute for Plant and Food Research, Palmerston North,
[2]Department of Molecular Medicine, College of Medicine, University of South Florida,
[3]Institute for Biological Instrumentation, Russian Academy of Sciences, Moscow,
[1]New Zealand
[2]USA
[3]Russia

1. Introduction

Proteins are important to the organisms and they control biochemical pathways within the cell. Each protein or group of proteins are responsible for various functions needed to maintain the living cell, including enzymatic catalysis; transporting or storing chemical compounds and energy; hormone regulation of many processes; maintaining structure of tissues and cells; antibody immune response; signal transduction by receptors and signalling proteins etc. The functions of proteins have been long related to their rigid three-dimensional (3D) structures in that a protein's biological function depended on its prior folding into a unique 3D structure. However, as it was discovered in recent years, not all biologically functional proteins fold spontaneously into globular structures. Some protein regions or entire proteins lack stable secondary and/or tertiary structures in solution yet possess crucial biological functions, and these disordered regions or proteins are key to understanding many biological processes such as transcriptional regulation, signalling and causes of diseases. In this chapter, we will focus on some basic concepts of intrinsically disordered proteins (IDPs), the recent progress of structural and functional studies of IDP, and the bioinformatics and experimental methods practically used for investigation of IDPs.

2. Intrinsically disordered proteins and their structural and functional studies

2.1 Concept of intrinsically disordered proteins (IDPs)

For over a hundred years, the structure-function relationships of proteins have been one of the central topics of protein science. In 1894, by observing the specificity of the enzymatic hydrolysis of glucosides, Fisher put forward the lock-and-key theory of protein functionality (Fischer, 1894) according to which protein functions are determined by their specific 3D structures. This lock-and-key model for protein binding was further reinforced by many experiments on the dependence of protein function on their 3D structures (Mirsky and Pauling, 1936; Wu, 1931) and was accepted widely to underlie almost all of the subsequent work and thinking (Phillips, 1986). The flood of protein 3D structures determined by modern structural biology through X-ray diffraction and nuclear magnetic resonance (NMR) spectroscopy has also highlighted a specific 3D structure as the necessary prerequisite for protein function (Berman et al., 2000). For those proteins with rigid structures, the amino acid sequence determines the protein's unique 3D structure and the sequence → structure → function paradigm have become paramount.

However, the generality of the sequence-structure-function paradigm was challenged by the observation (Karush, 1950) that the binding site of bovine serum albumin adopts a large number of configurations with similar energy levels. Upon interacting with a substrate, the best-fitting configuration was selected from the structural ensemble of bovine serum albumin, which was called configurational adaptability. Koshland further proposed this configurational adaptability as induced-fit process (Koshland, 1958), indicating that protein conformational changes are responsible for its function. At the time induced-fit was proposed, it was unclear whether the process of binding induced a new conformation or resulted in selection of the best-fit alternative from an ensemble of structures in equilibrium. The induced-fit theory was further supported by the fact that protein domains are capable of moving upon substrate binding, which were firstly observed in binding of glucose to yeast hexokinase and following numerous protein crystal structures (Bennett and Steitz, 1978; McDonald et al., 1979). One example of latter comes from crystal structures of folylpolyglutamate synthetase (FPGS) with and without substrates, a crucial enzyme to retain folic acids for normal cell growth (Sheng et al., 2000; Sun et al., 1998). An ATP-bound FPGS is activated by binding of folate as the second substrate, triggering a large closing movement of domains of FPGS that enables the enzyme to adopt a form for binding the third substrate, L-glutamate, and effect the addition of a polyglutamate tail to the folate (Sun et al., 2001). Thus, the induced-fit hypothesis not only covers structural accommodations within the binding pocket for binding a diverse set of related but structurally distinct molecules but also includes large domain movements upon ligand binding to gain deeper understanding of structure-function relationship of proteins (Koshland, 1994). The induced-fit mechanisms of protein binding are obviously involved in protein conformational changes with either small rearrangements of few interacting groups or larger domain-domain rearrangements between alternative 3D structures in multi-domain proteins.

In contrast to the views described above that function depends strictly on prior 3D structures, or on structural accommodations within a prior 3D structure, or on domain movements between different structures, it has been discovered that unfolded regions or entire proteins play crucial roles in protein functions (Bychkova et al., 1996; Daughdrill et al., 1997; Riek et al., 1996; Uversky et al., 1996; Uversky et al., 1997). This indicates the diversity of protein folding and functions, and raises intriguing questions about the role of protein disorder in biological processes. Disorder in either the binding protein and/or its partner prevents the presentation of rigid 3D structures that can be bound by other rigid complementary structures or shift between distinct states of rigid structures. These ruled out the lock-and-key and induced-fit mechanisms of binding. Obviously, the predominant sequence → 3D structure → function paradigm is no longer sufficient for many unfolded functional proteins, suggesting that a more comprehensive model is needed. In the late 1990's, studies of functional unfolded proteins emerged as a new research field of protein structure-function relationships. Several research groups simultaneously and independently made the important conclusion, after systematic research, that naturally disordered proteins, characterized by the lack of a well-defined 3D structure under physiological conditions and existing as highly dynamic ensembles of inter-converting structures, are not just rare exceptions but represent a new and very broad class of proteins with vital biological functions (Dunker et al., 2001; Dunker et al., 2000; Tompa, 2003; Uversky, 2002a; Uversky et al., 2000; Wright and Dyson, 1999). This important conclusion was reached from different starting points using different experimental and theoretical approaches, including bioinformatics, NMR spectroscopy, multi-parametric protein folding and misfolding studies and protein structural characterization. Since then, a new protein structure-function paradigm has been established to include the novel functions of disordered proteins. The discovery and characterization of functional unfolded proteins is one of the fastest growing areas of protein science and the literature of studies on these proteins has been increasing continually and has become especially rapid during the past decade (Dunker et al., 2007; Uversky, 2010). These functional unfolded proteins are known by different names, including intrinsically disordered, natively denatured, natively unfolded, intrinsically unstructured, natively disordered, inherently disordered and now are widely known as intrinsically disordered proteins (IDPs) among other names. Those proteins either forming crystals without partners or possessing ordered globular forms without partners in NMR experiments, will be termed here as "structured", "natively folded", or just "ordered".

2.2 The protein quartet model

Historically, the protein structure-function paradigm emphasized the role of a rigid 3D structure as being a necessary prerequisite to protein function. We now know from the cumulated experimental data on intrinsic disorder that the functional protein or protein region can exist as a structural ensemble, at either secondary or tertiary level. Both unfolded regions with a little elements of secondary structure (random coils) and collapsed tertiary structures with poorly packed side chains (molten globule-like) are included in the range of intrinsic disorder. These ideas were presented as the Protein Trinity Paradigm (Dunker et al., 2001; Dunker and Obradovic, 2001). The Protein Trinity Paradigm relates protein function to the three thermodynamic states of protein. In other words, the intracellular functional proteins or regions of such proteins can exist in any one of the three thermodynamic states, namely, ordered forms, molten globules, and random coils. Thus, a particular function is proposed to arise from any one of these states or a transition between any two of the states. According to this view, the native state of a protein is not just the ordered state, but any of the three states. The molten globule was initially discovered as an equilibrium structure observed in studies of protein denaturation in which the partially unfolded intermediates between the ordered state and the random coil were observed as the major species in urea, guanidine and pH titration studies. In these experiments, the protein converted from an ordered native state into a form having some liquid like characteristics with the side chains changing from rigid to non-rigid packing, while its secondary structure remains almost unchanged and the shape remains compact (Ohgushi and Wada, 1983; Ptitsyn and Uversky, 1994). The molten globule has been proposed to be responsible for biological functions. For example, molten globular state was reported to be involved in the process of translocation of proteins across membranes (Bychkova et al., 1988) and the transfer of retinal from its bloodstream carrier to its cell-surface receptor (Bychkova et al., 1992; Bychkova et al., 1998).

By summarizing a large number of experimental results on the conformational behaviours of IDPs, one of us further revealed that the extended disordered region or entire proteins did not possess uniform structural properties as random coils. They were split into two structurally different subclasses, intrinsic random coil-like and intrinsic pre-molten globule-like conformations (Uversky, 2002a). Proteins in pre-molten globule state are more compact than random coil, exhibiting some amount of residual secondary structure, but they are still essentially less dense than molten globule and ordered proteins. It was also noted that molten globule and pre-molten globule (as folding intermediates of globular proteins) might represent different phase states of protein, as they are separated by the first-order phase transition (Uversky, 1997; Uversky and Ptitsyn, 1994, 1996). These observations introduced the native pre-molten globule state of functional unfolded proteins, a new player on the protein functioning field. As ordered, molten globule, pre-molten globule, and random coil conformations possess clearly defined structural differences, the Protein Trinity Paradigm has been extended to the Protein Quartet Model with protein functions arising from four specific conformations (native ordered, molten globules, pre-molten globules and random coils) and transitions between any two of these states (Fig. 1). All of these four structurally defined protein native states can be characterized by using various experimental approaches and applications. For example, pre-molten globules with some residual secondary structure can be characterized by far-UV CD spectra as a typical disordered polypeptide chain with a pronounced minimum in the vicinity of 200 nm. See following sections for more information.

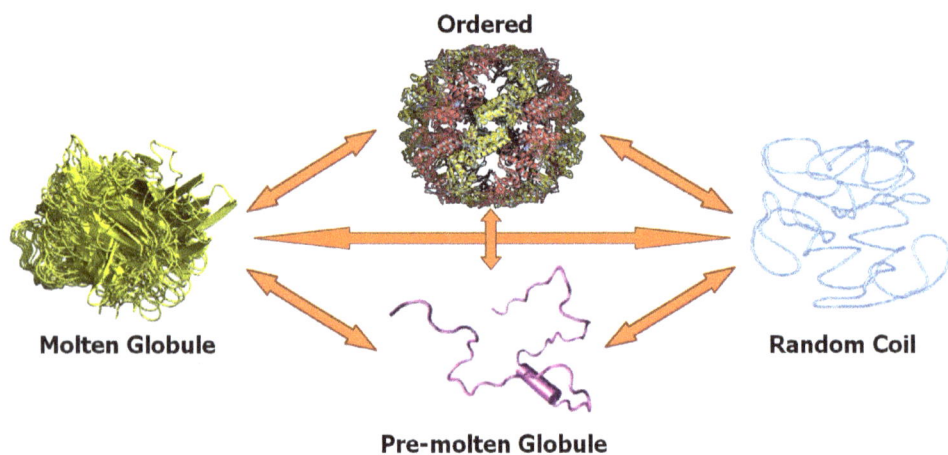

Fig. 1. The protein quartet model

2.3 Functional features of IDPs

Given that the intrinsic disorder represents an important structural class of proteins, and in order to meet the increasing interests in systemizing the crucial functions of IDPs, a database of disordered proteins (DisProt) has been created (Vucetic *et al.*, 2005). DisProt provides structural and functional information on proteins or regions that lack a rigid 3D structure under putatively native conditions. Verified by X-ray diffraction, NMR and CD spectra and several other biophysical techniques, each disordered protein included in the database is given the name, various aliases, accession code, amino acid sequence, location of the disordered region(s) and methods used for structural (disorder) characterization. Most entries list the biological function(s) of each disordered region or protein, if applicable. To date, there are 643 IDPs and 1375 intrinsically disordered regions (IDRs) listed in DisProt.

Among the rapidly increasing number of publications on IDPs, bioinformatics studies have predicted that about 25-30% of eukaryotic proteins are mostly disordered (Oldfield *et al.*, 2005a); more than 50% eukaryotic proteins have long regions of disorder (Dunker *et al.*, 2000; Oldfield *et al.*, 2005a); and more than 70% of signalling proteins and the vast majority of cancer associated proteins have long disordered regions (Iakoucheva *et al.*, 2002); 82-94% of transcription factors from three transcription factor datasets possess extended disordered regions (Liu *et al.*, 2006). IDPs are now widely accepted as ubiquitously existing in all kingdoms of life (Dunker *et al.*, 2000; Ward *et al.*, 2004). Since IDPs and IDRs have amazing conformational variability with a variety of functions, the terms "unfoldome" and "unfoldomics" have been recently introduced (Cortese *et al.*, 2005; Dunker *et al.*, 2007; Midic *et al.*, 2009). Unfoldome is attributed to a large set of functional IDPs and disordered regions within the proteome while Unfoldomics deals with both the identification of the set of proteins or regions in the unfoldome of a given organism and their functions, structures, interactions and evolution (Uversky *et al.*, 2009).

Literature search and comprehensive survey on functions of IDPs characterized by using different experiments suggested that IDPs or IDRs fall into broad functional classes including: (i) entropic chains activities stemming directly from disorder; (ii) molecular recognition via binding to other proteins or to nucleic acids; (iii) scavengers which store and/or neutralize small ligands; (iv) molecular assembly which assemble, stabilize and regulate large multi-protein complexes; (v) various protein modifications (acetylation, hydroxylation, ubiquitination, methylation, phosphorylation) and proteolysis etc. (Dunker *et al.*, 2002a; Dunker *et al.*, 2002b; Dyson and Wright, 2005b; Tompa, 2002; Uversky, 2010). Some illustrative biological functions of IDPs have also been collected in numerous literatures including cell division regulation, transcriptional and translational regulation, molecular chaperoning and cell signalling etc. (Dunker *et al.*, 2005; Dunker *et al.*, 2001; Radivojac *et al.*, 2007; Tompa, 2005; Wright and Dyson, 1999). Obviously, structural flexibility and plasticity originating from the lack of a rigid 3D structure probably represents a major functional advantage for IDPs, reflecting from the fact that IDPs or IDRs can interact with a broad range of binding partners including protein, membranes, nucleic acids, and small molecules (Oldfield *et al.*, 2008; Tompa and Csermely, 2004).

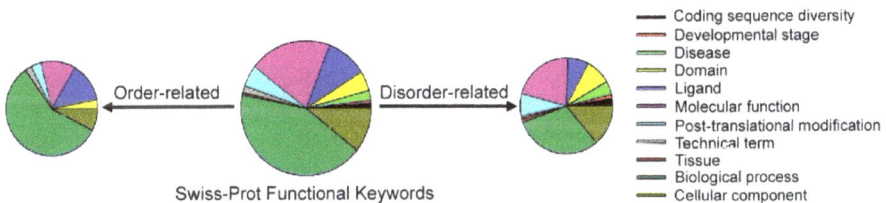

Fig. 2. Functional anthology of IDPs. Modified from Uversky et al. (2009) *Bmc Genomics*, 10 (Suppl 1): S7.

Specifically, a majority of the IDPs or IDRs characterized are involved in regulation, cell-signalling, and control pathways via interactions with multiple partners using high specificity/low affinity strategy, and such disordered regions often become folded upon binding to their partners, confirming that prior 3D structure is not required for molecular recognition (Dunker *et al.*, 2005; Dyson and Wright, 2002; Haynes *et al.*, 2006; Uversky *et al.*, 2005). The crucial role of IDPs in cell-signalling is further confirmed by the finding that eukaryotic proteomes which have developed extensive interaction networks possess higher frequency of IDPs than bacteria and archaea (Dunker *et al.*, 2000; Ward *et al.*, 2004). The close correlation between IDPs and cell-signalling implies that IDPs would play a critical role in protein interaction networks (see following section for details).

Although functional studies of the IDPs characterized experimentally have made significant progress in revealing the functional diversity of IDPs, they are bound to provide only an incomplete view due to limited number of known IDPs. By taking advantage of bioinformatics methods, statistical approaches utilizing a novel data mining tool have been used for a comprehensive study of functional roles of IDPs or IDRs from Swiss-Prot database containing over 200,000 proteins. In the mean time, at least one illustrative and experimentally validated example of functional disorder or order for the vast majority of functional keywords were found to support the bioinformatics analyses. These studies represent a functional anthology of IDPs or IDRs and provide researchers with a novel theoretical tool that could be used to strengthen the understanding of functional diversity of IDPs and protein structure-function relationships (Vucetic *et al.*, 2007; Xie *et al.*, 2007a; Xie *et al.*, 2007b). In these studies, it was shown that many protein functions are associated with long disordered regions; the 262 of 710 Swiss-Prot functional keywords were found to be strongly positively correlated with long IDRs; whereas 302 were strongly negatively correlated with such regions. Those Swiss-Prot functional keywords used in the analyses are associated with various biological processes, cellular components, domains, technical terms, developmental processes, coding sequence diversities, ligands, molecular function, post-translational modifications, tissue and diseases. When all of the functional keywords were classified into eleven functional categories, disorder-associated keywords were found for all eleven categories while order-associated keywords were found only for seven of the eleven categories (Vucetic *et al.*, 2007; Xie *et al.*, 2007b) (Fig. 2). Among them, coding sequence diversities, developmental processes, diseases and tissue functional keywords were exclusively strongly correlated with IDPs or IDRs. Therefore, the functional diversity provided by disordered regions complements functions of proteins with ordered structures, implying that IDPs and IDRs are characterized by a wide functional repertoire.

2.4 Protein-protein interactions involving IDPs

One of the most important functions of IDPs and IDRs is molecular recognition in regulatory and cell-signalling processes via protein-protein interactions (Dunker and Obradovic, 2001; Wright and Dyson, 1999). Protein - protein interactions are organized into complex networks which are central to many processes in regard to the physiology and function of cells. Studies of the protein interaction map proposed that most proteins interact with just a few partners and a small number of proteins interact with many partners. Such a small number of proteins, called hubs, represent a few highly connected nodes in the protein–protein interaction networks (Rual *et al.*, 2005; Stelzl *et al.*, 2005). Hubs can interact with multiple partners to connect various biological molecules in the network either simultaneously (party hubs) or at different times and locations (date hubs). It has been suggested that date hubs organize the proteome, connecting biological processes to each other, whereas party hubs act inside functional modules, forming scaffolds for various molecular machines or coordinated processes (Han *et al.*, 2004). With their ability to interact with multiple partners, hubs play a central role in various cellular biological processes by defining the properties of the protein interaction network (Barabasi and Oltvai, 2004). The high level of hub connectivity should be reflected in protein structures that render hub proteins the ability to carry out highly specific interactions with multiple, structurally diverse partners. Statistical investigation on protein interaction databases and bioinformatics studies have revealed that intrinsic disorder is a common feature of hub proteins from four eukaryotic interactomes and disordered domains confer hubs with the ability to interact with multiple structurally diverse partners in interaction networks (Dunker *et al.*, 2005; Haynes *et al.*, 2006; Patil and Nakamura, 2006). IDRs provide hubs the required binding promiscuity to interact with a large number of small molecules, proteins or nucleic acids. Alternatively, structured hubs bind to disordered regions in their many interaction partners (Oldfield *et al.*, 2008).

The abilities to interact with multiple partners (binding promiscuity) and to carry out binding-induced folding to accommodate diverse binding sites of different partners (binding plasticity) make IDPs and IDRs central in signalling and functional regulation of the cells (Uversky *et al.*, 2005). The p53 protein, regulating more than 150 genes and binding to over 100 partners (Zhao *et al.*, 2000), represents a typical example showing that intrinsic disorder is critical for function through binding promiscuity and binding plasticity (Oldfield *et al.*, 2008). Such protein interactions involving p53 known as one-to-many binding mode are illustrated in Fig. 3, in which the interactions with ten partners are mediated by protein regions experimentally confirmed as IDRs (Uversky *et al.*, 2009). Fig. 3 also indicates that protein binding sites can be predicted, using disorder predictor PONDR®VL-XT, to correspond to some short rigid regions (downward spikes) within predicted long regions of disorder. Actually, short rigid segments within long disordered regions are subject to bioinformatics screening to be the potential protein binding sites known as Molecular Recognition Features (MoRFs).

In addition to promote binding diversity by interacting with numerous partners, molecular recognition involving IDPs provides other important functional advantages over globular proteins with 3D structure for signalling and regulation: disordered regions can bind their partners with high specificity and low affinity. To permit specific recognition, disordered regions usually undergo binding-induced folding during protein interactions (Dyson and Wright, 2002, 2005b), involving a disorder-to-order transition in which IDPs or IDRs adopt a highly structured conformation upon binding to their biological partners. This binding-induced folding can occur for the whole IDPs, or large or short IDRs. It is known that a large decrease in conformational entropy due to folding of disordered regions in the disorder-to-order transition can uncouple specificity from binding strength (Dunker *et al.*, 2001; Schulz, 1979). With such a high specificity/low affinity, the regulatory interaction between an IDP and its partner is both highly specific and easily dispersed – activating and terminating a signal are equally important (Dunker *et al.*, 2002a). An IDP has been suggested to contain a "conformational preference" for the structure it will take upon binding (Fuxreiter *et al.*, 2004). This preferred conformation could be α-helix (α-MoRFs) (Cheng *et al.*, 2007), β-strand (β-MoRFs) or an irregular structure (ι-MoRFs) (Mohan *et al.*, 2006; Vacic *et al.*, 2007a). Fig. 3 shows these different conformational preferences in that a single intrinsic disordered region of p53 (residues 374 - 388 in the C-terminal regulatory domain) forms all three major secondary structure types in the bound state: α-helix when associating with S100ββ, a β-sheet with sirtuin and different irregular structures with CBP and cyclin A. The set of residues involved in these interactions exhibit a high extent of overlap along the sequence. However, p53 utilizes different residues for the interactions with four different binding partners, suggesting that the same intrinsic disordered region sequence is induced by the different partners in entirely different ways (Oldfield *et al.*, 2008).

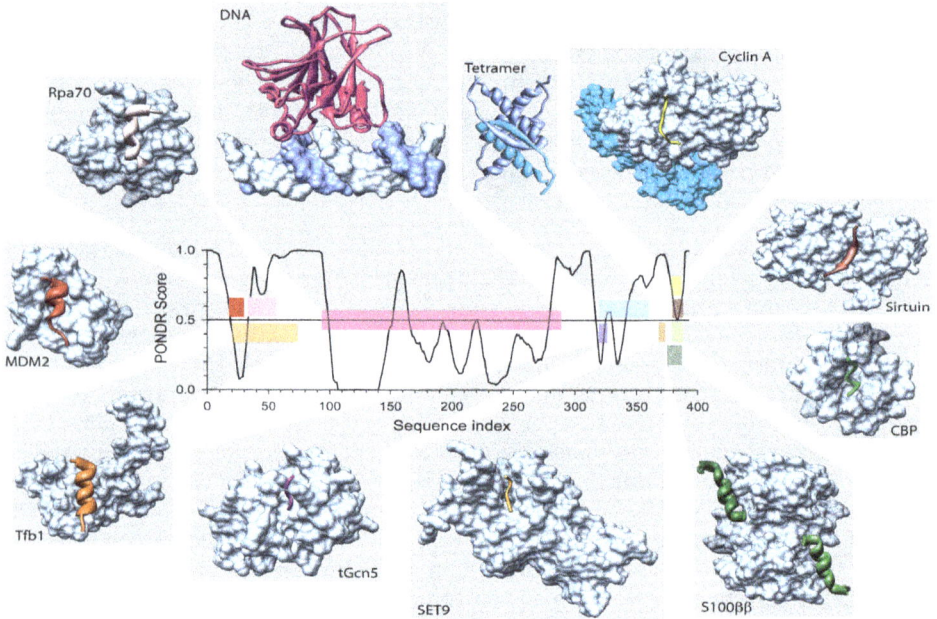

Fig. 3. The protein interactions involving p53. PONDR scores of intrinsic disorder was predicted by PONDR®VL-XT predictor. Residues with scores above 0.5 (threshold) are disordered and those below 0.5 are ordered. The N-domain (residues 1-100) and C-domain (residues 290-390) were predicted as disordered regions while the central DNA binding domain was ordered. The ten binding sites in both N- and C-domains of p53 are at or near the downward spikes in the plot of disorder scores. The complex structures containing various p53 binding regions are displayed around the predicted disorder pattern. In complexes, the structures of p53 segments bound to their partners are shown in different colours. And the same colours are used for the bars in the plot of disorder scores to indicate the positions of the segments in the sequence of p53. The Protein Data Bank IDs and partner names for complex structures are as follows: (1tsr DNA), (1q2d tGcn5), (3sak p53 (tet dom)), (1xqh set9), (1h26 cyclinA), (1ma3 sirtuin), (1jsp CBP bromo domain), (1dt7 s100ββ), (2gs0 Tfb1), (1ycr MDM2), and (2b3g rpa70). Reproduced from Uversky et al. (2009) *Bmc Genomics*, 10 (Suppl 1): S7.

3. Bioinformatics methods for predicting structures of IDPs

Bioinformatics has contributed greatly to the studies of IDPs. Driven by a rapidly increasing number of experimentally verified IDPs in late 1990's, the bioinformatics research of IDPs have promoted the correlation between protein sequence analyses and characterization of intrinsic disorder, and made it possible to investigate the intrinsic disorder nature of proteins from large databases such as interactomes, genomes and Swiss-Prot database. Bioinformatics analyses of IDPs have provided a conceptual framework for experimental studies of molecular function and protein-protein interactions (Sun *et al.*, 2011). Below we will focus on some basic sequence analysis tools for prediction of IDPs or IDRs.

3.1 Amino acids compositional profile of IDPs

The propensity for disorder is encoded in the peculiarities of protein amino acid sequences. By comparing the compositions of the disordered protein datasets with each other and with ordered protein datasets, it was found that IDPs are generally enriched in polar and charged residues and are depleted of hydrophobic residues except for proline (Dunker *et al.*, 2001; Uversky *et al.*, 2000). The relative fractional differences in composition for each amino acid residue between the studied set and a set of ordered proteins are calculated as (C_x-C_{order})/C_{order}, where C_x is the percentage of a given amino acid in the studied set, and C_{order} is the corresponding percentage in a set of ordered proteins. The compositional profile can be visualized by plotting this relative fractional difference in composition against each of twenty amino acids of protein (Dunker *et al.*, 2001; Vacic *et al.*, 2007b). Thus, in the studied set, negative peaks correspond to the amino acids which are depleted in comparison to the set of ordered proteins, and positive peaks indicate the amino acids which are enriched. Most IDPs are substantially depleted in amino acids W, C, F, I, Y, V, L, H, T and N (order-promoting residues) and enriched in amino acids K, E, P, S, Q, R, D and M (disorder-promoting residues). Amino acids A and G are neutral in regards to order and disorder (Radivojac *et al.*, 2007). For the order-promoting residues, the hydrophobic (I, L, and V) and aromatic amino acid residues (W, Y and F) normally form the hydrophobic core of an ordered globular protein. The order-promoting cysteine is known to have a significant contribution to the protein conformational stability via disulfide bond formation or being involved in coordination of different prosthetic groups. On the other hand, disorder-promoting residues (R, Q, S, E, D and K) are polar and charged, i.e. their abundance defines a large net charge of an IDP at physiological pH. Although disorder-promoting proline is hydrophobic, it is well known as a structure terminator (Romero *et al.*, 2001). These compositional biases of IDPs are characterized as low overall hydrophobicity and high net charge (Uversky *et al.*, 2000), and widely used as one of the criteria for IDPs prediction.

The DELLA proteins (DELLAs), a plant-specific protein family, function as repressors of gibberellin (GA)-responsive plant growth and are the key regulatory targets in the GA signalling pathway. The N-domains of DELLAs have been experimentally verified to be intrinsically disordered, and play an important role in molecular recognition (Sun *et al.*, 2008; Sun *et al.*, 2010). Similar to disordered proteins from the DisProt database (Sickmeier *et al.*, 2007), the N-domains of DELLAs showed an overall lack of order-promoting residues and enrichment in disorder-promoting residues, in particular S, M and D, a characteristic of IDPs (Fig. 4). A special feature of the N-domains of DELLAs was a depletion of K, Q and R residues, indicating that DELLAs are a special group of IDPs lacking these three disorder-promoting residues. This analysis can also be performed using a web Composition Profiler tool (http://www.cprofiler.org/) which automates composition profiling with graphical output (Vacic *et al.*, 2007b). There are four different background datasets available, including both intrinsic disordered and ordered datasets, as a comparison to the query sequences.

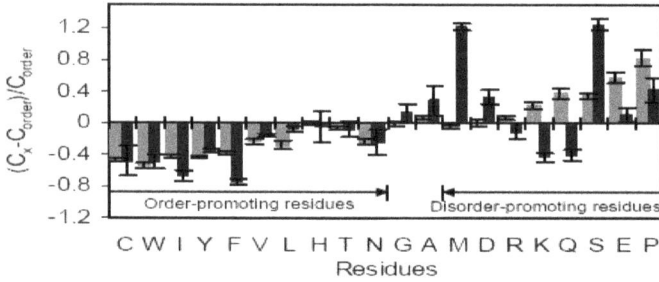

Fig. 4. The compositional profile of the N-domains of eight DELLA proteins (black bars) and disordered proteins of Dis-Pro database (grey bars) in comparison to the ordered globular proteins from the protein data bank. The eight DELLA proteins are from *Arabidopsis* (AtGAI, AtRGA, AtRGL1, AtRGL2 and AtRGL3), rice (SLR1), barley (SLN1) and wheat (RHT1).

3.2 Low sequence complexity of IDPs

The proteins with long disordered regions exhibit a close relationship with low sequence complexity. An investigation of Swiss-Prot database for both low sequence complexity and long disorder regions showed that nearly all the identified low-complexity segments are also predicted as disordered (Dunker *et al.*, 2001). In addition, IDPs all exhibit lower sequence complexity compared to, but partly overlapping with, the distribution of the sequence complexity for ordered proteins (Romero *et al.*, 2001). It was further revealed that IDPs or IDRs and low complexity sequences have similar compositional bias – more disorder-promoting residues (R, K, E, P, and S) and less order-promoting residues (C, W, Y, I, and V). Therefore the low sequence complexity is frequently accompanied by disorder, though not exclusively since low sequence complexity can sometimes occur in structurally ordered proteins (Romero *et al.*, 2001). Overall, simultaneous use of sequence complexity analysis and other disorder predictions will provide a better view of protein disorder.

As an example, plant-specific GRAS proteins, including DELLA subfamily, play critical roles in plant development and various signalling processes (Sun *et al.*, 2011; Sun *et al.*, 2010). One common feature of GRAS proteins is that all of the N-domains contain homopolymeric stretches of certain amino acid residues such as S, T, P, Q, G, D or A. Most of these amino acids are disorder-promoting residues and observed in low sequence complexity segments. By using an iterative algorithm for the complexity analysis of sequence (CAST) (Promponas et al., 2000), the segments with low complexity in all GRAS proteins are mostly located within the N-domains which have previously been proven both experimentally and theoretically to be intrinsically disordered (Sun et al., 2011; Sun et al., 2010).

3.3 Charge-hydrophobicity (CH) and cumulative distribution function (CDF) plots

IDPs, as shown in the compositional profile, are characterized to have low overall hydrophobicity and high net charge. The combination of low mean hydrophobicity and high net charge may represent a prerequisite under physiological conditions for lack of folding in some kinds of IDPs. Statistical analysis of both intrinsically ordered and disordered protein datasets resulted in a plot of the net charge of a protein against its mean hydropathy (CH-plot), showing that ordered and disordered proteins tend to occupy two different areas within the charge-hydrophobicity phase space, separated by a linear boundary line (Uversky et al., 2000): $<R> = 2.785 <H> - 1.151$, where the mean net charge $<R>$ of the protein is calculated as the absolute value of the difference between the numbers of positively charged and negatively charged residues divided by the total number of amino acids, the mean hydrophobicity $<H>$ is defined as the sum of the normalized hydrophobicity (Kyte and Doolittle approximation with a window size of 5 and normalization on the scale from 0 to 1) of all residues divided by the total number of residues minus 4. Figure 5A represents the original charge-hydrophobicity phase space; an IDP with a given mean net charge will most likely locate above the green boundary line. Further statistics with a wider range of IDPs has shown that the mean net charge and mean hydrophobicity of IDPs can be scattered over the charge-hydrophobicity phase space, and sometimes cross into the area of ordered proteins (Oldfield et al., 2005a). Therefore, an added boundary margin allowed the accuracy of the estimation to reach to 95%. As an example of CH-plot analysis, four of the N-domains of eight DELLAs fit into the disordered area but all of them are located within a boundary margin (Fig. 5B) (Sun et al., 2010).

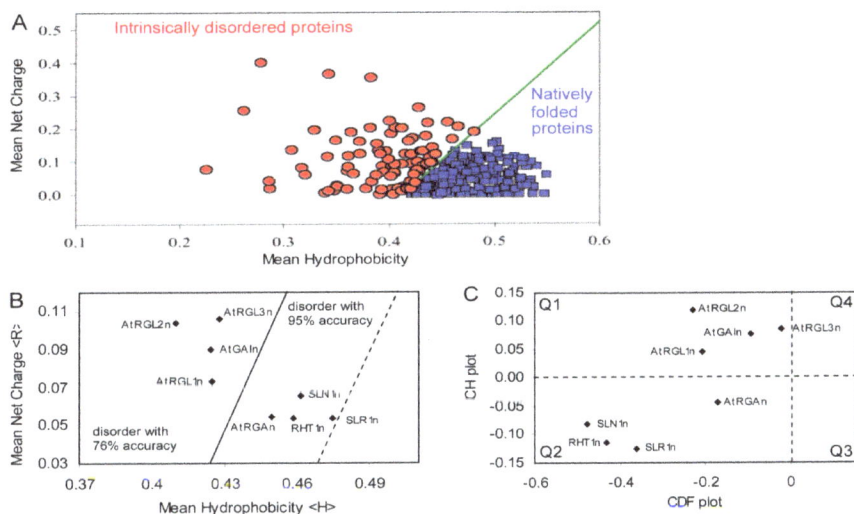

Fig. 5. (A) Mean net charge versus mean hydrophobicity plot for the set of 275 folded (blue squares) and 91 natively unfolded proteins (red circles). (B) Mean net charge versus mean hydrophobicity plot of the N-domains of eight DELLAs. A boundary margin of +0.045 (dotted) extends the disorder estimation accuracy to 95%. (C) Combined CDF-CH plot of the N-domains of eight DELLAs. Modified from Sun et al. (2010) J. Biol. Chem., 285, 11557-11571.

A combination of low hydrophobicity and high net charge as a prerequisite for IDPs can be explained from a physical viewpoint in that high net charge leads to charge-charge repulsion, and low hydrophobicity indicates less driving force for protein compaction. However, the CH-plot is a linear disorder classifier that takes into account only two parameters of the particular sequence - charge and hydrophobicity - and is predisposed to discriminate proteins with substantial amounts of extended disorder (random coils and pre-molten globules) from proteins with globular conformations (molten globule-like and rigid well-structured proteins) (Oldfield *et al.*, 2005a; Uversky *et al.*, 2000). Another binary disorder classifier, cumulative distribution function (CDF) analysis, discriminates all disordered conformations, including molten globules, from rigid well-folded proteins (Oldfield *et al.*, 2005a; Xue *et al.*, 2009). Therefore, simultaneous CDF-CH plot analysis gives a more accurate prediction for a wider range of sequences. CDF is a cumulated histogram of disordered residues at various disordered scores that are obtained from the disorder predictor PONDR-VSL2 (Peng *et al.*, 2006) (see next section for details). The cumulated histogram for structured proteins increases faster in the range of smaller disordered scores and then flattens at larger disordered scores. The cumulated histogram for disordered proteins increases slightly in the range of lower disordered scores but significantly at higher disordered scores. So, there is also a boundary line identified in the CDF plot. The distances to the boundary lines in both CH and CDF plots for a specific protein are further used to build up the CH-CDF plot. In the resulting CH-CDF plot, coordinates of each spot are calculated as a distance of the corresponding protein in the CH-plot from the boundary (as Y-coordinate in the CH-CDF plot) and an average distance of the respective cumulative distribution function (CDF) curve from the CDF boundary (as X-coordinate in the CH-CDF plot). Positive and negative Y values in the CH-CDF plot correspond to proteins predicted within CH-plot analysis to be intrinsically disordered or ordered, respectively. In contrast, positive and negative X values are attributed to proteins predicted within CDF analysis to be ordered or intrinsically disordered, respectively. Thus, the resultant quadrants of CDF-CH phase space correspond to the following expectations: Q1, proteins predicted to be disordered by both methods; Q2, proteins predicted to be disordered by CDFs but compact by CH-plots (i.e., putative molten globules); Q3, ordered proteins; Q4, proteins predicted to be disordered by CH-plots, but ordered by CDFs. All of the N-domains of eight DELLAs are, located in Q1 and Q2 quadrants (Fig. 5C), intrinsically disordered with different levels of compactness, which is consistent with physical and biological evidence (Sun *et al.*, 2010).

3.4 Prediction of intrinsic disorder and potential binding sites

Disorder prediction has been one of important approaches in IDPs and IDRs research. It is a powerful tool for study of IDPs, especially considering time and cost, compared to the experimental methods. Furthermore, it can be easily used to investigate large datasets such as proteome, interactome etc. So far, more than fifty predictors of disorder have been developed to evaluate intrinsic disorder of a given sequence on a per-residue basis (He *et al.*, 2009). These predictors utilize biased amino acid compositions of IDRs, various datasets derived from experiments and different computing techniques, and many of them are accessible on public servers. A partial list of various predictors can be found on DisProt website (http://www.disprot.org/predictors.php). In this chapter, we only focus on a series of **P**redictors **O**f **N**atural **D**isordered **R**egions (PONDRs).

A basic strategy for developing these predictors includes constructing training datasets of both ordered and disordered segments, selecting sequence attributes of order and disorder and applying a neural networks model in training. Predictor PONDR®VL-XT applies three different neural networks, one trained on **V**ariously characterized **L**ong disordered regions for internal region of the sequence and two trained on **X**-ray characterized **T**erminal disordered regions for N- and C-terminal regions ($\geq$ 5 amino acids) (Romero *et al.*, 2001). The PONDR®VL-XT has outputs from the first to the last residue in a sequence, and furthermore it provides the basis for CDF plot and potential binding site (MoRFs) predictions. PONDR®

VSL2 combines neural networks for both short (≤ 30 residues) and long disordered regions with each neural network trained by the dataset of that specific length. This predictor gives relatively higher accuracy of prediction in the PONDR series (Peng *et al.*, 2006). PONDR®VL3 applies ten neural networks and selects the final prediction by simple majority voting. The input features of these predictors are various sequence profiles. This predictor has higher accuracy in predicting longer disordered regions. All of these PONDR predictors have relatively high accuracy (> 80%), and the accuracy has been further improved by a meta-predictor PONDR-FIT that combines PONDR-VLXT, PONDR-VSL2, PONDR-VL3 and other three different predictors and is so far one of the predictors with higher accuracy of disorder prediction (Xue *et al.*, 2010).

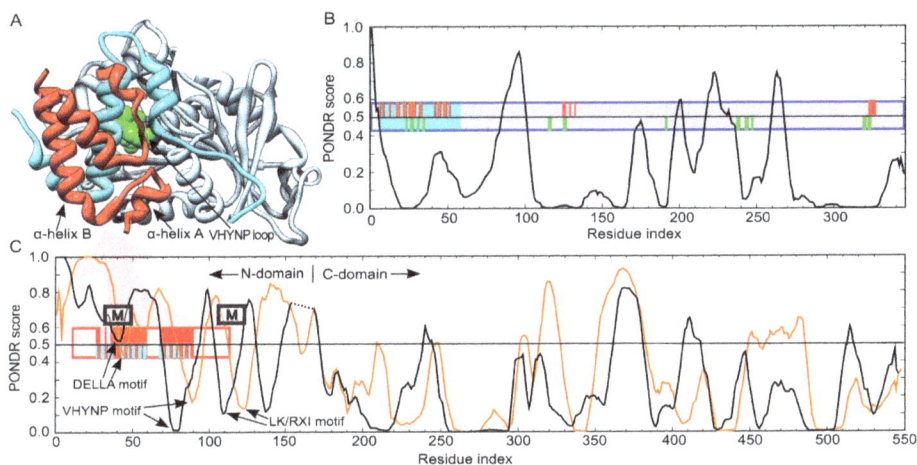

Fig. 6. PONDR disorder predictions for AtGID1a, AtGAI and AtRGL2. **(A)** X-ray crystal structure of AtGAIn-AtGID1a/GA₃ complex (PDB 2ZSH) with ribbon representation of AtGAIn (red), the N-terminal binding pocket of AtGID1a (cyan), GA receptor domain of AtGID1a (light blue) and GA₃ (green van der Waals surface) **(B)** Disorder prediction for AtGID1a **(C)** Disorder predictions for AtGAI (black line, its C-domain is shifted right to align with that of AtRGL2) and AtRGL2 (orange line). Disorder predictions were made with PONDR®VL-XT with a threshold 0.5 (≥ 0.5 for disorder and < 0.5 for order). Boxes indicate the fragments of AtGAIn and AtGID1a crystallized in the complex, with filled positions indicating region of defined density in the crystal structure. Ticks indicate residues that interact with the portion of the complex with the corresponding colours. Approximate positions of α-MoRFs predicted for most DELLA proteins are indicated by boxes labelled 'M'. Reproduced from Sun et al. (2010) *J. Biol. Chem.*, 285, 11557-11571.

As an example, PONDR®VL-XT was used to predict the disorders for two DELLAs (AtGAI and AtRGL2) together with the GA-receptor (AtGID1a) (Fig. 6). The PONDR scores of AtGID1a indicated a folded structure (Fig. 6B). The PONDR scores of AtGAI and AtRGL2, similar to each other, indicated that the C-domains of both DELLAs are dominated by ordered structures with most residues having PONDR score < 0.5, the threshold for order/disorder (Fig. 6C). In contrast, the N-domains of both DELLAs, AtGAIn and AtRGL2n, are clearly intrinsically disordered except for some short rigid segments corresponding to the DELLA, VHYNP and LK/RXI motifs (Fig. 6C, indicated by arrows).

These disorder predictions of the N-domains AtGAIn and AtRGL2n are consistent with the results from compositional profile (Fig. 4) and CH-CDF plot analysis (Fig. 5).

As discussed in Section 2.4 (protein-protein interactions involving IDPs), such short ordered segments within long disordered regions detected as downward spikes in disorder prediction using PONDR®VL-XT (Fig. 6C) are termed **M**olecular **R**ecognition **F**eatures (MoRFs). They are potential binding sites for protein interactions, and responsible for molecular recognition via a disorder-to-order transition upon binding to their interacting partners. By utilizing this specific sequence pattern of IDPs, a unique bioinformatics tool dedicated to the identification of MoRFs or potential protein-protein interaction sites in IDPs has been developed. α-MoRFs-I and its updated form α-MoRFs-II, the identifiers of α-helix forming MoRFs, are focused on short binding regions within long regions of disorder that are likely to form helical structure upon binding (Cheng *et al.*, 2007; Oldfield *et al.*, 2005 b). The α-MoRFs predictor defines a heuristic for binding-associated downward spikes and removes false positive predictions. It assigns relatively short segments (20 residues) that gain functionality through a disorder-to-order transition induced upon binding to a partner. The identifiers of β-sheet or irregular structure forming MoRFs are under development.

As an example, α-MoRFs have been predicted at or near the DELLA motif and the LK/RXI motif in the N-domains of DELLAs (Fig. 6C), suggesting that the DELLA motif is a binding site that undergoes disorder-to-α-helix transition upon binding to GID1 receptors. The formation of this α-helix has been confirmed and shown as α-helix A and α-helix B in the crystal structure of the AtGID1a-GA$_3$-AtGAIn ternary complex (Fig. 6A) (Murase *et al.*, 2008). Although the VHYNP motif appears as a large downward spike in the PONDR score pattern (Fig. 6C), it was not identified in our α-MoRFs prediction for any of the DELLAs. It is clearly shown in Fig. 6A that the VHYNP motif binds to AtGID1a, forming an irregular VHYNPSD loop. Therefore, it is ι-MoRFs rather than α-MoRFs involved in the binding-induced folding. The binding-induced folding of the DELLA and VHYNP motifs have also been supported by other biochemical evidence (Sun *et al.*, 2010). The LK/RXI motif was also identified in our α-MoRFs prediction. It is possibly a potential binding site of DELLAs for an unknown component in DELLA signalling pathway as this motif was not involved in interactions with GID1 receptors.

4. Experimental methods for investigation of IDPs

Having specific amino acid sequences, IDPs possess a number of distinctive structural characteristics that can be utilized for their identification. This includes but not limited to, sensitivity to proteolysis, disorder characteristics of CD and NMR spectroscopy, small-angle X-ray scattering, hydrodynamic measurement, dynamic light scattering, deuterium-hydrogen exchange mass spectrometry, Raman and infrared spectroscopy, monoclonal antibody based immunoassays and so on (Uversky, 2002a). While every method emphasizes different structural features of IDPs, simultaneous application of several techniques mentioned above will provide unambiguous evidence for the presence of varying degrees of unfoldedness in a given protein. Here, we will focus on some of frequently used techniques.

4.1 Hydrodynamic dimensions of IDPs

Hydrodynamic dimension is the most definite characteristic of the conformational state of a protein. According to Uversky (Uversky, 1993, 2002b), a protein molecule can exist in natively folded (NF), molten globule (MG), pre-molten globule (PMG), native coil (Coil) and completely unfolded states (U_{urea}) and is characterized by the following relationships of its Stokes radius (R_S) and theoretic molecular weight (MW^{Theo}): natively folded, log R_S^{NF} = 0.369×log (MW^{Theo}) – 0.254; molten globule, log R_S^{MG} = 0.334×log (MW^{Theo}) – 0.053; pre-molten globule, log R_S^{PMG} = 0.403×log (MW^{Theo}) – 0.239; native coil, log R_S^{Coil} = 0.493×log (MW^{Theo}) – 0.551; complete unfolded random coil in urea, log R_S^{Uurea} = 0.521×log (MW^{Theo}) – 0.649. Therefore, experimentally determined apparent molecular weight (MW^{App}) and Stokes radius (Rs^D) can be used to identify and classify IDPs into one of three subclasses: molten globule (MG)-like, pre-molten globule (PMG)-like or random-coil like, depending on their hydrodynamic characteristics.

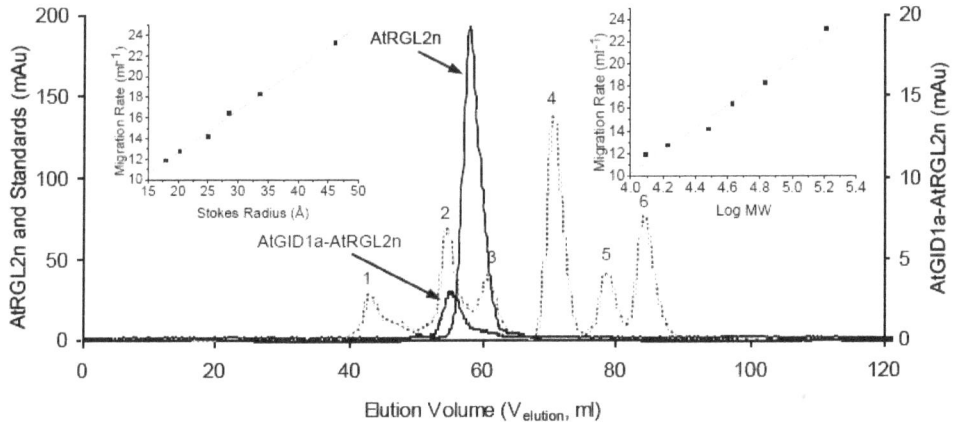

Fig. 7. Size-exclusion chromatography of AtRGL2n and AtGID1a-AtRGL2n complex. Six natively folded globular proteins (1. IgG₁, 158 kDa; 2. BSA, 67 kDa; 3. Ovalbumin, 43 kDa; 4. Carbonic Anhydrase, 30 kDa; 5. Myoglobin, 17 kDa; 6. Cytochrome c, 12.3 kDa corresponding to numbered peaks of hatched line) were used as standards to calibrate the Superdex 75-16/60 column. The resultant migration rate ($1000/V_{elution}$) plotted against Stokes radius (left inset) and molecular weight (right inset) were used to determine Stokes radius (Rs^D) and apparent molecular weight (MW^{App}) of DELLAs and the complex. As examples, the peaks of AtRGL2n and AtGID1a-AtRGL2n complex are shown in solid and dotted lines, respectively. Adopted from Sun et al. (2010) *J. Biol. Chem.*, 285, 11557-11571.

As an example, we investigated hydrodynamic dimensions of the N-domains of DELLAs using size-exclusion chromatography (Fig. 7) in which six natively folded globular proteins were used as standards to calibrate the Superdex 75-16/60 column (see insets). AtRGL2n and its ternary complex AtGID1a-GA₃-AtRGL2n were then applied on to the gel filtration column using the same buffer and flow rate. The determined Stokes radius (Rs^D) and apparent molecular weight (MW^{App}) of AtRGL2n and the ternary complex were calculated using migration rate ($1000/V_{elution}$). The MW^{App} of free AtRGL2n is approximately two to three times larger than its theoretical molecular weight (MW^{Theo}), indicating that it has an extended conformation with low compactness. The value of Rs^D reveals that free AtRGL2n belongs to the PMG-like IDPs. In contrast, the MW^{App} and Rs^D values show that the AtGID1a-GA₃-AtRGL2n complex becomes natively folded, supporting the hypothesis that intrinsic disordered AtRGL2n must undergo a binding-induced folding during the complexing of AtGID1a/GA$_3$ and AtRGL2n (Sun *et al.*, 2010).

4.2 CD spectra of IDPs

The far-UV circular dichroism (CD) spectra have been widely used for identification of IDPs due to its quick operation and easy accessibility. Contrary to α-helix and β-sheet, random coil displays specific shape of the far-UV CD spectrum, with a large negative ellipticity in the vicinity of 200 nm, low ellipticity at 190 nm and an ellipticity close to zero in the vicinity of 222 nm (Johnson, 1988; Kelly and Price, 1997). This is a very useful graphical criterion for identification of IDPs. Double wavelength statistics showed the far-UV CD spectra characteristics of IDPs: some random coil-like IDPs having averaged ellipticities at 200 nm (-18900 deg·cm^2·dmol^{-1}) and at 222 nm (-1700 deg·cm^2·dmol^{-1}) while some pre-molten globule-like IDPs having averaged ellipticities at 200 nm (-10700 deg·cm^2·dmol^{-1}) and at 222 nm (-3900 deg·cm^2·dmol^{-1}) (Uversky, 2002a).

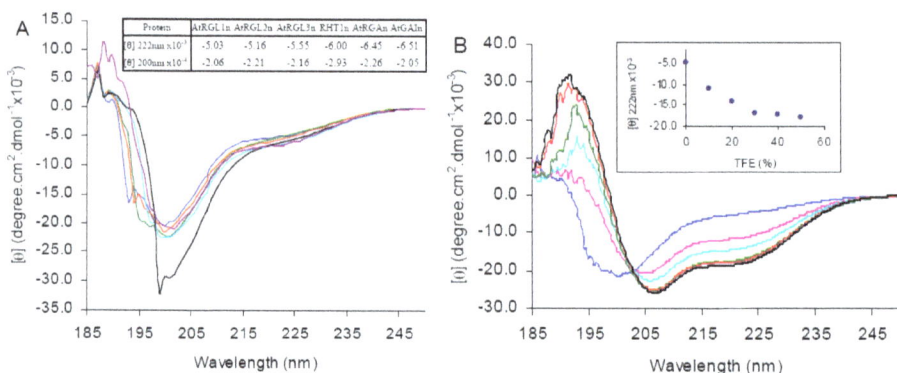

Fig. 8. Far-UV CD spectra of the N-domains of DELLAs **(A)** CD of AtRGL1n (blue), AtRGL2n (green), AtRGL3n (orange), AtRGAn (cyan), RHT1n (black) and AtGAln (purple). The ellipticity [θ] at both 200 nm and 222 nm are shown in inset **(B)** 2,2,2-trifluoroethanol (TFE)-induced folding of AtRGL2n with TFE percentage: 0% (blue), 10% (pink), 20% (cyan), 30% (green), 40% (red) and 50% (black). The ellipticity [θ] at 222 nm versus TFE is shown in inset. Adopted from Sun et al. (2010) *J. Biol. Chem.*, 285, 11557-11571.

As an example of using CD spectra to characterize IDPs, the far-UV CD spectra of the N-domains of six DELLAs (Fig. 8A) display a large negative ellipticity at 200 nm and low ellipticity at 190 nm, characteristic of proteins in a largely disordered conformation. The ellipticity values at 222 nm are close to that of pre-molten globule-like IDPs, on the other hand, the ellipticity values at 200 nm are close to that of random coil-like IDPs (Fig. 8A, inset table). Alternatively, the solvent 2,2,2-Trifluoroethanol (TFE) mimics the hydrophobic environment experienced by protein - protein interactions and has therefore been widely used as a probe for the propensity of IDPs to undergo an induced folding upon target binding (Dyson and Wright, 2002). To test the potential binding-induced folding of the N-domains of DELLAs, far-UV CD spectra of AtRGL2n were recorded in the presence of increasing concentrations of TFE (Fig. 8B). AtRGL2n showed an increased α-helicity upon the addition of TFE, as indicated by the characteristic peak at 192 nm and double minima at 208 and 222 nm. Most of the disorder-to-α-helix transition takes place in presence of 30% TFE, at which point the α-helical content reaches 51.5% as estimated from the ellipticity at 222 nm. The TFE results alone reveal a potential of AtRGL2n to form α-helices upon binding to its interacting partners (Sun *et al.*, 2010).

4.3 NMR spectra of IDPs

As discussed above, CD spectroscopy is a very useful technique for identification of IDPs. However, this technique aims to detect overall tendency of conformational ensemble of intrinsic disorder and the resultant spectroscopic characteristics reflect local structural propensities averaged over the whole protein molecule. It does not provide conformational information about local residual structures such as those retained in pre-molten globules. In past decade, NMR spectroscopy has rapidly developed into a key technique for studying conformational ensemble and dynamics of proteins in unfolded and partially folded states (Dyson and Wright, 2005a). NMR spectroscopy investigates both local and long-range conformational behaviour at atomic resolution on timescales varying over many orders of magnitude, it has been used together with small-angle X-ray scattering (SAXS) to characterize the conformational ensemble of IDPs (Wells *et al.*, 2008). The residual dipolar couplings (RDCs) of NMR spectra has become a powerful tool to describe quantitatively the level of local structure and transient long-range order in IDPs, see review (Jensen *et al.*, 2009) for details.

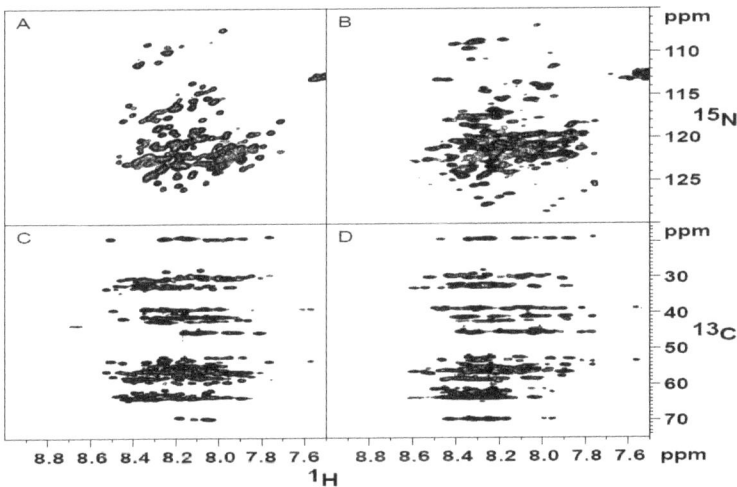

Fig. 9. NMR spectra showing chemical shift dispersion of AtRGL2n and AtRGAn. **(A)** and **(B)** are 1H-15N HSQC of AtRGL2n and AtRGAn, respectively. **(C)** and **(D)** show the 1H-13C plane of a CBCA(CO)NH experiment for AtRGL2n and AtRGAn, respectively. Reproduced from Sun et al. (2010) *J. Biol. Chem.*, 285, 11557-11571.

Here, we just show examples of characteristic NMR spectra of IDPs. The two-dimensional 15N, 1H-HSQC spectra and 1H-13C planes from a CBCA(CO)NH experiment have been collected for two N-domains of DELLAs, AtRGL2n and AtRGAn (Fig. 9). The narrow ranges of chemical shifts in the HSQC spectra for AtRGL2n and AtRGAn are characteristic of unstructured proteins (Fig. 9A and 9B, respectively). The CBCA(CO)NH planes (Fig. 9C and 9D) correlate the amide proton chemical shift with the C_α and C_β shifts of the previous residue, in which horizontal rows of peaks at chemical shifts typical of random coil can be seen. Neither of these rows displays a significant spread in ^{13}C shift values, suggesting a nearly uniform and therefore disordered environment. While the narrow range of ^{13}C shifts seen here is consistent with disorder, the chemical dispersion of AtRGAn (Fig. 9B) does appear to be slightly higher than that of AtRGL2n (Fig. 9A), indicating that AtRGAn may have relatively more local residual structures (Sun et al., 2010).

4.4 Deuterium / hydrogen exchange mass spectra

Deuterium/hydrogen exchange mass spectrometry (DHXMS) provides insights into protein structure and dynamics on a per amino acid basis. Compared to NMR spectra, the DHXMS technique requires much less time for sample preparation and can be routinely applied to larger proteins. The backbone amide hydrogens of proteins reversibly interchange with deuterium in D_2O solvent and the exchange rate of each amide hydrogen in a protein directly and precisely reports solvent accessibility to it, revealing the protein's conformational states on the scale of individual amino acids. Amide hydrogens that are involved in intramolecular H-bonds via secondary structures and/or permanently buried inside the protein are protected from the exchanging. Conversely, exchanging occurs readily at sites that are solvent-exposed and involved no hydrogen bond of secondary structures. Therefore, the exchange rates determined by mass spectrometry following the deuterium/hydrogen exchanging allow direct localization of structured or unfolded regions of the protein. DHXMS has been used in structural genomics to identify the disordered regions within large number of crystallographic targets (Pantazatos et al., 2004), in exploring protein conformational dynamics and the structural aspect of solution-phase proteins (Konermann et al., 2008).

As an example, Fig. 10 shows the hydrogen/deuterium exchange of AtRGL1n determined by MS. The MBP moiety of a well folded structure exhibited inaccessibility to solvent for most of the protein except the two terminal regions. In contrast, AtRGL1n showed instantly high exchange rate for the whole polypeptide chain, implying that free AtRGL1n without interacting partners is totally unfolded (Sheerin et al., 2011).

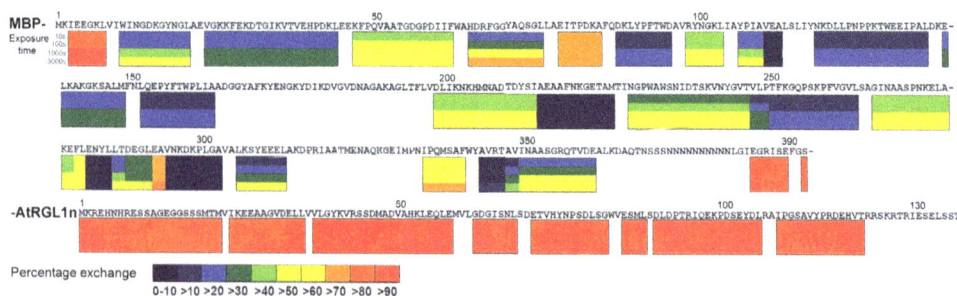

Fig. 10. The solvent accessibility of a maltose-binding protein (MBP) fused AtRGL1n. The exchange rates of backbone amide hydrogens were determined by MS following exposure to D_2O for 10-3000 s intervals. Adopted from Sheerin et al. (2011) Biochem. J., 435(3): 629-639.

4.5 Monoclonal antibody immunoassays for intrinsic-disorder based protein interactions

Monoclonal antibodies (mAb) have long been recognized and used as powerful molecular probes to monitor protein folding and conformational changes (Goldberg, 1991). The conformational specific mAb that recognizes conformation of epitopes (small groups of sequential or distal amino acids) allows a direct insight into the conformation of antigenic proteins at individual amino acids level. This method was utilized in conformational characterization of paired helical filaments (PHF) of tau protein, an IDP in its monomeric form and polymerizes through binding-induced folding into insoluble PHF, causing Alzheimer's disease (AD) and related tauopathies. A special mAb MN423 raised against the PHF core of tau protein recognizes three spatially close amino acid segments which reside on a nearly 90 amino acid-long polypeptide chain including the C-terminus. The disclosure of the spatial proximity of these segments represents constraints for intra-molecular folding of the PHF core, leading to propose the model for folding of tau polypeptide chain in the PHF core (Skrabana et al., 2006). The conformational specific mAb has often been used to monitor conformational changes of targeted proteins. Bax is a pro-apoptotic member of the B-cell lymphoma-2 (Bcl-2) family which are either IDP or contain IRDs that are critical to their function (Rautureau et al., 2010). Bax undergoes a conformational change triggered by a TNF-related apoptosis-inducing ligand (TRAIL), leading to effective induction of mitochondrial apoptosis (Sundararajan et al., 2001). This critical conformational change can be efficiently detected by using a mAb that recognizes only the three-dimensional epitope of conformationally changed Bax and was immobilized on a chip for surface plasmon resonance (SPR) detection (Kim et al., 2005).

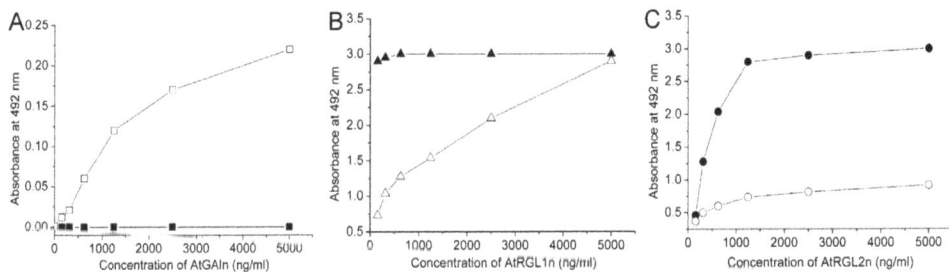

Fig. 11. Interaction of AtGID1 receptors extracted from *Arabidopsis quadruple*-DELLA mutant (*QUAD*) with the recombinant N-domains of DELLAs using mAb AD7 in sandwich immunoassays. All assays were carried out for AtGAIn (□), AtRGL1n (△), AtRGL2n (O) mixed with *QUAD* extract and AtGAIn (■), AtRGL1n (▲), AtRGL2n (●) mixed with extraction buffer only. Adopted from Sun et al. (2010) *J. Biol. Chem.*, 285, 11557-11571.

Furthermore, the conformational specific mAb has advantage over other techniques in monitoring conformational changes of antigenic proteins involved in a real time reaction system with multiple interacting components. This method has been successfully applied to gain an insight into the folding of DELLAs upon binding to the AtGID1, a GA receptor extracted from *Arabidopsis* tissue and binding to both the DELLA and VHYNP motifs of DELLAs (Sun et al., 2010). Both biophysical data and biological functions suggested that the *Arabidopsis* DELLA family may be further divided into two subgroups: AtGAI, AtRGA

(RGA-group) and AtRGL1, AtRGL2 and AtRGL3 (RGL-group). This classification was reinforced by different conformations of unbound *Arabidopsis* DELLAs with regard to the VHYNP motifs. The mAb AD7, conformational specific to the VHYNP motif, does not recognize unbound AtGAIn and AtRGAn but AtRGL1n, AtRGL2n and AtRGL3n (Fig. 11), implying the different conformations of VHYNP motif between these two subgroups in their unbound form. Furthermore, the ELISA assays of the recombinant N-domains (AtGAIn, AtRGL1n and AtRGL2n), AtGID1/GA$_3$ and mAb AD7 showed that the AtGID1 only partially blocks mAb AD7 from binding to both AtRGL1n and AtRGL2n (Fig. 11B, 11C). In contrast, binding of AtGID1/GA$_3$ to AtGAIn renders mAb AD7 binding to AtGID1/GA$_3$-AtGAIn complex (Fig. 11A). This indicates that the VHYNPSD loop in the unbound AtGAIn undergoes conformational changes induced by binding of AtGID1/GA$_3$, resulting in at least the partial conformational epitope recognized by mAb AD7.

5. Conclusion

As one of the fastest growing areas of protein science in the past decade, IDPs or IDRs lack secondary and/or tertiary structures yet possess crucial cellular functions under physiological conditions. Protein interactions involving IDPs or IRDs can trigger binding-induced folding of IDPs or IDRs via a disorder-to-order transition. Such important characteristics enable IDPs or IDRs to play critical roles in molecular recognition. The intrinsic disorder based protein interactions will be a key factor in elucidating mechanisms of biological processes and regulations such as disease, various signal transductions, plant growth and development, transcriptional regulation and so on.

6. References

Barabasi, A L and Oltvai, Z N (2004) Network biology: Understanding the cell's functional organization. *Nature Reviews Genetics*, 5, 101-U115. doi: 10.1038/nrg1272.

Bennett, W S and Steitz, T A (1978) Glucose-induced conformational change in yeast hexokinase. *Proc. Natl. Acad. Sci. U. S. A.*, 75, 4848-4852.

Berman, H M, Westbrook, J, Feng, Z, Gilliland, G, Bhat, T N, Weissig, H, Shindyalov, I N and Bourne, P E (2000) The Protein Data Bank. *Nucleic Acids Research*, 28, 235-242.

Bychkova, V E, Berni, R, Rossi, G L, Kutyshenko, V P and Ptitsyn, O B (1992) Retinol-binding protein is in the molten globule state at low pH. *Biochemistry*, 31, 7566-7571.

Bychkova, V E, Dujsekina, A E, Fantuzzi, A, Ptitsyn, O B and Rossi, G L (1998) Release of retinol and denaturation of its plasma carrier, retinol-binding protein. *Folding & Design*, 3, 285-291.

Bychkova, V E, Dujsekina, A E, Klenin, S I, Tiktopulo, E I, Uversky, V N and Ptitsyn, O B (1996) Molten globule-like state of cytochrome c under conditions simulating those near the membrane surface. *Biochemistry*, 35, 6058-6063.

Bychkova, V E, Pain, R H and Ptitsyn, O B (1988) The molten globule state is involved in the translocation of proteins across membranes. *Febs Letters*, 238, 231-234.

Cheng, Y G, Oldfield, C J, Meng, J W, Romero, P, Uversky, V N and Dunker, A K (2007) Mining alpha-helix-forming molecular recognition features with cross species sequence alignments. *Biochemistry*, 46, 13468-13477. doi: 10.1021/bi7012273.

Cortese, M S, Baird, J P, Uversky, V N and Dunker, A K (2005) Uncovering the unfoldome: Enriching cell extracts for unstructured proteins by acid treatment. *Journal of Proteome Research*, 4, 1610-1618. doi: 10.1021/pr050119c.

Daughdrill, G W, Chadsey, M S, Karlinsey, J E, Hughes, K T and Dahlquist, F W (1997) The C-terminal half of the anti-sigma factor, FlgM, becomes structured when bound to its target, sigma(28). *Nature Structural Biology*, 4, 285-291.

Dunker, A K, Brown, C J, Lawson, J D, Iakoucheva, L M and Obradovic, Z (2002a) Intrinsic disorder and protein function. *Biochemistry*, 41, 6573-6582. doi: 10.1021/bi012159+.

Dunker, A K, Brown, C J and Obradovic, Z (2002b) Identification and functions of usefully disordered proteins. *Unfolded Proteins*, 62, 25-49.

Dunker, A K, Cortese, M S, Romero, P, Iakoucheva, L M and Uversky, V N (2005) Flexible nets - The roles of intrinsic disorder in protein interaction networks. *Febs Journal*, 272, 5129-5148. doi: 10.1111/j.1742-4658.2005.04948.x.

Dunker, A K, Lawson, J D, Brown, C J, Williams, R M, Romero, P, Oh, J S, Oldfield, C J, Campen, A M, Ratliff, C R, Hipps, K W, Ausio, J, Nissen, M S, Reeves, R, Kang, C H, Kissinger, C R, Bailey, R W, Griswold, M D, Chiu, M, Garner, E C and Obradovic, Z (2001) Intrinsically disordered protein. *Journal of Molecular Graphics & Modelling*, 19, 26-59.

Dunker, A K and Obradovic, Z (2001) The protein trinity - linking function and disorder. *Nature Biotechnology*, 19, 805-806.

Dunker, A K, Obradovic, Z, Romero, P, Garner, E C and Brown, C J (2000) Intrinsic protein disorder in complete genomes. *Genome Inform Ser Workshop Genome Inform*, 11, 161-171.

Dunker, A K, Oldfield, C J, Meng, J W, Romero, P, Yang, J Y, Chen, J W, Vacic, V, Obradovic, Z and Uversky, V N (2007) The unfoldomics decade: an update on intrinsically disordered proteins. *Bmc Genomics*, 9. doi: 10.1186/1471-2164-9-s2-s1.

Dyson, H J and Wright, P E (2002) Coupling of folding and binding for unstructured proteins. *Current Opinion in Structural Biology*, 12, 54-60.

Dyson, H J and Wright, P E (2005a) Elucidation of the protein folding landscape by NMR. In *Nuclear Magnetic Resonance of Biological Macromolecules, Part C* Vol. 394. San Diego: Elsevier Academic Press Inc, pp. 299-+.

Dyson, H J and Wright, P E (2005b) Intrinsically unstructured proteins and their functions. *Nature Reviews Molecular Cell Biology*, 6, 197-208. doi: 10.1038/nrm1589.

Fischer, E (1894) Einfluss der configuration auf die wirkung der enzyme. *Ber. Dt. Chem. Ges.*, 27, 2985-2993.

Fuxreiter, M, Simon, I, Friedrich, P and Tompa, P (2004) Preformed structural elements feature in partner recognition by intrinsically unstructured proteins. *Journal of Molecular Biology*, 338, 1015-1026. doi: 10.1016/j.jmb.2004.03.017.

Goldberg, M E (1991) Investigating protein conformation, dynamics and folding with monoclonal-antibodies. *Trends in Biochemical Sciences*, 16, 358-362.

Han, J D J, Bertin, N, Hao, T, Goldberg, D S, Berriz, G F, Zhang, L V, Dupuy, D, Walhout, A J M, Cusick, M E, Roth, F P and Vidal, M (2004) Evidence for dynamically organized modularity in the yeast protein-protein interaction network. *Nature*, 430, 88-93. doi: 10.1038/nature02555.

Haynes, C, Oldfield, C J, Ji, F, Klitgord, N, Cusick, M E, Radivojac, P, Uversky, V N, Vidal, M and Iakoucheva, L M (2006) Intrinsic disorder is a common feature of hub proteins from four eukaryotic interactomes. *Plos Computational Biology*, 2, 890-901. doi: 10.1371/journal.pcbi.0020100.

He, B, Wang, K J, Liu, Y L, Xue, B, Uversky, V N and Dunker, A K (2009) Predicting intrinsic disorder in proteins: an overview. *Cell Research*, 19, 929-949. doi: 10.1038/cr.2009.87.

Iakoucheva, L M, Brown, C J, Lawson, J D, Obradovic, Z and Dunker, A K (2002) Intrinsic disorder in cell-signaling and cancer-associated proteins. *Journal of Molecular Biology*, 323, 573-584. doi: 10.1016/s0022-2836(02)00969-5.

Jensen, M R, Markwick, P R L, Meier, S, Griesinger, C, Zweckstetter, M, Grzesiek, S, Bernado, P and Blackledge, M (2009) Quantitative Determination of the Conformational Properties of Partially Folded and Intrinsically Disordered Proteins Using NMR Dipolar Couplings. *Structure*, 17, 1169-1185. doi: 10.1016/j.str.2009.08.001.

Johnson, W C (1988) Secondary structure of proteins through circular-dichroism spectroscopy. *Annual Review of Biophysics and Biophysical Chemistry*, 17, 145-166.

Karush, F (1950) Heterogeneity of the binding sites of bovine serum albumin. *J. Am. Chem. Soc.*, 72, 2705-2713.

Kelly, S M and Price, N C (1997) The application of circular dichroism to studies of protein folding and unfolding. *Biochimica Et Biophysica Acta-Protein Structure and Molecular Enzymology*, 1338, 161-185.

Kim, M, Jung, S O, Park, K, Jeong, E J, Joung, H A, Kim, T H, Seol, D W and Chung, B H (2005) Detection of Bax protein conformational change using a surface plasmon resonance imaging-based antibody chip. *Biochemical and Biophysical Research Communications*, 338, 1834-1838. doi: 10.1016/j.bbrc.2005.10.155.

Konermann, L, Tong, X and Pan, Y (2008) Protein structure and dynamics studied by mass spectrometry: H/D exchange, hydroxyl radical labeling, and related approaches. *Journal of Mass Spectrometry*, 43, 1021-1036. doi: 10.1002/jms.1435.

Koshland, D E (1958) Application of a theory of enzyme specificity to protein synthesis. *Proc. Natl. Acad. Sci. U. S. A.*, 44, 98-104.

Koshland, D E (1994) The key-lock theory and the induced fit theory. *Angew. Chem.-Int. Edit. Engl.*, 33, 2375-2378.

Liu, J G, Perumal, N B, Oldfield, C J, Su, E W, Uversky, V N and Dunker, A K (2006) Intrinsic disorder in transcription factors. *Biochemistry*, 45, 6873-6888. doi: 10.1021/bi0602718.

McDonald, R C, Steitz, T A and Engelman, D M (1979) Yeast hexokinase in solution exhibits a large conformational change upon binding glucose or glucose-6-phosphate. *Biochemistry*, 18, 338-342.

Midic, U, Oldfield, C J, Dunker, A K, Obradovic, Z and Uversky, V N (2009) Protein disorder in the human diseasome: unfoldomics of human genetic diseases. *Bmc Genomics*, 10. doi: 10.1186/1471-2164-10-s1-s12.

Mirsky, A E and Pauling, L (1936) On the structure of native, denatured and coagulated proteins. *Proc. Natl. Acad. Sci. U.S.A.*, 13, 739-766.

Mohan, A, Oldfield, C J, Radivojac, P, Vacic, V, Cortese, M S, Dunker, A K and Uversky, V N (2006) Analysis of molecular recognition features (MoRFs). *Journal of Molecular Biology*, 362, 1043-1059. doi: 10.1016/j.jmb.2006.07.087.

Murase, K, Hirano, Y, Sun, T P and Hakoshima, T (2008) Gibberellin-induced DELLA recognition by the gibberellin receptor GID1. *Nature*, 456, 459-464.

Ohgushi, M and Wada, A (1983) Molten-globule state - a compact form of globular-proteins with mobile side-chains. *Febs Letters*, 164, 21-24.

Oldfield, C J, Cheng, Y, Cortese, M S, Brown, C J, Uversky, V N and Dunker, A K (2005a) Comparing and combining predictors of mostly disordered proteins. *Biochemistry*, 44, 1989-2000. doi: 10.1021/bi047993o.

Oldfield, C J, Cheng, Y G, Cortese, M S, Romero, P, Uversky, V N and Dunker, A K (2005b) Coupled folding and binding with alpha-helix-forming molecular recognition elements. *Biochemistry*, 44, 12454-12470. doi: 10.1021/bi050736e.

Oldfield, C J, Meng, J, Yang, J Y, Yang, M Q, Uversky, V N and Dunker, A K (2008) Flexible nets: disorder and induced fit in the associations of p53 and 14-3-3 with their partners. *Bmc Genomics*, 9 Suppl 1, S1.

Pantazatos, D, Kim, J S, Klock, H E, Stevens, R C, Wilson, I A, Lesley, S A and Woods, V L (2004) Rapid refinement of crystallographic protein construct definition employing enhanced hydrogen/deuterium exchange MS. *Proc. Natl. Acad. Sci. U. S. A.*, 101, 751-756.

Patil, A and Nakamura, H (2006) Disordered domains and high surface charge confer hubs with the ability to interact with multiple proteins in interaction networks. *Febs Letters*, 580, 2041-2045. doi: 10.1016/j.febslet.2006.03.003.

Peng, K, Radivojac, P, Vucetic, S, Dunker, A K and Obradovic, Z (2006) Length-dependent prediction of protein intrinsic disorder. *Bmc Bioinformatics*, 7. doi: 10.1186/1471-2105/7/208.

Phillips, D C (1986) Development of concepts of protein-structure. *Perspectives in Biology and Medicine*, 29, S124-S130.

Promponas, V J, Enright, A J, Tsoka, S, Kreil, D P, Leroy, C, Hamodrakas, S, Sander, C and Ouzounis, C A (2000) CAST: an iterative algorithm for the complexity analysis of sequence tracts. *Bioinformatics*, 16, 915-922.

Ptitsyn, O B and Uversky, V N (1994) The molten globule is a 3rd thermodynamical state of protein molecules. *Febs Letters*, 341, 15-18.

Radivojac, P, Iakoucheva, L M, Oldfield, C J, Obradovic, Z, Uversky, V N and Dunker, A K (2007) Intrinsic disorder and functional proteomics. *Biophysical Journal*, 92, 1439-1456. doi: 10.1529/biophysj.106.094045.

Rautureau, G J P, Day, C L and Hinds, M G (2010) Intrinsically Disordered Proteins in Bcl-2 Regulated Apoptosis. *International Journal of Molecular Sciences*, 11, 1808-1824.

Riek, R, Hornemann, S, Wider, G, Billeter, M, Glockshuber, R and Wuthrich, K (1996) NMR structure of the mouse prion protein domain PrP(121-231). *Nature*, 382, 180-182.

Romero, P, Obradovic, Z, Li, X H, Garner, E C, Brown, C J and Dunker, A K (2001) Sequence complexity of disordered protein. *Proteins-Structure Function and Genetics*, 42, 38-48.

Rual, J F, Venkatesan, K, Hao, T, Hirozane-Kishikawa, T, Dricot, A, Li, N, Berriz, G F, Gibbons, F D, Dreze, M, Ayivi-Guedehoussou, N, Klitgord, N, Simon, C, Boxem, M, Milstein, S, Rosenberg, J, Goldberg, D S, Zhang, L V, Wong, S L, Franklin, G, Li, S M, Albala, J S, Lim, J H, Fraughton, C, Llamosas, E, Cevik, S, Bex, C, Lamesch, P, Sikorski, R S, Vandenhaute, J, Zoghbi, H Y, Smolyar, A, Bosak, S, Sequerra, R, Doucette-Stamm, L, Cusick, M E, Hill, D E, Roth, F P and Vidal, M (2005) Towards a proteome-scale map of the human protein-protein interaction network. *Nature*, 437, 1173-1178.

Schulz, G E (1979) Nucleotide Binding Proteins. In *Molecular Mechanism of Biological Recognition* (Balaban, M., ed. New York: Elsevier/North-Holland, pp. 79-94.

Sheerin, D J, Buchanan, J, Kirk, C, Harvey, D, Sun, X, Spagnuolo, J, Li, S, Liu, T, Woods, V A, Foster, T, Jones, W T and Rakonjac, J (2011) Inter- and intra-molecular interactions of Arabidopsis thaliana DELLA protein RGL1. *Biochemical Journal*, 435, 629-639. doi: 10.1042/bj20101941.

Sheng, Y, Sun, X L, Shen, Y, Bognar, A L, Baker, E N and Smith, C A (2000) Structural and functional similarities in the ADP-forming amide bond ligase superfamily: Implications for a substrate-induced conformational change in folylpolyglutamate synthetase. *Journal of Molecular Biology*, 302, 427-440. doi: 10.1006/jmbi.2000.3987.

Sickmeier, M, Hamilton, J A, LeGall, T, Vacic, V, Cortese, M S, Tantos, A, Szabo, B, Tompa, P, Chen, J, Uversky, V N, Obradovic, Z and Dunker, A K (2007) DisProt: the database of disordered proteins. *Nucleic Acids Research*, 35, D786-D793. doi: 10.1093/nar/gkl893.

Skrabana, R, Sevcik, J and Novak, M (2006) Intrinsically disordered proteins in the neurodegenerative processes: Formation of tau protein paired helical filaments and their analysis. *Cell. Mol. Neurobiol.*, 26, 1085-1097. doi: 10.1007/s10571-006-9083-3.

Stelzl, U, Worm, U, Lalowski, M, Haenig, C, Brembeck, F H, Goehler, H, Stroedicke, M, Zenkner, M, Schoenherr, A, Koeppen, S, Timm, J, Mintzlaff, S, Abraham, C, Bock, N, Kietzmann, S, Goedde, A, Toksoz, E, Droege, A, Krobitsch, S, Korn, B, Birchmeier, W, Lehrach, H and Wanker, E E (2005) A human protein-protein interaction network: A resource for annotating the proteome. *Cell*, 122, 957-968. doi: 10.1016/j.cell.2005.08.029.

Sun, X, Xue, B, Jones, W T, Rikkerink, E, Dunker, A K and Uversky, V N (2011) A functionally required unfoldome from the plant kingdom: intrinsically disordered N-terminal domains of GRAS proteins are involved in molecular recognition during plant development. *Plant Mol. Biol.*, 77, 205-223. doi: 10.1007/s11103-011-9803-z.

Sun, X L, Bognar, A L, Baker, E N and Smith, C A (1998) Structural homologies with ATP- and folate-binding enzymes in the crystal structure of folylpolyglutamate synthetase. *Proc. Natl. Acad. Sci. U. S. A.*, 95, 6647-6652.

Sun, X L, Cross, J A, Bognar, A L, Baker, E N and Smith, C A (2001) Folate-binding triggers the activation of folylpolyglutamate synthetase. *Journal of Molecular Biology*, 310, 1067-1078. doi: 10.1006/jmbi.2001.4815.

Sun, X L, Frearson, N, Kirk, C, Jones, W T, Harvey, D, Rakonjac, J, Foster, T and Al-Samarrai, T (2008) An E. coli expression system optimized for DELLA proteins. *Protein Expression and Purification*, 58, 168-174. doi: 10.1016/j.pep.2007.09.003.

Sun, X L, Jones, W T, Harvey, D, Edwards, P J B, Pascal, S M, Kirk, C, Considine, T, Sheerin, D J, Rakonjac, J, Oldfield, C J, Xue, B, Dunker, A K and Uversky, V N (2010) N-terminal Domains of DELLA Proteins Are Intrinsically Unstructured in the Absence of Interaction with GID1/Gibberellic Acid Receptors. *J. Biol. Chem.*, 285, 11557-11571. doi: 10.1074/jbc.M109.027011.

Sundararajan, R, Cuconati, A, Nelson, D and White, E (2001) Tumor necrosis factor-alpha induces Bax-Bak interaction and apoptosis, which is inhibited by adenovirus E1B 19K. *J. Biol. Chem.*, 276, 45120-45127.

Tompa, P (2002) Intrinsically unstructured proteins. *Trends in Biochemical Sciences*, 27, 527-533.

Tompa, P (2003) Intrinsically unstructured proteins evolve by repeat expansion. *Bioessays*, 25, 847-855. doi: 10.1002/bies.10324.

Tompa, P (2005) The interplay between structure and function in intrinsically unstructured proteins. *Febs Letters*, 579, 3346-3354. doi: 10.1016/j.febslet.2005.03.072.

Tompa, P and Csermely, P (2004) The role of structural disorder in the function of RNA and protein chaperones. *Faseb Journal*, 18, 1169-1175. doi: 10.1096/fj.04-1584rev.

Uversky, V N (1993) Use of fast protein size-exclusion liquid-chromatography to study the unfolding of proteins which denature through the molten globule. *Biochemistry*, 32, 13288-13298.

Uversky, V N (1997) Diversity of equilibrium compact forms of denatured globular proteins. *Protein and Peptide Letters*, 4, 355-367.

Uversky, V N (2002a) Natively unfolded proteins: A point where biology waits for physics. *Protein Sci.*, 11, 739-756. doi: 10.1110/ps.4210102.

Uversky, V N (2002b) What does it mean to be natively unfolded? *European Journal of Biochemistry*, 269, 2-12.

Uversky, V N (2010) The Mysterious Unfoldome: Structureless, Underappreciated, Yet Vital Part of Any Given Proteome. *Journal of Biomedicine and Biotechnology*, 2010, 568068. doi: 10.1155/2010/568068.

Uversky, V N, Gillespie, J R and Fink, A L (2000) Why are "natively unfolded" proteins unstructured under physiologic conditions? *Proteins-Structure Function and Genetics*, 41, 415-427.

Uversky, V N, Kutyshenko, V P, Protasova, N Y, Rogov, V V, Vassilenko, K S and Gudkov, A T (1996) Circularly permuted dihydrofolate reductase possesses all the properties of the molten globule state, but can resume functional tertiary structure by interaction with its ligands. *Protein Sci.*, 5, 1844-1851.

Uversky, V N, Narizhneva, N V, Ivanova, T V, Kirkitadze, M D and Tomashevski, A Y (1997) Ligand-free form of human alpha-fetoprotein: Evidence for the molten globule state. *Febs Letters*, 410, 280-284.

Uversky, V N, Oldfield, C J and Dunker, A K (2005) Showing your ID: intrinsic disorder as an ID for recognition, regulation and cell signaling. *Journal of Molecular Recognition*, 18, 343-384. doi: 10.1002/jmr.747.

Uversky, V N, Oldfield, C J, Midic, U, Xie, H B, Xue, B, Vucetic, S, Iakoucheva, L M, Obradovic, Z and Dunker, A K (2009) Unfoldomics of human diseases: linking protein intrinsic disorder with diseases. *Bmc Genomics*, 10. doi: 10.1186/1471-2164-10-s1-s7.

Uversky, V N and Ptitsyn, O B (1994) Partly folded state, a new equilibrium state of protein molecules - 4-state guanidinium chloride-induced unfolding of beta-lactamase at low-temperature. *Biochemistry*, 33, 2782-2791.

Uversky, V N and Ptitsyn, O B (1996) Further evidence on the equilibrium "pre-molten globule state": Four-state guanidinium chloride-induced unfolding of carbonic anhydrase B at low temperature. *Journal of Molecular Biology*, 255, 215-228.

Vacic, V, Oldfield, C J, Mohan, A, Radivojac, P, Cortese, M S, Uversky, V N and Dunker, A K (2007a) Characterization of molecular recognition features, MoRFs, and their binding partners. *Journal of Proteome Research*, 6, 2351-2366. doi: 10.1021/pr0701411.

Vacic, V, Uversky, V N, Dunker, A K and Lonardi, S (2007b) Composition Profiler: a tool for discovery and visualization of amino acid composition differences. *Bmc Bioinformatics*, 8. doi: 10.1186/1471-2105-8-211.

Vucetic, S, Obradovic, Z, Vacic, V, Radivojac, P, Peng, K, Iakoucheva, L M, Cortese, M S, Lawson, J D, Brown, C J, Sikes, J G, Newton, C D and Dunker, A K (2005) DisProt: a database of protein disorder. *Bioinformatics*, 21, 137-140. doi: 10.1093/bioinformatics/bth476.

Vucetic, S, Xie, H B, Iakoucheva, L M, Oldfield, C J, Dunker, A K, Obradovic, Z and Uversky, V N (2007) Functional anthology of intrinsic disorder. 2. Cellular components, domains, technical terms, developmental processes, and coding sequence diversities correlated with long disordered regions. *Journal of Proteome Research*, 6, 1899-1916. doi: 10.1021/pr060393m.

Ward, J J, Sodhi, J S, McGuffin, L J, Buxton, B F and Jones, D T (2004) Prediction and functional analysis of native disorder in proteins from the three kingdoms of life. *Journal of Molecular Biology*, 337, 635-645. doi: 10.1016/j.jmb.2004.02.002.

Wells, M, Tidow, H, Rutherford, T J, Markwick, P, Jensen, M R, Mylonas, E, Svergun, D I, Blackledge, M and Fersht, A R (2008) Structure of tumor suppressor p53 and its intrinsically disordered N-terminal transactivation domain. *Proc. Natl. Acad. Sci. U. S. A.*, 105, 5762-5767. doi: 10.1073/pnas.0801353105.

Wright, P E and Dyson, H J (1999) Intrinsically unstructured proteins: Re-assessing the protein structure-function paradigm. *Journal of Molecular Biology*, 293, 321-331.

Wu, H (1931) Studies on denaturation of proteins XIII. A theory of denaturation. *Chinese J. Physiol.*, 1, 219-234.

Xie, H B, Vucetic, S, Iakoucheva, L M, Oldfield, C J, Dunker, A K, Obradovic, Z and Uversky, V N (2007a) Functional anthology of intrinsic disorder. 3. Ligands, post-translational modifications, and diseases associated with intrinsically disordered proteins. *Journal of Proteome Research*, 6, 1917-1932. doi: 10.1021/pr060394e.

Xie, H B, Vucetic, S, Iakoucheva, L M, Oldfield, C J, Dunker, A K, Uversky, V N and Obradovic, Z (2007b) Functional anthology of intrinsic disorder. 1. Biological processes and functions of proteins with long disordered regions. *Journal of Proteome Research*, 6, 1882-1898. doi: 10.1021/pr060392u.

Xue, B, Dunbrack, R L, Williams, R W, Dunker, A K and Uversky, V N (2010) PONDR-FIT: A meta-predictor of intrinsically disordered amino acids. *Biochimica et Biophysica Acta*, 1804. doi: :10.1016/j.bbapap.2010.01.011.

Xue, B, Oldfield, C J, Dunker, A K and Uversky, V N (2009) CDF it all: Consensus prediction of intrinsically disordered proteins based on various cumulative distribution functions. *Febs Letters*, 583, 1469-1474. doi: 10.1016/j.febslet.2009.03.070.

Zhao, R B, Gish, K, Murphy, M, Yin, Y X, Notterman, D, Hoffman, W H, Tom, F, Mack, D H and Levine, A J (2000) Analysis of p53-regulated gene expression patterns using oligonucleotide arrays. *Genes & Development*, 14, 981-993.

12

Generation of Xylose-Fermenting
Saccharomyces Cerevisiae
by Protein-Engineering

Seiya Watanabe and Keisuke Makino
Ehime University
Kyoto University
Japan

1. Introduction

The utilization of ethanol produced from plant biomass (so-called "bioethanol"), which is derived from the fixation of atmospheric CO_2, as an industrial carbon source and car fuel is one of the most important research issues for the realization of a sustainable global environment. Currently, bioethanol is produced mainly from agricultural crop biomass, which, however, compete with food and animal feed. Alternatively, "lignocellulosic biomass", such as woods and agricultural residues, is an attractive feedstock, and consists of cellulose, hemicellulose, and lignin. In agricultural crops, corn for example, starch, consisting of cellulose (hexose polymers), accounts for 70% of the mass, and its biological fermentation is easy. On the other hand, in lignocellulosic biomass, hemicellulose, accounting for 30% of the mass, is composed of pentoses as well as hexoses such as xylose (and L-arabinose), which cannot be fermented by natural microorganisms.

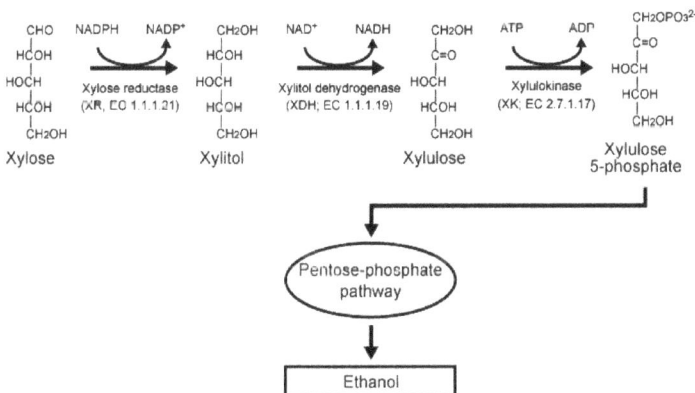

Fig. 1. Xylose metabolism in recombinant *S. cerevisiae*.

Yeasts, in particular, *Saccharomyces cerevisiae* have long been used for production of alcoholic beverages such as wine, beer and Japanese "sake", but the native strains also cannot ferment xylose as a carbon source. On the other hand, *S. cerevisiae* transforming with genes encoding to xylose reductase (XR; EC 1.1.1.21) and xylitol dehydrogenase (XDH; EC 1.1.1.9) (mainly from inherently xylose-metabolizing yeast, *Pichia stipitis*: PsXR and PsXDH) acquires the ability to ferment xylose to ethanol (Fig. 1), whereas produce the excretion of xylitol without the addition of a co-metabolizable carbon source. This is probably caused by several combined factors, and in particular, an intercellular redox imbalance due to a different coenzyme specificity of XR (with NADPH-preference) and XDH (with strict NAD$^+$-dependence) has been thought to be one of the main factors (Jeffries & Jin, 2004). Therefore, generation of novel XR or XDH mutants with modified coenzyme specificity by protein-engineering approach can solve this bottle neck problem.

2. Protein-engineering of PsXDH

XDH belongs to polyol dehydrogenase (PDH) subfamily in a medium-chain dehydrogenase/reductase (MDR) superfamily (Nordling et al., 2002), in which also contains sorbitol dehydrogenase (SDH; EC 1.1.1.14) and L-arabinitol 4-dehydrogenase from several organisms. Most PDHs catalyze strict NAD$^+$(H)-dependent interconversion between alcohols and their corresponding ketones or aldehydes. As well as other MDR enzymes, the coenzyme-binding mode is followed by a classical "Rossmann-fold", and one zinc atom at the catalytic site is necessary for enzyme activity (referred to as "catalytic zinc") (Fig. 2). Furthermore, some possess an additional second zinc atom, which is known as a "structural zinc", although its role has not yet been clarified.

2.1 Modification of coenzyme specificity toward NADP$^+$

2.1.1 Strategy of mutant design

Since NAD$^+$(H) only differs from NADP$^+$(H) in the phosphate group esterified at 2'-position of adenosine ribose, a limited number of amino acid residues interacting with these portions are the first candidates for protein engineering using site-directed mutagenesis. Most of these studies are Rossmann-fold type oxidoreductases, and "landmark" amino acid residues have already been proposed for the discrimination between NAD$^+$(H) and NADP$^+$(H) (Baker et al., 1992).

Generally, subunit structures in the MDR superfamily are comprised of two domains, namely a "coenzyme binding domain" within the intermediate segment and a "catalytic domain" consisting of N- and C-terminal segments (Fig. 2). The former possesses a similar β-α-β motif, centred around a highly conserved Gly-X-Gly-X-X-Gly sequence (where X is any amino acid; positions 9~11 in Table 1), in which several specific amino acid residues play a important role to discriminate between NAD$^+$(H) and NADP$^+$(H). The primary determinant of NAD$^+$(H) specificity is the presence of an aspartate residue which forms double-hydrogen bonds to both the 2'- and 3'-hydroxyl groups in the ribosyl moiety of NAD$^+$(H), and induces negative electrostatic potential to the binding site. Commonly, this residue in NADP$^+$(H)-dependent dehydrogenases is replaced by a smaller and uncharged residue such as glycine, alanine and serine, accompanied by the concurrent presence of an arginine residue that forms a positive-binding pocket for the 2'-phosphate group of NADP$^+$(H).

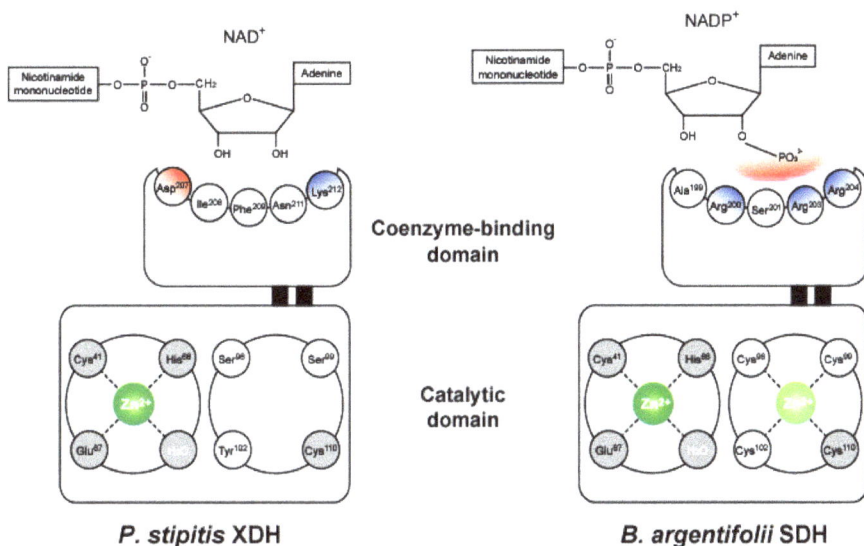

Colors of red and blue indicate positive- and negative-charge, respectively.
Dark- and light-green zinc atoms correspond to catalytic and structural zinc, respectively.
The gray-colored ligands are conserved between the two enzymes.

Fig. 2. Schematic diagram of PsXDH and BaSDH structures.

Bemisia argentifolii (whitefly) SDH (BaSDH) is the only NADP⁺(H)-dependent enzyme among the characterized PDHs (Banfield et al., 2001). Although the crystal structure of BaSDH-NADP⁺ complex has not yet been resolved, a phosphate ion from the crystallization buffer is found adjacent to Ala^{199}, Arg^{200} and Arg^{204}, in the apo-form structure (Fig. 1). The Ala^{199}-Arg^{200} is obviously homologous to Asp^{207}-Ile^{208} in PsXDH (positions 12 and 13 in Table 1) as described above. These aspartate and hydrophobic residues are conserved completely among other NAD⁺(H)-dependent PDHs.

2.1.2 D207A/I208R mutant

Therefore, D207A (Asp^{207}→Ala), I208R and AR (D207A/I208R) mutants were first constructed, functionally expressed as (His)$_6$-tagged enzymes in *Escherichia coli* and purified by Ni^{2+}-affinity chromatography, together with wild-type (WT) enzyme (Fig. 3) (Watanabe et al., 2005). Single substitutions produced a more positive effect on NADP⁺ kinetics than those of NAD⁺; their k_{cat}/K_m^{NADP} values showed an approximately 5~48-fold increase (Fig. 4). However, the mutant enzymes still preferred NAD⁺ to NADP⁺. Double mutation, AR, led to a moderate increase of k_{cat}/K_m^{NADP} compared to single mutations. In particular, there was a complete synergistic effect in the AR mutant; the increase in k_{cat}/K_m for AR was 250-fold, the same as the product of the ratios of the increase in k_{cat}/K_m for D207A and for I208R. On the other hand, k_{cat}/K_m^{NAD} values did not significantly change in the mutants. Overall, the double mutants improved the catalysis for NADP⁺, although k_{cat}/K_m^{NADP} (149 min⁻¹ mM⁻¹) did not reach the k_{cat}/K_m^{NAD} of WT (2,760 min⁻¹ mM⁻¹).

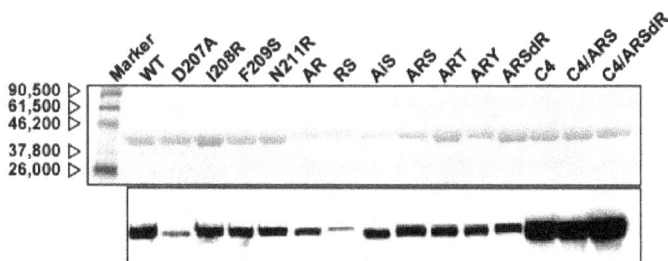

Fig. 3. Expression and purification of recombinant PsXDHs.

SDS-PAGE of purified enzyme (upper panel); intercellular expression level in *E. coli*, estimated by Western blot analysis (lower panel).

Enzyme	Source organisms	Coenzyme	Zn²⁺	Structural zinc loop			Rossmann-fold motif		
							1 1		111 1
				12 34 56		7 8	9 0 1		234 5
XDH	*P. stipitis*	NAD⁺	1	91	EPGIPSRFSDEYKSGHYNLCPH		181	VFGAGPVGLLAAAVAKTFGAKGVIVVDIFDNKL
XDH	*G. mastotermitis*	NAD⁺	1	93	EPGVPSRHSDEYKSGRYNLCPH		176	IYGAGPVGLLVAAVASAFGAESVTIIDLVESRL
XDH	*S. cerevisiae*	NAD⁺	1	94	EPGIPDRFSPEMKEGRYNLDPN		177	VFGAGPIGLLAGKVASVFGAADVVFVDLLENKL
XDH	*M. morganii*	NAD⁺	1	90	EPGIPDLQSPQSRAGIYNLDPA		177	VIGAGTIGIITQSALAG-GCSDVIICDVFDEKL
XDH	*A. oryzae*	NAD⁺	2	97	EPGIPCRRCEPCKEGKYNLCEK		179	VFGAGPVGLLCCAVARAFGSPKVIAVDIQKGRL
SDH	*B. argentifolii*	NADP⁺	2	91	EPGVPCRRCQFCKEGKYNLCPD		174	VIGAGPIGLVSVLAAKAYGAFV-VCTARSPRRL
SDH	*S. cerevisiae*	NAD⁺	1	93	EPGVPSRYSDETKEGRYNLCPH		177	VFGAGPVGLLTGAVARAFGATDVIFVDVFDNKL
SDH	Rat	NAD⁺	1	113	EPGVPREVDEYCKIGRYNLTPT		196	VCGAGPVGMVTLLVAKAMGAAQVVVTDLSASRL
SDH	*Callithrix* sp.	NAD⁺	1	95	EPGAPRETDEFCKTGRYNLSPT		178	VCGAGPIGLVTLLVAKAMGASQVVVTDLSAPRL
SDH	*S. pombe*	NAD⁺	2	92	EPGCVCRLCDYCRSGRYNLCPH		175	VMGCGTVGLLMMAVAKAYGAIDIVAVDASPSRV
SDH	Sheep	NAD⁺	1	92	QPGAPRQTDEFCKIGRYNLSPT		175	VCGAGPIGLVNLLAAKAMGAAQVVVTDLSASRL
SDH	Silkworm	NAD⁺	2	90	EPTQPCRSCELCKRGKYNLCVE		173	ILGAGPIGILCAMSAKAMGASKIILTDVVQSRL
SDH	Peach	NAD⁺	2	107	EPGISCWRCEQCKGGRYNLCPD		190	VIGAGPIGLVSVLSARAFGAARIVIVDVDDERL
SDH	Apple	NAD⁺	1	113	EPGVPREVDEYCKIGRYNLTPT		185	IVGAGPIGLVSVLAARAFGAPRIVIVDMDDRRL
SDH	*E. japonica*	NAD⁺	2	114	EPGISCKRCNLCKQGRYNLCRK		197	VVGAGPIGLVTLLAARAFGAPRIVIADVNDERL
SDH	Human	NAD⁺	1	94	EPGAPRENDEFCKMGRYNLSPS		177	VCGAGPIGMVTLLVAKAMGAAQVVVTDLSATRL
LAD1	*H. jecorina*	NAD⁺	2	116	EPNIICNACEPCLTGRYNGCEK		198	VCGAGPIGLVSMLCAAAAGACPLVITDISESRL

Table 1. Partial sequence alignment of the PDH subfamily enzymes including PsXDH.

2.1.3 D207A/I208R/F209S mutant

It is difficult to judge whether the position of 209 should be involved in the mutation sites, because there is homologous Ser residue regardless of NAD⁺(H)- and NADP⁺(H)-dependent enzyme(s) (position 14 in Table 1). In fact, the homologous position has not been targeted by site-directed mutagenesis for the reversal of coenzyme specificity in other NAD(P)⁺-dependent dehydrogenases. Very surprisingly, an additional introduction of F209S mutation into AR mutant leaded dramatic striking effects in kinetic constants for both NAD⁺ and NADP⁺ (Fig. 4) The k_{cat}/K_m^{NAD} dropped 15-fold compared to WT by an increase of K_m and a decrease of k_{cat}. On the other hand, k_{cat}/K_m^{NADP} increased dramatically (up to 4,100-fold), caused by a large decrease of K_m value and increase of k_{cat} value. It is noteworthy that k_{cat}/K_m^{NADP} (2,790 min⁻¹ mM⁻¹) was almost identical to that of WT for NAD⁺.

F209S single mutation itself and/or double mutation with D207A or I209R (AiS and RS mutants, respectively) gave no significant effect, compared with ARS mutant. What is important factor(s) for the change of the kinetics constants for NADP⁺ between AR and ARS? ARY (D207A/I208R/F209Y) and ART (D207A/I208R/F209T) mutants were constructed for analysis of the volumetric effect of hydroxyl side chain (Fig. 4). k_{cat}/K_m^{NADP}

value of ART mutant was almost the same as that of ARS. Furthermore, the k_{cat} values for NADP+ was reduced by an order of magnitude from ARS>ART>AR>ARY in agreement with the order of the side-chain volume. These results indicate that not only the hydroxyl group but also the relatively small volume group is required for the side chain of the residue at this position to make hydrogen bonds with amino moiety(s) of the side chain of neighbour amino acid residues (probably Arg[208] and Lys[212] in the ARS mutant) similar to BaSDH. Rosell et al. (2003) reported a study about several NAD+(H)-dependent mutants of a vertebrate NADP+(H)-dependent alcohol dehydrogenase (ADH) isozyme 8. For complete reversal of the coenzyme specificity toward NAD+(H), substitution of three amino acid residues at homologous positions of Asp[207]-Ile[208]-Phe[209] in PsXDH was necessary. These results indicate that, in some cases of dehydrogenase, an amino acid residue close to the "landmark" amino acid residues would also be effective for coenzyme recognition.

2.1.4 D207A/I208R/F209S/N211R mutant

Although ARS mutant shows a satisfied ratio of NADP+/NAD+ preference in terms of k_{cat}/K_m, 15.4, the thermostability decreased significantly, as described below. BaSDH possess Arg[203], while the basic residue is only rarely seen among NAD+(H)-PDHs involving PsXDH (position 15 in Table 1). A single mutation of N211R gave no effect on coenzyme specificity (Fig. 4). On the other hand, introduction of this mutation into ARS mutant (ARSdR) produced about a 2-fold decrease in the k_{cat}/K_m^{NAD} value mostly because of a dramatic decrease of K_m value compared to WT, whereas no additional change in the kinetics constant for NADP+ was observed. Finally, this best quadruple mutant showed over 4,500-fold higher k_{cat}/K_m^{NADP} and over 32-fold lower k_{cat}/K_m^{NAD} values, compared with WT; complete reversal coenzyme specificity was achieved. Furthermore, the N211R mutation compensated thermolability in ARS mutant (see below).

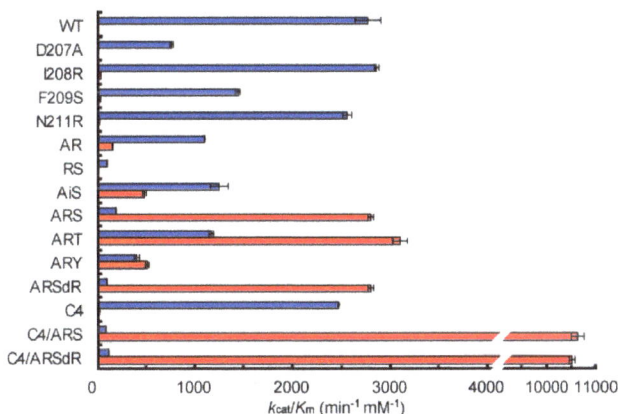

Fig. 4. Catalytic efficiency (k_{cat}/K_m) with NAD(P)+ for WT and mutant PsXDH.

2.1.5 Importance of reference enzyme(s)

Metzger & Hollenberg (1995) also attempted to identify a set of amino acid residues in PsXDH for specificity toward NAD+. They introduced the potential NADP+(H)-recognition sequence of *E. coli* glutathione reductase (no homology with PsXDH) and thermophilic ADH (~30% homology) into the homologous sequence in PsXDH, Asp[207]-Ile[208]-Phe[209]-Asp[210]-Asn[211]-Lys[212]. Among several mutants constructed, D207V/I208R/F209S/D210H mutant, which is very similar to the D207A/I208R/F209S (ARS) mutant, was not expressed in the host cell, in contrast of this study (see Fig. 3). Generally, there are a few sequence homologies among MDRs at the coenzyme recognition site except some specific residues, although the conformational fold is conserved. BaSDH, a reference enzyme in this study, is phylogenetically closer to PsXDH than glutathione reductase and/or ADH, supposing that the step-by-step acquirement of activity with NADP+ in PsXDH seems to mimic the natural evolutional process.

2.2 Thermostabilization

Protein stability is one of most important factors for application of industrial enzymes. Although XDH itself is not directly utilized for ethanol production, it is well known that there is significant relationship between the stability and intercellular expression level. Since there have been no reports about the XDH of thermophilic microorganisms, we cannot undertake site-directed mutagenesis based on the structural comparison of thermophilic XDH and mesophilic PsXDH and/or the random mutagenesis method. Alternative method for the thermostabilization should be developed.

2.2.1 Strategy of mutant design

In the BaSDH (Banfield et al., 2001), structural zinc atom is coordinated with four cysteine residues (Cys[96], Cys[99], Cys[102] and Cys[110]) as well as many other structural zinc-containing ADHs (a pattern of C-C-C-C) (Fig. 1 and positions 2, 4, 6 and 7 in Table 1). There are some substitution patterns of these four cysteines in the PDHs. Among them, enzymes with the pattern of D-S-M-D, R-D-C-S, R-D-C-T have been shown to have no structural zinc. These analyses suggested that four fully conserved cysteine residues are necessary for binding to the structural zinc atom, and that PsXDH having a pattern of S-S-Y-C, has no structural zinc atom. On the other words, it is likely that the substitutions of Ser[96], Ser[99] and Tyr[102] by cysteine residues may increase the thermostability due to introduction of the structural zinc; so-called C4 mutant (S96C/S99C/Y102C) (Watanabe et al., 2005).

2.2.2 Introduction of structural zinc

Atomic absorption spectroscopy analysis revealed that the WT and C4 mutant contain 1.0 and 1.9 mol zinc/mol subunit. Since the putative binding site for the catalytic zinc is conserved completely in PsXDHs (Cys[41], His[66] and Glu[67]), compared to other MDRs, this result should be interpreted as the acquirement of an additional zinc in C4 mutant. The C4 mutant displayed almost the same k_{cat}/K_m^{NAD} value; strict NAD+-dependent enzyme (Fig. 4). The thermostability of the C4 mutant was estimated from the T_m value by circular dichroism (CD) measurements (left panel in Fig. 5). The introduction of the structural zinc atom increased the thermostability of the C4 mutant (4.5°C increase of the T_m value compared to WT). Similar results were obtained by heat treatment of the enzyme; the enzyme activity of WT decreased to 15% after incubation for 10 min at 40°C, whereas no inactivation of the C4 mutant was detected by the same treatment (right panel in Fig. 5). The level of translational expression of the C4 mutant was approximately 3-fold higher than that of WT (Fig. 3), suggesting that the stability of the C4 mutant was also increased *in vivo*.

In cases of human ADHs (Jeloková et al., 1994) and bacterial phenylacetaldehyde reductase (Wang et al., 1999), it is impossible to remove structural zinc without loss of stable folding or enzyme activity. On the other hand, when ADHs from yeast (Magonet et al., 1992) and thermophilic archaea (Ammendola et al, 1992) are incubated with a chelating reagent, the structural zinc atom is selectively removed with no effect on the catalytic zinc or loss of enzyme activity, while the enzyme becomes very sensitive to treatment with heat and/or denaturant. These different results may reflect the role change of the structural zinc in the enzyme during evolution. Ancestral MDR enzyme had probably possessed this zinc by coordination of four cysteine ligands, which is absolutely necessary for both stability and enzyme activity (such as human ADH). Eventually, the role had been limited for maintaining the stability (such as yeast ADH), and subsequently no significant contribution, which leaded a loss of the structural zinc and random substitution of four cysteine ligands (such as PsXDH). Therefore, C4 mutation of PsXDH also seems to mimic the natural evolutionary process "reversely", as well as modification of coenzyme specificity toward NADP+.

In case of random mutagenesis method, single (or double) mutation with moderate effects on thermostability are first isolated, and these combination and/or further randomization are subsequently carried out. However, C4 mutation don't have to be generated by this method, because single and/or double mutations of S96C, S99C and Y102C give no effect to stabilize PsXDH. Since it is easy to identify whether a MDR enzyme contains a structural zinc atom by simple analysis of the amino acid sequence alignment, this novel strategy for protein stabilization would be applicable for other MDRs that contain no structural zinc atom natively.

2.2.3 Thermostabilization of NADP$^+$-dependent mutants

Modification of coenzyme specificity toward NADP+ in PsXDH resulted in a significant loss of thermostability (Fig. 5), which is also found in other several dehydrogenases/reductases. Therefore, the C4 mutation was combined with each ARS and ARSdR mutation, to construct the C4/ARS and C4/ARSdR mutant, respectively (Watanabe et al., 2005). Their translational expression levels in E. coli were almost the same as the C4 mutant (Fig. 3), estimating that the stability in vivo might increase compared to their parent enzymes, ARS and ARSdR. In fact, their T_m values were almost the same as that of the C4 mutant and, in particular, C4/ARS was dramatically stabilized by improving the T_m of 7.7°C compared to ARS (left panel in Fig. 5). Heat treatment of the enzymes provided similar results; after incubation for 10 min at 45°C, C4/ARS and C4/ARSdR maintained about half the activity, whereas no enzyme activity of ARS and ARSdR mutants was detected by the same treatment (right panel in Fig. 5). The zinc content of C4/ARS and C4/ARSdR mutants was 1.69 and 1.78 mol zinc/mol subunit, respectively, suggesting that their increased thermostability was due to the structural zinc atom introduced.

Unexpectedly, C4/ARS and C4/ARSdR mutants showed a further ~4-fold improved k_{cat}/K_m^{NADP} values, mainly caused by an increase in k_{cat}^{NADP} values (Fig. 4). Finally, their k_{cat}/K_m^{NADP} values reached much higher than the initial k_{cat}/K_m^{NAD} value of WT; ratio of NADP+/NAD+ preference in terms of k_{cat}/K_m, 142 and 101, respectively.

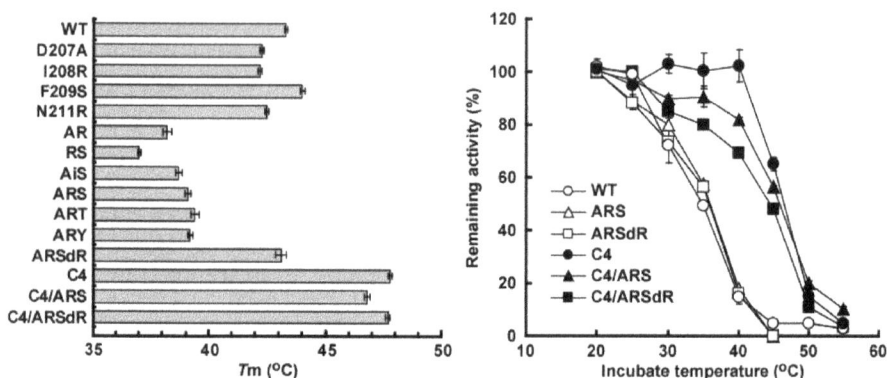

Fig. 5. Comparison of thermostability for WT and mutant PsXDH, estimated by CD analysis (left panel) and heat-inactivation (right panel).

Although protein stabilization generally confers rigidity on the protein and subsequently produces increased ligand affinity, there is no such change in the kinetic constants of C4/ARS and C4/ARSdR mutants. It is noteworthy that the C4 mutation site is separate from ARS/ARSdR sequentially and structurally (Fig. 1 and Table 1). These results suggest the possibility that thermostabilization can enhance the catalytic activity of enzymes, which modified the coenzyme specificity in other dehydrogenases/reductases.

2.2.4 Optimization of structural zinc-binding loop

If the hypothesis described in 2.2.2. is similar to evolutional process, the effect might have been also appeared neighbor region of four cysteine ligands, suggesting one possibility that the optimization leads to further stabilize C4 mutant. The structural zinc is bound within a protruding loop from the catalytic domain, whose corresponding amino acid sequence among all PDHs is completely or highly conserved regardless of the existence of the zinc atom; Glu^{91}, Gly^{93}, Lys^{103}, Gly^{105}, Tyr^{107}, Asn^{108} and Leu^{109} (numbering for PsXDH). Therefore, among amino acid residues except them, four positions of 95, 98, 101 and 112 in PsXDH C4 mutant (positions 1, 3, 5 and 8 in Table 1) were selected as site-directed mutation, to construct C4/P95S, C4/F98R, C4/E101F and C4/H112D mutants, respectively (Annaluru et al., 2007).

Among them, significant thermostabilization were found only in C4/F98R and C4/E101F mutants, compared with parent C4 mutant; higher T_m values of 4.2°C and 3.0°C, respectively. Similarly, after incubation for 10 min at 50°C, C4/F98R and C4/E101F maintained approximately 50% and 70% of the activity, whereas only 20% for C4 mutant. Therefore, based on the results, the C4/F98R/E101F mutant was constructed. Expectedly, the synergic effect of combination of two successive mutations on thermostabilization was observed in this mutant; the T_m value was 2.1°C and 3.3°C higher than C4/F98R and C4/E101F, respectively, and approximately 90% of activity remained after incubation for 10 min at 50°C. Finally, this quintuple mutation, S96C/S99C/Y102C/F98R/E101F, leaded to dramatically stabilize PsXDH WT by increase of 10.8°C in T_m.

The region of the structural zinc-binding loop is known to be involved in stabilizing the dimer–dimer interaction. In case of BaSDH, the arginine and phenylalanine residues at the homologous positions with Phe[98] and Glu[101] in PsXDH exist and participate in the interface interactions of two identical subunits. Since PsXDH is also homodimeric structure, thermostabilization from C4 to C4/F98R/E101F may be due to enhancement of intersubunit interaction rather than intrasubunit.

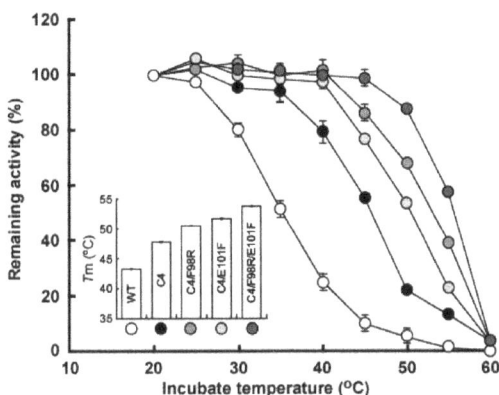

Fig. 6. Heat-inactivation curve of C4-derivative mutant and the T_m (inset).

3. Protein-engineering of PsXR

XR belongs to the aldo-keto reductase (AKR) superfamily, which is made up of 14 different families and approximately 120 members (Ellis, 2002). Their three-dimensional structure share a common $(\alpha/\beta)_8$ barrel fold, with a highly conserved coenzyme binding pocket at the C-terminus; no Rossmann-fold type binding mode, in contrast of XDH. Although most members of this superfamily show strong dependence on NADPH, a few members including XR (AKR2B5), 3α-hydroxysteroid dehydrogenase (AKR1C9) and 3-dehydroecdysone 3β-reductase (AKR2E1) can utilize both NADH and NADPH. To relax an intercellular redox imbalance between XR and XDH in *S. cerevisiae* cells, generation of XR mutant(s) with modified coenzyme specificity is alternative approach.

3.1 Modification of coenzyme specificity toward NADH

3.1.1 Strategy of mutant design

NADH-preferring PsXR mutants were designed by different three strategies based on the following observations (Watanabe et al., 2007).

1. In contrast of most XR including PsXR, XR from *Candida parapsilosis* (CpXR) shows 100-fold higher utilization of NADH than NADPH in terms of catalytic efficiency; it possesses a unique arginine residue in the coenzyme-binding pocket instead of Lys[270] in the PsXR (Table 2) (Lee et al., 2003).
2. Several site-directed mutagenetic and crystallographic studies for modifying coenzyme specificity toward NADH, using NADPH-preferring NAD(P)H-preferring XR from *Candida tenuis* (CtXR), were reported (Fig. 7) (Petschacher et al., 2005; Petschacher & Nidetzky, 2005). This fungal XR shows ~77 % sequential similarity to PsXR. The K274R/N276D mutations lead the highest ratio of NADH/NADPH preference in terms

of k_{cat}/K_m among the series of CtXR mutants investigated, and corresponds to K270R/R272D mutations in PsXR.

3. 2,5-Diketo-D-gluconic acid reductase, a member of the AKR superfamily (AKR5C1), also shows inherent NADPH specificity. However, mutations equivalent to K270G and R276H in PsXR leads to efficient NADH preference (Banta et al., 2002).

Enzyme	Source organisms	Coenzyme preference	2 2 2	2 2 3	2 2 4	2 7 0	2 7 1	2 7 2	2 7 6
XR	*P. stipitis*	NADPH	V	E	L	K	S	N	R
XR	*C. tropicalis*	NADPH	L	E	L	K	S	N	R
XR	*P. guilliermondii*	NADPH	V	E	L	K	S	N	R
XR	*P. tannophilus*	NADPH	L	E	L	K	S	T	T
XR	*K. lactis*	NADPH	L	E	L	K	S	S	R
XR	*A. niger*	NADPH	L	E	L	K	S	N	R
XR	*C. parapsilosis*	NADH	L	E	M	K	S	S	R
XR	*C. tenuis*	NADPH	V	E	M	K	S	L	R
XR(M)	*C. tenuis*	NADH	V	E	M	R	D	L	R
2,5-DKGRA	*Corynebacterium* sp.	NADPH	Y	D	L	K	S	V	R
2,5-DKGRA(M)	*Corynebacterium* sp.	NADH	Y	D	L	G	S	V	R
2,5-DKGRA(M)	*Corynebacterium* sp.	NADH	Y	D	L	K	S	V	H

"M" is a mutant enzyme.

Table 2. Partial sequence alignment of the AKR superfamily.

3.1.2 Characterization of mutants

Specific activities of the recombinant PsXR WT enzyme in the presence of NADH and NADPH were 7.2 and 15.7 U mg^{-1}, respectively; NADPH-preference specificity. Among several constructed single and/or multiple mutants, the high ratio of NADH/NADPH in k_{cat}/K_m was found in R276H and K270R/N272D mutants: 2.62 and 7.29, respectively. Their kinetic constants for NADH were almost same as those of WT, although K270R/N272D mutant showed approximately 5-fold higher K_m value with NADH. On the other hand, the k_{cat}/K_m^{NADP} values of R276H and K270R/N272D mutants decreased 28-fold and 370-fold, compared with WT, which were mainly due to 40-fold decrease of k_{cat} value and 1,120-fold increase K_m value, respectively. When the R276H mutation was introduced in the K270R/N272D mutant, no further improvement of NADH preference appeared. This may be due to antagonistic effect by the introduction of two arginine residues, which may also account for a similar result found in the corresponding triple mutant of CtXR, K274R/N276D/R280H (Petschacher et al., 2005; Petschacher & Nidetzky, 2005): the effect of steric hindrance on the 2'-phosphate group of NADP$^+$, generated by the introduction of the side chain of Arg^{270}, could be strongly reduced when side chain of Arg^{276} is substituted to the smaller side chain of a histidine.

Surprisingly, the greatest thermostabilization was found in the R276H single mutant (increase of 8.8^{o}C in T_m). Similar results were obtained by heat treatment of the enzyme: the activity of the WT enzyme decreased to only 4.1% after incubation for 10 min at 30°C, whereas mutants containing the R276H mutation maintained 92% of activity after the same treatment. These results were unexpected because, as observed in PsXDH (see above), the modification(s) of coenzyme specificity leads to a significant loss of thermostability.

3.2 Modification of coenzyme specificity toward NADPH

In CtXR, several limited numbers of amino acid residues commonly play a role to bind both to NADH and NADPH; Glu[223], Lys[270], Ser[271], Asn[272] and Arg[276] in PsXR (Fig. 7) (Kavanagh et al., 2002, 2003). Therefore, it is very surprisingly that the substitution of Glu[223]→Ala in PsXR leaded to eliminate NADH-dependent activity completely, although the k_{cat}/K_m^{NADPH} value was ~25% of WT (Khattab et al., 2011). Furthermore, since only Ser[271] is specific for NADPH-binding, the substitution to alanine may change the coenzyme specificity toward NADH over NADPH. However, the additional S271A mutation in E223A mutant compensated the decreased NADPH-dependent activity, and the k_{cat}/K_m^{NADPH} value (32.4 min^{-1} µM^{-1}) was almost comparable with that of WT (38.6 min^{-1} µM^{-1}).

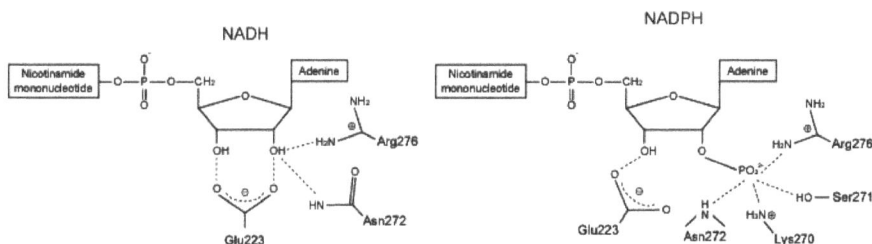

Fig. 7. Schematic diagram of putative interactions between PsXR and NAD(P)H.

4. Xylose-fermentation by S. cerevisiae using mutated XDH and/or XR

Although the native *S. cerevisiae* cannot ferment xylose, the *S. cerevisiae* transformed with the native genes encoding PsXR and PsXDH, a most potent recombinant strain, acquires the ability to ferment xylose to ethanol.

4.1 Screening of XDH and XR mutants

Each gene encoding PsXDH WT, ARS, ARSdR, C4, C4/ARS or C4/ARSdR was constitutively expressed in a laboratory *S. cerevisiae* strain together with PsXR WT gene (Watanabe et al., 2007b), whereas each gene encoding PsXR WT, K270R, R276H or K270R/N272D together with PsXDH WT gene (Watanabe et al., 2007ac); referred to as Y-X ("X" is a enzyme name). Fermentation by these recombinant yeasts was started with 5 g L^{-1} glucose and 15 g L^{-1} xylose in a minimal medium. Y-WT produced ethanol at 2.22 g L^{-1} and excluded xylitol at 1.42 g L^{-1}. Among the XDH mutants, Y-ARSdR produced the highest amount of ethanol at 2.97 g L^{-1} and excreted the lowest amount of xylitol at 0.16 g L^{-1}, whereas among the XR mutants, Y-R276H produced the highest amount of ethanol at 2.30 g L^{-1} and excreted the lowest amount of xylitol at 0.97 g L^{-1}. These results indicate that these protein-engineering approach are useful for the fermentation of xylose to ethanol using *S. cerevisiae*.

4.2 Which of modification between XDH and XR is better?

The fermentation by Y-ARSdR and Y-R276H was further subjected using a bioreactor under the same conditions as the preliminary experiment and compared with that by Y-WT. Y-

ARSdR produced ethanol at 7.02 g L^{-1} with a yield of 0.46 g of ethanol g^{-1} of total consumed sugars, whereas 5.94 g L^{-1} with a yield of 0.43 g of ethanol g^{-1} of total consumed sugars for Y-R276H (Fig. 2). When compared with Y-WT, Y-ARSdR and Y-R276H showed 86% and 52% decrease of unfavorable xylitol excretion with 41% and 20% increased ethanol production, respectively. Measurement of the intracellular coenzymes concentrations that the ratios of NADH/NAD$^+$ and NADPH/NADP$^+$ are more similar to those of strain harboring empty vectors (Y-vector) by an order of magnitude from Y-ARSdR>Y-R276H>Y-WT (Fig. 9). These results clearly indicate that utilization of NADP$^+$(H) by XDH and XR gives more positive effect to prevent xylitol accumulation than that of NAD$^+$(H), due to maintenance of redox balance in yeast cells.

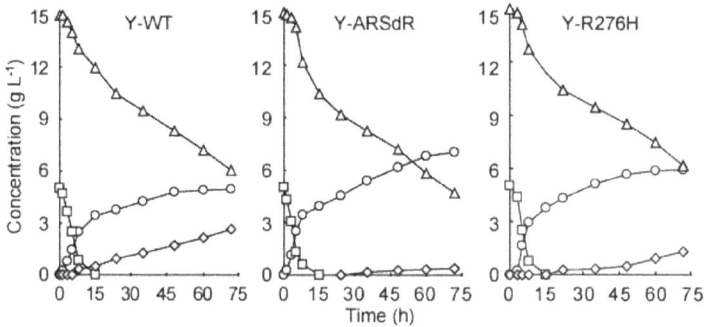

Fig. 8. Xylose fermentation by recombinant *S. cerevisiae* Y-WT, Y-ARSdR and Y-R276H.

Glucose (square); xylose (triangle); xylitol (rhomboid); ethanol (circle).

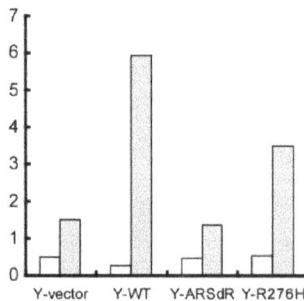

Fig. 9. Ratios of NADH/NAD$^+$ and NADPH/NADP$^+$ in recombinant *S. cerevisiae*.

4.3 Generation of more applicative *S. cerevisiae* strain

It is well known that overexpression of the *XKS1* gene encoding xylulokinase (XK) from *S. cerevisiae* (ScXK) is shown to aid xylose utilization, and that native strain can metabolize (ferment) "xylulose". A flocculent *S. cerevisiae* strain IR-2 was identified as high ability to ferment xylulose; a useful host strain for genetically engineering xylose fermentation (Matsushika et al, 2008ab). Therefore, PsXR, PsXDH (WT or ARSdR mutant) and ScXK genes were integrated on chromosome, and constitutively overexpressed; MA-R4 (vector control) and MA-R5 recombinant strains, respectively (Matsushika et al, 2009).

On a nutrient-rich medium containing 45 g L^{-1} xylose, MA-R4 and MA-R5 consumed 79% and 95% of the xylose at 33 h, respectively (Fig. 10); MA-R5 produced a maximum of 16.0 g L^{-1} ethanol at 48 h, while MA-R4 produced no more than 15.3 g L^{-1}. MA-R5 exhibited a 39% higher xylose consumption rate and a 39% higher ethanol production rate than MA-R4. These results suggest that protein-engineering of XDH is applicable to generate industrial recombinant *S. cerevisiae* capable to ferment both hexose and pentose at the same time.

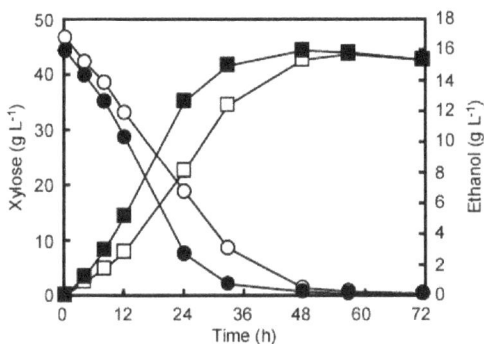

Xylose consumption (circles); ethanol production (squares).

Fig. 10. Xylose fermentation by recombinant *S. cerevisiae*, MA-R4 (open symbols) or MA-R5 (closed symbols).

5. Conclusion

A large body of literature has described the alteration of nicotinamide coenzyme specificity, by which several useful information have already been available. However, the full reversal of coenzyme specificity, in terms of having a mutant enzyme catalytically efficient as the wild type, has been rarely achieved. As illustrated in this chapter, in particular XDH, selection of reference enzyme with close phylogenetic relationship and continuous multiple site-directed mutagenesis including the known "landmark" amino acid residues were necessary for mutant design. Furthermore, it was shown that thermostabilization is dramatically achieved by carefully designed site-directed mutagenesis; introduction of metal ion and subsequently optimization of the binding region. These novel strategies would be very applicable for other dehydrogenase/reductase enzymes with Rossmann-fold type coenzyme-binding mode including a MDR superfamily.

Difference coenzyme specificity between XDH and XR is one of main factors for accumulating unfavourable by-product, xylitol, during xylose-fermentation by recombinant *S. cerevisiae*. In spite of this importance, a few approach based on protein-engineering have been reported for a long time. Recently, several metabolic engineering have been attempted to enhance the pentose-phosphate pathway and/or xylose-uptake; introduction and overexpression of other endogenous genes including xylulokinase (XK), transketolase (TKL1), transaldolase (TAL1) and several hexose-transporter (HXT1-7) genes (Jeffries & Jin, 2004). Combination of these approaches, the protein-engineered enzyme and useful industrial host strain (such as IR-2) should be considered to achieve more effective ethanol production from xylose by recombinant *S. cerevisiae*.

6. References

Ammendola, S., Raia, C.A., Caruso, C., Camardella, L., D'Auria, S., De Rosa, M. & Rossi, M. (1992). Thermostable NAD+-dependent alcohol dehydrogenase from *Sulfolobus solfataricus*. Gene and protein sequence determination and relationship to other alcohol dehydrogenases. *Biochemistry*, Vol.31, No.49, pp. 12514-12523, ISSN 0006-2960

Annaluru, N., Watanabe, S., Pack, S.P., Saleh, A.A., Kodaki, T. & Makino, K. (2007). Thermostabilization of *Pichia stipitis* xylitol dehydrogenase by mutation of structural zinc binding loop. *Journal of Biotechnology*, Vol.129, No.4, pp. 717-722, ISSN 0168-1656

Baker, P.J., Britton, K.L., Rice, D.W., Rob, A. & Stillman, T.J. (1992). Structural consequences of sequence patterns in the fingerprint region of the nucleotide binding fold. Implications for nucleotide specificity. *Journal of Molecular Biology*, Vol.228, No.2, pp. 662-671, ISSN 0022-2836

Banfield, M.J., Salvucci, M.E., Baker, E.N. & Smith, C.A. (2001). Crystal structure of the NADP(H)-dependent ketose reductase from *Bemisia argentifolii* at 2.3 Å resolution. *Journal of Molecular Biology*, Vol.306, No.2, pp. 239-250, ISSN 0022-2836

Banta, S., Swanson, B.A., Wu, S., Jarnagin, A. & Anderson, S. (2002). Alteration of the specificity of the cofactor-binding pocket of *Corynebacterium* 2,5-diketo-D-gluconic acid reductase A. *Protein Engineering*, Vo.15, No.2, pp. 131-140, ISSN 0269-2139

Ellis, E.M. (2002). Microbial aldo-keto reductases. *FEMS Microbiology Letters*, Vo.216, No.2, pp. 123-131, ISSN 0378-1097

Jeffries, T.W. & Jin, Y.S. (2004). Metabolic engineering for improved fermentation of pentoses by yeasts. *Applied Microbiology and Biotechnology*, Vol.63, No. 5, pp. 495-509, ISSN 0175-7598

Jeloková, J., Karlsson, C., Estonius, M., Jornvall, H. & Hoog, J.O. (1994). Features of structural zinc in mammalian alcohol dehydrogenase. Site-directed mutagenesis of the zinc ligands. *Europian Journal of Biochemistry*, Vol.225, No.3, pp. 1015-1019, ISSN 0014-2956

Kavanagh, K.L., Klimacek, M., Nidetzky, B. & Wilson, D.K. (2002). The structure of apo and holo forms of xylose reductase, a dimeric aldo-keto reductase from *Candida tenuis*. *Biochemistry*, Vol.41, No. 28, pp. 8785-8795, ISSN 0006-2960

Kavanagh, K.L., Klimacek, M., Nidetzky, B. & Wilson, D.K. (2003). Structure of xylose reductase bound to NAD+ and the basis for single and dual co-substrate specificity in family 2 aldo-keto reductases. *Biochemical Journal*, Vol.373, No.2, pp. 319-326, ISSN 0264-6021

Lee, J., Koo, B. & Kim, S. (2003). Cloning and characterization of the *xyl1* gene, encoding an NADH-preferring xylose reductase from *Candida parapsilosis*, and its functional expression in *Candida tropicalis*. *Applied and Environmental Microbiology*, Vol.69, No.10, pp. 6179-6188, ISSN 0099-2240

Magonet, E., Hayen, P., Delforge, D., Delaive, E. & Remacle, J. (1992). Importance of the structural zinc atom for the stability of yeast alcohol dehydrogenase. *Biochemical Journal*, Vol.287, No.2, pp. 361-365, ISSN 0264-6021

Matsushika, A., Inoue, H., Watanabe, S., Kodaki, T., Makino, K. & Sawayama, S. (2009). Efficient bioethanol production by recombinant flocculent *Saccharomyces cerevisiae* with genome-integrated NADP+-dependent xylitol dehydrogenase gene. *Applied and Environmental Microbiology*, Vol.75, No.11, pp. 3818-3822, ISSN 1098-5336

Matsushika, A., Watanabe, S., Kodaki, T., Makino, K., Inoue, H., Murakami, K. & Sawayama, S. (2008a). Expression of protein engineered NADP+-dependent xylitol dehydrogenase increases ethanol production from xylose in recombinant *Saccharomyces cerevisiae*. *Applied Microbiology and Biotechnology*, Vo.81, No.2, pp. 243-255, ISSN 1432-0614

Matsushika, A., Watanabe, S., Kodaki, T., Makino, K. & Sawayama, S. (2008b). Bioethanol production from xylose by recombinant *Saccharomyces cerevisiae* expressing xylose reductase, NADP+-dependent xylitol dehydrogenase, and xylulokinase. *Journal of Bioscience and Bioengineering*, Vol.105, No.3, pp. 296-299, ISSN 1389-1723

Metzger, M.H. & Hollenberg, C.P. (1995). Amino acid substitutions in the yeast *Pichia stipitis* xylitol dehydrogenase coenzyme-binding domain affect the coenzyme specificity. *Europian Journal of Biochemistry*, Vol.228, No.1, pp. 50-54, ISSN 0014-2956

Nordling, E., Jornvall, H. & Persson, B. (2002). Medium-chain dehydrogenases/reductases (MDR). Family characterizations including genome comparisons and active site modelling. *Europian Journal of Biochemistry*, Vol.269, No.17, pp. 4267-4276, ISSN 0014-2956

Petschacher, B., Leitgeb, S., Kavanagh, K. L., Wilson, D. K. & Nidetzky, B. (2005). The coenzyme specificity of *Candida tenuis* xylose reductase (AKR2B5) explored by site-directed mutagenesis and X-ray crystallography. *Biochemical Journal*, Vol.385, No.1, pp. 75-83, ISSN 1470-8728

Petschacher, B. & Nidetzky, B. (2005). Engineering *Candida tenuis* xylose reductase for improved utilization of NADH. Antagonistic effects of multiple side chain replacements and performance of site-directed mutants under simulated *in vivo* conditions. *Applied and Environmental Microbiology*, Vol.71, No.10, pp. 6390-6393, ISSN 0099-2240

Rosell, A., Valencia, E., Ochoa, W.F., Fita, I., Pares, X. & Farres, J. (2003). Complete reversal of coenzyme specificity by concerted mutation of three consecutive residues in alcohol dehydrogenase. *Journal of Biological Chemistry*, Vol.278, No.42, pp. 40573-40580, ISSN 0021-9258

Khattab, S.M., Watanabe, S., Saimura, M. & Kodaki, T. (2011). A novel strictly NADPH-dependent *Pichia stipitis* xylose reductase constructed by site-directed mutagenesis. *Biochemical and Biophysical Research Communications*, Vol.404, No.2, pp. 634-637, ISSN 1090-2104

Wang, J.C., Sakakibara, M., Matsuda, M. & Itoh, N. (1999). Site-directed mutagenesis of two zinc-binding centers of the NADH-dependent phenylacetaldehyde reductase from styrene-assimilating *Corynebacterium* sp. strain ST-10. *Bioscience, Biotechnology and Biochemistry*, Vol.63, No.12, pp. 2216-2218, ISSN 0916-8451

Watanabe, S., Pack, S.P., Saleh, A.A., Pack, S.P., Annaluru, N., Kodaki, T. & Makino, K. (2007a). The Positive effect of the decreased NADPH-preferring activity of xylose reductase from *Pichia stipitis* on the ethanol production using xylose-fermenting recombinant *Saccharomyces cerevisiae*. *Bioscience, Biotechnology and Biochemistry*, Vol.71, No.5, pp. 1365-1369, ISSN 0916-8451

Watanabe, S., Saleh, A.A., Pack, S.P., Annaluru, N., Kodaki, T. & Makino, K. (2007b). Ethanol production from xylose by recombinant *Saccharomyces cerevisiae* expressing protein engineered NADP+-dependent xylitol dehydrogenase. *Journal of Biotechnology,* Vol.130, No.3, pp. 316-319, ISSN 0168-1656

Watanabe, S., Saleh, A.A., Pack, S.P., Annaluru, N., Kodaki, T. & Makino, K. (2007c). Protein engineering of xylose reductase from *Pichia stipitis* for improved NADH-specificity and the efficient ethanol production from xylose in recombinant *Saccharomyces cerevisiae. Microbiology,* Vol.153, No.9, pp. 3045-3055, ISSN 1350-0872

Watanabe, S., Kodaki, T. & Makino, K. (2005). Complete reversal of coenzyme specificity of xylitol dehydrogenase and increase of thermostability by the introduction of structural zinc. *Journal of Biological Chemistry* Vol.280, No.11, pp. 10340-10349, ISSN 0021-9258

13

In Silico Engineering of Proteins That Recognize Small Molecules

Sushil Kumar Mishra, Gabriel Demo,
Jaroslav Koča and Michaela Wimmerová
CEITEC - Central European Institute of Technology, Masaryk University, Brno,
National Centre for Biomolecular Research, Faculty of Science, Masaryk University, Brno,
Czech Republic

1. Introduction

The ability of proteins to recognize other molecules in a highly selective and specific manner and to create supramolecular complexes has many biological implications. For example, interactions between receptor-ligand, antigen-antibody, DNA-protein, lectin-sugar are involved in many biologically important processes including transcription of genetic information, enzyme catalysis, transmission of nervous and hormonal signals, host recognition by microbes etc. Therefore, characterizing the structure and energy profile of such supramolecular complexes appears as a key factor in understanding biological function. This may have, in many cases, direct pharmacological consequences. The function of many proteins is driven by reversible binding to small molecules, with either activating or inhibitory effect over the protein's activity. Under these circumstances, it is clear that, in any drug design endeavor, where the goal is to find or build a small molecule that can regulate the function of a protein, it is absolutely essential to understand the stability and behavior of protein-ligand complexes.

However, there is also another kind of an approach to determine protein recognition ability and selectivity mechanisms. It is called protein engineering, and it is based on altering the affinity/selectivity of a protein by substituting some amino-acid residues by other ones in order to identify the most important residues and their specific contribution to the binding activity. Protein engineering is useful not only in the characterization of a protein's binding abilities, but also has applications in bioanalysis and biotechnology. For example, a protein may be engineered (i.e., modified by substituting amino acid residues) to bind specific carbohydrates on the cell surface, and subsequently be used as a marker for diseases characterized by such glycosylation. Another pharmacologically relevant event that can benefit from protein engineering is pathogen/host recognition. In this case, protein engineering may be employed, for instance, to mimic bacterial mutations that lead to multi-drug resistance, to understand their mode of action and to develop new antibacterial drugs. This is certainly a timely issue, as infectious diseases are a leading cause of death worldwide, and they are often connected with a drug resistance. A similar situation occurs in the case of viruses, where the high rate of mutation turns the protein of interest into a continuously moving target, making it tedious to develop drugs or vaccines, e.g., for HIV or influenza viruses.

Protein engineering is typically performed *in vitro,* with *in vivo* consequences and applications. In some cases, it may be very efficient to perform computer modeling and simulations before starting wet laboratory experiments. In such cases we are talking about *in silico* protein engineering, and the goal is to design appropriate mutations in a much faster and cheaper way. In this chapter, we will cover the majority of *in silico* approaches used for protein engineering. The chapter describes not only procedures involved in the *in silico* engineering process itself, but also the description what kind of information is necessary to be able to start *in silico* process. The chapter is composed of several sections. The first describes methods for 3D structure prediction, a necessary step to perform any *in silico* engineering, but not involved in the engineering itself. We further describe various approaches for *in silico* mutagenesis. Afterwards we introduce a number of techniques which enable the prediction of the preferred orientation of the ligand in the binding pocket, as well as the calculation of the binding free energy, again a technique not directly included in protein engineering itself, but necessary to perform it. Some successful examples of *in silico* protein engineering are also given. The whole process is schematically shown in the flowchart in Fig. 1.

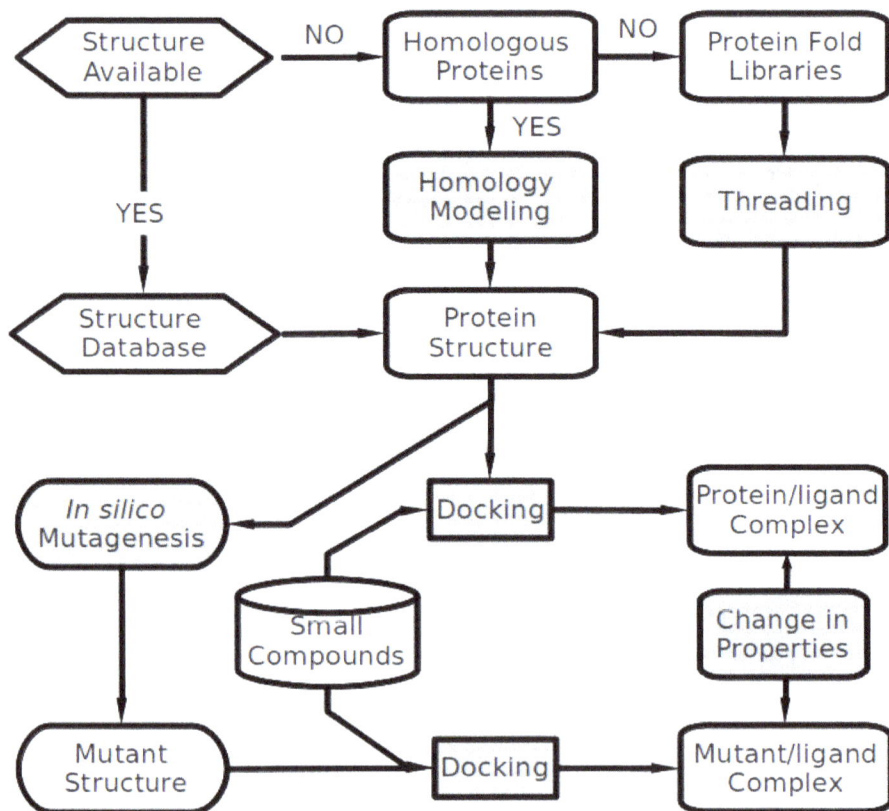

Fig. 1. Flowchart of steps performed within *in silico* protein engineering

2. 3D-Structure as the key prerequisite

A number of proteins that are involved in the cell recognition machinery bind small molecules. In this case, we call these proteins *receptors*, and the small molecules *ligands*. The 3D structure of a receptor is the starting point in the *in silico* protein engineering process. Current experimental methods for protein structure determination are very well established. If the experimental structure is not available, computational approaches are used to model the 3D structure of the receptor.

2.1 Experimental 3D-structure

The 3D protein structure can be obtained by X-ray crystallography or by NMR spectroscopy. Both methods allow to refine the atomic coordinates against experimental structural restraints and constraints. The final 3D model is obtained when the refinement statistics reach relevant global minimum values. The quality of a structure from X-ray crystallography or NMR spectroscopy is defined by experimental data, but the quality of the refined model is based on the interpretation of the model through the personal view of the scientist. In most cases, this freedom in model interpretation is the main source of uncertainty in the results obtained by refining approaches.

2.1.1 X-ray crystallography

The first protein structure determined by X-ray crystallography was solved in the late 1950s. Since that success, over 60 thousand X-ray crystal structures of proteins, nucleic acids and other biological macromolecules have been determined. X-ray crystallography is used to determine the arrangement of atoms in a crystal lattice. The procedure of the 3D structure obtaining is composed of four key steps (see Fig. 2). The crystal under investigation is placed in the way of beams of X-rays, which, upon collision with the crystal, are diffracted in a specific pattern based on the structure of the lattice. The diffraction pattern is used to compute the electron density map of the crystal, from which the mean positions of atoms in the crystal can be determined, together with other information. The resulting electron density map is an average electron density of all the molecules within the crystal. *Structure refinement* refers to the process by which structural models are fit to the information gained from the electron density map. During structure refinement, automated tools for chain tracing, side chain-building, ligand building and water detection are used. The structure refinement continues until the correlation between the diffraction data and the model reach a global minimum (Giacovazzo, 2002).

The atomic positions and their respective B-factors (Debye-Waller factors) can be refined to fit the observed diffraction data. The B-factor, also termed the temperature factor, describes the degree to which the electron density is spread out, accounting for thermal motions and reflecting the fluctuation of atoms about their average positions. Thus, for proteins, the B-factor allows for the identification of areas of large mobility, such as disordered loops, but it can also be the marker of errors in the process of model building (Yuan et al., 2005). The relative agreement of the structure with regard to the experimental data is measured by the R-factor and the "free" R-factor (R-free). The R-free is analogous to the R-factor, which is calculated from a subset (~5%) of reflections that were not included in the structure refinement. The value of R-free is monitored during the whole refinement process, and it prevents any over-refinement and over-interpretation of the data (Brunger, 1992).

X-Ray Crystallography

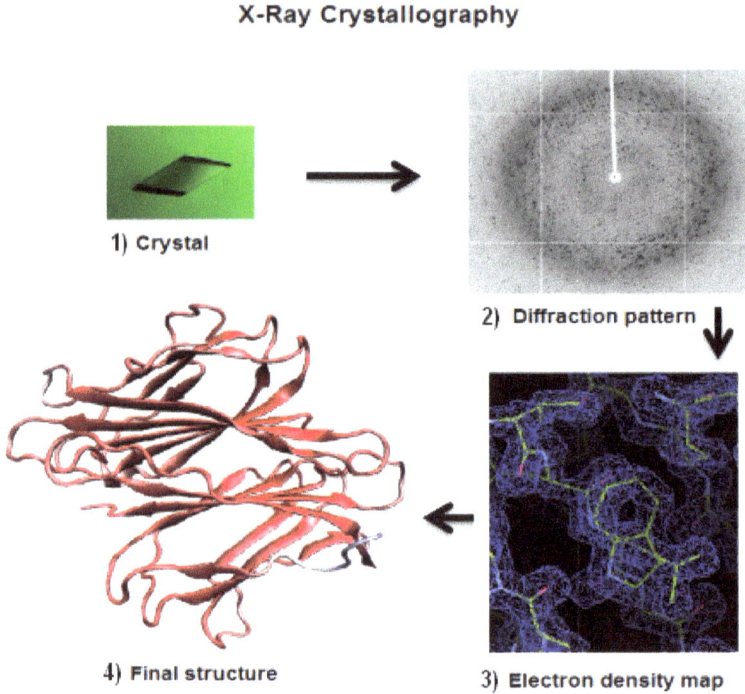

1) Crystal

2) Diffraction pattern

3) Electron density map

4) Final structure

Fig. 2. Four main steps to solve a protein structure by X-ray crystallography: (1) to crystallize the protein, (2) to collect the diffraction, (3) to calculate the electron density map, (4) to refine and validate the model of the structure of the protein

A number of factors contributes to the final quality of an X-ray structure. The first factor relates to the crystal characteristics and its diffraction properties, and is evaluated in terms of resolution. Here, the term *resolution* refers to the level of detail that can be inferred from the electron density map. For proteins, resolutions of less than 2.5 Å are considered meaningful, though the goal is to obtain resolutions of under 1.5 Å, where individual atoms can be clearly pinpointed from the electron density map. Most errors result from highly disordered areas in the electron density maps, like flexible loops of proteins. The electron density of atoms with high residual disorder is smeared in the electron density map, and is no longer detectable. Atoms that give weak scattering (i.e., diffraction of the X-ray beams), such as hydrogen, are normally invisible. Single atoms of protein side chains can be detected multiple times in an electron density map, because of multiple conformations of those respective residues (di Luccio & Koehl, 2011).

2.1.2 NMR spectroscopy

NMR spectroscopy is often the only way to obtain high resolution information on protein dynamics as well as on the protein structure in a solvent. NMR spectroscopy uses the magnetic properties of nuclei that possess a spin. To facilitate NMR experiments, it is necessary to isotopically label the protein with ^{13}C and ^{15}N (for ^1H there is no need to label the protein because this isotope has a natural abundance of 99.9%). The procedure is schematically pictured in Fig.3.

NMR

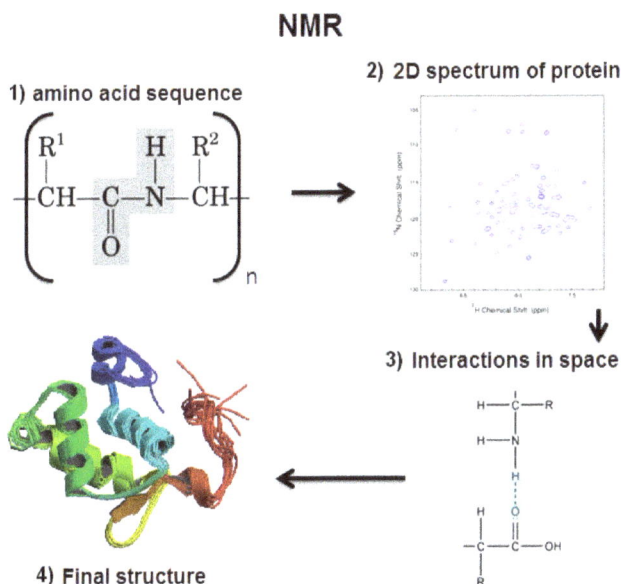

Fig. 3. For solving a protein structure by NMR in solution it is needed: 1) to know the amino acid sequence, 2) to measure the multidimensional spectra 3) to calculate the distances by NOE and J-coupling effects and 4) to refine and validate the 3D structure of the protein

The molecule of interest is placed in a strong magnetic field, and each of these nuclei is characterized by a unique resonance frequency, depending on the electron density of the local chemical environment (chemical shifts), but also on the combination of the local magnetic field and the external field. In the case of proteins, the number of nuclei involved can be large, therefore multidimensional experiments (2D, but also 3D and 4D experiments) are usually performed. The most important method for protein structure determination utilizes NOE (Nuclear Overhauser effect) experiments to measure the distances between pairs of atoms within the molecule that are not connected via chemical bonds (through-space coupling effects). Other NMR experiments are performed in order to measure the distances between pairs of atoms that are connected through chemical bonds (J-coupling). The goal is to assign the observed chemical shifts from multidimensional spectra to their specific atoms (nuclei) in the protein. All the values are then quantified and translated into angle and distance restrains. Most of these restraints correspond to ranges of possible values instead of precise constraints. These restraints are subsequently used to generate the 3D structure of the molecule by solving a distance geometry problem (Wüthrich, 1990, 2003).

The structure determination of macromolecules by NMR spectroscopy shares similarities with X-ray crystallography in terms of possible sources of errors. The errors in an NMR structure can result from an improper experiment setup, as well as from the human misinterpretation of the experimental data (Saccenti & Rosato, 2008). Molecular modeling techniques are used to generate a set of models for the protein structure that satisfy the obtained experimental restraints, as well as standard stereochemistry. Analogously to X-ray methods, the quality of NMR measurements affects the quality of the structures. The value of the root mean square (RMS) difference between each model and a "mean" structure defines the precision of a set of models for a protein structure. The quality of each model is evaluated by the number of the experimental restraints violations in the final model.

2.2 Homology modeling

Despite significant progress in X-ray crystallography and NMR spectroscopy, the structures of many biotechnologically and therapeutically relevant proteins remain undiscovered for various reasons. In such a case, homology modeling can be used to obtain their 3D structure. Homology modeling is a purely computational procedure that consists of building a protein model using a structural template, normally coming from proteins with a known structure. The procedure is composed of four key steps as seen in Fig. 4.

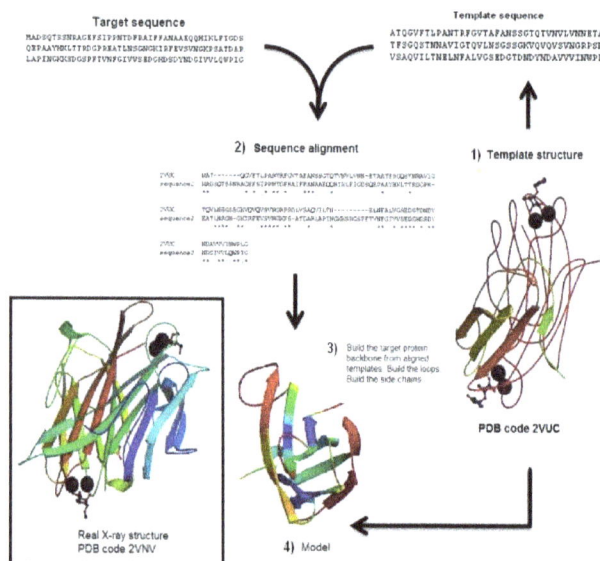

Fig. 4. Homology modeling consists of: 1) Identification of the template, 2) Alignment of the target sequence with the template sequence, 3) Building the target protein backbone, loops and side chains and 4) Refining and evaluating the final model.

Template selection and sequence alignment

An initial step for comparative modeling is to check whether there is any protein in the current PDB database having a similar sequence as the protein of interest. If so, the structure of this protein will be used as a template. The search for the template has to proceed using a sequence comparison algorithm that is able to identify the global sequence similarity (i.e., the degree to which the sequence of amino acids is conserved in the protein under investigation compared to the template protein). Homology modeling of a target protein sharing over 30% sequence identity with its template is expected to generate structural models whose accuracy is close to that of an experimental structure, however, Roessler and coworkers showed that even proteins sharing 40% of sequence identity can display different folds (Roessler, 2008).

The sequence of the protein with unknown structure is aligned against the sequence of the template protein, meaning that the sequences are arranged in such a way that the regions which contain the same amino acids in both proteins are superimposed. Then the C_α coordinates of the aligned residues from the template are copied over to the target protein in order to form the skeletal backbone (Nayeem et al, 2006). Commonly used alignment techniques are: standard pairwise sequence alignment, where only 2 sequences are compared at a time, or multiple sequence alignment, where more sequences are compared at a time and which is generally used when the target and template sequences belong to the same family. There are complex sequence alignment algorithms that optimize a score based on a substitution matrix and gap penalties. Most errors are caused by the sequence alignment technique. Errors appear frequently in the loop regions between secondary structures, as well as in regions where the sequence similarity is low. Structural alignment techniques are also available, which attempt to find areas of structural similarity between proteins. Recent techniques aim to use as much information as possible while performing the sequence alignment (amino-acid variation profiles, secondary structure knowledge, structural alignment data of known homologs) (Nayeem et al., 2006; Zhang, 2002).

Loop building

Loops participate in many biological events and contribute to functional aspects such as enzyme active sites formation or ligand-receptor recognition. The flexible nature of loops causes problems in the prediction of their conformation. Databases of loop conformations or modeling by *ab initio* methods are used in order to determine the proper structure of loops. In the database approach, a library of protein fragments is scanned for fragments whose length matches to the corresponding lenght of the modelled loop (for short loops) (di Luccio & Koehl, 2011; Zhang, 2002). The *ab initio* loop prediction approach relies on a conformational search guided by various scoring functions and is used for longer loops (Olson et al., 2008; van Vlijmen et al., 1997).

The side-chain positioning problem

Most of the side-chain positioning methods are based on rotamer libraries with discrete side-chain conformations. Rotamer libraries contain a list of all the preferred conformations of the side-chains of all twenty amino acids, along with their corresponding dihedral angles (Lovell, 2000). Side chain prediction techniques choose the best rotamer for each residue of the protein based on a score that includes both geometric and energetic constraints (combinatorial problem). The combinatorial problem is solved by heuristic techniques such as mean field theory, derivatives of the dead-end elimination theorem or Monte Carlo techniques (Vasquez, 1996).

Refinement and validation of the final model

When determining the structure of a protein by homology modeling, the last step is refining the model. However, it was shown that refining a structural model by energy minimization only (i.e., without experimental constraints) many times leads to structures that are different compared to those obtained by X-ray crystallography. To avoid such problems, several approaches can be applied including evolutionary derived distance constraints (Misura et al., 2006), the combination of molecular dynamics and statistical potentials (Zhu et al., 2008), adding a differentiable smooth statistical potential (Summa & Levitt, 2007) or considering the solvent effects (Chopra et al., 2008).

For the model validation step, scoring functions are used. These are functions based on statistical potentials, local side-chain and backbone interactions, residue environments, packing estimates, solvation energy, hydrogen bonding, and geometric properties. The validation of models can also come from experiments, and further later experimental constraints/restraints can be used to improve the accuracy of the respective models (di Luccio & Koehl 2011).

Generally, the quality of the homology model is dependent on the quality of the sequence alignment and of the template structure. The presence of alignment gaps (commonly called indels) in the target but not in the template complicates the model building process. In addition, it's very hard to deal with the gaps in the template structure (e.g, caused by the poor resolution of an X-ray structure). At 70% sequence identity between the model and the template, the root mean square deviation (RMSD) between the coordinates of the corresponding C^α atoms is typically ~1–2 Å. The RMSD can rise to 2–4 Å at 25% sequence identity. The errors are significantly higher in the loop regions, because of the increased flexibility in these areas, both in the target, as well as in the template. Errors in side chain packing and positioning increase with decreasing amino acid sequence identity, and are caused also by the fact that most side chains can exist in several conformations. These errors may be significant, and they imply that homology models must be utilized carefully. Nevertheless, homology models can be useful in reaching *qualitative* conclusions about the biochemistry of the query sequence (conserved residues can stabilize the folding, participate in binding small molecules or play a role in the interaction with another protein or nucleic acid) (di Luccio & Koehl 2011). The state of the art in homology modeling is assessed in a biannual large-scale experiment known as the Critical Assessment of Techniques for Protein Structure Prediction, or CASP. A particularly interesting example is provided by the application of homology modeling to virtual screening for GPCR (G-protein coupled receptor) antagonists (Evers & Klabunde, 2005).

Online portals, such as the Protein Structure Initiative (PSI) model portal (http://www.sbkb.org), or the Swiss-Model Repository (http://swissmodel.expasy.org), bring to the community a large database of models. The PSI model portal ((http://www.proteinmodelportal.org) currently provides 22.3 million comparative protein models for 3.8 million distinct UniProt entries with relevant validation data.

A variety of software is currently in use for homology modeling of protein structures:

GeneMine: Homology modeling in GeneMine (Lee & Irizarry, 2001) uses SegMod, a segment match modeling protocol (Levitt, 1992). The target sequence is divided into short segments. Corresponding structural fragments are taken from a structural database and then matched to the sequence. The fragments are then fitted onto the framework of the template structure. The program generates 10 independent models, from which an average model is constructed and stereochemically refined to minimize conformational repulsion.

DS MODELER: The protein homology modeling program DS MODELER (Accelrys Software Inc.) includes the software tool MODELLER (Sali & Blundell, 1993). MODELLER makes structure predictions based on distance restraints obtained from the template, from the database of crystal structures in the PDB, and from a molecular force field. Loops are generated *de novo*, by a process that incorporates knowledge-based potentials from known crystal structures.

ICM: The homology modeling option in ICM (Abagyan & Batalov, 1997) is completely automated. The template is used for matching the backbone, as well as the side chain conformations for the residues that are identical to the template. Loops are inserted from conformational databases with matching loop ends. The non-identical side chains are given the most preferred rotamer, and then optimized by torsional scan and minimization.

SWISS-MODEL: SWISS-MODEL is an automated protein structure homology modeling server accessible from the ExPASy Web server (Schwede et al., 2003). The input for SWISS-MODEL is a sequence alignment and a PDB file for the template. The homology model is constructed using the ProModII program (Peitsch, 1995). Model construction includes backbone and side chain building, loop building, validation of the quality and of the packing of the model. The model coordinates are returned in PDB format.

2.3 Threading

Threading is used to model the structure of a protein when no homologs with a known 3D structure are available. Protein threading is based on the idea that there is a limited number of different folds in nature (approximately 1000), and thus a new structure has a similar structural fold to those already deposited in the PDB. The threading approach is a specialized sub-class of fold recognition. It works by comparing a target sequence against a library of potential fold templates using energy potentials and/or other similarity scoring methods. The template with the lowest energy score (or highest similarity score) is then assumed to best fit the fold of the target protein. The procedure is composed of three key steps shown in Fig. 5.

Threading improves the sequence alignment sensitivity by introducing structural information (the secondary or tertiary structure of the targets) into the alignment. For example, some amino acids are preferred in helical secondary structure, some can appear more frequently in hydrophobic environments, *etc*. This different behavior of amino acids produces different secondary and tertiary structures of proteins, depending on what environment they are exposed to.

The earliest threading approach was the '3D profiles' method (Luthy et al., 1992), in which the structural environment at the position of each residue of the template is classified into 18 classes, based on the position status, local secondary structure and polarity.

Frequently used threading methods are based on the Profile Hidden Markov Model method (HMM) (Durbin, 1998). All the sequences in the database are clustered into a set of families. In an HMM algorithm, the target is represented by the predicted secondary structure, while the template structures are represented with the template's secondary structure patterns. The majority of current threading methods are based on *residue pairwise interaction energy* methods, where, in each step of the threading procedure, the alignment score is calculated by adding up all the pairwise interaction energies between each target residue and the template residues surrounding it.

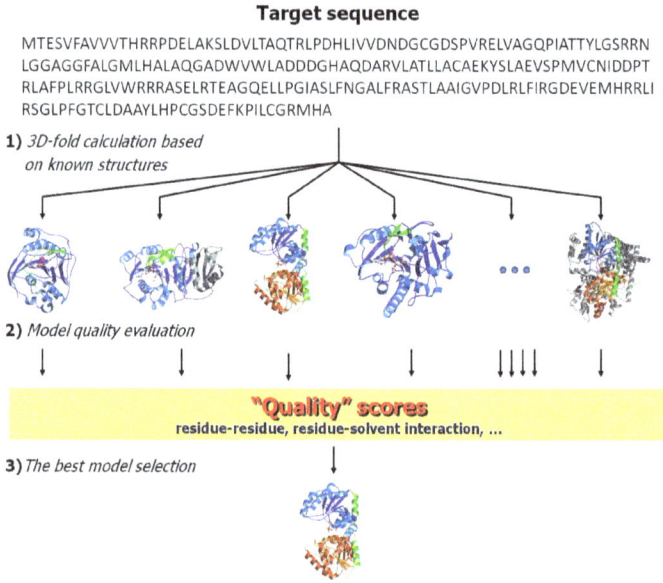

Target sequence

MTESVFAVVVTHRRPDELAKSLDVLTAQTRLPDHLIVVDNDGCGDSPVRELVAGQPIATTYLGSRRN
LGGAGGFALGMLHALAQGADWVWLADDDGHAQDARVLATLLACAEKYSLAEVSPMVCNIDDPT
RLAFPLRRGLVWRRRASELRTEAGQELLPGIASLFNGALFRASTLAAIGVPDLRLFIRGDEVEMHRRLI
RSGLPFGTCLDAAYLHPCGSDEFKPILCGRMHA

1) *3D-fold calculation based on known structures*

2) *Model quality evaluation*

"Quality" scores
residue-residue, residue-solvent interaction, ...

3) *The best model selection*

Fig. 5. Three main steps of threading: 1) The construction of a structure target database based on templates, 2) Calculation of the quality of each model and 3) Selecting the best model

Threading methods are not able to give a good sequence–structure alignment. The first reason is that the structure information has many approximations. Most of the threading methods use a *'frozen'* approximation. It means that the target residues are in the same environments as the template residues if they belong to the same structural fold. But, especially in loop regions, two homologous structures can have slightly different environments. Therefore, only conserved regions are used in threading (Madej et al., 1995).

A variety of threading software is available:

GenTHREADER is a fast and powerful protein fold recognition method (Jones, 1999a). It is used to make structural alignment profiles in the construction of the fold library. PSI-BLAST (Position-Specific Iterated - Basic Local Alignment Search Tool, Altschul et al., 1997) profiles, bidirectional scoring and secondary structures predicted by PSIPRED (Jones, 1999b), have also been incorporated into the modified protocol. Because of these implementations, the sensitivity and the accuracy of alignments is increased (McGuffin & Jones, 2003). New implementations for structure prediction on a genomic scale and for discriminating superfamilies from one another were added recently (Lobley, 2009).

3D-PSSM (Kelley et al., 2000) is using PSIPRED to predict the secondary structure of target proteins, and PSI-BLAST for sequence-profile alignments. The target profiles are aligned against 3D position-specific scoring matrices (PSSMs), which are generated for the templates within the fold library. For each template, PSI-BLAST is used to generate an initial 1D sequence based PSSM, which is then further enhanced using solvation potentials, secondary structures and structural alignments, resulting in a 3D-PSSM.

Phyre2 (Kelley & Sternberg, 2009) is a major update to the original Phyre server. It is designed to predict the 3D structure of a protein from its sequence. Phyre2 uses the alignment of hidden Markov models via HHsearch (Söding, 2005) in order to significantly improve the accuracy of the alignment, as well as the rate of detection of homologous regions. For regions that are not detectable by homology, *ab initio* folding simulations called Poing are used (Jefferys et al, 2010).

3. *In Silico* mutagenesis of proteins

The ultimate goal of protein engineering is to design a protein with novel properties, starting from existing proteins. Protein engineering in the field of recognition has been particularly successful in changing ligand specificity and binding affinity. Consequently, we are interested in changing the structure of a macromolecule in a predetermined way, such that we can affect its recognition ability. During the last years, the availability of computational and graphical tools, which allow to display and explore the three dimensional structures of proteins, has made *in silico* mutagenesis easier and more feasible.

Basically, two approaches are available - mutation of a single, or of multiple residues.

3.1 Performing *in silico* mutagenesis

The 3D structure of a protein molecule is generally stored as a text file which contains information about the chains, residues, atoms and atom types, atomic coordinates and their occupancy. Performing *in silico* protein mutagenesis basically means changing the lines of the text that encode the information about the residue being mutated, followed by a set of additional operations meant to properly integrate the mutated residue into the structure.

The mutation of one residue to another does not change anything in the backbone atoms. In addition, the protein side chains all start by the β carbon atom, which is the same for all the amino acids except for the glycine. Therefore, the single amino acid mutation is straightforward, since only the side chain atoms need to be changed. The most critical step is to check for steric clashes that may occur, especially when an amino acid with a short side chain is mutated into another one having a longer side chain. Moreover, the new amino acid may adopt several side chain orientations. This problem is handled using the concept of rotamers, which are defined as low energy side-chain conformations, and are sampled according to their occurrence in proteins. Computational chemistry tools are able to include all the possible side chain conformations by using rotamer libraries. Several molecular modeling platforms facilitate single point mutation using the concept of rotamers.

Some of commonly used software packages to perform single point or multiple point mutations at selected positions:

Swiss-Pdb Viewer: an application that allows to analyse several proteins at the same time (Guex & Peitsch, 1997). The proteins can be superimposed in order to deduce structural alignments and compare their active sites. Swiss-Pdb Viewer allows to browse a rotamer library for amino acids side chains. Amino acid mutations, H-bonds, angles and distances between atoms are easy to obtain.

Pymol: an open-source molecular visualization system (Schrodinger LLC). It can produce high quality 3D images of small molecules and biological macromolecules, such as proteins.

PyMol has a mutagenesis wizard to perform mutations. Several side chain orientations (rotamers) are possible. The rotamers are ordered according to their frequency of occurrence in proteins.

MODELLER: contains the routine 'mutate_model', which allows *in silico* side chain replacement, as well as modeling the final structure of the mutated protein. The routine introduces a single point mutation at a user-specified residue, and optimizes the mutant side chain conformation by conjugated gradient and a molecular dynamics simulation (Sali & Blundell, 1993).

Triton: a graphical interface for computer aided protein engineering. It implements the methodology of *in silico* site-directed mutagenesis to design new protein mutants with required properties, using the external program MODELLER mentioned above. The program allows to perform the one-, two- or multiple-point amino acid substitutions in a very user-friendly and automated way (Prokop et al, 2008). Output data can be easily visualized, written or organized as input files for any of the other computational chemistry modules that Triton interfaces. Routines to study enzyme kinetics and protein/ligand binding are available.

3.2 Alanine scanning mutagenesis

Alanine scanning mutagenesis is a method usually used to determine the contribution of a particular residue to protein function by mutating that residue into alanine. Alanine scanning involves substituting of a larger group of atoms with a smaller one. Alanine is the residue of choice because it removes the side chain beyond the β carbon of the amino acid in question, and, most importantly, because it does not alter the main-chain conformation (Wells, 1991). Additionally, it does not impose extreme electrostatic or steric strain in the system. Glycine would also cancel the contribution of the side chain, but could introduce conformational flexibility into the protein backbone, and therefore is not commonly used.

Alanine-shaving is the process of making multiple simultaneous alanine mutations and can be helpful, e.g., in investigating the cooperativity between side chains (Bogan & Thorn, 1998). Cooperativity can be detected by multiple mutation cycles (Carter, 1986), in which the free energy change caused by the simultaneous mutations at selected residue positions in a protein is compared with the sum of the free energy changes associated with single mutations at each of the selected positions. This technique has also been used experimentally (Bogan & Thorn 1998).

4. Qualitative and semi-quantitative approaches to evaluate the recognition ability of proteins

Molecular recognition can be viewed as the ability of a certain biomacromolecule to interact preferentially with a particular target molecule. A necessary prerequisite for any *in silico* protein engineering approach is the ability to evalute how strong the recognition is. In biological systems, the process of recognition, governed by non-covalent interactions, results in the formation of a complex, where one biomacromolecule interacts with another biomacromolecule or a small molecule. Modern computer modeling and simulation methods, such as docking or free energy calculations, make it possible to study the molecular recognition process between two molecules *in silico*. Evaluation of the recognition ability of biomacromolecules is performed in two steps: (i) docking the small molecule into the biomacromolecule and (ii) analyzing the interactions and factors that determine the binding affinity.

4.1 Principles of molecular docking

Molecular docking is a widely-used computational tool for the study of molecular recognition, which aims to predict the preferred binding orientation of one molecule to another when bound together in a stable complex. Docking can be performed between two proteins, a protein and a small molecule, a protein and an oligonucleotide or between an oligonucleotide and a small molecule. We use the terms *receptor* and *ligand* to describe the role of binding partners in docking. *Receptor* denotes the system we are docking to (most commonly a protein), while *ligand* denotes the molecule being docked (drug-like compounds, peptide, carbohydrate, etc.). The docking product is commonly referred to as the *complex*. Inside the complex, the position of the ligand relative to the receptor is called the *binding mode*. The space within the receptor where binding modes are explored is commonly known as the *search (or grid) space*.

As already mentioned, receptor-ligand docking programs usually run in two primary parts. The first stage is *searching the grid space* and it leads to the generation of possible binding modes of the ligand within the predefined search space in the receptor. The second stage of docking is *scoring*, and it refers to the process of quantifying the binding strength of each mode of binding using a function called a *scoring function*. We describe each stage in the following pages.

4.2 Receptor site characterization

In the process of docking, the first issue is where to dock the ligand, i.e., how to define a search space on the receptor where the search will be performed. If the 3D structure and the binding site of the receptor is known, the search space is defined within and around this binding site. However, it can happen that the 3D structure of the receptor has not been solved, or there is no experimental evidence indicating a possible region for the ligand binding. In this case, it is recommended to do a prior identification of the binding site by using specialized tools such as PASS (Brady & Stouten 2000), Q-sitefinder (Laurie, 2005), ICM Pocketfinder (An J, 2005) etc. The ligand binding site prediction itself is a complex and tedious problem, thus we will not discuss the details here (for further reading see: Huang & Zou, 2010; Yuriev et al. 2011).

If no prior identification of the binding site is done, it is indeed possible to define the search space around the whole receptor. This approach is known as *blind docking*. A library of ligands is docked into the receptor in order to get an idea of the potential binding regions. The reliability of the blind docking results highly depends on the correct prediction of the binding regions, and represents a compromise between speed and accuracy.

4.3 Sampling protein and ligand conformational flexibility in docking

The main docking operations focus on the ligand. However, during the docking, several preliminary assumptions need to be made about the receptor flexibility. About 85% of proteins undergo conformational changes upon ligand binding, mainly movements in the essential binding site residues (Najmanovich et al., 2000). Therefore, performing accurate molecular docking is quite difficult, because of the many possible conformational states of both the biomacromolecule, and the ligand flexible areas. Depending on how conformational flexibility is handled during the docking, we distinguish between two classes. The *rigid body docking* method handles both binding partners as rigid bodies. The bond angles, bond lengths and torsion angles of the docking partners are not modified at any stage of the docking. By contrast, in *flexible docking* procedures, binding partners are considered as flexible molecules. This kind of procedure allows the specified atom or group of atoms to acquire the preferred position upon binding. Flexible docking is further categorized into two types: *flexible ligand* docking, where only the conformation of the ligand changes during the docking, and *flexible receptor docking*, where both the conformation of the ligand and the conformation of the receptor can change.

4.4 Sampling conformational and configurational space

Search space where we sample the structural arrangement of two molecules without changing the conformation of any of the molecule is called configurational search space. This term can be used for search space on rigid docking. Whereas in flexible docking, we search for the configurations of the system with two molecules, each of them being able to adopt several conformations. The configurational and conformational search is done via a set of algorithms that sample all the desired degrees of freedom of the ligand in order to find the correct binding mode. The set of operations performed to improve a binding mode is often referred to as *optimization*. Optimization is a difficult problem in docking, because it requires successful conformational search combined with an effective global sampling across the entire range of possible docking orientations.

There are basically three general categories of such algorithms, based on shape matching, systematic search and stochastic search, respectively.

Shape Matching is an approach based on the geometrical overlap between two molecules. The algorithm first generates a "negative image" of the binding site starting from the molecular surface of the receptor, which consists of a number of overlapping spheres of varying radii. The ligand is placed into the binding site using the surface complementarity approach, i.e., the molecular surface of the ligand has to attain maximum close surface contacts to the molecular surface of the binding site of the protein. To do this, ligand atoms are matched to the sphere centres of the negative image. The ligand can then be oriented in the binding site by performing least squares fitting of the ligand atom positions to the sphere centres. The degree of shape complementarity is measured by a certain score function. Maximizing this score function leads to the docked configuration. Note that this is not the function used in the second docking stage, though that one is also referred to as score or scoring function. Examples of docking programs which are based on this approach are DOCK (Kuntz et al., 1982), FRED (McGann et al., 2003) and MS-DOCK (Sauton et al., 2008).

Systematic Search algorithms try to explore all the conformational degrees of freedom of the ligand and combine them with the search on the system with the receptor. Depending on the way how the search is carried out, there are three main subclasses of systematic search algorithms.

A-*Systematic or pseudosystematic search*, where a huge number of poses are generated by rotating all the rotatable bonds by a given interval (in degrees). These poses are then filtered by using some geometrical and chemical constraints. The remaining poses are subjected to more accurate optimization. This hierarchal sampling method is currently used by the Glide (Friesner et al., 2004) and FRED (McGann et al., 2003) docking programs.

B-*Fragmentation methods* divide the ligand into small fragments (both rigid and flexible). First, a rigid core fragment is placed into the active site. Then, the more flexible fragments are sequentially linked by covalent bonds by using the "place-and-join" approach. Currently, docking programs like LUDI (Böhm, 1992), DOCK (Ewing & Kuntz, 1997), FlexX (Rarey et al., 1996) and eHiTs (Zsoldos et al., 2006) provide this methodology.

C-*Database or conformational ensemble* methods use an ensemble of pre-generated ligand conformations to deal with ligand flexibility, which is then combined with a search for proper receptor/ligand orientation. Databases or libraries of conformations can be generated within the docking program or separately, using other programs such as OMEGA (OpenEye Scientific, NM). FLOG (Miller et al., 1994) is a typical software using this methodology, but some other programs like MS-DOCK (Sauton et al., 2008) and Q-Dock (Brylinski & Skolnick, 2008) also offer this approach.

Random or ***stochastic*** methods are also available. They attempt to sample the space by making random changes to the receptor/ligand system. Whether a geometry change is accepted or rejected is decided using a predefined probability function. This may result in non-reproducible results, even if the docking is repeated with the same parameters. There are mainly four types of stochastic search algorithms.

A. *Monte Carlo (MC)* is used for a large set of optimization problems, ranging from economics, mathematics to nuclear physics or even regulating the flow of traffic. In docking, the ligand is first placed into the binding site of the receptor, and this binding mode is scored. A new geometry is generated by applying random changes to the rotatable bonds or the position of the ligand with respect to the receptor. The new binding mode is then scored. If the score of the new binding mode is better than that of the old one, this change is accepted. Otherwise, a probability (P) to accept the change is calculated as $P \approx \exp(-\Delta E / K_b T)$. Here ΔE is the change in score, K_b is Boltzmann's constant and T is the absolute temperature of the system. A random number (r), between 0 and 1, is generated, and if $r < P$, the change is accepted. After such an evaluation, another random change is applied to the ligand and the whole procedure is repeated until a reasonable number of orientations is obtained. AutoDock (Morris et al. 1998), ICM (Abagyan et al. 1994) and QXP (McMartin & Bohacek, 1997) are key examples of programs that use MC-based optimization procedures.

B. Genetic Algorithms (GA) are based on ideas derived from natural evolution, such as mutation, crossover, inheritance and selection. To solve the optimization problem, GAs simulate the survival of the fittest among individuals over consecutive generations. Each geometry of the ligand with respect to the protein is defined by a set of state variables called genes. Genes describe the translation, rotation and orientation of the ligand. A full set of a ligand's state variables is referred to as the genotype, whereas the phenotype is represented by the atomic coordinates. Genetic operations such as mutation, crossover, inheritance and selection are applied to the population until the fitness criterion is fulfilled.

Some of the most popular programs like AutoDock (Morris et al., 1998), GOLD (Jones et al., 1995, 1997), and Lead finder (Stroganov et al., 2008) include GA or hybrid approaches to find the optimal orientation of the ligand.

C. Tabu search (TS) is a meta-heuristic approach where a local search is combined with storing a list of previously considered geometries, along with a probability criterion, which ensures that only a new geometry will be sampled further. A random change is only accepted if the RMSD between the new conformation and any of the previously sampled geometries is greater than a threshold. The programs PRO_LEADS (Baxter et al., 1998) and PSI-DOCK (Pei et al., 2006) are TS based software.

D. Particle Swarm optimization (PSO) is one of the evolutionary computational techniques inspired by the social behaviour. SO exploits the population of individual to probe the premising region of search space. The population is called *swarm* and the individuals are called *particles*. These algorithms maintain a population of geometries by modeling swarm intelligence, a concept referring to the collective behaviour of otherwise fully independent particles. A number *particles* is randomly set into motion through this space. At each iteration, they observe the fitness of themselves and their neighbours and emulate successful neighbours (those whose current position represents a better solution to the problem than theirs) by moving towards them. The major advantage of PSO, compared with GA, is its relative simplicity and quick convergence. Examples of docking programs that use swarm optimization are SODOCK (Chen et al. 2007), Tribe-PSO (Chen et al., 2006), PSO@AutoDOck (Namasivayam & Günther, 2007).

4.5 Scoring ligand poses

Once a reasonable set of receptor/ligand geometries has been generated, ranking these modes is the second critical aspect of the docking procedure. To recognize the true binding modes from all the geometries, the binding affinity is scored using scoring functions, i.e., each binding mode is analysed by a set of equations and compared to the other binding modes. If the search algorithms predict a "correct" binding mode but the scoring function fails to rate this as a top scoring orientation, then the suggested output will be a false negative binding mode. Therefore, scoring functions should be able to distinguish between a true binding mode and all other modes explored. However, using a rigorous scoring function for several hundreds of binding modes is computationally expensive. Hence, computationally feasible empirical scoring functions are commonly used by all available docking software. Numerous scoring functions developed and evaluated so far can be grouped into three basic categories.

A. *Force field based*: A force field is a way to express the potential energy of the system by using a mathematical function and a set of parameters. A basic functional form of a force field encapsulates both bonded terms (between atoms that are linked by a covalent bond) and non-bonded terms (also called " non-covalent "). Non-bonded terms describe van der Waals and long range electrostatic forces. The generic equations (1-3) for force fields such as in AMBER (Weiner & Kollman, 1981) or CHARMM (Brooks et al., 1983), are expressed as:

$$V(r^N) = V(r^N)_{Bonded} + V(r^N)_{Non-bonded} \qquad (1)$$

$$V(r^N)_{Bonded} = \sum_{bonds} \frac{k_i}{2}(r_i - r_{eq})^2 + \sum_{angles} \frac{k_\theta}{2}(\theta_i - \theta_{eq})^2 + \sum_{torsions} \frac{V_n}{2}(1 + \cos(n\varphi - \varphi_0)) \qquad (2)$$

$$V\left(r^N\right)_{Non-bonded} = \frac{q_i q_j}{D r_{ij}} + \sum_{i=1}^{N} \sum_{j=i+1}^{N} \left(4\varepsilon_{ij} \left[\left(\frac{\sigma_{ij}}{r_{ij}}\right)^{12} - 2 \left(\frac{\sigma_{ij}}{r_{ij}}\right)^{6} \right] \right) \tag{3}$$

where r^N denotes geometry of the system, $V(r^N)$ is its potential energy, r_i and r_{eq} are the actual and equilibrium bond lengths, respectively, for the bond i, the θ_i and θ_{eq} is the same for bond angles, the φ and φ_0 is the same for dihedral angles, q_i and q_j are partial charges on the atom i and j, respectively; r_{ij} is distance of atoms i and j, D is dielectric constant and the remaining symbols are force field parameters.

Force field based scoring functions calculate the binding score as a sum of individual contributions made by various interactions in the bound complex. Force field based scoring functions commonly used in docking software mainly use non-bonded and torsion terms. The binding process normally takes place in water, so the desolvation energies of the ligand and the protein are sometimes taken into account implicitly. Since hydrogen bonding is one of the dominating interactions for the majority of complexes, some of the docking software, like AutoDock (Morris et al., 2009) and G-Score (Kramer et al., 1999), include a separate term for the treatment of hydrogen bonding.

B. *Empirical scoring functions*: Empirical based scoring functions, as in the case of force field methods, calculate the binding score of a complex as a sum of several weighted empirical energy terms that account for various types of non-bonded interactions. However, as opposed to the force field methods, empirical based scoring functions are much less systematic and general. The final score ΔG is calculated as a sum of weighted empirical energy terms, $\Delta G = \sum W_i * \Delta G_i$, where ΔG_i represents individual empirical energy terms, such as vdW energy, electrostatic energy, hydrogen bonding, desolvation, hydrophobicity, entropy etc., while W_i is the corresponding weight coefficient for a particular energy term, determined by linear fitting to an experimental data set. A set of X-ray receptor ligand complexes and their corresponding experimental binding energies are usually used as training data to calculate the weight coefficients by regression analysis. Due to the simple nature of the equation, these methods are computationally much more efficient compared to force field based methods. However, there are also significant drawbacks. General applicability of these functions is strongly dependent on the experimental data set used for their parametrization. It is not reliable to use such a scoring function for a data set that is structurally different from the training set. Glidescore (Halgren et al. 2004), LigScore (Krammer et al., 2005), and X-Score (Wang et al., 2002) are examples of software using empirical scoring functions.

C. *Knowledge based scoring Functions*: Knowledge based scoring functions use the sum of the potential of mean force (PMF) between the protein and the ligand, using data derived from 3D structure databases. These scoring functions are based on capturing the protein ligand atom pair frequency of occurrence in the structural database. It is assumed that each interaction type between a protein atom of type *i* and a ligand atom of type *j*, found at a certain distance r_{ij}, has an interaction free energy $A(r)$, which is defined by an inverse Boltzmann relation (Eq. 4).

$$A(r) = -K_b T \ln\left[\rho(r) / \rho^*(r)\right] \tag{4}$$

where K_b is the Boltzmann constant, T is the absolute temperature, $\rho(r)$ is the density of occurrence of the atom pair at distance r in the training set and $\rho^*(r)$ is this density in a reference state where the atomic interactions are zero.

The advantage of knowledge based scoring functions over empirical scoring functions is that there is no fitting to the experimental free energy of the complexes in the training set, whereas solvation and entropic effects are included implicitly. It should be noted that knowledge based scoring functions are used to reproduce the experimental structures rather than to predict binding energies. They can identify non-binders on their own or in combination with some other docking software during virtual screening. Since not all the possible interactions can be inferred from the crystal structure, these scoring functions may not be so robust and accurate, but they usually offer a good balance between speed and accuracy.

4.6 Techniques to improve the performance of scoring functions

Consensus scoring: This is a combination of the information obtained from different scores. The approach is helpful in balancing out the error of individual scoring functions, thus improving the probability of finding an appropriate solution. Several published studies show that combining the scores from different methods performs better than considering only the individual scores. MultiScore (Terp et al., 2001) and X-Score (Wang et al., 2002) are the most popular examples using consensus scoring.

Clustering: We often find an incorrect geometry with a slightly more favorable binding score than the correct geometry. However, these incorrect geometries are found with a very low frequency (~1-2%) when multiple docking experiments are performed. Thus, RMSD based clustering of all the docking solutions can be performed. To get the correct pose, the best energy conformation from the most populated cluster should be chosen.

4.7 Description of some commonly used docking programs

Table 1.1 summarizes the main features, license type and source for the most popular docking programs. We further provide a more detailed description of a few selected pieces of docking software. We would like to state that these methods are not necessarily the most accurate ones, but they are definitely the most widely used and the most cited in the docking community.

AutoDock: AutoDock3 (Morris et al., 1998) and AutoDock4 (Morris et al., 2009) are force field based docking programs which have been widely used for the automated docking of small molecules, such as peptides, enzyme inhibitors and other ligands, into macromolecules, such as proteins, enzymes and nucleic acids. AutoDock offers optimization procedures like simulated annealing, genetic algorithm (GA) for global searching, a local search (LS) method to perform energy minimization, or a combination of both (GALS) for getting the accurate docked complex. The scoring function used in AutoDock is inspired by the MD programs AMBER, CHARMM or GROMOS, it includes terms for the Lennard-Jones potential, Coulombic electrostatic potential, hydrogen bonding, partial entropic contribution, desolvation upon binding and a hydrophobic effect. The scaling parameters for these terms were derived from a set of 30 protein-ligand complexes.

Software	Ligand sampling methods[a]	Receptor sampling methods[a,b]	Scoring function[c]	Solvation scoring[b,d]	License type[e]	Source
AutoDock3	SA, GA	NA	MM+ED	DDS, DS	FAS	(Morris et al., 1998)
AutoDock4	SA, GA	SE	MM+ED	DDS, DS	FAS	(Morris et al., 2009)
AutoDock Vina	CB	CB	ML	NA	OPS	(Trott et al., 2009)
DOCK6	IC	SE	MM	DDD/GB/PB	FAS	(Kuntz et al., 1982)
ICM	MC	MC	MM+KB	DDD,PBE,DS	CPL	(Abagyan et al., 1994)
Glide	CE+MC	TOS	MM+ED	DS	CPL	(Halgren et al., 2004)
GOLD	GA	NA	MM+ED	NA	CPL	(Jones et al., 1995, 1997)
FlexX/FlexE	IC	SE	MM+ES	NA	CPL	(Rarey et al., 1996)

[a]Sampling methods can be Genetic Algorithm (GA), Conformational Expansion (CE), Monte Carlo (MC), Simulated Annealing (SA), Molecular Dynamics (MD), Incremental Construction (IC), Merged Target Structure Ensemble (SE), a combination of GA, SA and MC (CB), and Torsional Search (TOS); see Section 4.4 for more information. [b]If the package does not accommodate this option, the symbol NA (not available) is used. [c]Scoring functions can be Empirical (ES), Knowledge Based (KB) or force field (MM) based; see Section 4.5 for more information. [d]The accuracy of the scoring function can be improved using implicit solvent models. Solvation scoring can be done using Distance-Dependent Dielectric (DDD), Poisson Boltzmann Dielectric (PBE), a parameterized desolvation term (DS), Generalized Born (GB), and linearized Poisson Boltzmann (PB) equations. The license type can be [e]Freely available (FAL), Open Source (OPS) and Commercial Paid License (CPL) for academic users only.

Table 1. Details of Commonly Used Docking Software.

The advantage of AutoDock4 over AutoDock3 is that it allows receptor flexibility, and also an improved new force field is used to calculate the binding energy. The force field of AutoDock4 includes a new intramolecular term, and a full desolvation model for desolvating polar and charged atoms. AutoDock facilitates the clustering of all the docked orientations by defining a root mean square tolerance, which can also be used to find the potential binding regions. It was seen that the lowest energy structure in the most populated cluster successfully reproduces the crystal structure.

AutoDock Vina (Trott & Olson, 2009): It is a new generation of docking software (referred to as Vina) from the Molecular Graphics Lab, the developer of the other versions of AutoDock. It is a user friendly, open source piece of software, capable of predicting binding modes with better accuracy, while it is significantly faster than AutoDock4. It uses a combination of optimization algorithms, such as the genetic algorithm, swarm optimization and simulated annealing, to place the ligand in the binding site. The scoring function used in Vina is more based on machine learning rather than directly on a force field. Similarly to AutoDock4, it allows receptor flexibility.

The philosophy behind the development of Vina was to make the software easy to use, so most of the parameters used during docking are set by default, reducing the possibility of making manual mistakes. A further speed up in docking is achieved by multithreading. Thus, overall, Vina is very suitable for docking a large set of different compounds.

DOCK: The program package DOCK (Kuntz et al., 1982, currently version 6.4) basically works in a few subsequent steps. First, the program "sphgen" is employed in order to identify a binding site and to generate spheres within the active site. Secondly, the program "grid" is used to generate scoring grids. Then the last program "DOCK" matches the sphere with the ligand atoms and uses the scoring grid to evaluate the ligand orientation. It constructs the ligand in the binding site step by step using the Anchor-and-Grow algorithm. Initially, the rigid anchor fragment of the receptor is placed at a selected position, and then is gradually enlarged by adding the flexible fragments of the ligand. An additional extension to DOCK allows rescoring the docked configuration of the ligand using several secondary scoring functions.

ICM: The Internal Coordinates Mechanics (ICM) software is a set of modules for various purposes, such as visualization, chemical drawing and editing, homology modeling, docking and virtual screening (Abagyan et al., 1994). The ICM-Docking and chemistry module performs flexible ligand docking in a grid based receptor field. The scoring function used in ICM primarily accounts for electrostatics, van der Waals, hydrogen bonds, and the hydrophobic term. ICM needs the protein structure to be converted into an ICM object before docking. It provides a simple, object based GUI which can be used for docking. The binding site can be defined by entering the binding site residues, using the graphical selection tool, or the implemented icmPocketFinder function. It generates receptor maps within the defined boundary, which are further used in the docking. It is necessary to define the initial position where sampling will begin. ICM facilitates interactive, as well as batch docking. In interactive docking, one ligand is docked at a time in the foreground, whereas batch ligand docking runs in the background and is thus ideal for large scale docking jobs and virtual screening of huge ligand libraries. ICM offers an attractive feature to visualize and browse the docking results, and scan the hit compounds.

BALLDock/SLICK: BALLDock/SLICK (Kerzmann et al., 2008) is specially designed for docking of carbohydrate like compounds, with applications in carbohydrate based drug design. Molecular docking of protein-carbohydrate complexes needs some special attention because of the special features of such interactions, such as the unusual flexibility of carbohydrates, stacking interactions with aromatic amino acids and a high number of hydrogen bonds involved in binding. Protein carbohydrate interactions are strongly influenced by CH···π interactions, which are mostly ignored in the commonly available scoring functions and are considered in BALLDock/SLICK. This docking program uses genetic algorithms to search the configurations of the ligand within the defined search space, and the scoring function SLICK is used to calculate the binding score of the docked conformations. Kerzmann et al. compared the performance of BALLDock with FlexX on a set of 22 lectins and sugar-binding proteins complexed with carbohydrate. FlexX achieved good results but still did not reach the predictive accuracy of BALLDock/SLICK.

TRITON: We previously introduced our in house graphical tool TRITON, and mentioned its use for homology modeling and mutagenesis. Another functionality of TRITON relates to *in silico* engineering of protein-ligand binding properties (Prokop et al., 2008). The program can be used as a graphical user interface for the docking software AutoDock3, AutoDock4 and Vina. It enables the user to do common pre-docking tasks, like creating a project directory, reading structures, manipulating structures, calculating various types of charges and finally preparing input files for docking. Docking wizards make the job easy for new users, where the step by step procedure decreases the possibility of missing any of the docking parameters. It includes and offers certain optimized docking parameters that can be used as basic starting points if the user is not sure about certain docking parameters used in the AutoDock suite of programs. TRITON also includes parameters for ions taken from case specific studies, so it is easy to handle ions during the docking. Another important feature of TRITON is that it facilitates interactive analysis and visualization of the docking results.

5. Free energy calculation

As discussed above, various docking software can successfully predict the correct binding mode of the ligand into the receptor (Taylor et al., 2002; Warren et al., 2006). However, the previously described empirical scoring functions, which are based on a single receptor/ligand structure, do not provide accurate enough predictions of the binding free energy (ΔG), the key quantity characterizing the strength of the receptor/ligand interaction. To tackle this problem, molecular dynamics (MD) or Monte Carlo (MC) based methods for free energy calculation were developed in the mid 1980s (Jorgensen & Ravimohan, 1985). These methods, formally rooted in statistical thermodynamics, are now frequently used to compute receptor/ligand binding free energy. The methods use molecular mechanics force fields and Newtonian physics to evaluate the dynamics of the system. In the case of MD, we follow the evolution of the dynamics of the system in time. The dynamics allows the system to accommodate various protein side chains as well as ligand conformations, and also ligand configurations with respect to the protein. Simulations are usually performed in the bound state. Here, we will discuss the methods most commonly used to evaluate the binding free energy between the receptor and the ligand, namely Free Energy Perturbation (FEP), Thermodynamic Integration (TI), and Molecular Mechanics Poisson-Boltzmann Surface Area (MM-PBSA). We will also give some notes about the combined molecular mechanics/quantum mechanics (QM/MM) techniques.

5.1 Free Energy Perturbation (FEP) and Thermodynamic Integration (TI)

The FEP and TI approaches for free energy calculation are based on statistical thermodynamics and are generally formulated not to calculate the absolute value of the free energy, but always a relative value, i.e., the free energy difference, ΔG, between two equilibrium states. This is of a great importance, since for *in silico* mutagenesis applications we always need only relative values.

The FEP and TI free energy calculations are carried out using a thermodynamic cycle. Such a cycle, adapted for *in silico* mutagenesis purposes, is shown in Fig. 6. It involves a mutation of either the receptor alone, or the receptor/ligand complex (start state) into another state (end state), where the receptor is mutated. The simulation can be performed in either implicit or explicit solvent. The final calculated quantity is $\Delta\Delta G$. This number will tell us whether the mutated protein (P_M) exhibits higher or lower affinity to the ligand L compared to the wild type protein (P_W).

As the start and the end state can be arbitrarily different, these calculations are sometimes referred to as *computational alchemy*.

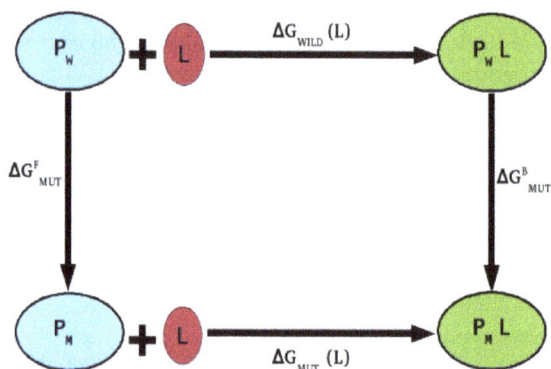

Fig. 6. Thermodynamic cycle for calculating the relative binding free energies of a ligand L to mutated system (P_M). ΔG^F_{mut} is free energy change between the wild type and mutated receptor, ΔG^B_{mut} is free energy change between the wild type receptor/ligand complex and the mutated receptor/ligand complex, $\Delta G_{wild}(L)$ and $\Delta G_{mut}(L)$ are binding free energies for the wild type receptor/ligand and mutated receptor/ligand complexes, respectively.

As the free energy is the state function, Eq's (5 and 6) must hold.

$$\Delta G_{wild}(L) + \Delta G^B_{mut} = \Delta G^F_{mut} + \Delta G_{mut}(L) \tag{5}$$

$$\Delta\Delta G = \Delta G_{mut}(L) - \Delta G_{wild}(L) = \Delta G^B_{mut} + \Delta G^F_{mut}(L) \tag{6}$$

The **FEP** calculations are based on the Zwanzig's formula (Zwanzig, 1954) to calculate the free energy difference ΔG between two states (see Eq. 7).

$$\Delta G^{FEP} = G_B - G_A = -k_B T \ln \left\langle \exp\left(\frac{V_B - V_A}{k_B T}\right)\right\rangle_A \tag{7}$$

where k_B is the Boltzmann constant, T is the absolute temperature, $<>_A$ denotes the MD or MC ensemble average over a simulation run for state A, V_A and V_B are the potential energies of state A and B, respectively. In general, an ensemble is an average set of systems, that are identical in all respect apart from the dynamics of the atom (k/a ensemble), considered all at once, each of which represents a possible state that the real system might be in. The potential energy difference can be averaged over an ensemble generated using the start and end state potential function for the forward and backward process, respectively.

The goal is to obtain the convergence of the values resulted from Eq. 7 within a reasonable time. It is assumed that the relevant geometries sampled on the potential energy of state A have a considerable overlap with those of state B.

The transition of state A into state B may also yield high energy geometries in the complex because of steric clashes with the neighbouring atoms. To overcome this issue, transition is done via many non-physical intermediate states that are usually constructed as a linear combination of the potential calculated for the start and end state. The potential energy of an intermediate state between A and B is given as shown in Eq. 8,

$$V_\lambda = (1 - \lambda)V_A + \lambda V_B \tag{8}$$

where λ varies from 0 to 1. This state is a hypothetical mixture of states A and B: when $\lambda=0$, $V_\lambda=V_A$, and when $\lambda=1$, $V_\lambda=V_B$. Therefore, the transformation of state A into state B is done smoothly, by changing the values of the parameter λ in small increments, $d\lambda$. In practice, the free energy difference between the states A and B is computed by summing over all the intermediate states along the λ variable (Eq. 9).

$$\Delta G = \sum_i dV \lambda i \tag{9}$$

This approach of breaking down the transitions into multiple smaller steps shares similarity with another approach used to compute free energy, namely Thermodynamic Integration (TI) (Kirkwood, 1935). **TI** is based on integrating a different equation from statistical thermodynamics, where the free energy difference between two states is obtained by integrating the derivative of the mixed potential function over λ (Eq. 10).

$$\Delta G^{TI} = \int_0^1 \left\langle \frac{\delta V(\lambda)}{\delta \lambda} \right\rangle_\lambda d\lambda \approx \sum_i w_i \left\langle \frac{\delta V(\lambda)}{\delta \lambda} \right\rangle_{\lambda_i} \tag{10}$$

In this case, the mixed potential $V(\lambda)$ is defined numerically by evaluating the linear interpolation between the potential function of the start and end state, respectively.

In principle, both FEP and TI should give the same results, as the free energy is a state function.

The relative binding free energy difference $\Delta\Delta G$ between the wild type protein P_W and its mutant P_M can easily be calculated from Eq's. 5 and 6 (for denotation see Fig. 6), and where ΔG^F_{mut} and ΔG^B_{mut} are calculated using the above described FEP or TI methods for the free and bound state, respectively.

As mentioned before, with the FEP or TI approach, the free energy associated with the two unphysical paths $P_W \rightarrow P_M$ (mutation in the free state) and P_W (L) $\rightarrow P_M$ (L) (mutation in the bound state) is calculated by sampling the degrees of freedom of the free protein or the complex using molecular dynamics (MD) or Monte Carlo (MC) methods. At regular intervals, the atoms of the residue which is being mutated are replaced by atoms of the residue which is desired at that place, and the potential energy along the paths is recorded. This quantity, averaged over the complete simulation, gives a proper free energy change ΔG_{mut}. However, the convergence of the free energies is a first critical issue in the accurate calculation of the binding free energy. This requires exhaustive sampling of the system, which is much more time consuming than docking or normal MD simulations. Moreover, the mutation may cause steric clashes with the neighbouring atoms, which makes the sampling issue even more complicated.

5.2 Molecular mechanics poisson-boltzmann surface area (MM-PBSA)

Another approach well suited for estimating the binding free energy of molecular complexes and their mutants is the Molecular Mechanics Poisson-Boltzmann Surface Area (MM-PBSA) method (Srinivasan et al., 1998). The MM-PBSA approach was initially used to study the stability of nucleotide fragments, but also to compute the relative or absolute binding free energy of protein-ligand complexes. Later extensions (see Kollman et al., 2000; Hou et al., 2011) have enabled employing the method for free energy calculation in *in silico* mutagenesis approaches, which is helpful in making predictions for protein engineering. Unlike FEP and TI, MM-PBSA is an endpoint method that calculates binding free energy without consideration of any intermediate state.

The MM-PBSA approach is used to calculate the free energy change ΔG_{bind} upon ligand binding according to equation 11. It combines the molecular mechanical energies with the continuum solvent approaches, and approximates the average of each state in order to calculate the binding free energy.

$$\Delta G_{bind} = <G_{complex}> - \left(<G_{receptor}> + <G_{ligand}>\right) \tag{11}$$

The single terms are defined by Eq's 12-14.

$$G_X = H - TS = E_{MM} + G_{sol} - TS \tag{12}$$

$$E_{MM} = E_{internal} + E_{electrostatic} + E_{vdw} \tag{13}$$

$$G_{sol} = G_{PB/GB} + G_{SA} \tag{14}$$

where X stands for the complex, receptor or ligand, T is the absolute temperature. The E_{MM}, G_{sol} and S are the gas phase molecular mechanics energy, solvation free energy and entropy, respectively. The E_{MM} includes several energy terms: $E_{Internal}$ for bond, angle and dihedral contributions, $E_{electrostatics}$ for coulomb interactions and E_{vdw} for van der Waals energies. G_{sol} is the sum of electrostatic solvation energy and non-polar contribution to the solvation free energy. The electrostatic contribution to the solvation free energy is calculated by solving either the linearized Poisson Boltzman (PB) or Generalized Born (GB) equation, while the non-polar contribution is estimated from the solvent accessible surface area (Connolly, 1983). If the solvation free energies are computed from the Generalized Born (GB) model, the method is termed also MM-GBSA. The last term, TS, includes the solute entropy S, which is usually calculated by quasi-harmonic analysis of the snapshots using normal mode analysis (Srinivasan et al., 1998).

Ideally, this approach is based on post-processing molecular dynamics trajectories. The free energy contributions are calculated for each component of the system (protein, ligand and the complex) from the snapshots taken from MD trajectories. In order to get the binding free energy of a ligand, two alternatives are used (see Fig. 7). The first is a multi trajectory approach, where we use the trajectories from three separate molecular dynamics simulations (on the complex, receptor and ligand). Snapshots of each component (protein, ligand and complex), taken from their corresponding simulation trajectories, are used to calculate the free energy terms. Note that this approach takes into account the influence of conformational changes upon binding on the final binding free energy.

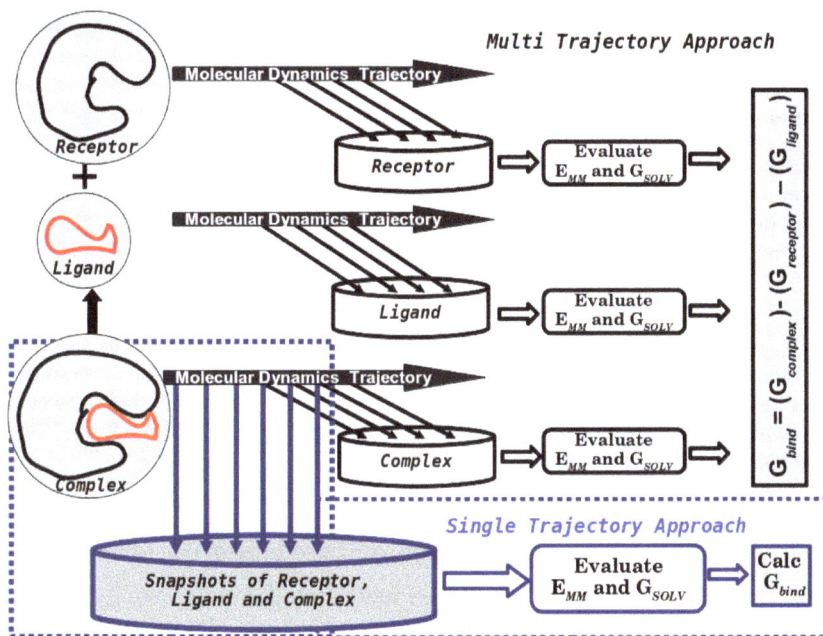

Fig. 7. Diagram for MM-PBSA and MM-GBSA calculations on a solvated complex. Single trajectory approach is surrounded by a blue dotted line.

In the second approach, molecular dynamics simulations are run on the complex only, in order to reduce noise and cancel out the errors in the simulations. Conformational snapshots for the receptor alone and the ligand alone are extracted from the MD simulation of the complex by removing the respective binding partner from the complex. Therefore, it is assumed that the structure of the receptor and ligand is the same in the bound and the free state, and no major conformational changes occur upon binding. In this approach, $E_{Internal}$ is canceled out between the complex, protein and ligand, which reduces the noise in calculations.

In principle, the first approach of running three independent molecular dynamics simulations of three species is more accurate than the single trajectory approach. In practice though, the multi trajectory approach seems not to be used extensively. This is understandable, since there is no proper way to get the convergence of E_{MM} values for the receptor within reasonable computational time. Hence, the regular implementation of this method is usually based on the second approach, where only the MD trajectory of the complex is used to compute the binding free energy.

A fundamental issues associated with the MM-PBSA approach is entropy calculation. The normal mode analysis (NMA) approach is usually employed to calculate entropy. However, this approach overestimates the loss of entropy upon ligand binding. In order to get meaningful absolute binding free energies, the entropy contribution must be determined in a consistent fashion. The best approach is to compute the relative binding free energies of a series of similarly sized ligands, where the entropy contribution is expected to cancel out

(Massova & Kollman, 1999). The *in silico* mutants of a protein are also expected not to have a significant change in entropic contribution to the binding.

5.3 Alanine scanning mutagenesis using MM-PBSA

In the *in silico* mutagenesis section we discussed the basis of the methodology of performing single point or multiple alanine mutations using computational approaches. Here we will only show how to use alanine scanning mutagenesis coupled with the MM-PBSA approach. We are distinguishing between two complementary problems of mutagenesis where binding free energy is calculated by the MM-PBSA approach. The first refers to the change in binding free energy upon alanine mutation at any location. This can be solved using the previously described single trajectory MM-PBSA approach on two systems, the wild type and the mutant. Molecular dynamics simulations of two systems (ligand complexed with the wild type and with the alanine mutant, respectively) are run under the same conditions. These two different trajectories are subjected to the MM-PBSA calculation previously described. The change in binding free energy upon mutation is now the difference between the binding free energy of the mutant and that of the wild type. In principle, this approach is accurate and recommended because it samples the conformational changes of the system upon mutation and takes into account their effect on the change in free energy.

The second issue refers to the individual contribution of each residue to the binding. MM-PBSA was first used in this respect in a study where a single MD simulation was used to compute the individual contribution of each residue to the binding in protein–protein complexes. Snapshots of mutants are generated from a single molecular dynamics trajectory of a wild type system. Mutations are performed by removing side chain atoms beyond the β carbon of the amino acids under investigation (Massova & Kollman 1999). These snapshots are used for binding free energy calculations by the MM-PBSA approach. The approach used in alanine scanning mutagenesis is depicted in Figure 8.

On the one hand, this is not a very accurate way to get the change in free energy upon alanine mutation. On the other hand, it is very fast, as the mutations can be performed at any location without running the molecular dynamics simulation of the mutant system. Therefore, once we have the MD trajectory of the wild type, a possible primary scan for all the locations could be done in minimal computational time. Since the method uses the MD trajectory of the wild type to create the mutants, it is assumed that the receptor/ligand complex adopts the same geometry upon the mutation. This is a limiting factor, as mutations of the residues around the ligand binding site can substantially affect the binding geometry. Nevertheless, it is expected that this approach can estimate the free energy contribution made by a particular residue compared to the wild type system (Moreira et al., 2007). This approach is mainly recommended for finding the hot spots in protein-protein interactions. Hot spots are residues which make a contribution of about 2.0 Kcal/mol to the total binding free energy of the system, and are very important from the protein engineering point of view because they can be used as key points to alter the protein's recognition ability. Alanine scanning also gives an idea about the residues which are close to the binding region, but do not contribute substantially to the binding energy. These locations in the protein can be used to make the binding stronger if a residue with favourable properties is placed at that location. The MM-PBSA approach is quite fast. The calculations for hundreds of ligands and hundreds of mutants are feasible using high performance computing facilities. One must mention here that these approaches are approximate, and the relevant predictions should be verified using FEP or TI before experimental trials.

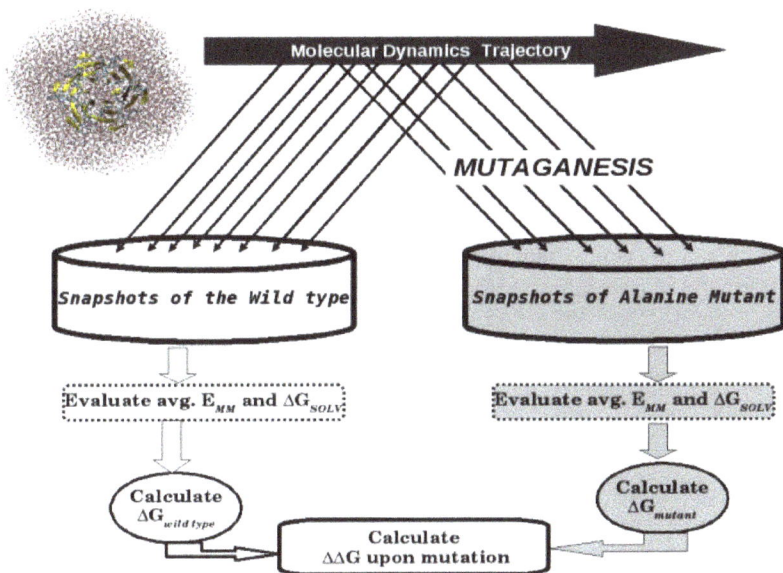

Fig. 8. showing a single trajectory alanine scanning mutagenesis approach used with MM-PBSA or MM-GBSA calculations.

5.4 Hybrid Quantum Mechanics/Molecular Mechanics (QM/MM) approaches

A combination of quantum mechanics and molecular mechanics (QM/MM), accompanied by the increasing computational power of modern parallel and vector-parallel platforms, has brought a real breakthrough in the simulation of large systems. Here we describe the current QM/MM strategies used for quantifying the binding energy of complexes involved in molecular recognition.

The seminal contribution made by Warshel et al. in 1976 marks the beginning of the QM/MM era (Warshel & Levitt, 1976). In brief, to model large biomolecules one uses a QM method to model the active region (originally substrates and co-factors of an enzymatic reaction), and an MM method for the treatment of the surroundings (e.g., protein and solvent). The QM/MM approaches are relatively new in the field of molecular docking. A few years ago, a combined QM/MM docking approach for the investigation of protein ligand complexes was presented for the first time, and very promising results were obtained by combining the fast docking technique with the subsequent QM/MM optimization of the docked structure (Beierlein & Clark, 2003). Later, in an attempt to develop a docking algorithm which can predict poses accurately for the cases where the conventional approach fails, QM/MM calculations were integrated in the scoring phase (Cho et al., 2005). A protein-ligand docking study of 40 complexes investigated through QM/MM based docking calculations suggests that the use of fixed charges during the docking exhibits on-trivial errors. Therefore, polarization of the QM region is suggested to be crucial for docking studies. It was found that including also some protein atoms in the QM region, along with the ligand atoms, increases the success rate of QM/MM docking procedures (Cho & Rinaldo, 2009).

There are also examples in literature where a QM/MM approach was used to calculate the binding free energy. For example, Gräter and coworkers evaluated the performance of a QM/MM approach combined with MM-PBSA to obtain the protein/ligand binding free energy for a set of 47 benzamidine derivatives binding to trypsin. The QM/MM-PBSA methods reproduced the experimental binding energy well, with a root-mean-square (RMS) error of 1.2 kcal/mol (Gräter et al., 2005). Later, QM charge densities were used to solve the PB equation in a test case of binding of balanol and its derivative to the protein linase A (Wang & Wong, 2007). Even if this approach is not being used very frequently in the field of binding free energy calculation, the availability of packages (e.g. Amber Tools 1.5) that facilitate such a QM/MM-PBSA calculation in protein/ligand complexes, along with recent developments, is expected to make the QM/MM-PBSA method more user friendly.

Nowadays, all the statistical mechanics techniques to determine free energy differences through sampling, e.g., TI, umbrella sampling, or FEP are being used in conjunction with semiempirical QM/MM methods (Chung et al., 2009; Tuttle, 2010). The continuous increase in computer power has played an essential role in the development of these methods. QM/MM methods are expected to be especially important in the field of molecular recognition for systems where ions are present, i.e., in the area of metalloproteins.

6. Case studies

We present a few studies where *in silico* protein engineering was successfully used to study molecular recognition. We compare our results with other recently published studies on altering the binding specificity of a receptor by using *in silico* mutagenesis. The citations particular to the studied systems and to the methods mentioned below are omitted here, and can be found in the respective papers.

6.1 Engineering of the PA-IIL lectin to understand its sugar preference

Lectins are proteins of non-immune origin that recognize carbohydrates with high specificity and affinity. They belong to a large family of proteins whose unifying feature is the ability to decode the information stored in the glycome. Lectins are involved in a diverse set of biological processes, such as cell-cell recognition, differentiation, signalling or the adhesion of infectious agents to host cells. Many of these functions are connected with the recognition of specific saccharide structures on the cell surface. Carbohydrate-binding proteins play a key role also in host cell recognition by pathogens, as their specific adhesion to the host cell tissue is the first stage of their infectivity. Thus, lectins from pathogens represent a primary target for anti-adhesion therapy, having a great potential in the field of drug design.

The selected study (Adam et al. 2007, 2008) is based on the *in silico* protein engineering of the protein PA-IIL, a lectin from an opportunistic human pathogenic bacterium *Pseudomonas aeruginosa*, which causes lethal complications in cystic fibrosis patients. PA-IIL is a tetrameric lectin characterized by an unusually high (micromolar) affinity to L-fucose, which is atypical in protein-carbohydrate binding. Lectins homologous to PA-IIL later identified in other microorganisms, display high sequence and structure similarity, but strongly differ from each other in terms of sugar preference. For example, the lectin RS-IIL from *Ralstonia solanacearum* strongly prefers D-mannose over L-fucose. Three amino acid residues, at positions 22–23–24, were identified as the key residues that describe the relationship between structure and binding specificity for these lectins, and were named the "specificity binding loop". Given the capital relevance of this loop, *in silico* approaches were applied to understand the precise role of the specificity loop in the sugar binding preference.

saccharide	Expt[a]	AD3[b]	DOCK (Std)[b]	DOCK (Amber)[b]
PA-IIL				
Me-α-L-Fuc	−8.71	−10.7	−40.01	−32.22
Me-α-L-Gal	−8.09	−9.93	−40.02	−28.92
α-L-Fuc	−6.94	−9.33	−35.92	−27.56
Me-α-D-Man	−5.97	−9.23	−43.50	−26.88
S22A				
Me-α-D-Man	−7.58	−10.47	−32.60	−22.30
Me-α-L-Fuc	−7.39	−9.26	−30.60	−23.47
α-L-Fuc	−6.84	−8.96	−28.81	−19.85
Me-α-L-Gal	−6.60	−9.49	−31.70	−20.34
S23A				
Me-α-L-Fuc	−9.04	−10.49	−37.90	−20.82
Me-α-L-Gal	−8.11	−10.79	−38.83	−19.86
α-L-Fuc	−7.32	−8.98	−35.43	−18.81
Me-α-D-Man	−5.84	−9.15	−39.95	−16.80
G24N				
Me-α-L-Fuc	−9.19	−9.65	−39.84	−28.19
Me-α-L-Gal	−8.06	−9.83	−38.42	−22.26
α-L-Fuc	−7.18	−9.29	−37.10	−22.48
Me-α-D-Man	−5.96	−9.15	−45.03	−25.16

Table 2. Experimental (Expt) and calculated energies of monosaccharides binding to PA-IIL and its mutants obtained from AutoDock3 and DOCK (all values in Kcal/mol). AD3 stands for AutoDoc3 binding energies, DOCK (std) for energies from inbuilt evaluation of the DOCK software and DOCK (Amber) for energies from DOCK reevaluated by AMBER. Saccharides used: α-L-Fuc (α-L-Fucopyranose), Me-α-L-Fuc (Me-α-L-fucopyranoside), Me-α-L-Gal (Me-α-L-galactopuranoside), Me-α-D-Man (Me-α-D-mannopyranoside). Values taken from [a]Adam et al, 2007 and [b]Adam et al, 2008

The dimeric structure of PA-IIL was used as a template structure for the homology modeling of three single-point mutants (S22A, S23A, and G24N matching amino acids in RS-IIL) of PA-IIL using our *in house* developed software TRITON, interfaced with MODELLER. In order to understand the role of a particular mutation with respect to sugar preference, different monosaccharides were docked into PA-IIL and its mutants using AutoDock3 and DOCK. Since PA-IIL has two Ca^{++} ions in the binding site, which mediate the sugar binding, the effect of their charge on the docking energy was also evaluated. A formal charge on Ca^{++} equal to 1.8 and 2.0 gave results in good agreement with experiment. The value of 1.8 was chosen as a compromise, because Ca^{++} surrounded by several negatively charged oxygen atoms adopts a smaller charge in reality (about 1.5, see Mitchell et al, 2005).

The docking simulations produced a series of binding energies for the possible complexes of saccharides bound to PA-IIL and its mutants. The results can be seen in Table 1.

Overall, the docking results from AutoDock3 confirm that PA-IIL has higher preference for Me-α-L-Fuc (-10.7 Kcal/mol) over Me-α-D-Man (−9.23 Kcal/mol), and the sugar preference switches from Me-α-L-Fuc (-9.26 Kcal/mol) to Me-α-D-Man (-10.47 Kcal/mol) upon the S22A mutation. Docking inside two other mutants S23A and G24N also shows the order of preference again similar to what can be observed experimentally. Qualitatively, DOCK overestimates the binding energy in more cases than AudoDock3. Compared to experimental results, the AutoDock3 results reproduced the experimental order of saccharide preference to a large extent. The authors conclude that the automated docking methods are capable of identifying preference trends, and, therefore, using *in silico* approaches in pre-planning the *in vitro* mutations can help to identify the best potential candidates for mutagenesis.

6.2 Double mutant avian H5N1 virus hemagglutinin

The study (Das et al., 2009) shows how the free energy perturbation approach is used to compute the binding affinity of hemagglutinin (HA) to sialylated glycan epitops. A typical influenza infection, caused by avian influenza A viruses (H1N1, H2N2, H3N2 and H5N1 subtypes), requires binding of the viral surface glycoprotein hemagglutinin (HA) to sialylated glycans present on the host cell surface in order to initiate the infection. A change in the binding specificity of the HAs from α-2,3 (common in avians) to α-2,6-linked (common in human) sialylated glycans is expected to facilitate transmission of the virus from avians to humans. Therefore, molecular recognition of the particular glycans, considered as a key point for such infections, was inspected using mutagenesis studies.

HAs are homo trimers, with each monomer comprising of two subunits. The Receptor Binding Domain (RBD) of HAs, formed by basically 3 loops, requires at minimum two mutations to switch receptor specificity from avian to human. It is also known that hemagglutinin H1 changes its specificity from human to avian epitopes after two mutations (D190E and D225E). The authors were interested in finding whether a double mutation in hemagglutinin H5 enable it to recognize the human analog, as it is seen for the H1 HA subtype.

The authors used *in silico* approaches to interpret and predict the critical mutations responsible for HA-receptor binding. In order to achieve this, the change in relative binding affinity of H5 HA to sialylated glycans upon mutation was calculated through free energy perturbation approaches. The change in binding energy due to a mutation is evaluated using a thermodynamic cycle (see Fig. 6), where $\Delta\Delta G_{bind}$ is calculated from the free energy change caused by the same mutation for the bound and free states respectively. This simulation was performed over 22 λ points, where each window was simulated for 0.3 ns. Therefore, a total of 66-ns of simulation were performed for each mutation in order to get proper sampling. The authors claim that before analyzing the effect of novel mutations on the H5 HA receptor, they validated their protocol by comparing the calculated binding affinities against experimental data for other mutants.

The authors conclude that the FEP calculations are in a fairly good agreement with the glycan array data, which was available for only a few H5 HA mutants. Most of the evaluated mutations resulted either in no change, or in weak binding affinity to α-2,6-linked sialylated glycans compared to α-2,3. They identified that a double mutation (V135S and A138S) in H5 HA enhances the specificity towards α-2,6-linked sialylated glycans: $\Delta\Delta G_{bind}=$ -2.56+-0.73 Kcal/mol for the human receptor, compared to $\Delta\Delta G_{bind}=$ 0.84+-1.02 Kcal/mol for the avian receptor. To validate the results, the authors repeated the calculations for the same mutants on H5 HA obtained from a different isolate, which also revealed a substantial increase in the binding affinity for the human receptor. In order to understand the forces behind the recognition, they performed a free energy component analysis and saw that the electrostatic interactions are the driving forces for change in binding specificity upon mutation.

Thus, this study used computational approaches to provide valuable insight into the molecular recognition of glycans. This is another example where *in silico* protein engineering approaches were used as a complementary tool to interpret and understand molecular recognition.

6.3 Structural basis of NR2B-selective antagonist recognition

The third example (Mony et al., 2009) gives a detailed characterization of the ifenprodil binding site in the NMDA receptor (NMDAR) by both *in silico* and *in vitro* approaches. The NMDA receptor is an ionotropic glutamate receptor, which serves as the predominant molecular device for controlling synaptic plasticity and memory function. Therefore, controlled activation of the NMDA receptor is of great interest as a potential therapeutic target.

In order to stop receptor overactivation, several NMDAR competitive antagonists were developed in the 1980s. However, these compounds failed in clinical trials because of their inability to discriminate between the various NMDAR subtypes, and caused a generalized inhibition. In the study we report here, the authors used the most promising NMDAR antagonist at that time, ifenprodil, and its derivatives, in order to characterize the ifenprodil binding site using both computational and experimental approaches. The ifenprodil binding site on NMDAR was mapped on NR2B subunit's N-terminal domain (NR2B NTD), and the authors were able to describe the structural determinants responsible for the high-affinity binding of ifenprodil on the NR2B subunit.

A homology modeled structure of NR2B NTD was generated using the sequence to structure alignment functionality within the comparative modeling tool of MODELER 9.0. The ifenprodil was docked into the modeled structure using LigandFit. During the docking, the structure of the protein was kept rigid and 20 conformers of the ligand were subjected to energy minimization in the molecular modeling tool CHARMM. A 1 ns MD simulation of the minimized structures was used to generate the pharmacophore model of the system. In this case the *in silico* approach was used to get a clear picture of the system before extensive experimental validation was achieved by site directed mutagenesis.

Docking showed that ifenprodil adopts a unique and well defined orientation in the central crevice of the NR2B NTD. Based on the *in silico* model, site directed mutagenesis proved 5 NR2B NTD residues (Thr76, Asp77, Asp206, Tyr231, Val262) are essential for the high affinity ifenprodil binding and receptor inhibition. The proposed model of ifenprodil binding to NR2B NTD shared some similarities with a previously proposed model, which had had no experimental validation (Mirienelli et al., 2007). The authors suggest that the differences in the models could be caused by the use of different sequence alignment for the loops situated in binding cleft. However this study showed that a suitable combination of *in silico* approaches can provide a good picture of what we can expect before starting any kind of experiment.

7. Concluding remarks

We have shown in this chapter how *in silico* protein engineering can be used in the field of molecular recognition. The particular steps one has to go through when using these techniques were described. They comprise of 3D structure determination, *in silico* mutagenesis, docking as the first approximation of the binding affinity, and, finally, accurate calculation of the binding free energy.

It should be highlighted that, in many cases, *in silico* approaches provide information complementary to that obtained by experimental approaches. A number of such methods have been implemented and are available in specialized software packages. Therefore users can test the different tools easily and select the ones able to perform well for the particular system they are interested in. We have provided also a brief list of the most frequently used computer programs for the particular tasks described. It is probably fair to say that *in silico* approaches are mostly useful for the visualization and intelligent design of protein engineering projects. As the computer power increases and software products become more and more sophisticated, it is highly probable that *in silico* protein engineering of proteins recognizing small molecules will become an even more useful tool in the future.

8. Acknowledgment

The authors thank the Ministry of Education of the Czech Republic (Contracts MSM0021622413, ME08008) and Czech Science Foundation (Contracts 303/09/1168, P207/10/0321, 301/09/H004) for financial support. This work was further supported by the project "CEITEC - Central European Institute of Technology" (CZ.1.05/1.1.00/02.0068) from European Regional Development Fund.

9. References

Abagyan, RA. & Batalov, S. (1997). Do aligned sequences share the same fold ?. *Journal of Molecular Biology*, Vol. 273, No. 1, pp. 355-368.

Abagyan, R.; Totrov, M. & Kuznetsov, D. (1994). ICM-A new method for protein modeling and design: applications to docking and structure prediction from the distorted native conformation. *Journal of Computational Chemistry*, Vol. 15, No. 5, pp. 488-506.

Adam, J.; Pokorná, M.; Sabin, C,; Mitchell, EP.; Imberty, A. & Wimmerová, M. (2007). Engineering of PA-IIL lectin from pseudomonas aeruginosa - unravelling the role of the specificity loop for sugar preference. *BMC Structural Biology*, Vol. 7:36.

Adam, J.; Kríz, Z.; Prokop, M.; Wimmerová, M & Koca, J. (2008). In silico mutagenesis and docking studies of *Pseudomonas aeruginosa* PA-IIL lectin predicting binding modes and energies. *Journal of Chemical Information and Modeling*, Vol. 48, No. 11, pp. 2234-2242.

Altschul, S., Madden, T., Schaffer, A., Zhang, J., Zhang, Z., Miller, W., and Lipman, D. (1997) Gapped BLAST and PSI-BLAST: a new generation of protein database search programs. *Nucleic Acids Research*, 25, 3389-3402

An, J.; Totrov, M. & Abagyan, R. (2005). Pocketome via comprehensive identification and classification of ligand binding envelopes. *Molecular & Cellular Proteomics: MCP*, Vol. 4, No. 6, pp. 752-761.

Baxter, CA.; Murray, CW.; Clark, DE.;Westhead, DR. & Eldridge, MD. (1998). Flexible docking using tabu search and an empirical estimate of binding affinity. *Proteins*, Vol. 33, No. 3, pp. 367-382.

Beierlein, F.; Lanig, H.; Schurer, G.; Horn, AHC. & Clark T. (2003). Quantum mechanical/molecular mechanical (QM/MM) docking: an evaluation for known test systems. *Molecular Physics*, Vol. 101, No. 15, pp. 2469-2480.

Bogan, AA. & Thorn, KS. (1998). Anatomy of hot spots in protein interfaces. *Journal of Molecular Biology*, Vol. 280, No. 1, pp. 1-9.

Böhm, HJ. (1992). The computer program LUDI: a new method for the de novo design of enzyme inhibitors. *Journal of Computer-Aided Molecular Design*, Vol. 6, No. 1, pp. 61-78.

Brady, GP. & Stouten, PF. (2000). Fast prediction and visualization of protein binding pockets with pass. *Journal of Computer-Aided Molecular Design*, Vol. 14, No. 4, pp. 383-401.

Brooks, B.; Bruccoleri, R.; Olafson, B.; States, D.; et al. (1983). CHARMM: a program for macromolecular energy, minimization, and dynamics calculations. *Journal of Computational. Chemistry*, Vol. 4, No. 2, pp. 187-217.

Brunger, AT. (1992). Free R value: a novel statistical quantity for assessing the accuracy of crystal structures. *Nature*, Vol. 355, No. 6359, pp. 472-475.

Brylinski, M. & Skolnick, J. (2008). Q-Dock: low-resolution flexible ligand docking with pocket-specific threading restraints. *Journal of Computational Chemistry*, Vol. 29, No. 10, pp. 1574-1588.

Carter, P. (1986). Site-directed mutagenesis. *Biochemical Journal*, Vol. 237, No. 1, pp. 1-7.

Chen, HM.; Liu, BF.; Huang, HL.; Hwang, SF. & Ho, SY (2007). SODOCK: swarm optimization for highly flexible protein–ligand docking. *Journal of Computational Chemistry*, Vol. 28, No. 2, pp. 612-623.

Chen, K.; Li, T. & Cao, T. (2006). Tribe-pso: a novel global optimization algorithm and its application in molecular docking. *Chemometrics and Intelligent Laboratory Systems*, Vol. 82, No. 1-2, pp. 248-259.

Cho, AE.; Guallar, V.; Berne, BJ. & Friesner, R. (2005). Importance of accurate charges in molecular docking: quantum mechanical/molecular mechanical (QM/MM) approach. *Journal of Computational Chemistry*, Vol. 26, No. 9, pp. 915-931.

Cho, AE. & Rinaldo, D. (2009). Extension of QM/MM docking and its applications to metalloproteins. *Journal of Computational Chemistry*, Vol. 30, No. 16, pp. 2609-2616.

Chopra, G.; Summa, CM. & Levitt, M. (2008). Solvent dramatically affects protein structure refinement. *Proceedings of the National Academy of Sciences*, Vol. 105, No. 51, pp. 20239 - 20244.

Chung, JY.; Hah, JM. & Cho, AE. (2009). Correlation between performance of QM/MM docking and simple classification of binding sites. *Journal of Chemical Information and Modeling*, Vol. 49, No. 10, pp. 2382-2387.

Connolly, ML. (1983). Analytical molecular surface calculation. *Journal of Applied Crystallography*, Vol. 16, No. 5, pp. 548-558.

Das, P.; Li, J.; Royyuru, AK. & Zhou, R. (2009). Free energy simulations reveal a double mutant avian H5N1 virus hemagglutinin with altered receptor binding specificity. *Journal of Computational Chemistry*, Vol. 30, No. 11, pp. 1654-1663.

Durbin, R. et al., (1998). *Biological sequence analysis: probabilistic models of proteins and nucleic acids*, Cambridge University Press, ISBN: 9780521629713, United Kingdom.

Evers, A. & Klabunde, T. (2005). Structure-based drug discovery using GPCR homology modeling: successful virtual screening for antagonists of the alpha 1A Adrenergic Receptor. *Journal of Medicinal Chemistry*, Vol. 48, No. 4, pp. 1088-1097.

Ewing, TJA. & Kuntz, ID. (1997). Critical evaluation of search algorithms for automated molecular docking and database screening. *Journal of Computational Chemistry*, Vol. 18, No. 9, pp. 1175-1189.

Friesner, RA, Banks, JL, Murphy, RB, Halgren, TA, et al. (2004). Glide: a new approach for rapid, accurate docking and scoring. 1. method and assessment of docking accuracy. *Journal of Medicinal Chemistry*, Vol. 47, No. 7, pp. 1739-1749.

Giacovazzo, C. (2002). *Fundamentals of crystallography*, Oxford University Press. ISBN: 0198509588, USA.

Gräter, F.; Schwarzl, SM.; Dejaegere, A.; Fischer, S. & Smith, JC. (2005). Protein/ligand binding free energies calculated with Quantum Mechanics/Molecular Mechanics. *The Journal of Physical Chemistry B*, Vol. 109, No. 20, pp. 10474-10483.

Guex, N. & Peitsch, MC. (1997). Swiss-model and the Swiss-PDB Viewer: an environment for comparative protein modeling. *Electrophoresis*, Vol. 18, No. 15, pp. 2714-2723.

Halgren, TA.; Murphy, RB.; Friesner, RA.; Beard, HS. et al. (2004). Glide: a new approach for rapid, accurate docking and scoring. 2. enrichment factors in database screening. *Journal of Medicinal Chemistry*, Vol. 47, No. 7, pp. 1750-1759.

Hou, T.; Wang, J.; Li, Y. & Wang, W. (2011). Assessing the performance of the MM/PBSA and MM/GBSA methods. 1. the accuracy of binding free energy calculations based on molecular dynamics simulations. *Journal of Chemical Information and Modeling*, Vol. 51, No. 1, pp. 69-82.

Huang, SY. & Zou, X. (2010). Advances and challenges in protein-ligand docking. *International Journal of Molecular Sciences*, Vol. 11, No. 8, pp. 3016-3034.

Jefferys, BR.; Lawrence AK. & Sternberg, MJE. (2010). Protein folding requires crowd control in a simulated cell. *Journal of Molecular Biology*, Vol. 397, No. 5, pp. 1329-1338.

Jones, DT. (1999a) Protein secondary structure prediction based on position-specific scoring matrices *Journal of Molecular Biology*, Vol. 292, No. 2, pp. 195-202.

Jones, DT. (1999b) Genthreader: an efficient and reliable protein fold recognition method for genomic sequences *Journal of Molecular Biology*, Vol. 287, No. 4, pp. 797-815.

Jones, G.; Willett, P. & Glen, RC.; (1995). Molecular recognition of receptor sites using a genetic algorithm with a description of desolvation. *Journal of Molecular Biology*, Vol. 245, No. 1, pp. 43-53.

Jones, G.; Willett, P.; Glen, RC.; Leach, AR. & Taylor, R. (1997). Development and validation of a genetic algorithm for flexible docking. *Journal of Molecular Biology*, Vol. 267, No. 3, pp. 727-748.

Jorgensen, WL. & Ravimohan, C. (1985). Monte carlo simulation of differences in free energies of hydration. *The Journal of Chemical Physics*, Vol. 83, No. 6, p. 3050.

Kelley, LA.; MacCallum, RM. & Sternberg, MJ. (2000). Enhanced genome annotation using structural profiles in the program 3D-PSSM. *Journal of Molecular Biology*, Vol. 299, No. 2, pp. 499-520.

Kelley, LA. & Sternberg, MJE. (2009). Protein structure prediction on the web: a case study using the phyre server. *Nature Protocols*, Vol. 4, No. 3, pp. 363-371.

Kerzmann, A,; Fuhrmann, J.; Kohlbacher, O. & Neumann, D. (2008). BALLDock/SLICK: a new method for protein-carbohydrate docking. *Journal of Chemical Information and Modeling*, Vol. 48, No. 8, pp. 1616-1625.

Kirkwood, J. (1935). Statistical mechanics of fluid mixtures. *Journal of Chemical Physics*, Vol. 3, No. 5, pp. 300-313.

Kollman, P A,; Massova, I.; Reyes, C.; Kuhn, B. et al. (2000). Calculating structures and free energies of complex molecules: combining molecular mechanics and continuum models. *Accounts of Chemical Research*, Vol. 33, No. 12, pp. 889-897.

Kramer, B.; Rarey, M. & Lengauer, T. (1999). Evaluation of the flexx incremental construction algorithm for protein-ligand docking. *Proteins*, Vol. 37, No. 2, pp. 228-241.

Krammer, A.; Kirchhoff, PD.; Jiang, X.; Venkatachalam, CM. & Waldman, M. (2005). LigScore: a novel scoring function for predicting binding affinities. *Journal of Molecular Graphics & Modelling*, Vol. 23, No. 5, pp. 395-407.

Kuntz, ID.; Blaney, JM.; Oatley, SJ.; Langridge, R. & Ferrin, TE. (1982). A geometric approach to macromolecule-ligand interactions. *Journal of Molecular Biology*, Vol. 161, No. 2, pp. 269-288.

Laurie, ATR. (2005). Q-Sitefinder: an energy-based method for the prediction of protein-ligand binding sites. *Bioinformatics*, Vol. 21, No. 9, pp. 1908-1916.

Lee, C. & Irizarry, K. (2001). The genemine system for genome/proteome annotation and collaborative data mining *IBM Systems Journal*, Vol. 40, No. 2, pp. 592-603.

Levitt, M. (1992). Accurate modeling of protein conformation by automatic segment matching. *Journal of Molecular Biology*, Vol. 226, No. 2, pp. 507-533.

Lobley, A., Sadowski, M.I. & Jones, D.T. (2009). pGenTHREADER and pDomTHREADER: New methods for improved protein fold recognition and superfamily discrimination. *Bioinformatics*, 25, 1761-1767.

Lovell, SC.; Word, JM.; Richardson, JS. & Richardson, DC. (2000). The penultimate rotamer library. *Proteins: Structure, Function, and Bioinformatics*, Vol. 40, No. 3, pp. 389-408.

di Luccio, E & Koehl, P. (2011). A quality metric for homology modeling: the h-factor. *BMC Bioinformatics*, Vol. 12, p. 48.

Luthy, R.; Bowie, JU. & Eisenberg, D. (1992). Assessment of protein models with three-dimensional profiles. *Nature*, Vol. 356, No. 6364, pp. 83-85.

Madej, T.; Gibrat, JF. & Bryant, SH. (1995). Threading a database of protein cores. *Proteins*, Vol. 23, No. 3, pp. 356-369.

Massova, I. & Kollman, PA. (1999). Computational alanine scanning to probe protein–protein interactions: a novel approach to evaluate binding free energies. *Journal of the American Chemical Society*, Vol. 121, No. 36, pp. 8133-8143.

McGann, MR.; Almond, HR.; Nicholls, A.; Grant, JA. & Brown, FK. (2003). Gaussian docking functions. *Biopolymers*, Vol. 68, No. 1, pp. 76-90.

McGuffin, LJ. & Jones, DT. (2003). Improvement of the genthreader method for genomic fold recognition. *Bioinformatics*, Vol. 19, No. 7, pp. 874 -881.

McMartin, C. & Bohacek, RS. (1997). QXP: powerful, rapid computer algorithms for structure-based drug design. *Journal of Computer-Aided Molecular Design*, Vol. 11, No. 4, pp. 333-344.

Miller, MD.; Kearsley, SK.; Underwood, DJ. & Sheridan, RP. (1994). FLOG: a system to select 'quasi-flexible' ligands complementary to a receptor of known three-dimensional structure. *Journal of Computer-Aided Molecular Design*, Vol. 8, No. 2, pp. 153-174.

Misura, KMS.; Chivian, D.; Rohl, CA.; Kim, DE. & Baker, D. (2006). Physically realistic homology models built with rosetta can be more accurate than their templates. *Proceedings of the National Academy of Sciences*, Vol. 103, No. 14, pp. 5361 -5366.

Mitchell, EP.; Sabin, C.; Šnajdrová, L.; Pokorná, M.; Perret, S.; Gautier, C.; Hofr, C.; Gilboa-Garber, N.; Koča, J.; Wimmerová, M. & Imberty, A. (2005). High affinity fucose binding of *Pseudomonas aeruginosa* lectin PA-IIL: 1.0 Å resolution crystal structure of the complex combined with thermodynamics and computational chemistry approaches. *Proteins: Structure, Function, and Bioinformatics*, Vol. 58, No. 3, pp.735-746.

Mony, L.; Krzaczkowski, L.; Leonetti, M.; Le Goff, A. et al. (2009). Structural basis of NR2B-selective antagonist recognition by n-methyl-d-aspartate receptors. *Molecular Pharmacology*, Vol. 75, No. 1, pp. 60-74.

Moreira, IS.; Fernandes, PA. & Ramos, MJ. (2007a). Computational alanine scanning mutagenesis--an improved methodological approach. *Journal of Computational Chemistry*, Vol. 28, No. 3, pp. 644-654.

Morris, GM.; Goodsell, DS.; Halliday, RS.; Huey, R.; et al. (1998). Automated docking using a lamarckian genetic algorithm and an empirical binding free energy function. *Journal of Computational Chemistry*, Vol. 19, No. 14, pp. 1639-1662.

Morris, GM.; Huey, R.; Lindstrom, W.; Sanner, MF.; et al. (2009). AutoDock4 and AutoDockTools4: automated docking with selective receptor flexibility. *Journal of Computational Chemistry*, Vol. 30, No. 16, pp. 2785-2791.

Najmanovich, R,; Kuttner, J,; Sobolev, V. & Edelman, M. (2000). Side-chain flexibility in proteins upon ligand binding. *Proteins: Structure, Function, and Genetics*, Vol. 39, No. 3, pp. 261-268.

Namasivayam, V. & Günther, R. (2007). PSO@autodock: a fast flexible molecular docking program based on swarm intelligence. *Chemical Biology & Drug Design*, Vol. 70, No. 6, pp. 475-484.

Nayeem, A.; Sitkoff, D. & Krystek Jr., S. (2006). A comparative study of available software for high-accuracy homology modeling: from sequence alignments to structural models. *Protein Science*, Vol. 15, No. 4, pp. 808-824.

Olson, MA.; Feig, M. & Brooks, CL. (2008). Prediction of protein loop conformations using multiscale modeling methods with physical energy scoring functions. *Journal of Computational Chemistry*, Vol. 29, No. 5, pp. 820-831.

OpenEye Scientific Software, Santa Fe, New Mexico. http://www.eyesopen.com/

Pei, J.; Wang, Q.; Liu, Z.; Li, Q.; et al. (2006). PSI-Dock: towards highly efficient and accurate flexible ligand docking. *Proteins: Structure, Function, and Bioinformatics*, Vol. 62, No. 4, pp. 934-946.

Peitsch, M.C. (1995). ProMod: Automated knowledge-based protein modelling tool. *PDB Quarterly Newsletter*, 72, 4.

Prokop, M.; Adam, J.; Kříž, Z.; Wimmerová, M. & Koča, J. (2008). Triton: a graphical tool for ligand-binding protein engineering. *Bioinformatics*, Vol. 24, No. 17, pp. 1955-1956.

Pymol. The PyMOL Molecular Graphics System, Version 1.3, Schrödinger, LLC. http://www.pymol.org/

Rarey, M.; Kramer, B.; Lengauer, T. & Klebe, G. (1996). A fast flexible docking method using an incremental construction algorithm. *Journal of Molecular Biology*, Vol. 261, No. 3, pp. 470-489.

Roessler, CG.; Hall, BM.; Anderson, WJ.; Ingram, WM.; et al. (2008). Transitive homology-guided structural studies lead to discovery of cro proteins with 40% sequence identity but different folds," *PNAS*, Vol. 105, No. 7, pp. 2343-2348.

Saccenti, E. & Rosato, A. (2008). The war of tools: how can NMR spectroscopists detect errors in their structures ?. *Journal of Biomolecular NMR*, Vol. 40, No. 4, pp. 251-261.

Sali, A. & Blundell, TL. (1993). Comparative protein modelling by satisfaction of spatial restraints. *Journal of Molecular Biology*, Vol. 234, No. 3, pp. 779-815.

Sauton, N.; Lagorce, D.; Villoutreix, BO. & Miteva, MA. (2008). MS-Dock: accurate multiple conformation generator and rigid docking protocol for multi-step virtual ligand screening. *BMC Bioinformatics*, Vol. 9, No. 1, p. 184.

Schwede, T.; Kopp, J.; Guex, N. & Peitsch, MC. (2003). Swiss-model: an automated protein homology-modeling server. *Nucleic Acids Research*, Vol. 31, No. 13, pp. 3381-3385.

Söding, J. (2005). Protein homology detection by HMM–HMM comparison. *Bioinformatics*, Vol. 21, No. 7, pp. 951-960.

Srinivasan, J.; Thomas EC.; Piotr, C.; Kollman, PA. & Case, DA. (1998). Continuum solvent studies of the stability of DNA, RNA, and phosphoramidate–DNA helices. *Journal of the American Chemical Society*, Vol. 120, No. 37, pp. 9401-9409.

Stroganov, OV.; Novikov, FN.; Stroylov, VS.; Kulkov, V. & Chilov, GG. (2008). Lead finder: an approach to improve accuracy of protein–ligand docking, binding energy estimation, and virtual screening. *Journal of Chemical Information and Modeling*, Vol. 48, No. 12, pp. 2371-2385.

Summa, CM. & Levitt, M. (2007). Near-native structure refinement using in vacuo energy minimization. *Proceedings of the National Academy of Sciences*, Vol. 104, No. 9, pp. 3177-3182.

Taylor, RD.; Jewsbury, PJ. & Essex, JW. (2002). A review of protein-small molecule docking methods. *Journal of Computer-Aided Molecular Design*, Vol. 16, No. 3, pp. 151-166.

Terp, GE.; Johansen, BN.; Christensen, IT. & Jørgensen, FS. (2001). A new concept for multidimensional selection of ligand conformations (multiselect) and multidimensional scoring (multiscore) of protein-ligand binding affinities. *Journal of Medicinal Chemistry*, Vol. 44, No. 14, pp. 2333-2343.

Trott, O. & Olson, AJ. (2010). AutoDock Vina: improving the speed and accuracy of docking with a new scoring function, efficient optimization, and multithreading. *Journal of Computational Chemistry*, Vol. 31, No. 2, pp. 455-461.

Tuttle, T. (2010). Applications of QM/MM in inorganic chemistry. *Spectroscopic Properties of Inorganic and Organometallic Compounds*, pp. 87-110, Royal Society of Chemistry, Cambridge, ISBN: 9781849730853

Vásquez, M. (1996). Modeling side-chain conformation. *Current Opinion in Structural Biology*, Vol. 6, No. 2, pp. 217-221.

van Vlijmen, HWT. & Karplus, M. (1997). PDB-based protein loop prediction: parameters for selection and methods for optimization. *Journal of Molecular Biology*, Vol. 267, No. 4, pp. 975-1001.

Wang, M. & Wong, CF. (2007). Rank-ordering protein-ligand binding affinity by a Quantum Mechanics/Molecular Mechanics/Poisson-Boltzmann-Surface Area model. *The Journal of Chemical Physics*, Vol. 126, No. 2, pp. 026101.

Wang, R.; Lai, L. & Wang, S. (2002). Further development and validation of empirical scoring functions for structure-based binding affinity prediction. *Journal of Computer-Aided Molecular Design*, Vol. 16, No. 1, pp. 11-26.

Warren, GL.; Andrews, CW.; Capelli, AM.; Clarke, B.; et al. (2006). A critical assessment of docking programs and scoring functions. *Journal of Medicinal Chemistry*, Vol. 49, No. 20, pp. 5912-5931.

Warshel, A. & Levitt, M. (1976). Theoretical studies of enzymic reactions: dielectric, electrostatic and steric stabilization of the carbonium ion in the reaction of lysozyme. *Journal of Molecular Biology*, Vol. 103, No. 2, pp. 227-249.

Weiner, PK. & Kollman, PA. (1981). AMBER: Assisted Model Building with Energy Refinement. a general program for modeling molecules and their interactions. *Journal of Computational Chemistry*, Vol. 2, No. 3, pp. 287-303.

Wells, JA. (1991). Systematic mutational analyses of protein-protein interfaces. *Methods in Enzymology*, Vol. 202, pp. 390-411.

Wuthrich, K.; (1990). Protein structure determination in solution by nmr spectroscopy. *Journal of Biological. Chemistry*, Vol. 265, No. 36, pp. 22059-22062.

Wuthrich, K. (2003). NMR studies of structure and function of biological macromolecules. *Journal of Biomolecular NMR*, Vol. 27, No. 1, pp. 13-39.

Yuan, Z.; Bailey, TL. & Teasdale, RD. (2005). Prediction of protein b-factor profiles. *Proteins: Structure, Function, and Bioinformatics*, Vol. 58, No. 4, pp. 905-912.

Yuriev, E.; Agostino, M. & Ramsland, PA. (2011). Challenges and advances in computational docking: 2009 in review. *Journal of Molecular Recognition: JMR*, Vol. 24, No. 2, pp. 149-164.

Zhang, H. (2002). Protein Tertiary Structures: Prediction from Amino Acid Sequences, *Encyclopedia of Life Sciences*, Macmillan Publishers Ltd, Nature Publishing Group, England.

Zhu, J.; Fan, H.; Periole, X.; Honig, B. & Mark, AE. (2008). Refining homology models by combining replica-exchange molecular dynamics and statistical potentials. *Proteins: Structure, Function, and Bioinformatics*, Vol. 72, No. 4, pp. 1171-1188.

Zsoldos, Z.; Reid, D.; Simon, A.; Sadjad, BS. & Johnson, AP. (2006). eHITS: an innovative approach to the docking and scoring function problems. *Current Protein & Peptide Science*, Vol. 7, No. 5, pp. 421-435.

Zwanzig, R. (1954). High-temperature equation of state by a perturbation method. i. nonpolar gases. *The Journal of Chemical Physics*, Vol. 22, No. 8, pp. 1420-1426.

Permissions

The contributors of this book come from diverse backgrounds, making this book a truly international effort. This book will bring forth new frontiers with its revolutionizing research information and detailed analysis of the nascent developments around the world.

We would like to thank Pravin T P Kaumaya, for lending his expertise to make the book truly unique. He has played a crucial role in the development of this book. Without his invaluable contribution this book wouldn't have been possible. He has made vital efforts to compile up to date information on the varied aspects of this subject to make this book a valuable addition to the collection of many professionals and students.

This book was conceptualized with the vision of imparting up-to-date information and advanced data in this field. To ensure the same, a matchless editorial board was set up. Every individual on the board went through rigorous rounds of assessment to prove their worth. After which they invested a large part of their time researching and compiling the most relevant data for our readers. Conferences and sessions were held from time to time between the editorial board and the contributing authors to present the data in the most comprehensible form. The editorial team has worked tirelessly to provide valuable and valid information to help people across the globe.

Every chapter published in this book has been scrutinized by our experts. Their significance has been extensively debated. The topics covered herein carry significant findings which will fuel the growth of the discipline. They may even be implemented as practical applications or may be referred to as a beginning point for another development. Chapters in this book were first published by InTech; hereby published with permission under the Creative Commons Attribution License or equivalent.

The editorial board has been involved in producing this book since its inception. They have spent rigorous hours researching and exploring the diverse topics which have resulted in the successful publishing of this book. They have passed on their knowledge of decades through this book. To expedite this challenging task, the publisher supported the team at every step. A small team of assistant editors was also appointed to further simplify the editing procedure and attain best results for the readers.

Our editorial team has been hand-picked from every corner of the world. Their multi-ethnicity adds dynamic inputs to the discussions which result in innovative outcomes. These outcomes are then further discussed with the researchers and contributors who give their valuable feedback and opinion regarding the same. The feedback is then collaborated with the researches and they are edited in a comprehensive manner to aid the understanding of the subject.

Apart from the editorial board, the designing team has also invested a significant amount of their time in understanding the subject and creating the most relevant covers. They scrutinized every image to scout for the most suitable representation of the subject and create an appropriate cover for the book.

The publishing team has been involved in this book since its early stages. They were actively engaged in every process, be it collecting the data, connecting with the contributors or procuring relevant information. The team has been an ardent support to the editorial, designing and production team. Their endless efforts to recruit the best for this project, has resulted in the accomplishment of this book. They are a veteran in the field of academics and their pool of knowledge is as vast as their experience in printing. Their expertise and guidance has proved useful at every step. Their uncompromising quality standards have made this book an exceptional effort. Their encouragement from time to time has been an inspiration for everyone.

The publisher and the editorial board hope that this book will prove to be a valuable piece of knowledge for researchers, students, practitioners and scholars across the globe.

List of Contributors

Eduard V. Bocharov, Konstantin V. Pavlov, Pavel E. Volynsky, Roman G. Efremov and Alexander S. Arseniev
Shemyakin-Ovchinnikov Institute of Bioorganic Chemistry RAS, Russia

Burcu Turanli-Yildiz, Ceren Alkim and Z. Petek Cakar
Istanbul Technical University (ITU), Dept. of Molecular Biology and Genetics, Research Center (ITU-MOBGAM), Istanbul, Turkey
ITU Dr. Orhan Ocalgiray Molecular Biology, Biotechnology and Genetics, Research Center (ITU-MOBGAM), Istanbul, Turkey

Emel Ordu
Yıldız Technical University, Turkey

Nevin Gül Karagüler
Istanbul Technical University, Turkey

Junko Tanaka, Hiroshi Yanagawa and Nobuhide Doi
Department of Biosciences and Informatics, Keio University, Japan

Jennifer D. Stone and David M. Kranz
Department of Biochemistry, University of Illinois at Urbana-Champaign, Urbana, USA

Yiyuan Yin, Min Mo and Roy A. Mariuzza
Institute for Bioscience and Biotechnology Research, University of Maryland, Rockville, USA

David L. Donermeyer, Paul M. Allen and K. Scott Weber
Department of Pathology and Immunology, Washington University School of Medicine, St. Louis, USA

Roy A. Mariuzza
Department of Cell Biology and Molecular Genetics, University of Maryland, College Park, USA

Steven Jacobs and Karyn O'Neil
Centyrex Venture, Johnson & Johnson, USA

Liudmila Lysenko and Nina Nemova
Institute of Biology, KarRC of Russian Academy of Science, Russian Federation

Paul D. Riggs
New England Biolabs, U.S.A.

Shu-Qun Liu, Xing-Lai Ji, Yan Tao, Ke-Qin Zhang and Yun-Xin Fu
Laboratory for Conservation and Utilization of Bio-Resources & Key Laboratory for Southwest Biodiversity, Yunnan University, Kunming, P. R. China

Shu-Qun Liu and Xing-Lai Ji
Sino-Dutch Biomedial and Information Engineering School, Northeastern University, Shenyang, P. R. China

Yue-Hui Xie
Teaching and Research Section of computer, Department of Basic Medical, Kunming Medical College, Kunming, P. R. China

De-Yong Tan
School of Life Sciences, Yunnan University, Kunming, P. R. China

Yun-Xin Fu
Human Genetics Center, School of Public Health, The University of Texas Health Science Center, Houston, Texas, USA

Aijun Wang, Natalie Winblade Nairn, Marcello Marelli and Kenneth Grabstein
Allozyne, USA

Xiaolin Sun and William T. Jones
The New Zealand Institute for Plant and Food Research, Palmerston North, New Zealand

Vladimir N. Uversky
Department of Molecular Medicine, College of Medicine, University of South Florida, USA
Institute for Biological Instrumentation, Russian Academy of Sciences, Moscow, Russia

Seiya Watanabe and Keisuke Makino
Ehime University, Kyoto University, Japan

Sushil Kumar Mishra, Gabriel Demo, Jaroslav Koča and Michaela Wimmerová
CEITEC - Central European Institute of Technology, Masaryk University, Brno, National Centre for Biomolecular Research, Faculty of Science, Masaryk University, Brno, Czech Republic

www.ingramcontent.com/pod-product-compliance
Lightning Source LLC
Chambersburg PA
CBHW070718190326
41458CB00004B/1023